Biodiversity and Ecosystem Functioning in Naturally and Experimentally Assembled Communities

Biodiversity and Ecosystem Functioning in Naturally and Experimentally Assembled Communities

Editors

Daniel Puppe
Panayiotis Dimitrakopoulos
Baorong Lu

MDPI • Basel • Beijing • Wuhan • Barcelona • Belgrade • Manchester • Tokyo • Cluj • Tianjin

Editors
Daniel Puppe
Leibniz Centre for
Agricultural Landscape
Research (ZALF)
Müncheberg, Germany

Panayiotis Dimitrakopoulos
University of the Aegean
Mytilene, Greece

Baorong Lu
Fudan University
Shanghai, China

Editorial Office
MDPI
St. Alban-Anlage 66
4052 Basel, Switzerland

This is a reprint of articles from the Special Issue published online in the open access journal *Biology* (ISSN 2079-7737) (available at: https://www.mdpi.com/journal/biology/special_issues/biodiversity_ecosystem_communities).

For citation purposes, cite each article independently as indicated on the article page online and as indicated below:

LastName, A.A.; LastName, B.B.; LastName, C.C. Article Title. *Journal Name* **Year**, *Volume Number*, Page Range.

ISBN 978-3-0365-8098-2 (Hbk)
ISBN 978-3-0365-8099-9 (PDF)

Cover image courtesy of Baorong Lu

Contents

About the Editors

Daniel Puppe

Daniel Puppe is a researcher at the Leibniz Centre for Agricultural Landscape Research (ZALF) in Müncheberg, Germany. As a biologist, he is generally interested in all forms of life. However, in his research he is focused on organisms that form biogenic silica structures (e.g., plant phytoliths, testate amoeba shells, or sponge spicules) and the role of biogenic silica for biogeochemical silicon cycling in terrestrial ecosystems and agricultural landscapes. With his research, Daniel aims to find environmentally friendly and sustainable strategies in order to deal with the challenges that global change poses to (agro)ecosystem functioning. He trained at the TU Brunswick (diploma in biology) and the BTU Cottbus-Senftenberg (PhD) in Germany. Daniel is a member of the German Soil Science Society (DBG) and the German Society for Protozoology (DGP) andserves as reviewer for science funding associations (German Research Foundation, National Science Centre of Poland) and international peer-reviewed journals, and is author of various scientific publications (https://orcid.org/0000-0003-3140-0528).

Panayiotis G. Dimitrakopoulos

Panayiotis G. Dimitrakopoulos is Professor of Functional Ecology at the Department of Environment of the University of the Aegean, Greece. He obtained his Ph.D. degree from the University of the Aegean in 2001. His postdoctoral research was carried out in the University of Zurich, Switzerland, and was funded from the European Science Foundation (LINKECOL program). His research focuses on functional plant ecology, community ecology, biodiversity conservation, and conservation policy. He has served as President of the Hellenic Ecological Society (2012–2014), member of the "Natura 2000" National Committee (2010–2020), and Honorary Visiting Senior Fellow at Anglia Ruskin University, Cambridge (2019–2022). He serves as Section Editor-in-Chief of Ecology in Biology and is a member of the editorial board of six journals and referee for more than 60 journals. He has published more than 85 papers in peer-reviewed journals (https://orcid.org/0000-0002-8374-4392).

Baorong Lu

Baorong Lu is a professor of evolutionary biology and genetic resources at the School of Life Sciences, Fudan University, China. He received his Ph.D. degree from the Swedish University of Agricultural Sciences (SLU) in 1993. He was promoted as an associated professor in SLU (since 1994) and a professor (since 1995) in the Chinese Academy of Sciences. He worked as a Germplasm Specialist (senior scientist) in the International Rice Research Institute during 1995–2000 until he joined Fudan University in the end of 2000. He received the Distinguished Young Fellow Award by the National Science Foundation of China in 2001. He has served as the member of the Chinese National Biosafety Committee since 2002. He is actively involved in research and teaching on botany, evolutionary biology, cytogenetics, biodiversity conservation, biotechnology, and the biosafety assessment of genetically engineered crops, particularly the issue of transgene flow and its ecological/environmental consequences. He has published more than 300 academic articles in peer-review journals and these articles have been cited more than 12,000 times. He has also served as the Editor-in-Chief, Associate Editor or an Editorial Member of several international and national journals.

Preface to "Biodiversity and Ecosystem Functioning in Naturally and Experimentally Assembled Communities"

In short, the term "ecosystem functioning" describes the interactions between life (i.e., plants, animals, and microorganisms) and the abiotic environment. While the significance of plants, animals, and microorganisms for ecosystem functioning is well known (e.g., maintaining biogeochemical processes), the role of biodiversity (e.g., abundance and distribution of organisms) in maintaining ecosystem functioning remains largely unknown. As biodiversity is severely threatened by global change, studying the consequences of biodiversity loss for ecosystem functioning is more topical than ever. In this context, each individual study on biodiversity and ecosystem functioning (BEF) can contribute to the overall picture of the underlying relationships. This book comprises a collection of review and research articles covering a broad range of BEF relationships in naturally and experimentally assembled communities. We thank all authors for their contributions (a short summary is given in the *Editorial*).

"It is the range of biodiversity that we must care for—the whole thing—rather than just one or two stars." (Sir David Attenborough).

Daniel Puppe, Panayiotis Dimitrakopoulos, and Baorong Lu
Editors

biology

MDPI

Editorial

Biodiversity and Ecosystem Functioning in Naturally and Experimentally Assembled Communities

Daniel Puppe [1,*], Panayiotis G. Dimitrakopoulos [2,*] and Baorong Lu [3,*]

1 Leibniz Centre for Agricultural Landscape Research (ZALF), 15374 Müncheberg, Germany
2 Department of Environment, University of the Aegean, 81132 Mytilene, Greece
3 Department of Ecology and Evolutionary Biology, School of Life Sciences, Fudan University, Shanghai 200333, China
* Correspondence: daniel.puppe@zalf.de (D.P.); pdimi@env.aegean.gr (P.G.D.); brlu@fudan.edu.cn (B.L.)

Citation: Puppe, D.; Dimitrakopoulos, P.G.; Lu, B. Biodiversity and Ecosystem Functioning in Naturally and Experimentally Assembled Communities. *Biology* **2023**, *12*, 835. https://doi.org/10.3390/biology12060835

Received: 5 June 2023
Accepted: 6 June 2023
Published: 9 June 2023

Numerous studies have proved that biodiversity and ecosystem functioning (BEF) are closely linked. In this context, interactions between biodiversity and ecosystem functions, such as biomass production, trophic transfer through plants and animals, and pollination, have been in the focus of research [1–3]. Based on this, we now know that biodiversity promotes biomass production and pollination success, for example. However, regarding other BEF relationships, the drawing of general conclusions is hampered by the fact that different studies often show inconsistent results. Thus, for a better understanding of BEF relationships, detailed research on the underlying mechanisms is urgently needed. This knowledge is crucial for harmonizing research findings and to craft policies for the conservation of biodiversity, as it is severely threatened by global warming and other anthropogenic environmental changes (e.g., pollution, resource depletion, or species invasions).

In this Special Issue, two review and thirteen research articles dealing with a broad range of BEF relationships are published. The first review article provides a comprehensive overview of the current knowledge of autotroph–herbivore interactions in marine ecosystems, using macroalgae as a role model to describe the effects of macroalgal properties (e.g., nutrient composition or defense mechanisms) on herbivore feeding behavior [4]. The second review article presents METAL (MacroEcological Theory on the Arrangement of Life), which uses niche-environment interactions to explain phenomena such as phenology, biogeographical shifts, and community arrangement/reorganization [5]. The research articles in this Special Issue comprise studies on various forms of life (from microbiota, such as bacteria and fungi, to macrobiota, including tulips, monkeys, or giant pandas) in different terrestrial and aquatic ecosystems (e.g., grasslands, rivers, or marine hydrothermal vents), as well as in farming systems. In the first research article, wheat–faba-bean mixtures were studied to shed light on the phenomenon whereby increased diversity (e.g., represented by cereal–legume mixtures) in crop systems is often associated with higher yields and ecological sustainability compared to monocultures [6]. Let et al. [7] analyzed the effects of environmental pollution on the biodiversity (measured by abundances) of the larvae of aquatic insects, i.e., mayflies (Ephemeroptera), stoneflies (Plecoptera), and caddisflies (Trichoptera), in a river in central Bohemia (Czech Republic). The third research article deals with plant diversity and its relationships with soil microbes, i.e., bacteria and fungi, in semi-arid grassland in China, highlighting the critical role of soil microbe diversity in regulating soil multifunctionality [8]. Arunrat et al. [9] compared the microbial diversity, community composition, and functional structure of bacterial communities between rice–fish co-cultures and rice monoculture farming systems in Thailand to identify environmental driving factors of bacterial compositions in these systems. Man et al. [10] investigated microbes in *Sphagnum*-dominated peatlands in central China to unravel the linkages between different types of *Sphagnum* and the diversity and ecological functions of *Sphagnum*-associated microbiomes. In the sixth research article, the effects of village development on the habitat quality of the Yunnan snub-nosed monkey (*Rhinopithecus bieti*) in south China were evaluated to derive strategies for the protection of the habitats of these endangered monkeys [11]. Smeti et al. [12] performed a field study across nine rivers in Greece to analyze biodiversity–biomass relationships in benthic diatoms, which have been largely under-studied until now. Jia et al. [13] analyzed the distribution and microhabitat selection of the giant panda (*Ailuropoda melanoleuca*) in China to develop practical habitat conservation and management measures. In the ninth research article, rice straw and stubble burning in paddy fields in central Thailand was examined to assess changes in soil bacterial communities and soil properties after burning [14]. In the tenth research article, the plasticity of the grass *Aeluropus lagopoides* in saline habitats of Saudi Arabia was studied to better understand the interactions between the morphological parameters of plants and their environment [15]. Bilias et al. [16] analyzed the natural nutrient status and rhizosphere fungal morphotypes of wild-growing tulips in Greece to gain deeper insights into the adaptation of tulips to the environment. Cai et al. [17] examined crop-weed introgression (i.e., the transfer of genetic material from one species into the gene pool of another species) promoted by the change in rice cultivars to better understand weedy plant evolution and human influences in agroecosystems. Last but not least, Gheibzadeh et al. [18] studied alpha, beta, and gamma carbonic anhydrases (i.e., metalloenzymes that catalyze the hydration of carbon dioxide) in the thermophilic microbiome of marine hydrothermal vents to detect horizontal gene transfer, which is discussed as an important tool in maintaining microbial biodiversity in these harsh ecosystems.

All the articles in this Special Issue provide in-depth discussions on various aspects of BEF relationships in naturally and experimentally assembled communities. The articles in this Special Issue will help to substantially deepen our understanding of BEF interactions and to elaborate biodiversity conservation in a changing world. Therefore, the collection of these articles is of great interest not only for scientists in this field, but also for students, university teachers, and policymakers, as well as those who are interested in BEF in general.

Conflicts of Interest: The authors declare no conflict of interest.

References

1. Loreau, M.; Naeem, S.; Inchausti, P.; Bengtsson, J.; Grime, J.P.; Hector, A.; Hooper, D.U.; Huston, M.A.; Raffaelli, D.; Schmid, B.; et al. Biodiversity and ecosystem functioning: Current knowledge and future challenges. *Science* **2001**, *294*, 804–808. [CrossRef] [PubMed]
2. Tilman, D.; Isbell, F.; Cowles, J.M. Biodiversity and ecosystem functioning. *Annu. Rev. Ecol. Evol. Syst.* **2014**, *45*, 471–493. [CrossRef]
3. van der Plas, F. Biodiversity and ecosystem functioning in naturally assembled communities. *Biol. Rev.* **2019**, *94*, 1220–1245. [CrossRef] [PubMed]
4. Cheng, A.; Lim, W.Y.; Lim, P.-E.; Amri, A.Y.; Poong, S.-W.; Song, S.-L.; Ilham, Z. Marine Autotroph-Herbivore Synergies: Unravelling the Roles of Macroalgae in Marine Ecosystem Dynamics. *Biology* **2022**, *11*, 1209. [CrossRef] [PubMed]
5. Beaugrand, G. Towards an understanding of large-scale biodiversity patterns on land and in the sea. *Biology* **2023**, *12*, 339. [CrossRef] [PubMed]
6. Ajal, J.; Weih, M. Nutrient accumulation pattern in mixtures of wheat and faba bean is strongly influenced by cultivar choice and co-existing weeds. *Biology* **2022**, *11*, 630. [CrossRef] [PubMed]
7. Let, M.; Černý, J.; Nováková, P.; Ložek, F.; Bláha, M. Effects of trace metals and municipal wastewater on the Ephemeroptera, Plecoptera, and Trichoptera of a stream community. *Biology* **2022**, *11*, 648. [CrossRef] [PubMed]
8. Li, Z.; Liu, X.; Zhang, M.; Xing, F. Plant Diversity and Fungal Richness Regulate the Changes in Soil Multifunctionality in a Semi-Arid Grassland. *Biology* **2022**, *11*, 870. [CrossRef] [PubMed]
9. Arunrat, N.; Sansupa, C.; Kongsurakan, P.; Sereenonchai, S.; Hatano, R. Soil Microbial Diversity and Community Composition in Rice–Fish Co-Culture and Rice Monoculture Farming System. *Biology* **2022**, *11*, 1242. [CrossRef] [PubMed]
10. Man, B.; Xiang, X.; Zhang, J.; Cheng, G.; Zhang, C.; Luo, Y.; Qin, Y. Keystone Taxa and Predictive Functional Analysis of Sphagnum palustre Tank Microbiomes in Erxianyan Peatland, Central China. *Biology* **2022**, *11*, 1436. [CrossRef] [PubMed]
11. Zhu, S.; Li, L.; Wu, G.; Liu, J.; Slate, T.J.; Guo, H.; Li, D. Assessing the Impact of Village Development on the Habitat Quality of Yunnan Snub-Nosed Monkeys Using the INVEST Model. *Biology* **2022**, *11*, 1487. [CrossRef]
12. Smeti, E.; Tsirtsis, G.; Skoulikidis, N.T. Geology Can Drive the Diversity–Ecosystem Functioning Relationship in River Benthic Diatoms by Selecting for Species Functional Traits. *Biology* **2023**, *12*, 81. [CrossRef]
13. Jia, W.; Yan, S.; He, Q.; Li, P.; Fu, M.; Zhou, J. Giant Panda Microhabitat Study in the Daxiangling Niba Mountain Corridor. *Biology* **2023**, *12*, 165. [CrossRef] [PubMed]

14. Arunrat, N.; Sereenonchai, S.; Sansupa, C.; Kongsurakan, P.; Hatano, R. Effect of Rice Straw and Stubble Burning on Soil Physicochemical Properties and Bacterial Communities in Central Thailand. *Biology* **2023**, *12*, 501. [CrossRef] [PubMed]
15. Assaeed, A.M.; Dar, B.A.; Al-Doss, A.A.; Al-Rowaily, S.L.; Malik, J.A.; Abd-ElGawad, A.M. Phenotypic Plasticity Strategy of Aeluropus lagopoides Grass in Response to Heterogenous Saline Habitats. *Biology* **2023**, *12*, 553. [CrossRef] [PubMed]
16. Bilias, F.; Karagianni, A.-G.; Ipsilantis, I.; Samartza, I.; Krigas, N.; Tsoktouridis, G.; Matsi, T. Adaptability of Wild-Growing Tulips of Greece: Uncovering Relationships between Soil Properties, Rhizosphere Fungal Morphotypes and Nutrient Content Profiles. *Biology* **2023**, *12*, 605. [CrossRef] [PubMed]
17. Cai, X.-X.; Wang, Z.; Yuan, Y.; Pang, L.-H.; Wang, Y.; Lu, B.-R. Crop–Weed Introgression Plays Critical Roles in Genetic Differentiation and Diversity of Weedy Rice: A Case Study of Human-Influenced Weed Evolution. *Biology* **2023**, *12*, 744. [CrossRef] [PubMed]
18. Gheibzadeh, M.S.; Manyumwa, C.V.; Tastan Bishop, Ö.; Shahbani Zahiri, H.; Parkkila, S.; Zolfaghari Emameh, R. Genome Study of α-, β-, and γ-Carbonic Anhydrases from the Thermophilic Microbiome of Marine Hydrothermal Vent Ecosystems. *Biology* **2023**, *12*, 770. [CrossRef]

biology

MDPI

Review

Marine Autotroph-Herbivore Synergies: Unravelling the Roles of Macroalgae in Marine Ecosystem Dynamics

Acga Cheng [1], Wai Yin Lim [1,2], Phaik-Eem Lim [1,2], Affendi Yang Amri [2], Sze-Wan Poong [2], Sze-Looi Song [2,3,*] and Zul Ilham [1,4,*]

1 Institute of Biological Sciences, Faculty of Science, Universiti Malaya, Kuala Lumpur 50603, Malaysia
2 Institute of Ocean and Earth Sciences, Universiti Malaya, Kuala Lumpur 50603, Malaysia;
3 Institute for Advanced Studies, Universiti Malaya, Kuala Lumpur 50603, Malaysia
4 Department of Biological and Environmental Engineering, College of Agriculture and Life Sciences, Cornell University, Ithaca, NY 14850, USA
* Correspondence: szelooi@um.edu.my (S.-L.S.); ilham@um.edu.my (Z.I.); Tel.: +60-37967-4014 (Z.I.)

Simple Summary: Invasive species are a leading hazard to marine ecosystems worldwide, coupled with climate change. Tackling the emerging biodiversity threat to maintain the ecological balance of the largest biome in the world has now become a pivotal part of the Sustainable Development Goals (SDGs). Marine herbivores are generally regarded as biological agents that restrict invasive species, and their efficiency depends on their dietary habits, especially the autotrophs they eat. Many researchers have found contradicting findings on the effects of nutritional attributes and novelty of autotrophs on herbivore eating behaviour. In light of the scattered literature on the mechanistic basis of autotroph-herbivore interactions, we provide a comprehensive review to fill knowledge gaps about synergies based on macroalgae, an important group of photosynthetic organisms in the marine biome that interact strongly with generalist herbivores. We also analyse macroalgal defence measures against herbivores, underlining unique features and potential roles in maintaining marine ecosystems. The nutritional qualities, shape, and novelty of autotrophs can alter herbivore feeding behaviour. Future research should explore aspects that can alter marine autotroph-herbivore interactions to resolve inconsistent results of specific features and the uniqueness of the organisms involved.

Abstract: Species invasion is a leading threat to marine ecosystems worldwide, being deemed as one of the ultimate jeopardies for biodiversity along with climate change. Tackling the emerging biodiversity threat to maintain the ecological balance of the largest biome in the world has now become a pivotal part of the Sustainable Development Goals (SDGs). Marine herbivores are often considered as biological agents that control the spread of invasive species, and their effectiveness depends largely on factors that influence their feeding preferences, including the specific attributes of their food–the autotrophs. While the marine autotroph-herbivore interactions have been substantially discussed globally, many studies have reported contradictory findings on the effects of nutritional attributes and novelty of autotrophs on herbivore feeding behaviour. In view of the scattered literature on the mechanistic basis of autotroph-herbivore interactions, we generate a comprehensive review to furnish insights into critical knowledge gaps about the synergies based largely on the characteristics of macroalgae; an important group of photosynthetic organisms in the marine biome that interact strongly with generalist herbivores. We also discuss the key defence strategies of these macroalgae against the herbivores, highlighting their unique attributes and plausible roles in keeping the marine ecosystems intact. Overall, the feeding behaviour of herbivores can be affected by the nutritional attributes, morphology, and novelty of the autotrophs. We recommend that future research should carefully consider different factors that can potentially affect the dynamics of the marine autotroph-herbivore interactions to resolve the inconsistent results of specific attributes and novelty of the organisms involved.

Citation: Cheng, A.; Lim, W.Y.; Lim, P.-E.; Yang Amri, A.; Poong, S.-W.; Song, S.-L.; Ilham, Z. Marine Autotroph-Herbivore Synergies: Unravelling the Roles of Macroalgae in Marine Ecosystem Dynamics. *Biology* **2022**, *11*, 1209. https://doi.org/10.3390/biology11081209

Academic Editors: Daniel Puppe, Panayiotis Dimitrakopoulos and Baorong Lu

Received: 28 May 2022
Accepted: 26 July 2022
Published: 12 August 2022

Keywords: autotroph-herbivore interactions; feeding behaviour; macroalgae; marine herbivores; nutrient acquisition

1. Introduction

In recent decades, mounting evidence suggests that biological invasions by invasive (also called alien or non-native) species are a growing threat to global biodiversity, and is exacerbated by climate warming [1,2]. Globalization, the transformation of technological regimes and expansions of transportation networks which modify the marine habitats are other recognized drivers behind the rapid shifting of invasive species across a broad geographical range [3–5]. In a narrower sense, species invasions can adversely influence the dynamics of specific communities, particularly concerning the extirpation of native species [6,7] and the reduction of species richness [8]. Climate, recipient communities, and invaders are considered the prime determinants of invasion impacts, with the characteristics of recipient communities being the most critical determinant [9]. The mechanisms of invasion impact on the diversity of native species, however, are still not well understood and in fact, previous findings on invasion consequences for species richness have been contradictory; viz., either positive, negative, neutral, or multifarious impacts [9]. This invasion paradox has led to many controversial debates over the past two decades [10]. The diversity and impact of invasive species on marine ecosystems are extensively covered in a recent review by Salimi et al. (2021) [11].

Studies on aquatic ecosystems showed that the interactions between marine herbivores and various plants and/or algae (hereinafter referred to as the "autotrophs") could reduce or even prevent the detrimental impacts of species invasions [12,13]. Lyons and Scheibling [14] reported that the establishment of the invasive green algae *Codium fragile* was enhanced by sea urchin food preference for kelps under increased water temperature and wave action, leading to increased herbivore pressure on local kelp stands. By and large, generalist marine herbivores such as most fishes and sea urchins that feed on autotrophs are common biological control agents that suppress the establishment and abundance of invasive species in the recipient communities [15,16]. It has been reported that the feeding (or grazing) preferences of the herbivores can determine the relationship between native or invasive autotrophs [17,18]. Recent findings also suggested that mechanisms underlying autotroph palatability could help resolve the inconsistent results of novelty [19,20].

Since the 1980s, efforts have been undertaken to understand the foraging behaviour of generalist marine herbivores [21–24]. Their selective foraging behaviour, which aims chiefly to regulate their nutritional needs for growth, fecundity, and performance [25,26], has been found to exert a profound impact on the biological structure of many marine ecosystems [27]. As such, theoretic insights on the nutritional relationships between herbivores and autotrophs will assist in the control and management of invasive species [28–30]. Generalist herbivores have also been found to make their food selection based on autotroph palatability, which depends primarily on their other unique attributes including, among others, secondary metabolites, morphology and physical stress [31–35]. Significant research has been devoted to examining the role and importance of some secondary metabolites in the survival and adaptation of autotrophs [36–38], but less attention has been paid to dissecting the value of their other attributes that may also influence the preferences of herbivores, i.e., whether to feed on native or invasive plants, or both [12]. It is worth noting that autotrophic characteristics may have the opposite effect on autotroph-herbivore interactions in controlled experimental studies where herbivores are restricted to a single autotroph species than in effects seen in field studies where herbivores are free to move around and cause natural autotroph damage. Future research examining the significance of autotroph features in interactions between autotrophs and herbivores must therefore carefully take into account the context in which the relationships have been observed [19].

In this study, relevant research papers are selected based on the pre-determined key search criteria (autotroph-herbivore interactions, feeding behaviour, macroalgae, marine herbivores, nutrient acquisition) and accessed via a reliable online database of reputable journals. One hundred and seven papers have been validated and reviewed comprehensively. At present, the available literature on the interactions between autotroph palatability and their various attributes on the nutritional ecology of marine herbivores is scattered and fragmentary [39–41]. It is also important to note that the chemical compounds in autotrophs that attract or deter their feeders are not well-addressed [42,43]. This review provides an overview of the unique attributes of macroalgae, which generally refers to the primary marine autotrophs and photosynthetic eukaryotes other than terrestrial plants that are capable of interacting strongly with marine herbivores [44]. Several key factors that can influence the feeding preferences of herbivores will be discussed, providing a better understanding of the interactions between autotrophs and herbivores that can potentially increase the ecological resilience of the marine biome. These are in line with the global trends in the marine sciences and sustainable development challenges, particularly in achieving the SDGs.

2. Marine Algae and Their Unique Attributes

Algae are the ultimate source of nutrients and energy for other organisms living in aquatic ecosystems. Although not considered plants, algae are photosynthetic in nature and produce over 70% of the global oxygen content [45–47]. Algae are also effective at sequestering carbon by converting almost 50% of the atmospheric carbon dioxide into organic molecules that build essential cellular constituents and intensify their energy production [48–52]. Macroalgae, being the most important primary producers in the oceans, house a wide range of nutritional quality within and among groups which often influences their palatability to herbivores [25,53]. For the most part, the proteins in macroalgae contain important amino acids, particularly the ones that cannot be synthesized by the animal body [54,55]. Animal hosts can thus obtain all these essential amino acids through symbiosis with the algae [56]. A variety of macroalgae reproduce either exclusively sexually or asexually, whilst some species demonstrate an alternation of generations involving both reproductive strategies in succession [57–59]. The following subsections discuss the unique characteristics and ecological relationships of each major group of macroalgae, including red algae (Rhodophytes), brown algae (Phaeophytes), and green algae (Chlorophytes) [51,52]. Figure 1 depicts the three major groups of macroalgae and examples of their common species.

2.1. Red Algae (Division Rhodophyta)

The first group is the eukaryotic red algae, or the Rhodophytes, comprising more than 6000 species of primarily marine algae ranging from microscopic to macroscopic in size [60,61]. These algae store their energy as a specialized polysaccharide, known as floridean starch, and their cell walls are made of unique cellulose and polysaccharides, such as agars and carrageenan galactans [62–64]. However, some other red algae may adopt sulfated mannans or neutral xylans as the main cell wall components rather than carrageenans [63]. Their photosynthetic pigments include chlorophylls *a* and *d*, while their accessory pigments are carotenoids, phycobilins, and xanthophyll [60,65,66] (Table 1). Some notable examples of red algae are, among others, filamentous species like *Pleonosporum* spp. and coralline algae like *Porolithon* spp., which contribute significantly to the building of tropical reefs and thalloid species. It is worth noting that the red algae have no flagellated cells or cells with any vestigial structure of flagellation [20]. Irish moss (*Chondrus crispus* Stackhouse), also known as the carrageen moss, is an example of an economically important red alga which has been used to bind proteins together to stabilize and add texture to various foods and beverages like ice cream, yogurt, and deli meats [67,68]. Another economically and nutritionally important species of red algae is nori (*Porphyra umbilicalis* Kützing); a high-protein and high-fibre algae which is commonly used in Japanese cuisine as an ingredient to wrap sushi [69]. *Porphyra* was proved to have the greatest protein content (ca. 35%) among the marine macroalgae, while some members of the brown algae in the order Laminariales have the lowest content (ca. 7%) [70,71].

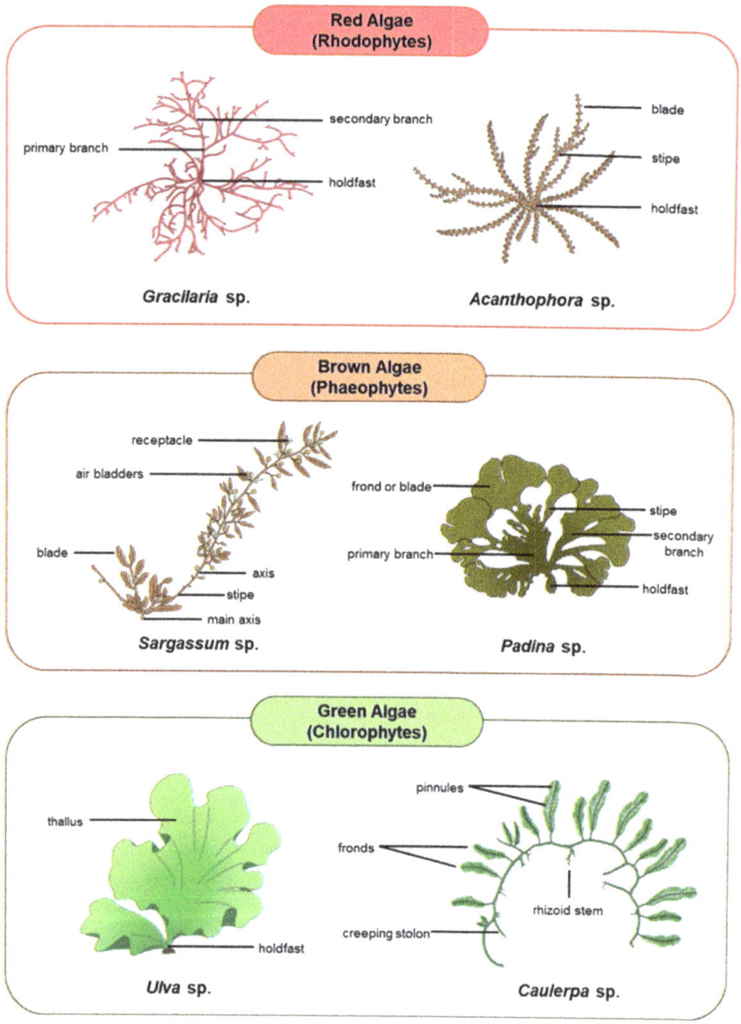

Figure 1. Major groups of macroalgae and examples of their common species.

Table 1. Major groups of macroalgae and their attributes.

Major Groups	Pigments	Cell Wall	Storage Components
Red algae (Rhodophytes)	Chlorophyll *a* (*d* in some Florideophyceae), R- and C- phycocyanin, allophycocyanin, R- and B-phycoerythrin, Alpha- and Beta-carotene, xanthophylls	Cellulose, xylans, galactan, alginate in corallinaceae	Floridean starch
Brown algae (Phaeophytes)	Chlorophyll *a*, *c*, Beta-carotene, fucoxanthin, xanthophylls	Cellulose, alginic acid, fucoidan	Laminaran, mannitol
Green algae (Chlorophytes)	Chlorophyll *a*, *b*, Alpha-, Beta- and Gamma- carotene, xantophylls	Cellulose, hydroxyproline glucosides, xylans, mannans, absent wall, calcified in some	Starch, oil

2.2. Brown Algae (Division Chromophyta)

In contrast to other algal groups, brown algae or the Phaeophytes are mostly developed from a secondary endosymbiosis event which involved a non-photosynthetic eukaryote and a unicellular red alga. Resultantly, brown algae exhibit several morphological and metabolic features that make them the most complex macroalgae [72]. Phaeophytes are mostly macroscopic in size, inclusive of the giant kelp (*Macrocystis pyrifera* (Linnaeus) C.Agardh), which can grow up to 10 m in length [73]. Most of the approximately 1800 species of brown algae live in the marine environment, especially in cool temperate waters located in both the Northern and Southern Hemispheres [74,75]. Fucans and alginates are the specific polysaccharides compounds, which can be found in the cell wall of brown algae [72]. Generally, brown algae consist of three distinctly recognizable parts–the holdfast, stipe, and leaf-like blades [76]. The holdfast is a root-like structure at the bottom, which is often joined by a stipe to one or more leaf-like blades depending on the species. The blades serve as the primary surface for important processes including photosynthesis and nutrient exchange in the algae [77,78]. Although photosynthesis takes place predominantly in the blades, it is crucial that the stipe has the adequate length to place the blades sufficiently close to the light source. Alternatively, algae can absorb sufficient light by swelling the body (thallus) or increasing their growth rate [79]. The photosynthetic pigments in brown algae are chlorophylls *a* and *c*, and their accessory pigments include carotenoids and xanthophylls [80] (Table 1). Fucoxanthin contains brown-coloured pigment and the unique xanthophyll in brown algae which gives them their characteristic dark colour [81]. Unlike red algae, most of the brown algae have two flagella which help them achieve locomotion [82]. Some examples of brown algae include the rockweeds (*Ascophyllum* spp. and *Fucus* spp.) and the giant kelps (*Macrocystis* sp.). These algae usually contain laminarin and mannitol, storage sugars which can be fermented to make alcohol [83]. Some brown algae possess the ability to take up certain important substances from seawater. For instance, the iodine concentration in an edible kelp, kombu, can be thousands of times as great in the cells of the species as in its surrounding water [84].

2.3. Green Algae (Division Chlorophyta)

On the other hand, green algae or the Chlorophytes are generally more closely related to the higher plants in comparison to brown and red algae, in particular their chloroplast structure [85,86]. The cell walls of most species of green algae are built mainly by cellulose, with some incorporation of glycans (hemicelluloses) [87]. Their photosynthetic pigments in the chloroplast are chlorophylls *a* and *b*, while their accessory pigments are carotenoids and xanthophylls, found in embryophytes [87] (Table 1). Green algae comprise of 9000 to 12,000 species, with the majority of them occurring in freshwater rather than the marine environments [86,87]. Most green algae are microscopic, except for a small number of species in some specific genera such as those in *Cladophora* which are multicellular and macroscopic [87–89]. The unicellular genera *Chlamydomonas* and *Chlorella* are some common examples of green algae in both marine and freshwater ecosystems worldwide, which consist of species that disperse in a wide range of habitats [90]. An example of more complex green algae includes *Volvox*, which forms large hollow-spherical colonies that consist of thousands of cells [91]. The green algae *Ulva* spp., *Caulerpa* spp., *Enteromorpha* spp., and *Codium* spp. are commonly used as a food source for humans. The *Ulva* spp., known generally as sea lettuce, are extensively consumed in many Asian countries especially in Japan, China, and the Republic of Korea [86,92]. Access to nitrogen is one of the major limiting factors in the growth of green algae on the grounds that most of them thrive in shallow water [93]. Nevertheless, the increased runoff of fertilizer-related nitrogen into the oceans, mainly from agriculture has created favourable conditions for the growth of green algae and also other groups of algae in the past few decades [94]. According to Lee (2018), the majority of green algae form zoogametes, which are motile flagellated gametes [20]. The review by Moreira et al. (2021) details how macroalgae from various divisions differ in their flagellal construction, orientation, and life cycle in general [89].

3. At a Glance: Key Defence Strategies of Marine Macroalgae against Herbivores

The base of a marine food web is dominated by photosynthetic autotrophs, notably macroalgae and microalgae (phytoplankton), which are the main producers of half of the Earth's oxygen and also the organic carbon required by all marine animals to survive [95–98]. The next level of the marine food web is made up of herbivores, from small zooplankton to larger animals (such as herbivorous fishes and manatees) that eat up a huge number of macroalgae [99–101]. Figure 2 illustrates a simplified conceptual model of the interaction between different levels of the marine food web, including macroalgae, herbivore, and predator, with the non-native autotroph having direct defences against herbivores. The interactions between herbivores and macroalgae are indeed one of the key drivers of marine ecosystem dynamics, gaining increasing scientific attention in recent decades (Table 2). Unfortunately, the synergies are currently being altered by climate change, affecting macroalgae growth rates and phenology, expression of chemical defenses, and herbivore behaviour and metabolism [102–111]. These macroalgae, however, have developed a variety of defence mechanisms to help them avoid herbivory and ensure their survival and abundance, as discussed in Sections 3.1 and 3.2.

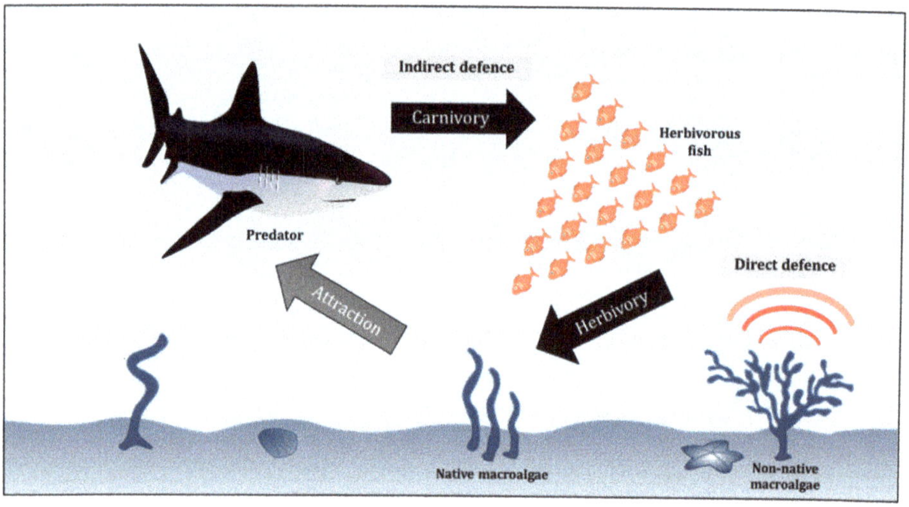

Figure 2. Conceptual model of macroalgae-herbivore-predator interaction.

Table 2. Key findings in autotroph-herbivore studies conducted since 2000.

Location	Autotroph(s)	Herbivore(s)	Key Findings	Reference
Australia	Algal turfs	Herbivorous fishes (Acanthuridae, Scaridae and Siganidae)	Fish response mechanisms to habitat-specific differences in food production remain unclear	[112]
Caribbean and Brazil	Macroscopic algae	Herbivorous fishes (Acanthuridae and Scaridae)	Temperature-related feeding processes are most likely involved in the distribution patterns of herbivores	[113]
Caribbean-Florida	Sea grass beds	Herbivorous fishes (Acanthuridae, Scaridae, and Pomacentridae)	Robust herbivorous fish assemblages can limit reefs from further macroalgal domination	[114]
Colombia	Macroalgae	Herbivorous fishes (Gobiidae, Pomacentridae, Labridae, Mugilidae, Labrisomidae, Gobiesocidae and Muraenidae)	Small crustacean prey items dominated the diets of most species. Macroalgae and diatoms consumption by a significant number of species was also observed	[115]
Caribbean	Algal turfs	Herbivorous fishes (Acanthuridae and Scaridae)	Herbivores in promoting reef recovery and resilience may depend on their feeding preferences, abundance, and biomass	[116]

Table 2. *Cont.*

ocation	Autotroph(s)	Herbivore(s)	Key Findings	Reference
ed Sea	Macroalgae	Sea urchins and herbivorous fish	Herbivores as a crucial top-down factor in controlling both benthic algal biomass and composition	[117]
apan	Algal beds (kelp)	Herbivorous fishes (Acanthuridae and Scaridae)	The importance of temperature-mediated fish herbivory in limiting the development of kelp populations in southern Japan is confirmed	[118]
Mediterranean Sea	Algae	Herbivorous fishes (Acanthuridae)	Expansion of tropical rabbitfishes poses a major threat to shallow water Mediterranean ecosystems	[118]
Portugal	Seagrass	Mesograzers (Amphipod and isopod)	Intraspecific variation should not be ignored when classifying a single seagrass species with respect to herbivory vulnerability. Seagrass structural traits confer mechanical resistance	[119]
Baltic Sea	Phytoplankton	Predatory zooplankton	Role of zooplankton filter feeders in controlling the development of phytoplankton	[120]
Malaysia	Macroalgae	Herbivorous fish (Chanidae)	Feeding behaviour of a herbivore could be influenced by the nutritional quality, morphology, and geography of the autotrophs	[39]

3.1. Physical Defences

A multitude of scientific research indicated that ocean warming has caused ecological impacts on various marine flora and fauna species across the globe, with a range of species marching away from their native homes in search of cooler climes [120–133]. With heatwaves sweeping through oceans twice as much as they did in the early 1990s, many biodiversity hotspots around the world are on the verge of imminent collapse. Many marine algae exhibit morphological plasticity that allows them to thrive in diverse habitats with various environmental pressures [134]. The study conducted by Diaz-Pulido et al. [134] showed that the morphology of different species of brown algae (*Padina boergesenii*) was significantly affected not only by herbivory but also by climatic and oceanographic factors, and this suggested that algal response to herbivory could also be a seasonal process [135–139]. Populations from more variable environments are considered to be more plastic [140], and algal phenotypic plasticity is potentially another pivotal mechanism that enables algae to respond to either fluctuating environments [141] or species invasion [142]. According to Fordyce [143], ecological interactions mediated by phenotypic plasticity, which are typical in nature, depend heavily on the morphological responses of the interacting organisms.

Hard encrusting calcified algae are common in the tropics where grazing is severe. Calcification of the coralline algal thallus is thought to have evolved as an adaptation to protect reproductive structures from herbivory by developing a multi-layered thallus in which reproductive structures are sunken beneath the calcareous surface cells and are thus protected from grazer access [144]. The calcified thalli may also decrease digestibility and in herbivorous herbivores (such as crabs) cause wear of chelipeds, mandibles, and the teeth of the gastric mill [145]. Increasing anthropogenic CO_2 emissions have led to elevated oceanic pCO_2 which may impact the structural integrity and protective function of the calcified thallus by decreasing calcification rates, and thus increasing the vulnerability of the coralline algae to bioerosion and grazing by excavating herbivores such as sea urchins and parrotfishes [146]. Non-calcifying macroalgae, on the other hand, typically use thallus toughness or mechanical strength as means of physical defence [136]. It is important to note that herbivore foraging is not essentially detrimental to the marine autotrophs. For example, limpets and chitons reportedly encourage coralline growth by regularly removing algal epiphytes from the surface of the coralline algae, which is necessary to avoid eventual overgrowing and killing of the coralline crust [147].

3.2. Chemical Defences

The chemical strategies of defence against herbivore are complex and generally assigned to two defence mechanisms–the direct and indirect defences [103,109]. In response to herbivory, direct defences are biologically mediated by autotroph chemistry and thus these defences can change the biological functions of the herbivores, including their feeding patterns, growth, and survival. In contrast, indirect defences against herbivory depend upon other species such as the natural enemies of the herbivores [110,111] (Figure 2). The chemical ecology of macroalgae has been widely elucidated in various regions and habitats, focusing primarily on herbivore offence and oxidative burst responses, which are chemical defences activated against pathogens and biofuels (Potin, 2008) [121–123]. An enormous diversity of secondary metabolites is regularly produced by autotrophs in response to herbivory in aquatic ecosystems [124,125]. Marine algae are known to be a viable source of specialized metabolites that play a crucial role in the ecosystem and climate functioning [126,127]. Tropical macroalgal taxa have been reported to produce a higher diversity of metabolites compared to their temperate counterparts, dominated by halogenated metabolites, terpenoids, acetogenins, and phenolics [127,128]. Mainly regulated by developmental, genetic, and environmental factors, these metabolites play diverse ecological functions in macroalgae, from being deterrent against herbivores to defenders to fight against specific pathogens and competitors for space with other marine organisms [127,129,130].

Over the past decade, considerable attention has been devoted to understanding the interactions between algal halogenated compound production and the environment, which includes global and anthropogenic climate changes [124,131]. Given that macroalgae produce a range of halogenated secondary metabolites, particularly chlorinated and brominated compounds that are predominant in red (90%) and green (7%) macroalgae, many studies have been conducted using these macroalgae to aid biosorption of pollutants in both industry and agriculture [124,132]. It is worth noting that halogenation of macroalgal components is involved in chemical defence mechanisms because halogenated metabolites are often associated with antibacterial, antifungal, antibacterial, and antioxidant properties [132,133].

4. Does Nutrient Acquisition in Algae Determine the Feeding Preferences of Marine Herbivores?

Palatability can be broadly defined as the characteristics and conditions of autotrophs that stimulate the animal to feed on them [33]. These include their structure, physical, and chemical attributes [148]. It has been long recognized that macronutrient composition influences palatability and foods that are higher in fat and protein content usually have higher palatability in terrestrial and aquatic ecosystems [149,150]. In marine communities, the preference and performance of the herbivores often relate directly to the nutritional value of algae or some other autotrophs, which is driven mostly by the protein and nitrogen content [151–155]. For example, several studies on the high-value marine abalone (Haliotis asinine) suggested that their diverse preferences are primarily influenced by the protein and nitrogen content of macroalgae [156–172]. Table 3 shows some examples of studies involving the interactions between marine autotrophs and herbivores based on the herbivore nutrient acquisition since the 2010s.

Living organisms require nitrogen to synthesize amino acids, the basic building blocks of protein that serve essential functions in virtually all biological processes [153]. Many previous studies have pointed out that low nitrogen consumption is associated with reduced food intake in generalist marine herbivores [39,148,160–163]. The study by Barile, Lapointe and Capo [156] on California sea hare (*Aplysia californica*) showed that this herbivorous gastropod preferred to feed on gracilarioid algae (*Gracilaria ferox*) with high levels of nitrogen. The lack of preference for protein-enriched algae (i.e., high-nutrient algae) can be explained by the compensatory feeding behaviour of some herbivore species. Previous studies reported the optimal growth rate and adequate intakes of limiting nutrients by testing the consumption rates of different herbivores on the low nutritional quality of algae foods [39,43,164,165]. According to Bradley et al. (2021), herbivores may avoid species that are less palatable or have lower nutritional value, which may affect their distribution and abundance [164]. However, the distinct nutritional drivers underlying the feeding preferences of specific marine herbivores are still a frontier that needs to be further explored.

Table 3. Examples of studies involving marine autotroph-herbivore interactions based on herbivore nutrient acquisition since the 2010s.

Nutrient	Marine Autotroph(s)	Marine Herbivore(s)	Ref(s)
Protein	Bull kelp (*Durvillaea antarctica*)	Talitrid amphipod (*Orchestoidea tuberculate*)	[25]
	Blade tissue of bull kelp (*Durvillaea antarctica*)	Talitrid amphipod (*Orchestoidea tuberculate*)	[167]
	Red algae (*Asparagopsis taxiformis*)	Abalone (*Haliotis asinina*)	[151]
	Grey weed (*Lessonia nigrescens*)	Talitrid amphipod (*Orchestoidea tuberculate*)	[48]
	Green seaweeds	White-spotted rabbitfish (*Siganus canaliculatus*)	[168]
	Brown algae (*Sargassum* spp.)	Marine isopod (*Idotea baltica*), periwinkle (*Littorina littorea*), and green sea urchin (*Psammechinus miliaris*)	[29]
	Epiphytic red algae	Butterfish (*Odax pullus*)	[26]
	Bull kelp (*Durvillea antarctica*)	Sea snail (*Diloma nigerrima*)	[157]
Nitrogen	Sea grapes (*Caulerpa racemosa*)	Purple sea urchin (*Paracentrotus lividus*)	[169]
	Brown forkweed (*Dictyota dichotoma*)	Long-spined sea urchin (*Diadema antillarum*) and herbivorous fishes	[170]
	Brown algae (*Sargassum yezoense*)	Sea urchin (*Hemicetrotus pulcherrimus*)	[160]
	Apical portions of brown algae fronds (*Sargassum* spp.)	Parrotfish (*Sparisoma aurofrenatum* and *Sparisoma chrysopterum*)	[171]
	Green algae (*Ulva* spp.)	Purple sea urchin (*Paracentrotus lividus*)	[163]
	Marine macroalgal species near Malaysian waters	Milkfish (*Chanos chanos*)	[39]
Carbon	Sea grapes (*Caulerpa racemosa*)	Purple sea urchin (*Paracentrotus lividus*)	[169]
	Macrophyte species in Northwestern Europe	Ringed China-mark (*Parapoynx stratiotata*	[12]
	Seagrass (*Cymodocea nodosa*)	Purple sea urchin (*Paracentrotus lividus*)	[163]
Phosphorus	Macrophyte species in Northwestern Europe	Ringed China-mark (*Parapoynx stratiotata*	[12]
	Apical portions of brown algae fronds (*Sargassum* spp.)	Surgeonfish (*Acanthurus coeruleus*) and parrotfish (*Sparisoma rubripinne* and *Sparisoma chrysopterum*)	[171]
	Marine macroalgal species near Malaysian waters	Milkfish (*Chanos chanos*)	[39]
Total phenolic	Bull kelp (*Durvillaea antarctica*)	Talitrid amphipod (*Orchestoidea tuberculate*)	[25]
	Bladder wrack (*Fucus vesiculosus*)	Flat periwinkle (*Littorina obtusata*)	[172]
	Bull kelp (*Durvillaea antarctica*)	Talitrid amphipod (*Orchestoidea tuberculate*)	[25]
	Marine macroalgal species near Malaysian waters	Milkfish (*Chanos chanos*)	[39]
Secondary metabolites	Bull kelp (*Durvillaea antarctica*)	Talitrid amphipod (*Orchestoidea tuberculate*)	[25]
	Brown algae (*Sargassum yezoense*)	Sea urchin (*Hemicetrotus pulcherrimus*)	[160]
	Brown algae (*Sargassum muticum*)	Periwinkle (*Littorina littorea*), and green sea urchin (*Psammechinus miliaris*)	[29]

5. Conclusions

One of the SDGs is uniquely dedicated to life below water, which is to conserve and sustainably use the oceans, seas, and marine resources for sustainable development. Nevertheless, maintaining the ecological balance of the largest biome in the world has become more challenging, especially when human-induced climate change continues to rapidly affect the diversity of marine life in an adverse way. To tackle the major threats to biodiversity, such as invasive species and habitat loss, it is worthwhile to dive into the unknown interactions between autotrophs and herbivores–those organisms that rule the base of the food chain. Theoretic insights on the synergies of autotrophs and herbivores in the marine biome are therefore crucial in controlling and managing invasive species, which almost always do more harm than good. A more concerted effort to test the major hypotheses in invasion biology, for example, the biotic resistance and enemy release hypotheses, is required to ensure the sustainability of the current marine ecosystems. Future research that aims to develop theories of marine ecology should be carefully designed, looking into various factors that can potentially affect the dynamics of different trophic levels within one or several food webs, including geographic variation and important attributes of the organisms involved. We strongly recommend the integration of evolutionary novelty theory with autotroph attributes and novelty in future studies to provide a better understanding of the consequences of biological invasions in the marine biome.

Author Contributions: Conceptualization, A.C. and Z.I.; data curation, W.Y.L.; funding acquisition, A.C., S.-L.S. and Z.I.; investigation, A.C., Z.I. and W.Y.L.; methodology, A.C. and W.Y.L.; project administration, A.C. and Z.I.; resources, A.C. and Z.I.; software, Z.I.; supervision, A.C.; validation, A.C. and Z.I.; visualization, A.Y.A., P.-E.L., S.-W.P. and S.-L.S.; writing—original draft, A.C., Z.I. and W.Y.L.; writing—review and editing, A.C., Z.I., S.-L.S. and W.Y.L. All authors have read and agreed to the published version of the manuscript.

Funding: This work was supported by Universiti Malaya [Grant Numbers: H-5620009 and ST007-2021]. The funders had no role in the preparation of the manuscript.

Institutional Review Board Statement: Not applicable.

Informed Consent Statement: Not applicable.

Data Availability Statement: Not applicable.

Acknowledgments: All the authors would like to acknowledge all the financial supports for this research project and the collaboration works between all the authors. A.C. would like to thank Usaha Fadzilat (M) Sdn. Bhd. for providing information on herbivorous fishes in Malaysia. Z.I. is Fulbright Visiting Research Scholar at Cornell University during the preparation of this work.

Conflicts of Interest: The authors declare no conflict of interest.

References

1. Thuiller, W.; Richardson, D.M.; Midgley, G.F. Will climate change promote alien plant invasions? In *Biological Invasions*; Springer: Berlin, Germany, 2008; pp. 197–211.
2. Seebens, H.; Essl, F.; Dawson, W.; Fuentes, N.; Moser, D.; Pergl, J.; Pyšek, P.; van Kleunen, M.; Weber, E.; Winter, M.; et al. Global trade will accelerate plant invasions in emerging economies under climate change. *Glob. Chang. Biol.* **2015**, *21*, 4128–4140. [CrossRef]
3. Pyšek, P.; Jarošík, V.; Hulme, P.E.; Kühn, I.; Wild, J.; Arianoutsou, M.; Bacher, S.; Chiron, F.; Didžiulis, V.; Essl, F.; et al. Disentangling the role of environmental and human pressures on biological invasions across Europe. *Proc. Natl. Acad. Sci. USA* **2010**, *107*, 12157–12162. [CrossRef]
4. Richardson, D.M.; Holmes, P.M.; Esler, K.J.; Galatowitsch, S.M.; Stromberg, J.C.; Kirkman, S.P.; Pyšek, P.; Hobbs, R.J. Riparian vegetation: Degradation, alien plant invasions, and restoration prospects. *Divers. Distrib.* **2007**, *13*, 126–139. [CrossRef]
5. Seebens, H.; Gastner, M.T.; Blasius, B. The risk of marine bioinvasion caused by global shipping. *Ecol. Lett.* **2013**, *16*, 782–790. [CrossRef]
6. Didham, R.K.; Tylianakis, J.M.; Hutchison, M.A.; Ewers, R.M.; Gemmell, N.J. Are invasive species the drivers of ecological change? *Trends Ecol. Evol.* **2005**, *20*, 470–474. [CrossRef]
7. Gilbert, B.; Levine, J.M. Plant invasions and extinction debts. *Proc. Natl. Acad. Sci. USA* **2013**, *110*, 1744–1749.
8. Winter, M.; Schweiger, O.; Klotz, S.; Nentwig, W.; Andriopoulos, P.; Arianoutsou, M.; Basnou, C.; Delipetrou, P.; Didžiulis, V.; Hejda, M.; et al. Plant extinctions and introductions lead to phylogenetic and taxonomic homogenization of the European flora. *Proc. Natl. Acad. Sci. USA* **2009**, *106*, 21721–21725. [CrossRef]
9. Dong, L.J.; Yu, H.W.; He, W.M. What determines positive, neutral, and negative impacts of Solidago canadensis invasion on native plant species richness? *Sci. Rep. UK* **2015**, *5*, 16804.

10. Stohlgren, T.J.; Rejmánek, M. No universal scale-dependent impacts of invasive species on native plant species richness. *Biol. Lett.* **2014**, *10*, 20130939. [CrossRef]
11. Salimi, P.A.; Creed, J.C.; Esch, M.M.; Fenner, D.; Jaafar, Z.; Levesque, J.C.; Montgomery, A.D.; Salimi, M.A.; Edward, J.K.P.; Raj, K.D.; et al. A review of the diversity and impact of invasive non-native species in tropical marine ecosystems. *Mar. Biodivers. Rec.* **2021**, *14*, 11. [CrossRef]
12. Grutters, B.M.C.; Roijendijk, Y.O.A.; Verberk, W.; Bakker, E. Plant traits and plant biogeography control the biotic resistance provided by generalist herbivores. *Funct. Ecol.* **2017**, *31*, 1184–1192. [CrossRef]
13. Parker, J.D.; Burkepile, D.E.; Hay, M.E. Opposing Effects of Native and Exotic Herbivores on Plant Invasions. *Science* **2006**, *311*, 1459–1461. [CrossRef]
14. Lyons, D.; Scheibling, R. Context-dependant survival of the invasive seaweed *Codium fragile* ssp. *tomentosoides* in kelp bed and urchin barren habitats off Nova Scotia. *Aquat. Biol.* **2008**, *2*, 17–27. [CrossRef]
15. Cebrian, E.; Ballesteros, E.; Linares, C.; Tomas, F. Do native herbivores provide resistance to Mediterranean marine bioinvasions? A seaweed example. *Biol. Invasions* **2010**, *13*, 1397–1408. [CrossRef]
16. Seastedt, T.R. Biological control of invasive plant species: A reassessment for the Anthropocene. *New Phytol.* **2015**, *205*, 490–502.
17. Joshi, J.; Vrieling, K. The enemy release and EICA hypothesis revisited: Incorporating the fundamental difference between specialist and generalist herbivores. *Ecol. Lett.* **2005**, *8*, 704–714. [CrossRef]
18. Shea, K.; Chesson, P. Community ecology theory as a framework for biological invasions. *Trends Ecol. Evol.* **2002**, *17*, 170–176.
19. Münzbergová, Z.; Skuhrovec, J. Data on Herbivore Performance and Plant Herbivore Damage Identify the Same Plant Traits as the Key Drivers of Plant–Herbivore Interaction. *Insects* **2020**, *11*, 865. [CrossRef]
20. Lee, R.E. *Phycology*; Cambridge University Press: Cambridge, UK, 2018; pp. 510–546.
21. Behmer, S.T.; Simpson, S.J.; Raubenheimer, D. Herbivore foraging in chemically heterogeneous environments: Nutrients and secondary metabolites. *Ecology* **2002**, *83*, 2489–2501.
22. Martinez, A.S.; Byrne, M.; Coleman, R.A. What and when to eat? Investigating the feeding habits of an intertidal herbivorous starfish. *Mar. Biol.* **2016**, *163*, 166. [CrossRef]
23. Senft, R.L.; Coughenour, M.B.; Bailey, D.W.; Rittenhouse, L.R.; Sala, O.; Swift, D.M. Large Herbivore Foraging and Ecological Hierarchies. *BioScience* **1987**, *37*, 789–799. [CrossRef]
24. Wahl, M.; Hay, M. Associational resistance and shared doom: Effects of epibiosis on herbivory. *Oecologia* **1995**, *102*, 329–340. [CrossRef]
25. Duarte, C.; Navarro, J.; Acuña, K.; Gómez, I. Feeding preferences of the sandhopper *Orchestoidea tuberculata*: The importance of algal traits. *Hydrobiologia* **2010**, *651*, 291–303. [CrossRef]
26. Johnson, J.S.; Clements, K.D.; Raubenheimer, D. The nutritional basis of seasonal selective feeding by a marine herbivorous fish. *Mar. Biol.* **2017**, *164*, 201. [CrossRef]
27. Taylor, D.I.; Schiel, D.R. Algal populations controlled by fish herbivory across a wave exposure gradient on southern temperate shores. *Ecology* **2010**, *91*, 201–211. [CrossRef]
28. Sagerman, J.; Enge, S.; Pavia, H.; Wikström, S.A. Low feeding preference of native herbivores for the successful non-native seaweed *Heterosiphonia japonica*. *Mar. Biol.* **2015**, *162*, 2471–2479. [CrossRef]
29. Schwartz, N.; Rohde, S.; Hiromori, S.; Schupp, P.J. Understanding the invasion success of *Sargassum muticum*: Herbivore preferences for native and invasive *Sargassum* spp. *Mar. Biol.* **2016**, *163*, 181. [CrossRef]
30. Thomas, M.B.; Reid, A.M. Are exotic natural enemies an effective way of controlling invasive plants? *Trends Ecol. Evol.* **2007**, *22*, 447–453. [CrossRef]
31. Chavanich, S.; Harris, L.G. The influence of macroalgae on seasonal abundance and feeding preference of a subtidal snail, *lacuna vincta* (montagu) (littorinidae) in the gulf of maine. *J. Molluscan Stud.* **2002**, *68*, 73–78. [CrossRef]
32. Molis, M.; Scrosati, R.A.; El-Belely, E.; Lesniowski, T.J.; Wahl, M. Wave-induced changes in seaweed toughness entail plastic modifications in snail traits maintaining consumption efficacy. *J. Ecol.* **2015**, *103*, 851–859. [CrossRef]
33. Pennings, S.C.; Siska, E.L.; Bertness, M.D. Latitudinal differences in plant palatability in Atlantic coast salt marshes. *Ecology* **2001**, *82*, 1344–1359.
34. Rodríguez, A.; Clemente, S.; Hernández, J.C.; Brito, A.; García, I.; Becerro, M.A. Nutritional, structural and chemical defenses of common algae species against juvenile sea urchins. *Mar. Biol.* **2017**, *164*, 127. [CrossRef]
35. Sudatti, D.B.; Fujii, M.; Rodrigues, S.V.; Turra, A.; Pereira, R.C. Prompt induction of chemical defenses in the red seaweed *Laurencia dendroidea*: The role of herbivory and epibiosis. *J. Sea Res.* **2018**, *138*, 48–55. [CrossRef]
36. Ianora, A.; Boersma, M.; Casotti, R.; Fontana, A.; Harder, J.; Hoffmann, F.; Pavia, H.; Potin, P.; Poulet, S.A.; Toth, G. New trends in marine chemical ecology. *Estuaries Coasts* **2006**, *29*, 531–551. [CrossRef]
37. Nylund, G.M.; Enge, S.; Pavia, H. Costs and Benefits of Chemical Defence in the Red Alga Bonnemaisonia hamifera. *PLoS ONE* **2013**, *8*, e61291. [CrossRef]
38. Maschek, J.A.; Baker, B.J. The Chemistry of Algal Secondary Metabolism. In *Algal Chemical Ecology*; Springer: Berlin, Germany, 2008; pp. 1–24. [CrossRef]
39. Yin, L.W.; Eem, L.P.; Amri, A.Y.; Looi, S.S.; Cheng, A. Exploring the role of macroalgal traits on the feeding behaviour of a generalist herbivore in Malaysian waters. *Bot. Mar.* **2020**, *63*, 407–417. [CrossRef]

40. Boyer, K.E.; Fong, P.; Armitage, A.R.; Cohen, R.A. Elevated nutrient content of tropical macroalgae increases rates of herbivory in coral, seagrass, and mangrove habitats. *Coral Reefs* **2004**, *23*, 530–538. [CrossRef]
41. Clements, K.D.; Raubenheimer, D.; Choat, J.H. Nutritional ecology of marine herbivorous fishes: Ten years on. *Funct. Ecol.* **2009**, *23*, 79–92. [CrossRef]
42. Wong, P.K.; Liang, Y.; Liu, N.Y.; Qiu, J.-W. Palatability of macrophytes to the invasive freshwater snail Pomacea canaliculata: Differential effects of multiple plant traits. *Freshw. Biol.* **2010**, *55*, 2023–2031. [CrossRef]
43. Machado, G.B.; Leite, F.P.; Sotka, E.E. Nutrition of marine mesograzers: Integrating feeding behavior, nutrient intake and performance of an herbivorous amphipod. *PeerJ* **2018**, *6*, e5929. [CrossRef]
44. Cock, J.M.; Coelho, S.M. Algal models in plant biology. *J. Exp. Bot.* **2011**, *62*, 2425–2430. [CrossRef]
45. Gislason, S. *Air and Breathing*; Environmed Research Inc.: Vancouver, BC, Canada, 2018; Volume 5.
46. Greenbaum, E.; Guillard, R.R.L.; Sunda, W.G. Hydrogen and Oxygen Photoproduction by Marine Algae. *Photochem. Photobiol.* **1983**, *37*, 649–655. [CrossRef]
47. Souvorov, A.V. *Marine Ecologonomics: The Ecology and Economics of Marine Natural Resources Management*, 1st ed.; Elsevier Science: Amsterdam, The Netherlands, 1999; Volume 6.
48. Arrigo, K.R. Carbon cycle: Marine manipulations. *Nature* **2007**, *450*, 491.
49. Chung, I.K.; Beardall, J.; Mehta, S.; Sahoo, D.; Stojkovic, S. Using marine macroalgae for carbon sequestration: A critical appraisal. *J. Appl. Phycol.* **2010**, *23*, 877–886. [CrossRef]
50. Moreira, D.; Pires, J.C. Atmospheric CO2 capture by algae: Negative carbon dioxide emission path. *Bioresour. Technol.* **2016**, *215*, 371–379. [CrossRef]
51. Bocanegra, A.; Bastida, S.; Benedi, J.; Ródenas, S.; Sánchez-Muniz, F.J. Characteristics and Nutritional and Cardiovascular-Health Properties of Seaweeds. *J. Med. Food* **2009**, *12*, 236–258. [CrossRef]
52. Makkar, H.P.S.; Tran, G.; Heuzé, V.; Giger-Reverdin, S.; Lessire, M.; Lebas, F.; Ankers, P. Seaweeds for livestock diets: A review. *Anim. Feed Sci. Technol.* **2016**, *212*, 1–17. [CrossRef]
53. Rothäusler, E.; Macaya, E.; Molis, M.; Wahl, M.; Thiel, M. Laboratory experiments examining inducible defense show variable responses of temperate brown and red macroalgae. *Rev. Chil. Hist. Nat.* **2005**, *78*, 603–614. [CrossRef]
54. Dawczynski, C.; Schubert, R.; Jahreis, G. Amino acids, fatty acids, and dietary fibre in edible seaweed products. *Food Chem.* **2007**, *103*, 891–899. [CrossRef]
55. Wang, J.T.; Douglas, A.E. Essential amino acid synthesis and nitrogen recycling in an alga-invertebrate symbiosis. *Mar. Biol.* **1999**, *135*, 219–222. [CrossRef]
56. Douglas, A.E. Host benefit and the evolution of specialization in symbiosis. *Heredity* **1998**, *81*, 599.
57. Barsanti, L.; Gualtieri, P. *Algae: Anatomy, Biochemistry, and Biotechnology*, 2nd ed.; CRC Press: Boca Raton, FL, USA, 2014.
58. Frenkel, J.; Vyverman, W.; Pohnert, G. Pheromone signaling during sexual reproduction in algae. *Plant J.* **2014**, *79*, 632–644. [CrossRef]
59. Clifton, K.E.; Clifton, L.M. The Phenology of Sexual Reproduction by Green Algae (Bryopsidales) on Caribbean Coral Reefs. *J. Phycol.* **1999**, *35*, 24–34. [CrossRef]
60. Gantt, E.; Grabowski, B.; Cunningham, F.X. Antenna Systems of Red Algae: Phycobilisomes with Photosystem II and Chlorophyll Complexes with Photosystem I. In *Light-Harvesting Antennas in Photosynthesis*; Springer: Dordrecht, The Netherlands, 2003; pp. 307–322. [CrossRef]
61. Masarin, F.; Cedeno, F.R.P.; Chavez, E.G.S.; De Oliveira, L.E.; Gelli, V.C.; Monti, R. Chemical analysis and biorefinery of red algae *Kappaphycus alvarezii* for efficient production of glucose from residue of carrageenan extraction process. *Biotechnol. Biofuels* **2016**, *9*, 122. [CrossRef]
62. Michel, G.; Helbert, W.; Kahn, R.; Dideberg, O.; Kloareg, B. The structural bases of the processive degradation of ι-carrageenan, a main cell wall polysaccharide of red algae. *J. Mol. Biol.* **2003**, *334*, 421–433.
63. Usov, A.I. Polysaccharides of the red algae. In *Advances in Carbohydrate Chemistry and Biochemistry*; Elsevier: Amsterdam, The Netherlands, 2011; Volume 65, pp. 115–217.
64. Vreeland, V.; Kloareg, B. Cell wall biology in red algae: Divide and conquer. *J. Phycol.* **2000**, *36*, 793–797.
65. Schubert, N.; García-Mendoza, E.; Pacheco-Ruiz, I. Carotrnoid composition of marine red algae. *J. Phycol.* **2006**, *42*, 1208–1216.
66. Squires, A.H.; Moerner, W.E. Direct single-molecule measurements of phycocyanobilin photophysics in monomeric C-phycocyanin. *Proc. Natl. Acad. Sci. USA* **2017**, *114*, 9779–9784. [CrossRef]
67. Rioux, L.-E.; Beaulieu, L.; Turgeon, S.L. Seaweeds: A traditional ingredients for new gastronomic sensation. *Food Hydrocoll.* **2017**, *68*, 255–265. [CrossRef]
68. Trius, A.; Sebranek, J.G.; Lanier, T. Carrageenans and their use in meat products. *Crit. Rev. Food Sci. Nutr.* **1996**, *36*, 69–85.
69. Blouin, N.A.; Brodie, J.A.; Grossman, A.C.; Xu, P.; Brawley, S.H. *Porphyra*: A marine crop shaped by stress. *Trends Plant Sci.* **2011**, *16*, 29–37. [CrossRef]
70. Fleurence, J. Seaweed proteins: Biochemical, nutritional aspects and potential uses. *Trends Food Sci. Technol.* **1999**, *10*, 25–28. [CrossRef]
71. Fleurence, J.; Morançais, M.; Dumay, J. Seaweed proteins. In *Proteins in Food Processing*; Elsevier: Amsterdam, The Netherlands, 2018; pp. 245–262.

72. Cock, J.M.; Sterck, L.; Rouzé, P.; Scornet, D.; Allen, A.; Amoutzias, G.; Anthouard, V.; Artiguenave, F.; Aury, J.-M.; Badger, J.H.; et al. The Ectocarpus genome and the independent evolution of multicellularity in brown algae. *Nature* **2010**, *465*, 617–621. [CrossRef]
73. Westermeier, R.; Patiño, D.J.; Müller, H.; Müller, D.G. Towards domestication of giant kelp (Macrocystis pyrifera) in Chile: Selection of haploid parent genotypes, outbreeding, and heterosis. *J. Appl. Phycol.* **2009**, *22*, 357–361. [CrossRef]
74. Aven, J.A.R.; Johnston, A.M.; Kübler, J.E.; Orb, R.E.K.; Cinroy, S.G.M.; Andley, L.I.L.H.; Crimgeour, C.H.M.S.; Alker, D.I.W.; Beardall, J.; Layton, M.N.C.; et al. Seaweeds in Cold Seas: Evolution and Carbon Acquisition. *Ann. Bot.* **2002**, *90*, 525–536. [CrossRef]
75. Wei, N.; Quarterman, J.; Jin, Y.-S. Marine macroalgae: An untapped resource for producing fuels and chemicals. *Trends Biotechnol.* **2013**, *31*, 70–77. [CrossRef]
76. Wernberg, T.; Thomsen, M.S. The effect of wave exposure on the morphology of Ecklonia radiata. *Aquat. Bot.* **2005**, *83*, 61–70. [CrossRef]
77. Koehl, M.A.R.; Silk, W.K.; Liang, H.; Mahadevan, L. How kelp produce blade shapes suited to different flow regimes: A new wrinkle. *Integr. Comp. Biol.* **2008**, *48*, 834–851. [CrossRef]
78. Stewart, H.L.; Carpenter, R.C. The Effects of Morphology and Water Flow on Photosynthesis of Marine Macroalgae. *Ecology* **2003**, *84*, 2999–3012. [CrossRef]
79. Toohey, B.; Kendrick, G.A.; Wernberg, T.; Phillips, J.C.; Malkin, S.; Prince, J. The effects of light and thallus scour from Ecklonia radiata canopy on an associated foliose algal assemblage: The importance of photoacclimation. *Mar. Biol.* **2004**, *144*, 1019–1027. [CrossRef]
80. Bidigare, R.R. Photosynthetic pigment composition of the brown tide alga: Unique chlorophyll and carotenoid derivatives. In *Novel Phytoplankton Blooms*; Springer: Berlin/Heidelberg, Germany, 1989; pp. 57–75.
81. Maria, A.G.; Graziano, R.; Nicolantonio, D.O. Anti-Obesity Activity of the Marine Carotenoid Fucoxanthin. *Mar. Drugs* **2015**, *13*, 2196–2214. [CrossRef]
82. Fu, G.; Nagasato, C.; Oka, S.; Cock, J.M.; Motomura, T. Proteomics Analysis of Heterogeneous Flagella in Brown Algae (Stramenopiles). *Protist* **2014**, *165*, 662–675. [CrossRef]
83. Horn, S.J.; Aasen, I.M.; Østgaard, K. Production of ethanol from mannitol by *Zymobacter palmae*. *J. Ind. Microbiol. Biotechnol.* **2000**, *24*, 51–57.
84. Amachi, S. Microbial Contribution to Global Iodine Cycling: Volatilization, Accumulation, Reduction, Oxidation, and Sorption of Iodine. *Microbes Environ.* **2008**, *23*, 269–276. [CrossRef]
85. Patron, N.J.; Keeling, P.J. Common evolutionary origin of starch biosynthetic enzymes in green and red algae1. *J. Phycol.* **2005**, *41*, 1131–1141. [CrossRef]
86. Kılınç, B.; Cirik, S.; Turan, G.; Tekogul, H.; Koru, E. *Seaweeds for Food and Industrial Applications*; IntechOpen: London, UK, 2013.
87. Lewis, L.A.; McCourt, R.M. Green algae and the origin of land plants. *Am. J. Bot.* **2004**, *91*, 1535–1556. [CrossRef]
88. Pulz, O.; Gross, W. Valuable products from biotechnology of microalgae. *Appl. Microbiol. Biotechnol.* **2004**, *65*, 635–648. [CrossRef]
89. Moreira, A.; Cruz, S.; Marques, R.; Cartaxana, P. The underexplored potential of green macroalgae in aquaculture. *Rev. Aquac.* **2021**, *14*, 5–26. [CrossRef]
90. Pinheiro, C.; Azevedo, J.; Campos, A.; Loureiro, S.; Vasconcelos, V. Absence of negative allelopathic effects of cylindrospermopsin and microcystin-LR on selected marine and freshwater phytoplankton species. *Hydrobiologia* **2012**, *705*, 27–42. [CrossRef]
91. Nozaki, H.; Mahakham, W.; Heman, W.; Matsuzaki, R.; Kawachi, M. A new preferentially outcrossing monoicous species of Volvox sect. Volvox (Chlorophyta) from Thailand. *PLoS ONE* **2020**, *15*, e0235622. [CrossRef]
92. García-Casal, M.N.; Ramirez, J.; Leets, I.; Pereira, A.C.; Quiroga, M.F. Antioxidant capacity, polyphenol content and iron bioavailability from algae (*Ulva* sp., *Sargassum* sp. and *Porphyra* sp.) in human subjects. *Br. J. Nutr.* **2008**, *101*, 79–85.
93. Li, Y.; Horsman, M.; Wang, B.; Wu, N.; Lan, C.Q. Effects of nitrogen sources on cell growth and lipid accumulation of green alga *Neochloris oleoabundans*. *Appl. Microbiol. Biotechnol.* **2008**, *81*, 629–636. [CrossRef]
94. Michalak, A.M.; Anderson, E.J.; Beletsky, D.; Boland, S.; Bosch, N.S.; Bridgeman, T.B.; Chaffin, J.D.; Cho, K.; Confesor, R.; Daloğlu, I.; et al. Record-setting algal bloom in Lake Erie caused by agricultural and meteorological trends consistent with expected future conditions. *Proc. Natl. Acad. Sci. USA* **2013**, *110*, 6448–6452. [CrossRef]
95. Maberly, S.C.; Pitt, J.-A.; Davies, P.S.; Carvalho, L. Nitrogen and phosphorus limitation and the management of small productive lakes. *Inland Waters* **2020**, *10*, 159–172. [CrossRef]
96. Bindoff, N.L.; Cheung, W.W.L.; Kairo, J.G.; Arístegui, J.; Guinder, V.A.; Hallberg, R.; Hilmi, N.; Jiao, N.; Karim, M.S.; Levin, L.; et al. Changing Ocean, Marine Ecosystems, and Dependent Communities. In *The Ocean and Cryosphere in a Changing Climate: Special Report of the Intergovernmental Panel on Climate Change, Intergovernmental Panel on Climate*; Cambridge University Press: Cambridge, UK, 2022; pp. 447–588.
97. Jorda, G.; Marbà, N.; Bennett, S.; Santana-Garcon, J.; Agusti, S.; Duarte, C.M. Ocean warming compresses the three-dimensional habitat of marine life. *Nat. Ecol. Evol.* **2019**, *4*, 109–114. [CrossRef]
98. Sorte, C.J.B.; Williams, S.L.; Carlton, J.T. Marine range shifts and species introductions: Comparative spread rates and community impacts. *Glob. Ecol. Biogeogr.* **2010**, *19*, 303–316. [CrossRef]
99. Wabnitz, C.C.C.; Lam, V.W.Y.; Reygondeau, G.; Teh, L.C.L.; Al-Abdulrazzak, D.; Khalfallah, M.; Pauly, D.; Palomares, M.L.D.; Zeller, D.; Cheung, W.W.L. Climate change impacts on marine biodiversity, fisheries and society in the Arabian Gulf. *PLoS ONE* **2018**, *13*, e0194537. [CrossRef]

100. Currin, C.; Newell, S.; Paerl, H. The role of standing dead Spartina alterniflora and benthic microalgae in salt marsh food webs: Considerations based on multiple stable isotope analysis. *Mar. Ecol. Prog. Ser.* **1995**, *121*, 99–116. [CrossRef]
101. Chapman, R.L. Algae: The world's most important "plants"—An introduction. *Mitig. Adapt. Strat. Glob. Chang.* **2010**, *18*, 5–12. [CrossRef]
102. Burkepile, D.E.; Parker, J.D. Recent advances in plant-herbivore interactions. *F1000Research* **2017**, *6*, 119. [CrossRef]
103. Jormalainen, V.; Honkanen, T. Macroalgal Chemical Defenses and Their Roles in Structuring Temperate Marine Communities. In *Algal Chemical Ecology*; Springer: Berlin/Heidelberg, Germany, 2008; pp. 57–89. [CrossRef]
104. Kinlan, B.P.; Gaines, S.D. Propagule dispersal in marine and terrestrial environments: A community perspective. *Ecology* **2003**, *84*, 2007–2020. [CrossRef]
105. Wallentinus, I. Introduced Marine Algae and Vascular Plants in European Aquatic Environments. In *Invasive Aquatic Species of Europe. Distribution, Impacts and Management*; Springer: Dordrecht, The Netherlands, 2002; pp. 27–52. [CrossRef]
106. Van Donk, E.; Ianora, A.; Vos, M. Induced defences in marine and freshwater phytoplankton: A review. *Hydrobiologia* **2010**, *668*, 3–19. [CrossRef]
107. Zamzow, J.; Amsler, C.; McClintock, J.; Baker, B. Habitat choice and predator avoidance by Antarctic amphipods: The roles of algal chemistry and morphology. *Mar. Ecol. Prog. Ser.* **2010**, *400*, 155–163. [CrossRef]
108. Sakanishi, Y.; Tanaka, K.; Kasai, H.; Tanaka, J. Characterization of thallus mechanical and physiological traits of tropical fucoids: A preliminary study. *Phycol. Res.* **2020**, *68*, 208–215. [CrossRef]
109. War, A.R.; Paulraj, M.G.; Ahmad, T.; Buhroo, A.A.; Hussain, B.; Ignacimuthu, S.; Sharma, H.C. Mechanisms of plant defense against insect herbivores. *Plant Signal. Behav.* **2012**, *7*, 1306–1320. [CrossRef]
110. Arimura, G.-I.; Matsui, K.; Takabayashi, J. Chemical and Molecular Ecology of Herbivore-Induced Plant Volatiles: Proximate Factors and Their Ultimate Functions. *Plant Cell Physiol.* **2009**, *50*, 911–923. [CrossRef]
111. Dicke, M.; van Poecke, R.M.; de Boer, J.G. Inducible indirect defence of plants: From mechanisms to ecological functions. *Basic Appl. Ecol.* **2003**, *4*, 27–42.
112. Russ, G.R.; Alcala, A.C. MARINE RESERVES: RATES AND PATTERNS OF RECOVERY AND DECLINE OF PREDATORY FISH, 1983–2000. *Ecol. Appl.* **2003**, *13*, 1553–1565. [CrossRef]
113. Floeter, S.R.; Behrens, M.D.; Ferreira, C.E.L.; Paddack, M.J.; Horn, M.H. Geographical gradients of marine herbivorous fishes: Patterns and processes. *Mar. Biol.* **2005**, *147*, 1435–1447. [CrossRef]
114. Paddack, M.J.; Cowen, R.K.; Sponaugle, S. Grazing pressure of herbivorous coral reef fishes on low coral-cover reefs. *Coral Reefs* **2006**, *25*, 461–472. [CrossRef]
115. Castellanos-Galindo, G.A.; Giraldo, A. Food resource use in a tropical eastern Pacific tidepool fish assemblage. *Mar. Biol.* **2008**, *153*, 1023–1035. [CrossRef]
116. Kopp, D.; Bouchon-Navaro, Y.; Louis, M.; Mouillot, D.; Bouchon, C. Juvenile Fish Assemblages in Caribbean Seagrass Beds: Does Nearby Habitat Matter? *J. Coast. Res.* **2010**, *26*, 1133–1141.
117. Jessen, C.; Wild, C. Herbivory effects on benthic algal composition and growth on a coral reef flat in the Egyptian Red Sea. *Mar. Ecol. Prog. Ser.* **2013**, *476*, 9–21. [CrossRef]
118. Vergés, A.; Steinberg, P.D.; Hay, M.E.; Poore, A.G.B.; Campbell, A.H.; Ballesteros, E.; Heck, K.L., Jr.; Booth, D.J.; Coleman, M.A.; Feary, D.A.; et al. The tropicalization of temperate marine ecosystems: Climate-mediated changes in herbivory and community phase shifts. *Proc. R. Soc. B Biol. Sci.* **2014**, *281*, 20140846. [CrossRef]
119. Martínez-Crego, B.; Arteaga, P.; Tomas, F.; Santos, R. The Role of Seagrass Traits in Mediating Zostera noltei Vulnerability to Mesograzers. *PLoS ONE* **2016**, *11*, e0156848. [CrossRef]
120. Kornijów, R.; Karpowicz, M.; Ejsmont-Karabin, J.; Nawrocka, L.; De Eyto, E.; Grzonkowski, K.; Magnuszewski, A.; Jakubowska, A.; Wodzinowski, T.; Woźniczka, A. Patchy distribution of phyto- and zooplankton in large and shallow lagoon under ice cover and resulting trophic interactions. *Mar. Freshw. Res.* **2020**, *71*, 1327–1341. [CrossRef]
121. Amsler, C.D. Induced defenses in macroalgae: The herbivore makes a difference. *J. Phycol.* **2001**, *37*, 353–356. [CrossRef]
122. Borell, E.M.; Foggo, A.; Coleman, R.A. Induced resistance in intertidal macroalgae modifies feeding behaviour of herbivorous snails. *Oecologia* **2004**, *140*, 328–334. [CrossRef]
123. Hemmi, A.; Jormalainen, V. Geographic covariation of chemical quality of the host alga Fucus vesiculosus with fitness of the herbivorous isopod *Idotea baltica*. *Mar. Biol.* **2003**, *145*, 759–768. [CrossRef]
124. Cabrita, M.T.; Vale, C.; Rauter, A.P. Halogenated Compounds from Marine Algae. *Mar. Drugs* **2010**, *8*, 2301–2317. [CrossRef]
125. Pavia, H.; Toth, G.B.; Åberg, P. Optimal defense theory: Elasticity analysis as a tool to predict intraplant variation in defenses. *Ecology* **2002**, *83*, 891–897.
126. Paul, C.; Pohnert, G. Production and role of volatile halogenated compounds from marine algae. *Nat. Prod. Rep.* **2011**, *28*, 186–195. [CrossRef]
127. Gaubert, J.; Payri, C.E.; Vieira, C.; Solanki, H.; Thomas, O.P. High metabolic variation for seaweeds in response to environmental changes: A case study of the brown algae *Lobophora* in coral reefs. *Sci. Rep.* **2019**, *9*, 993. [CrossRef]
128. Baker, B.J.; Amsler, C.D.; McClintock, J.B. Macroalgal Chemical Defenses in Polar Marine Communities. In *Algal Chemical Ecology*; Amsler, C.D., Ed.; Springer: Berlin/Heidelberg, Germany, 2008; pp. 91–103. [CrossRef]
129. Kooke, R.; Keurentjes, J.J.B. Multi-dimensional regulation of metabolic networks shaping plant development and performance. *J. Exp. Bot.* **2011**, *63*, 3353–3365. [CrossRef]

130. Dell, C.; Hay, M.E. Induced defence to grazing by vertebrate herbivores: Uncommon or under-investigated? *Mar. Ecol. Prog. Ser.* **2016**, *561*, 137–145.
131. Jerković, I.; Marijanović, Z.; Roje, M.; Kuś, P.M.; Jokić, S.; Čož-Rakovac, R. Phytochemical study of the headspace volatile organic compounds of fresh algae and seagrass from the Adriatic Sea (single point collection). *PLoS ONE* **2018**, *13*, e0196462. [CrossRef]
132. Nielsen, B.; Maneein, S.; Farid, A.; Milledge, J. The Effects of Halogenated Compounds on the Anaerobic Digestion of Macroalgae. *Fermentation* **2020**, *6*, 85. [CrossRef]
133. Küpper, F.C.; Miller, E.P.; Andrews, S.J.; Hughes, C.; Carpenter, L.J.; Meyer-Klaucke, W.; Toyama, C.; Muramatsu, Y.; Feiters, M.C.; Carrano, C.J. Emission of volatile halogenated compounds, speciation and localization of bromine and iodine in the brown algal genome model Ectocarpus siliculosus. *JBIC J. Biol. Inorg. Chem.* **2018**, *23*, 1119–1128. [CrossRef]
134. Diaz-Pulido, G.; McCook, L.J.; Larkum, A.W.; Lotze, H.K.; Raven, J.A.; Schaffelke, B.; Smith, J.E.; Steneck, R.S. Vulnerability of macroalgae of the Great Barrier Reef to climate change. *Phycologia* **2007**, *46*, 131–136.
135. Lewis, S.M.; Norris, J.N.; Searles, R.B. The Regulation of Morphological Plasticity in Tropical Reef Algae by Herbivory. *Ecology* **1987**, *68*, 636–641. [CrossRef]
136. Bittick, S.J.; Clausing, R.J.; Fong, C.R.; Fong, P. Bolstered physical defences under nutrient-enriched conditions may facilitate a secondary foundational algal species in the South Pacific. *J. Ecol.* **2016**, *104*, 646–653. [CrossRef]
137. Hay, M. The Functional Morphology of Turf-Forming Seaweeds: Persistence in Stressful Marine Habitats. *Ecology* **1981**, *62*, 739–750. [CrossRef]
138. Yñiguez, A.; McManus, J.; Collado-Vides, L. Capturing the dynamics in benthic structures: Environmental effects on morphology in the macroalgal genera *Halimeda* and *Dictyota*. *Mar. Ecol. Prog. Ser.* **2010**, *411*, 17–32. [CrossRef]
139. Charrier, B.; Le Bail, A.; de Reviers, B. Plant Proteus: Brown algal morphological plasticity and underlying developmental mechanisms. *Trends Plant Sci.* **2012**, *17*, 468–477. [CrossRef]
140. Schaum, C.E.; Collins, S. Plasticity predicts evolution in a marine alga. *Proc. R. Soc. B Biol. Sci.* **2014**, *281*, 20141486.
141. Graham, M.H.; Vasquez, J.A.; Buschmann, A.H. Global ecology of the giant kelp Macrocystis: From ecotypes to ecosystems. *Oceanogr. Mar. Biol.* **2007**, *45*, 39.
142. Yun, H.Y.; Molis, M. Comparing the ability of a non-indigenous and a native seaweed to induce anti-herbivory defenses. *Mar. Biol.* **2012**, *159*, 1475–1484. [CrossRef]
143. Fordyce, J.A. The evolutionary consequences of ecological interactions mediated through phenotypic plasticity. *J. Exp. Biol.* **2006**, *209*, 2377–2383. [CrossRef]
144. Steneck, R.S. Escalating herbivory and resulting adaptive trends in calcareous algal crusts. *Paleobiology* **1983**, *9*, 44–61. [CrossRef]
145. Kennish, R.; Williams, G.A.; Lee, S.Y. Algal seasonality on an exposed rocky shore in Hong Kong and the dietary implications for the herbivorous crab *Grapsus albolineatus*. *Mar. Biol.* **1996**, *125*, 55–64. [CrossRef]
146. Johnson, M.D.; Carpenter, R.C. Ocean acidification and warming decrease calcification in the crustose coralline alga Hydrolithon onkodes and increase susceptibility to grazing. *J. Exp. Mar. Biol. Ecol.* **2012**, *434*, 94–101.
147. Borowitzka, M.A.; Vesk, M. Ultrastructure of the corallinaceae. I. The vegetative cells of *Corallina officinalis* and *C. cuvierii*. *Mar. Biol.* **1978**, *46*, 295–304. [CrossRef]
148. Van Alstyne, K.L.; Pelletreau, K.N.; Kirby, A. Nutritional preferences override chemical defenses in determining food choice by a generalist herbivore, *Littorina sitkana*. *J. Exp. Mar. Biol. Ecol.* **2009**, *379*, 85–91. [CrossRef]
149. Schai-Braun, S.C.; Reichlin, T.S.; Ruf, T.; Klansek, E.; Tataruch, F.; Arnold, W.; Hackländer, K. The European Hare (*Lepus europaeus*): A Picky Herbivore Searching for Plant Parts Rich in Fat. *PLoS ONE* **2015**, *10*, e0134278. [CrossRef]
150. Cebrian, J.; Shurin, J.B.; Borer, E.; Cardinale, B.J.; Ngai, J.T.; Smith, M.D.; Fagan, W.F. Producer Nutritional Quality Controls Ecosystem Trophic Structure. *PLoS ONE* **2009**, *4*, e4929. [CrossRef]
151. Angell, A.R.; Pirozzi, I.; De Nys, R.; Paul, N.A. Feeding Preferences and the Nutritional Value of Tropical Algae for the Abalone Haliotis asinina. *PLoS ONE* **2012**, *7*, e38857. [CrossRef]
152. Lemoine, N.P.; Giery, S.T.; Burkepile, D.E. Differing nutritional constraints of consumers across ecosystems. *Oecologia* **2014**, *174*, 1367–1376. [CrossRef]
153. Wu, G.; Bazer, F.W.; Dai, Z.; Li, D.; Wang, J.; Wu, Z. Amino Acid Nutrition in Animals: Protein Synthesis and Beyond. *Annu. Rev. Anim. Biosci.* **2014**, *2*, 387–417. [CrossRef]
154. Shepherd, S.A.; Steinberg, P.D. Food preferences of three abalone species with a review of the food of abalone. In *Abalone of The World: Biology, Fisheries and Culture*; Shepherd, S.A., Tegner, M.J., Guzman del Proo, S.A., Eds.; Blackwell Scientific: Oxford, UK, 1992; pp. 169–181.
155. Adin, R.; Riera, P. Preferential food source utilization among stranded macroalgae by Talitrus saltator (Amphipod, Talitridae): A stable isotopes study in the northern coast of Brittany (France). *Estuar. Coast. Shelf Sci.* **2003**, *56*, 91–98. [CrossRef]
156. Barile, P.J.; E Lapointe, B.; Capo, T.R. Dietary nitrogen availability in macroalgae enhances growth of the sea hare *Aplysia californica* (Opisthobranchia: Anaspidea). *J. Exp. Mar. Biol. Ecol.* **2004**, *303*, 65–78. [CrossRef]
157. Quintanilla-Ahumada, D.; Quijón, P.A.; Navarro, J.M.; Pulgar, J.; Duarte, C. Living on a trophic subsidy: Algal quality drives an upper-shore herbivore's consumption, preference and absorption but not growth rates. *PLoS ONE* **2018**, *13*, e0196121.
158. Renaud, S.M.; Luong-Van, J.T. Seasonal variation in the chemical composition of tropical Australian marine macroalgae. *Proc. Eighteenth Int. Seaweed Symp.* **2006**, *1*, 155–161. [CrossRef]

159. Bleakley, S.; Hayes, M. Algal Proteins: Extraction, Application, and Challenges Concerning Production. *Foods* **2017**, *6*, 33. [CrossRef]
160. Endo, H.; Suehiro, K.; Kinoshita, J.; Agatsuma, Y. Combined Effects of Temperature and Nutrient Enrichment on Palatability of the Brown Alga *Sargassum yezoense* (Yamada) Yoshida & T. Konno. *Am. J. Plant Sci.* **2015**, *6*, 275–282. [CrossRef]
161. Hauxwell, J.; McClelland, J.; Behr, P.J.; Valiela, I. Relative Importance of Grazing and Nutrient Controls of Macroalgal Biomass in Three Temperate Shallow Estuaries. *Estuaries* **1998**, *21*, 347–360. [CrossRef]
162. Vergés, A.; Alcoverro, T.; Romero, J. Plant defences and the role of epibiosis in mediating within-plant feeding choices of seagrass consumers. *Oecologia* **2010**, *166*, 381–390. [CrossRef]
163. Jiménez-Ramos, R.; Brun, F.G.; Egea, L.G.; Vergara, J.J. Food choice effects on herbivory: Intra-specific seagrass palatability and inter-specific macrophyte palatability in seagrass communities. *Estuarine, Coast. Shelf Sci.* **2018**, *204*, 31–39. [CrossRef]
164. Bradley, D.J.; Boada, J.; Gladstone, W.; Glasby, T.M.; Gribben, P.E. Sublethal effects of a rapidly spreading native alga on a key herbivore. *Ecol. Evol.* **2021**, *11*, 12605–12616. [CrossRef]
165. Duarte, C.; Acuña, K.; Navarro, J.M.; Gómez, I.; Jaramillo, E.; Quijón, P. Variable feeding behavior in *Orchestoidea tuberculata* (Nicolet 1849): Exploring the relative importance of macroalgal traits. *J. Sea Res.* **2014**, *87*, 1–7. [CrossRef]
166. Lyons, D.A.; Scheibling, R.E. Effect of dietary history and algal traits on feeding rate and food preference in the green sea urchin *Strongylocentrotus droebachiensis*. *J. Exp. Mar. Biol. Ecol.* **2007**, *349*, 194–204. [CrossRef]
167. Duarte, C.; Acuña, K.; Navarro, J.M.; Gómez, I. Intra-plant differences in seaweed nutritional quality and chemical defenses: Importance for the feeding behavior of the intertidal amphipod *Orchestoidea tuberculata*. *J. Sea Res.* **2011**, *66*, 215–221. [CrossRef]
168. You, C.; Zeng, F.; Wang, S.; Li, Y. Preference of the herbivorous marine teleost Siganus canaliculatus for different macroalgae. *J. Ocean Univ. China* **2014**, *13*, 516–522. [CrossRef]
169. Tomas, F.; Box, A.; Terrados, J. Effects of invasive seaweeds on feeding preference and performance of a keystone Mediterranean herbivore. *Biol. Invasions* **2010**, *13*, 1559–1570. [CrossRef]
170. Chan, A.; Lubarsky, K.; Judy, K.; Fong, P. Nutrient addition increases consumption rates of tropical algae with different initial palatabilities. *Mar. Ecol. Prog. Ser.* **2012**, *465*, 25–31. [CrossRef]
171. Shantz, A.; Ladd, M.; Burkepile, D. Algal nitrogen and phosphorus content drive inter- and intraspecific differences in herbivore grazing on a Caribbean reef. *J. Exp. Mar. Biol. Ecol.* **2017**, *497*, 164–171. [CrossRef]
172. Cacabelos, E.; Olabarria, C.; Incera, M.; Troncoso, J.S. Do grazers prefer invasive seaweeds? *J. Exp. Mar. Biol. Ecol.* **2010**, *393*, 182–187.

biology

MDPI

Review

Towards an Understanding of Large-Scale Biodiversity Patterns on Land and in the Sea

Grégory Beaugrand

CNRS, Univ. Littoral Côte d'Opale, Univ. Lille, UMR 8187 LOG, F-62930 Wimereux, France; gregory.beaugrand@univ-lille.fr

Simple Summary: Among such questions as the origin of the universe or the biological bases of consciousness, understanding the origin and arrangement of planetary biodiversity is one of the 25 most important scientific enigmas according to the American journal *Science* (2005). This review presents a recent theory called the 'macroecological theory on the arrangement of life' (METAL). METAL proposes that biodiversity is strongly influenced by the climate and the environment in a deterministic manner. This influence mainly occurs through the interactions between the environment and the ecological niche of species *sensu* Hutchinson (i.e., the range of species tolerance when several factors are taken simultaneously). The use of METAL in the context of global change biology has been presented elsewhere. In this review, I explain how the niche–environment interaction generates a mathematical constraint on the arrangement of biodiversity, a constraint called the great chessboard of life. The theory explains (i) why biodiversity is generally higher toward low-latitude regions, (ii) why biodiversity peaks at the equator in the terrestrial realm and why it peaks at mid-latitudes in the oceans, and finally (iii) why there are more terrestrial than marine species, despite the fact that life first appeared in the marine environment.

Abstract: This review presents a recent theory named 'macroecological theory on the arrangement of life' (METAL). This theory is based on the concept of the ecological niche and shows that the niche-environment (including climate) interaction is fundamental to explain many phenomena observed in nature from the individual to the community level (e.g., phenology, biogeographical shifts, and community arrangement and reorganisation, gradual or abrupt). The application of the theory in climate change biology as well as individual and species ecology has been presented elsewhere. In this review, I show how METAL explains why there are more species at low than high latitudes, why the peak of biodiversity is located at mid-latitudes in the oceanic domain and at the equator in the terrestrial domain, and finally why there are more terrestrial than marine species, despite the fact that biodiversity has emerged in the oceans. I postulate that the arrangement of planetary biodiversity is mathematically constrained, a constraint we previously called 'the great chessboard of life', which determines the maximum number of species that may colonise a given region or domain. This theory also makes it possible to reconstruct past biodiversity and understand how biodiversity could be reorganised in the context of anthropogenic climate change.

Keywords: biodiversity; climate; theory; metal; determinism; randomness; biogeography; bioclimatology

Citation: Beaugrand, G. Towards an Understanding of Large-Scale Biodiversity Patterns on Land and in the Sea. *Biology* **2023**, *12*, 339. https://doi.org/10.3390/biology12030339

Academic Editors: Daniel Puppe, Panayiotis Dimitrakopoulos, Baorong Lu and Ettore Randi

Received: 5 November 2022
Revised: 13 February 2023
Accepted: 14 February 2023
Published: 21 February 2023

1. Introduction

The discipline of biology covers all living systems, from the simplest organic molecules (molecular biology) to large biomes (biogeography), by crossing many organisational levels, such as cells, tissues, organs, species, biocoenoses, and ecosystems [1]. It is essentially a science of complexity (Box 1) [2,3]. Since the origin of life, whether on Earth or elsewhere [4], biological systems have constantly evolved to adapt to their environment [5]. Species have emerged or died at gradual or sometimes more sudden rates, apparent balances punctuated by periods when changes occur relatively quickly [6]. The variety of species is not only perceptible from a morphological or anatomical point of view but is also reflected in many life history traits (size, growth, lifespan) that influence reproduction and individual survival [7]. The diversity exhibited by the living is almost inexhaustible, and evolutionary tinkering may obscure a form of intelligibility that researchers aim to discover [8,9]. It is a subtle mix between chance and necessity [10]: chance because diversity finds its origin in the genetic variability maintained by mutations and intra and inter-chromosomal mixing, and necessity because there are fundamental limits, whether physical, genetic, physiological, or ecological.

Box 1. Complexity in ecology and the scientific approach we have adopted to consider it within the framework of METAL.

Complexity in biology
+ Innumerable actors and factors.
+ All elements are interconnected and interdependent.
+ Multiple actions and feedbacks at different organizational levels and spatio-temporal scales.
+ Nonlinearity (threshold effect, hysteresis).
+ Emergence of new properties that are difficult to predict from properties of the parts.

How to deal with this complexity within the framework of METAL theory?
+ The system is complex but it can be simplified at certain organizational levels (consideration of emergent properties) and to some spatio-temporal scales (i.e. at the largest scales).
+ At some organizational levels and spatio-temporal scales, the laws influencing the arrangement of biodiversity are simple.
+ Non-linearity can be overcome (e.g. the concept of niche considers elegantly the non-linear responses of species to environmental fluctuations).
+ The use of ecological properties at the organizational level and at the relevant spatio-temporal scales enables one to unify phenomena, patterns of variability and biological events that govern the arrangement of biodiversity.
+ Their unification gives a high level of coherence to the phenomena and observed events and improves their understanding and predictability.

The origin and evolution of biodiversity are now better known. Charles Darwin, and neo-Darwinism, laid the solid theoretical foundations [11–15]. However, there remains a fundamental question to be resolved. How is biodiversity and the species that compose it organised on our planet and how is the abundance or number of species modified in space and time [16]? These questions are fundamental because biodiversity strongly influences the functioning of ecosystems and thus regulates services such as atmospheric carbon dioxide sequestration, but also provisioning services, i.e., the exploitation of ecosystems [17–20]. Moreover, to understand how anthropogenic climate change will affect individuals, species, and biocoenoses, the essential prerequisite is (i) to understand how these biological systems are naturally organised and (ii) to identify cardinal factors and mechanisms responsible for the alterations to then anticipate the modifications caused by environmental changes.

In this review, I present the macroecological theory on the arrangement of life (METAL), a theory that proposes that biodiversity is strongly influenced by the climatic and environmental regime in a deterministic manner (https://biodiversite.macroecologie.climat.cnrs.fr; accessed on 1 February 2023). This influence mainly occurs through the interactions between the ecological niche of species *sensu* Hutchinson (i.e., the range of a species tolerance when several factors are taken simultaneously) and the climate and environment [17]. The niche–environment interaction is therefore a fundamental interaction in ecology that enables one to predict and unify (i) at a species level, local changes in abundance, species phenology, and biogeographic range shifts, and (ii) at a community level, the arrangement of biodiversity in space and time as well as long-term community/ecosystem shifts, including regime shifts [21–30]. This theory offers a way to make testable ecological and biogeographical predictions to understand how life is organised and how it responds to global environmental changes [26]. More specifically, I show how METAL helps in understanding (i) why there are more species at low latitudes than at the poles, (ii) why the peak of biodiversity is located at mid-latitudes in the oceanic domain and at the equator in the terrestrial domain, and (iii) finally, why there are more terrestrial than marine species, despite the fact that biodiversity has emerged in the oceans. METAL has not been tested on prokaryotes (Bacteria and Archaea) yet because the species concept is fuzzy in this group, being replaced by the concept of operational taxonomic unit (i.e., taxa defined by molecular data analysis) [31,32]. Moreover, the ecological niche of prokaryotes can be more diverse and extreme, especially for Archaea [33,34], and their geographical ranges can be wide [35]. Therefore, all ecological principles examined in this review are only relevant for eukaryotes.

2. Patterns of Variability in Nature

For millennia, humans have detected recurring patterns of variability in nature or cycles [17,36–41]. The multitude of environments that our planet conceals forces clades to adapt to the local conditions, a process that rapidly fills the niche space [42]. Biogeographic studies have provided compelling evidence that some species are present only in tropical environments, while others are exclusively found in temperate or polar regions [41,43–46]. For example, Figure 1 shows that the spatial distribution of marine zooplankton (here copepod crustaceans) exhibits distinct patterns of variability [47], some species are present in the cold Labrador current (*Calanus glacialis*), others essentially along the European continental shelf (*Candacia armata*) or in the waters of the north (*Paraeuchaeta norvegica*) or the south (*Clausocalanus* spp.) of the North Atlantic, or finally at the transition between these waters along the North Atlantic Current (*Metridia lucens*).

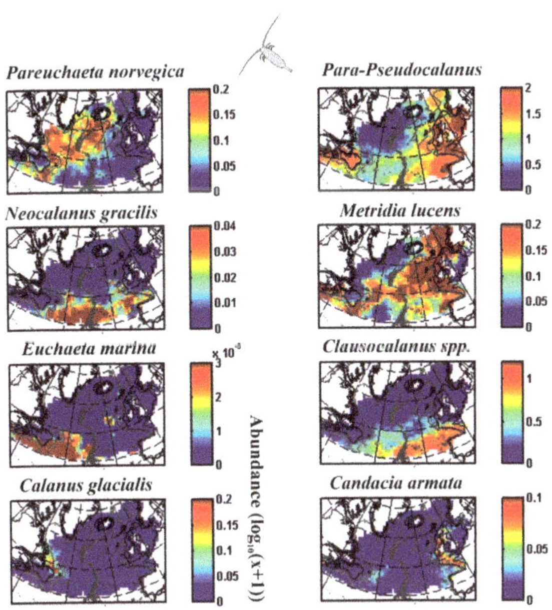

Figure 1. Mean spatial distribution of some marine copepods (planktonic marine crustaceans). Maximum abundance values are in red and zero abundances are in dark blue. The absence of colour corresponds to an absence of sampling. Some copepods are present in the icy or cold waters of the North Atlantic Ocean (*Pareuchaeta norvegica* or *Calanus glacialis*). Others occur in subtropical waters (*Clausocalanus* spp., *Neocalanus gracilis* and *Euchaeta marina*). The *Para-Pseudocalanus* group is present in temperate waters, *Metridia lucens* at the limit between temperate and cold waters, and *Candacia armata* mainly south of the European continental slope. These examples show that the distribution of species is not random on a large scale and that there are therefore control mechanisms. Redrawn, from Beaugrand and colleagues [47].

Bioclimatologists and ecologists have noted the existence of cycles where periods of high abundance alternate with periods of low abundance or even absence [37,40,48–50]. In temperate ecosystems (e.g., the North Sea), some species flourish in the spring; we speak of spring phenology. Others bloom in the summer; we then speak of summer phenology [49,50]. The presence of these recurrent patterns of variability in space or time suggests the existence of control mechanisms, whether autogenic or allogenic [50].

3. The Difficult Identification of Patterns in Ecology

It is more difficult than it seems to identify these patterns of variability in nature, and sometimes—especially on a small scale—it may seem that there are no rules governing the arrangement of biodiversity [17]. Take the example of the simulated distribution of individuals from a fictitious species in a hypothetical region (Figure 2).

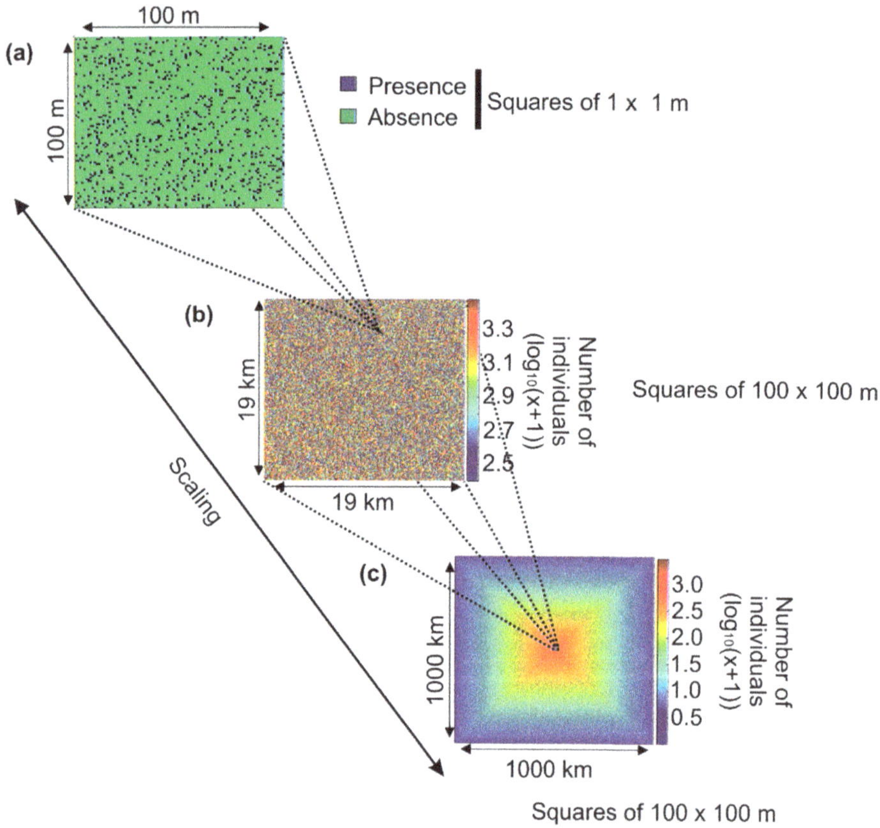

Figure 2. Hypothetical distribution of a species from the scale of a region of 100 × 100 m to a scale of 1000 × 1000 km. (**a**) On a local scale (100 × 100 m), the presence of individuals of the same species (blue squares, 1 × 1 m square) seems random. (**b**) On a more regional scale (19 × 19 km), the number of individuals is counted in each 100 × 100 m square. The density of individuals in the target region still seems random, although this density is between 2.4 and 3.5 (in decimal logarithmic scale). (**c**) On a large scale (1000 × 1000 km), a pattern of variability is clearly observed and the abundance of the species is greater towards the centre of the geographical domain. The transition from small to large scale is called scaling.

If we identify the number of individuals in an imaginary geographical square of 100 × 100 m, the distribution of individuals in this square appears random because no pattern of variability is identifiable (Figure 2a; the blue squares 1 × 1 m represent an individual). If we now cover an area of 19 × 19 km (there are 19,000/100 squares of 100 × 100 m in the figure, i.e., 190 × 190 = 36,100 squares), the distribution of the total number of individuals in each square of 100 × 100 m remains unintelligible (Figure 2b). Now imagine that we can examine the distribution of the number of individuals of this same species in a large region of 1000 × 1000 km (there are in Figure 2c 1,000,000/100 squares of 100 × 100 m, i.e., 10,000 × 10,000 = 100 million): we now see a pattern of variability emerging. The species is more abundant towards the centre of the region (Figure 2c). Looking at the pattern by taking height, that is to say, from a local to a large spatial scale, allows us to precisely identify the contours of the spatial distribution of this fictitious species (Box 1). An ecologist, who often studies biological systems on a small scale, may conclude that there are no detectable patterns of variability and that, therefore, the distribution of individuals is random and does not obey any rules (Box 1). On the other hand, a biogeographer may conclude that there is a structure, which implies the existence of underlying control mechanisms. The problem arises if researchers from these different disciplines extrapolate their results from the small to the large scale or inversely. In such a case, an ecologist may conclude that there are no principles governing the spatial distribution of a species and a biogeographer may establish certain predictions that are likely to be challenged on smaller scales. We touch here on the burning problem of scaling at the origin of so much controversy [51–53]. Referring to the analogy of an ecological theatre made by Hutchinson [54], Wiens [53] said, "to understand the drama, we must view it on the appropriate scale". Note that this phenomenon is also observed along the time dimension. It is therefore essential in the construction of all theories to specify its limits according to the spatio-temporal scales one considers [55].

4. Towards a Better Understanding of Principles of Biodiversity Organisation and Climate Change Biology

METAL (macroecological theory on the arrangement of life) has recently been proposed to connect a large number of phenomena observed in biogeography (spatial distribution of species, communities and biodiversity), ecology (phenology, gradual or abrupt changes in communities or biodiversity), paleoecology (past distribution of species, communities and biodiversity) and bioclimatology (biogeographic and phenological shifts, temporal changes in abundance and biodiversity at local or regional scales) [17,21,22,24,29,30,49,50,56] (https://biodiversite.macroecologie.climat.cnrs.fr; accessed on 1 February 2023).

The unification of these phenomena is obtained by using the concept of the ecological niche of Hutchinson [57,58], which constitutes the elementary macroscopic brick of the theory, giving meaning and coherence to all phenomena, patterns of variability or events cited above (Figure 3). METAL considers the fundamental niche (i.e., the niche without the influence of species interaction), and current models do not explicitly include the influence of biotic interaction yet [25,29]. The niche can be divided into five components: (i) climatic, (ii) physico-chemical, (iii) substrate, or trophic with (iv) dietary and (v) resource concentration components [23,25,49]. It integrates all environmental conditions where a species' individual can ensure its homeostasis, grow, and reproduce. A species' niche therefore includes phenotypic plasticity, encompassing polyphenism and reaction norm (i.e., a species niche integrates the niches of all individuals of that species).

Therefore, the niche–environment interaction is considered to be a fundamental interaction in biology that explains and unifies a large number of patterns observed in ecology, biogeography and climate change biology [26]. This occurs because the genome controls many processes at infraspecific organisational levels (e.g., molecular processes) that affect physiological and morphological traits that in turn influence individual performance and fitness and finally determine the ecological niche of a species [50] (Figure 3). The use of the niche makes it possible (i) to implicitly consider these infraspecific processes without having to model them and (ii) to integrate the emergence of new biological properties impossible to anticipate from the property of the individual parts when crossing one or several organisational levels (here from the molecular to the specific level)(Box 1) [59,60].

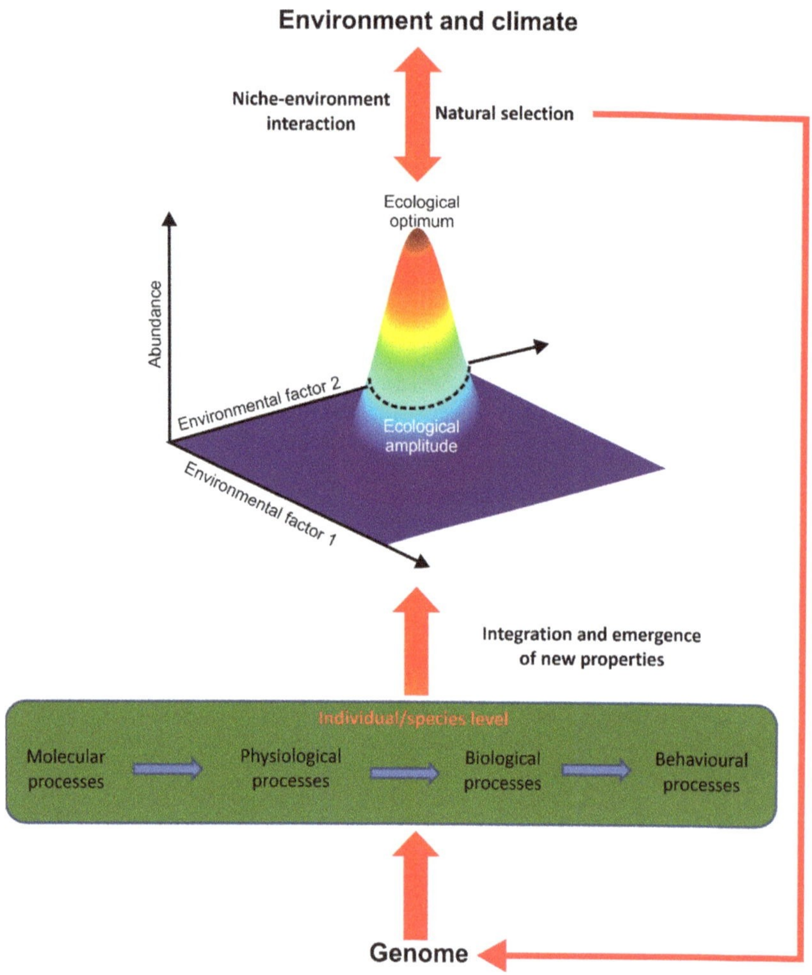

Figure 3. The concept of the ecological niche, the elementary macroscopic brick of METAL. The ecological niche of a species is quantified by simultaneously considering all the ecological factors that influence its abundance. The concept is therefore multidimensional. The ecological optimum represents the values of the ecological parameters for which the maximum abundance is observed. Ecological amplitude is the degree of ecological valence that a species tolerates. Put simply, it is the width of the ecological niche. The use of the ecological niche within METAL makes it possible to integrate molecular, physiological, biological and behavioural processes controlled in part by the genome and the environment. Such processes are impossible to model for all living species on our planet using a reductionist approach. Moreover, the concept of niche makes it possible to consider the emergence of new properties at a specific organisational level. The niche–environment (including climatic) interaction makes it possible to explain, unify and predict a large number of patterns observed in ecology, paleoecology, biogeography and climate change biology. The niche–environment interaction affects the species genome through processes involved in natural selection.

Also known as species distribution models (SDMs) or bioclimatic envelope models [61–63], METAL integrates ecological niche models (ENMs) in its framework. ENMs primarily focus on the realised niche, which is based on past or contemporary spatial distribution and some key environmental (including climatic) variables. They then use the realised niche to project the likely distribution of a species in the past, present or future.

ENMs have been extensively applied to project future species spatial distributions in the context of global climate change [61,64–69]. METAL provides a robust scientific baseline for ENMs and shows that this niche approach can be extended to explain many different phenomena at different organisational levels and spatio-temporal scales [22,70].

The niche–environment interaction is crucial for explaining, unifying and predicting a very large number of phenomena, patterns of variability or biological events observed in nature (Figure 4a) [26]. At an individual organisational level, the niche–environment interaction controls a large number of physiological and behavioural responses, such as the phenomena of thermotaxis and chemotaxis (Figure 4b) [17]. At a population organisational level, the niche–environment interaction controls species' phenology and long-term changes in local abundance, including the arrival or extirpation of individuals of a species in a given area (Figure 4c) [49,50]. At a specific organisational level, the niche–environment interaction controls the distributional range of a species and even its extinction (Figure 4d) [23].

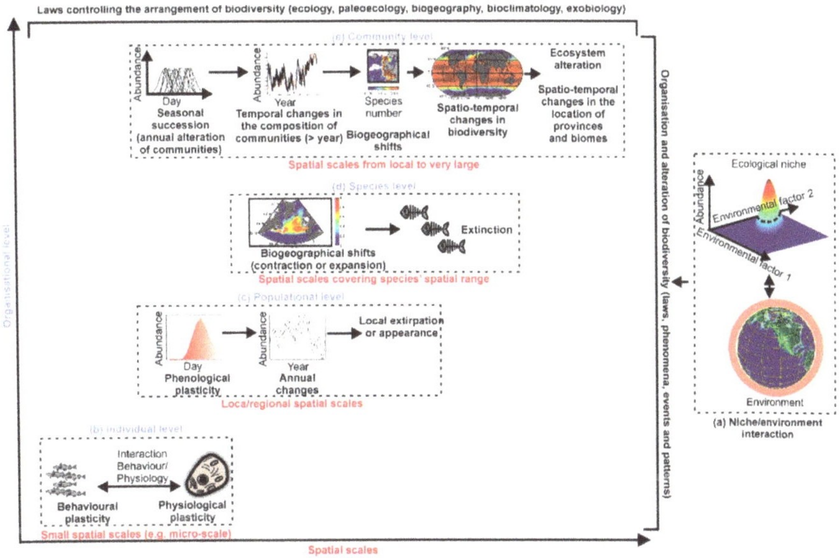

Figure 4. The niche–environment interaction and its influences on the arrangement of biological systems from the individual to the community organisational level and from the micro-scale to the mega-scale. (**a**) Niche–environment interaction. Organisational levels of individual (**b**), population (**c**), species (**d**) and community (**e**). Since the arrangement of communities affects the environment of their habitat, the influence of the niche–environment interaction on the community is also exerted on ecosystems and ecotones, provinces and biomes. In black (bold): phenomena, patterns of variability and biological events. Only the main ones are represented here. In blue (bold): organisational level. In red (bold): spatial scales.

At a community level, the niche–environment interaction helps in understanding how communities are formed and modified, thus providing a theoretical basis for synecology and phytosociology (Figure 4e) [49,50]. The theory explains the seasonal succession observed in the marine planktonic environment, the gradual or abrupt modifications in communities, the biogeographical changes of biocoenoses (or assemblages), their contractions or expansions, and their eventual disappearance (Figure 4e) [23,28]. We can thus explain and anticipate major biological changes but also understand how biodiversity is organised and how it can be altered in the context of climate change [27,56].

Note, however, that human activities now influence a large number of processes that can interfere with the niche–environment interaction. For example, the extinction of a species or its long-term changes can be explained by anthropogenic pressures such as fishing, land use, hunting and pollution, pressures that are not further mentioned here [25,71–73]. A METAL model has recently considered fishing pressure and the niche together to explain the long-term changes in cod spawning stock biomass in the North Sea since the beginning of the 1960s [25].

The use of METAL in the context of climate change biology has been presented elsewhere [26]. In this review, I show how the niche–environment interaction generates a mathematical constraint on the large-scale arrangement of biodiversity and explains why there are more species on land than in the marine realm. To make progress on these questions, the scientific community continues to collect and inventory species and to study their biology [74,75]. A study suggested that the number of terrestrial and marine species could be 8,740,000 and 2,210,000, respectively [76]. Because the scientific team estimated that 1,233,500 species had been inventoried in the terrestrial environment and 193,756 in the marine environment (bottom and surface), this suggests that between 9% (marine) and 14% (terrestrial) of species have been named and described so far. (Note, however, that there exist many estimations in the scientific literature [75,77–79].) Meanwhile, ecologists continue to investigate the multiple interactions of these species with the environment, including the climate, but also biotic interactions, an essential prerequisite for understanding their spatial distributions (biogeography), their temporal patterns of reproduction (phenology) and their fluctuations from seasonal to centenary and millennial, as well as changes occurring on a geological time scale (ecology, palaeoecology and bioclimatology) [56,80–86]. With such a poor fundamental knowledge on species' biology, how can we understand how factors and processes affect large-scale biodiversity patterns and design models to reconstruct them?

5. Large-Scale Biodiversity Patterns

5.1. A Brief Overview of the Main Hypotheses or Theories That Have Attempted to Explain Large-Scale Biodiversity Patterns

Why do some regions of the globe have more species than others? Among such scientific questions as the origin of life, the biological basis of consciousness or the composition of the universe, this question was cited as one of the 25 most important enigmas by the American journal *Science* in 2005 [16,87]. Indeed, for most taxonomic groups, it has been noticed that warm regions contain a higher number of species than polar regions [41,88–92]. Biogeographers generally speak of latitudinal biodiversity gradients to describe the large-scale biodiversity patterns observed in nature. The plural is important because the gradient may be different from one taxonomic group to another and from one domain to another (e.g., terrestrial and marine) [41]. For example, a maximum is obtained at the equator for a large number of terrestrial taxonomic groups, while it is rather subtropical for most oceanic taxonomic groups [41,88]. Although the existence of this biogeographical pattern has been known since Alexander von Humboldt in 1807 while he was in Central America and Charles Darwin after the return of the second expedition of the HMS *Beagle* in 1836, and that many hypotheses have been formulated for decades, no consensus has been reached [93–101].

What causes the latitudinal gradient in biodiversity, whether on land or in the sea, has been a topic of debate for decades, and more than 20 hypotheses or theories have been proposed [41,101–109]. It is beyond the scope of the present paper to review and discuss all of them, and below I only briefly review the main hypotheses or theories. While some authors have propounded that the biodiversity gradients are related to the larger area of the tropical belts [96,110], others have proposed null models of biodiversity, such as the neutral theory of biodiversity and biogeography [99] and the mid-domain effect (MDE) [111,112]. Moreover, it has been suggested that time is an important factor because speciation needs it to operate [90,113–117]. The tropics may assemble more species over a longer time period because they are more climatically stable than higher latitudes [95], and studies have provided evidence that the tropics are both a species cradle (higher origination rates) and a museum (more long-term climatic stability) [118,119]; See Vasconcelos and colleagues [120], however. Some studies have suggested that richer taxa have quicker diversification rates [121] and the metabolic theory of ecology predicts that the molecular clock is affected by body mass and temperature through metabolism [122].

Another popular hypothesis invokes the positive role of energy on biodiversity [95,123,124]. The energy hypothesis is frequently divided into two [123]: (i) exosomatic energy, where climatic factors such as temperature, precipitation and photosynthetically active radiation positively affect biodiversity, and (ii) endosomatic energy, which is the level of energy contained in the biomass that affects individuals and therefore the number of species. The latter hypothesis may be tested by using chlorophyll concentration or primary production [17]. Climate stability has also been invoked to explain the higher biodiversity in the tropics [125], along with magnitude, severity and frequency of environmental perturbations that are thought to limit species richness in temperate and polar regions [126,127]. In space, environmental heterogeneity promotes higher biodiversity [128,129]. For example, island species richness is positively correlated with habitat diversity [129]. The niche-assembly theory posits that there is more species richness in the tropics because there are more ecological niches, the niche being defined in term of resources [130]. Some hypotheses have invoked biotic interaction as a cause of speciation and therefore high species richness [131]. For example, Emerson and Kolm have provided evidence that the proportion of endemic species in an island covaries positively with biodiversity, suggesting that species richness increases speciation [17,131–133]; see, however, [134,135]. The argument seems tautological to some authors in terms of the search for the primary cause of these large-scale biodiversity patterns [17,95].

Perhaps the most compelling hypotheses are those that invoke an environmental control of biodiversity, such as environmental stability or energy availability [88,136,137]. Climatic hypotheses have been frequently proposed because large-scale biodiversity patterns correlate well with environmental parameters [88,137]. Among hypotheses, it has been suggested that global climate change may have shaped the large-scale patterns of biodiversity prevailing on Earth today because most clades originated in warm habitats, as temperatures have been predominantly warm during its history [138]. This hypothesis is known as tropical niche conservatism (TNC) [139]. Temperature has often been suggested to explain large-scale patterns in the distribution of marine organisms [88,140,141]. However, the exact mechanisms (e.g., metabolic theory of ecology [98]) by which the parameter may influence large-scale biodiversity patterns remain uncertain [100–102]. Finally, many authors have also suggested that many causes or factors interact to shape large-scale biodiversity patterns [30,142,143].

5.2. Modelling Biodiversity in METAL

Understanding the spatio-temporal arrangement of biodiversity on a large scale requires the development of numerical models where biological, environmental and climatic knowledge are put into equations [27,29,30,100]. In the context of the application of METAL, the fundamental bases of the biodiversity model are simple [100]. A large number of fictitious species is generated. Each fictitious species (called hereafter a pseudospecies) has unique physiological preferences that define their ecological (fundamental) niches, that is to say, their responses to climatic and environmental variability [23]. We can initially consider a simple niche, considering only the bioclimatic dimensions temperature and water availability (here precipitation). Temperature is an essential factor controlling the physiology of all species living on our planet and precipitation is a proxy for water availability, a variable just as important as temperature for terrestrial species. These climatic dimensions are fundamental, and many studies have underlined their importance [46,88,140].

Figure 5 shows an example with two marine pseudospecies, one being more eurythermal (i.e., tolerating a greater range of thermal variation) and the other being more stenothermal (i.e., tolerating a smaller range of thermal variation). In this example, we see that the stenothermal pseudospecies is characterised by a more limited range and lower abundance than the more eurythermal pseudospecies [21,23].

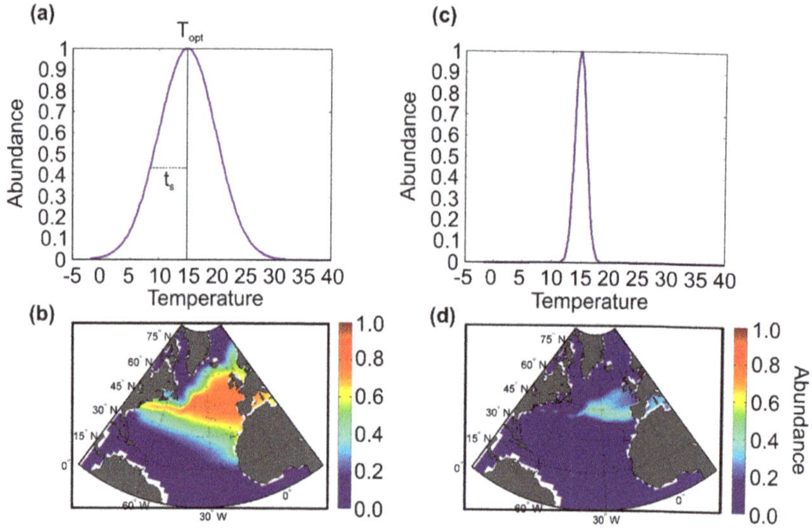

Figure 5. Idealised relationship between the ecological niche of a marine species and its spatial distribution. In this example, the ecological niche is a thermal niche with a Gaussian distribution characterised by two parameters: the optimum temperature and the thermal amplitude (parameter close to the standard deviation). The optimum temperature (T_{opt}) is 15 °C for the two fictitious niches (**a**,**c**). The thermal amplitude (t_S) is higher for (**a**) than (**c**). The spatial distribution is wider and the abundance of the species higher when the species has a thermal niche with a large thermal amplitude (**b**,**d**). In reality, the niche of a species is multidimensional. From Beaugrand and colleagues [23].

We can thus create a multitude of pseudospecies by varying the optimum and the ecological amplitude (i.e., niche breadth) of each niche dimension. Figure 6 shows the creation of marine pseudospecies from a simple Gaussian thermal niche [21,23]. Note that different types of niches can be used: from rectangular to trapezoidal [25,100] and from logistic to beta distribution [25,27], symmetrical or asymmetrical [25], parametric or nonparametric [23]. Moreover, the niche can be multidimensional [144], including nutrients, solar radiation or mixed-layer depth for phytoplankton, bathymetry and sediment types for fish, soil pH and composition for plants [25,65,66,144–146]. So far, most METAL simulations have been based on niches that vary between 0 (i.e., absence of a species for a given environmental regime) and 1 (i.e., highest abundance, or presence in case of a rectangular niche). Therefore, all species can reach the same level of maximum abundance. Although this assumption may possibly hold for a clade composed of species with a similar size, this is not so for a group that exhibits large size variability (e.g., mammals) [147–149]. Note, however, that this assumption does not affect biodiversity when the selected indicator is species richness (see below).

Different thermal optima and amplitudes are used [21,23]. In this example, when distributional ranges originating from one thermal niche are spatially separated, it is considered that they represent different species; therefore, one niche can give several species, in agreement with Buffon's law, also known as the first principle of biogeography [41]. We see that thermal niches with lower thermal amplitudes give more species, although they exhibit smaller distributional ranges (Figure 6, left maps vs. right maps). Figure 5 shows that there is a relationship between the average abundance of a species and its area of distribution, a relationship already demonstrated empirically by Brown [150]. We extended this relationship by indicating that there is a positive link between the ecological amplitude of a species, its average abundance and its distribution area [23] (Figures 5 and 6). These relationships hold for species of the same size [147,148] and trophic guild.

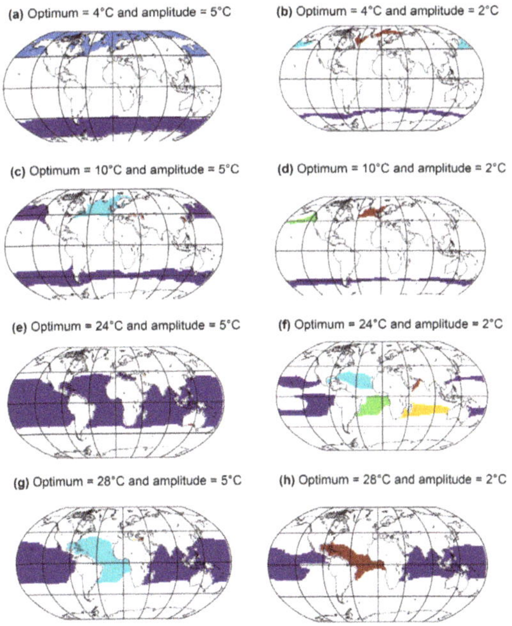

Figure 6. Different types of spatial distribution of marine species generated from thermal niches by varying the thermal optimum and amplitude. The different colours on the map represent different species generated from the same thermal niche. The same niche can give rise to several species if and only if individuals from different species cannot meet (allopatric speciation). Niches with a low thermal amplitude generate more species (e.g., (**a**,**b**) and (**e**,**f**)). The current location of continents at the equator and in the northern latitudes allows more species to form by allopatric speciation. Methods, from Beaugrand and colleagues [29].

Examples from Figure 6 show that a niche can lead to more pseudospecies in the Northern than in the Southern Hemisphere (Figure 6b–d). This is due to the current location of continents that act as a barrier against gene flux, triggering more allopatric speciation in the Northern than the Southern Hemisphere (towards high latitudes). When the thermal amplitude is larger, the pseudospecies are more eurygraph and a single niche leads to fewer pseudospecies, e.g., only one in Figure 6a in each hemisphere. Moreover, the current configuration (i.e., south to north configuration) of the continents also enables more pseudospecies to emerge in the tropics (Figure 6g–h), especially when the pseudospecies are stenoecious and therefore stenograph (Figure 6e,f). Note that parapatric and sympatric speciations are not accounted for in this example. Allopatric speciation is thought to be a widespread mode of speciation in the marine environment, despite more evidence that other modes of speciation might also play a role [151]. Parapatric speciation is thought to be possible in the ocean [152–154]. Clinal parapatric speciation has been suggested for salps and some benthic species [151,155]. Sympatric speciation might also be frequent for marine invertebrates [156].

To reproduce the large-scale arrangement of biodiversity, we can build a model that first creates millions of niches, which then allow pseudospecies to establish themselves in a given region as long as environmental fluctuations are suitable [17,29,30,100]. The principle of the model is simple. It starts to create a large number of niches where both optima and amplitudes with respect to temperature only for the marine realm and both temperature and precipitation for the terrestrial realm vary. Many niches (i.e., with all possible optima and amplitudes), which can also overlap, are created (i) for temperature between −1.8 °C and 44 °C in both realms and (ii) for precipitation between 0 and 3000 mm in the terrestrial realm only. The full procedure is described in [29]. At the end of the procedure, there are a maximum of 101,397 and 94,299,210 niches in the marine and terrestrial realms, respectively. About 25% and 1% of these niches are chosen randomly to perform the simulations in the marine and terrestrial realms, respectively [29]. The use of fictitious niches and species is especially useful, since we have only inventoried 9% of marine and 14% of terrestrial biodiversity and we know little about the biology of most species (see Section 5.1). A niche can give rise to several pseudospecies if individuals from different regions never come into contact (e.g., Figure 6f) [29]. Pseudospecies are gradually colonising the terrestrial and marine environment (surface and bottom). During the simulations, the species organize themselves into communities and the biodiversity. More precisely, here the number of species in a given region is reproduced.

Beaugrand and colleagues [29] used this approach to model the biodiversity of the terrestrial and marine realm, including the surface and the bottom of neritic and oceanic regions (Figure 7). These numerical experiments (or simulations) correctly reconstruct large-scale biodiversity patterns as they are observed nowadays for a large number of taxonomic groups in the terrestrial and marine environment (e.g., crustaceans, fish, cetaceans, plants, birds) [29]. The biodiversity maps for the ocean floors (Figure 7c,f) remain provisional, as few observations have been made to date to confirm these predictions [29]. The model also reproduces well past biodiversity patterns of the Last Glacial Maximum and the mid-Pliocene (e.g., foraminifera), as well as the Ordovician (e.g., acritarchs) [27,56].

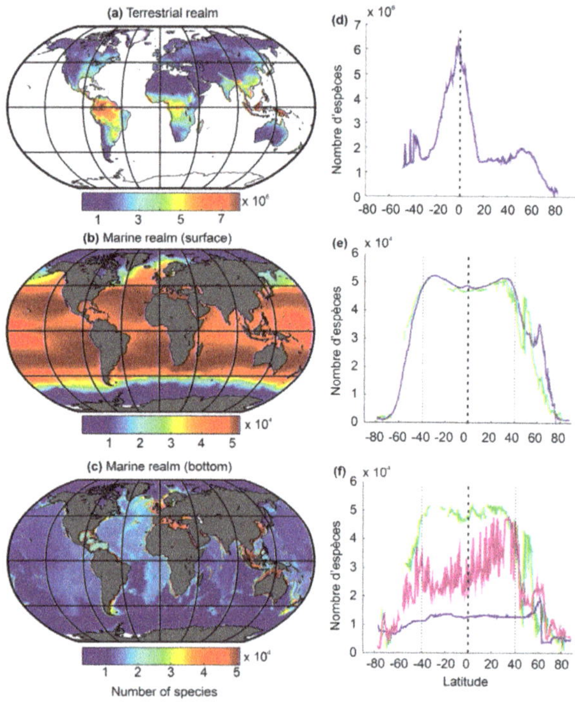

Figure 7. Average distribution of biodiversity (i.e., number of species) in terrestrial (**a**,**d**) and marine (**b**,**e**) surface biodiversity and (**c**,**f**) benthic biodiversity reconstituted from a bioclimatic model derived from METAL [29,100]. (**d**–**f**) The curves show the latitudinal gradient of biodiversity observed for each environment. (**e**) The blue curve reflects the latitudinal biodiversity of the oceanic regions (bathymetry above 200 m) and the green curve reflects the latitudinal biodiversity of the continental-shelf regions (bathymetry below 200 m). (**f**) The curve in green reflects the latitudinal biodiversity of the continental shelf (bathymetry lower than 200 m), the curve in blue reflects that of the deep regions (bathymetry higher than 2000 m), and that in magenta reflects the latitudinal biodiversity of the continental slope (bathymetry between 200 and 2000 m). From Beaugrand and colleagues [29].

6. The Great Chessboard of Life

The reconstruction of large-scale biodiversity patterns observed in nature is possible because the niche–climate interaction generates a mathematical constraint on the maximum number of species that can establish in a given region [30]. We have named this constraint the great chessboard of life (Figure 8) [30]. This particular chessboard has a number of geographical squares (i.e., wide squares in Figure 8) that correspond to different regions (marine or terrestrial). Note that these geographical squares are limited on the figure (i.e., 6 × 8 = 48 squares), but should be higher to correctly represent the variety of environments, e.g., one for every degree of latitude and longitude (i.e., 180 latitudes × 360 longitudes = 64,800 squares). Each square on the chessboard is composed of sub-squares (i.e., the narrow squares in Figure 8), which represent the number of climatic niches that determines the maximum number of species that can colonise a square (i.e., a region or a wide square). Only one species can establish in a sub-square (i.e., a climatic niche) of the chessboard according to the competitive exclusion principle of Gause [157], thereby the more sub-squares (L) in a given region, the higher the maximum number of species that an area may contain (Figure 8). S is the number of species that a square (i.e., an area) actually contains. Therefore, L represents a fundamental limit (what I call here a mathematical constraint) on species richness, even if the actual number can still vary according to other processes (see below). The different pieces on the chessboard (e.g., king, queen, pawn) symbolize the different biological properties of the species (e.g., their differences in terms of life history traits, such as reproduction). Q represents niche saturation, with Q = (S/L) × 100. A saturation of 100% means that all niches or potential species that a square may contain are occupied. Biological (degree of clade origination) and climatic (repeated Pleistocene glaciations) causes influence the percentage of saturation of the niches in each geographical square so that there remains a degree of valence on the number of species present on the great chessboard of life [30].

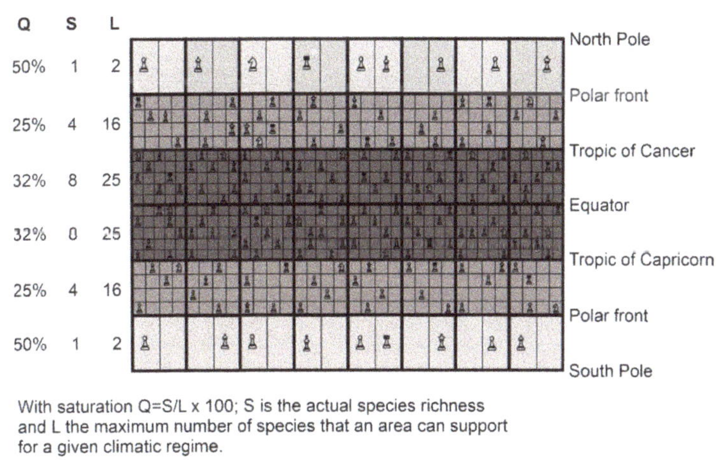

Figure 8. The great chessboard of life that illustrates the mathematical influence on current large-scale biodiversity patterns in the marine realm. Each square on the chessboard, which represents a region, is composed of sub-squares, which represent the number of climatic niches that determines the maximum number of species that can colonise a square (i.e., a region). The different pieces on the chessboard (e.g., king, queen, pawn) symbolize the different biological properties of the species (e.g., their differences in terms of life history traits, such as reproduction). Note that it is also applicable on land. From Beaugrand and colleagues [30].

The number of maximum niches fixes an upper limit on the number of species that can colonise a given region by speciation or immigration [30]. Few species can colonise areas located towards the minimum (e.g., −1.8 °C) and maximum limits (e.g., 44 °C) of temperature and precipitation (e.g., 0 and 3000 mm for precipitation) [29]. The choice of these minimum and maximum values in the METAL models is therefore important because it affects the results [100]. In marine polar areas, corresponding nowadays to the lowest limit of temperature (close to −1.8 °C), the number of species that can establish are fundamentally limited by L, since two species having the same niche cannot coexist at the same time and in the same place [157]. On the chessboard (Figure 8), the number of sub-squares is two in the wide square between the polar front and North Pole.

At low latitudes, since the theoretical upper limits are not observed nowadays (i.e., the upper limit for temperature is frequently fixed to 44 °C; for a justification of the threshold, see [100]), terrestrial biodiversity is maximum at the equator and marine biodiversity in subtropical regions (Figure 7) [27,29,56,100]. (I will come back to this point in Section 7). The great chessboard of life therefore suggests that there is more species richness in regions where there are more ecological niches (*sensu* Hutchinson [58]), providing compelling evidence for the niche-assembly hypothesis, although the niche in this hypothesis has been usually defined in terms of resources [130] (see Section 5.1).

The biogeographical constraints (i.e., low and high number of niches in high and low latitudes, respectively) imposed by the chessboard on biodiversity may be quickly detectable because clade diversification (not implemented in this METAL model) takes place relatively rapidly on a geological time scale [158]. Moreover, in the marine environment, taxa such as plankton have high dispersal capabilities [44,159,160] and may rapidly conform to the chessboard. Niche saturation (i.e., the number of observed species on the theoretical number of available niches) may help measuring the degree of conformity of the different taxonomic groups on the chessboard [30]. Although it is commonly assumed that niche saturation increases towards the equator (i) because evolutionary rates are thought to increase from cold to warm regions [122,161] and (ii) because of the presence of strong climate-induced environmental perturbations in extra-tropical regions that limit species richness [116,162], niche saturation is frequently highest towards the poles [30]. These apparently counterintuitive results suggest that the few sub-squares (i.e., the climatic niches) available on each region (i.e., wide square) of the chessboard are frequently occupied in polar regions (Figure 8). This means that low polar biodiversity should not always be attributed to low diversification rates (origination *minus* extinction) [114,115], but rather to a smaller maximum number of species' niches (i.e., the parameter L on the chessboard in Figure 8) at saturation that locally limits biodiversity. This low number of niches over polar regions, and inversely the high number of niches equatorwards, originating from the niche–environment interaction, represent a mathematical constraint on the arrangement of biodiversity. Although there remains a great degree of freedom on the type and number of species that can establish in a region (e.g., origination and diversification of a clade), this number cannot exceed a threshold set by the niche–climate interaction.

In the marine realm, large-scale patterns of niche saturation differ among taxonomic groups, which suggests the existence of a particular chessboard for each group that might originate from taxon-specific diversification history [163,164]. For example, pinnipeds, which exhibit an inversed latitudinal biodiversity gradient, originate from Arctoid carnivores 25–27 Ma in the cold regions of the North Pacific [165]. Place of origin and time of emergence may therefore blur large-scale biodiversity patterns imposed by the great chessboard of life. Moreover, life history traits of each group make the great chessboard of life specific to a given taxonomic group, which sometimes explains the lack of universality of large-scale biodiversity patterns (Figure 8) [30].

The rate of diversification remains an important parameter because it determines the degree of niche occupation in a given geographical cell. Indeed, when polar regions are excluded, niche saturation of most groups (e.g., plankton and fish) but mammals is higher over permanently stratified regions [147]. Moreover, many clades should exhibit latitudinal biodiversity gradients towards the equator because their probability of emergence should be higher in the tropics, where there are more available niches, and palaeontological data have provided compelling evidence of greater rates of origination for tropical clades [166]—the hypothesis of tropical niche conservatism [139].

Beaugrand and colleagues suggest that the total number of species on the chessboard diminishes with organismal complexity [30], which can be explained by basic ecological and evolutionary processes. Endosomatic energy decreases from primary producers to top predators as a consequence of the second law of thermodynamics, decreasing the number of individuals and thereby species richness and niche saturation from producers to higher trophic levels [17,41]. Positive relationships between numbers of individuals and species richness has often been proposed to explain large-scale biodiversity patterns, e.g., the productivity theory [167], the area hypothesis [168], and the unified neutral theory of biodiversity and biogeography [99]. In addition to diminishing the number of individuals [169], a larger body also increases generation time [147], which slows down evolution [122,169]. Therefore, the likelihood that a taxon exhibits a large-scale biodiversity pattern different from the one imposed by the great chessboard of life is greater when its mean niche saturation is lower. This is especially the case for marine mammals. Pinnipeds, which have a low degree of niche saturation (<1% [30]), exhibit a pattern that does not conform to the great chessboard of life [30]. Note that it is possible to do specific simulations to account for the biology of a specific clade or taxonomic group, such as euphausiids, fish, coral reefs, or mangroves [29,145].

7. Differences in Latitudinal Biodiversity Gradients between the Terrestrial and the Marine Domains

A biodiversity peak is observed at the equator in the terrestrial realm and around the subtropical regions in the marine realm (Figure 7a–d). This distinction is related to the differential influence of atmospheric pressure fields in the marine and terrestrial realms [5,17,170,171]. Indeed, the high-pressure centres (i.e., the large-pressure high linked to the descending branches of the Hadley and Ferrel cells) provide climatic stability and heat, which increase surface biodiversity in the marine environment (Figure 9) [17]. However, above the continents, the high-pressure centres strongly limit precipitation, and biodiversity is therefore very low due to the lack of water availability [5,17,170,171]. The maximum values of biodiversity are reached in regions where precipitation is regular (towards the equator) and decreases when moving away from the influence of the intertropical convergence zone (ITCZ) [5], i.e., when it occurs a few weeks a year (monsoon areas). Finally, the cold ocean floors do not show a typical biodiversity gradient but a very homogeneous biodiversity pattern, except in high-latitude regions, where biodiversity decreases slightly, and over seamounts, where it is higher (Figure 7e–f).

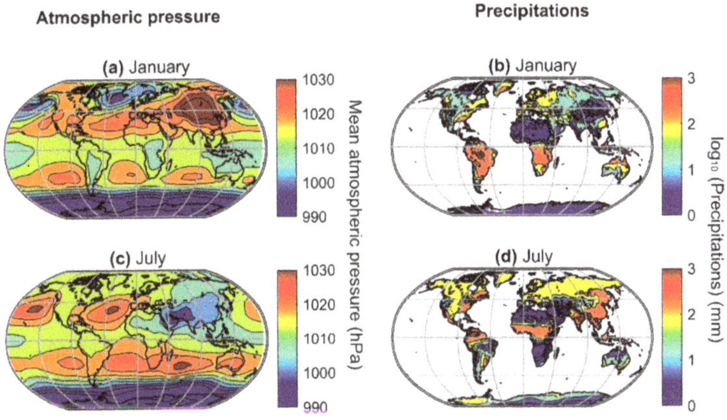

Figure 9. Average positions of the planet's major pressure highs and lows and their influences on average precipitation in January (**a**,**b**) and July (**c**,**d**). Atmospheric pressure (in hPa) (**a**,**c**). Precipitation (in mm) (**b**,**d**). From Beaugrand [17].

The chessboard of life reorganizes when climate changes, which makes it dynamic from small to large temporal scales (i.e., geological scales) [30]. Indeed, large-scale biodiversity patterns are not stable over time [172], and METAL suggests that they were sometimes very different from those currently observed [56]. For example, during a cold period at the end of the Ordovician (510 million years ago), a very significant contrast probably existed between tropical biodiversity and the biodiversity of high-latitude regions (Figure 10a). Conversely, during the warm period of Stage 4 of the Cambrian (510 million years), the latitudinal gradient of biodiversity was probably reversed (Figure 10b). The current latitudinal gradient of biodiversity, characterised by a more or less regular increase in biodiversity from the poles to the equator, has therefore probably not always been observed since the appearance of eukaryotes on our planet [56]. Mannion and colleagues [172] also proposed that the current latitudinal gradient of biodiversity has not been a permanent feature through the Phanerozoic. They suggested that a biodiversity peak occurred during cold icehouse climatic regimes, whereas temperate peaks (or flattened gradients) were observed during warmer greenhouse regimes. In the context of current climate change, the use of METAL also suggests that the contrast between regions of low and high biodiversity may diminish towards the end of the century because a rise in temperature over permanently stratified regions (e.g., tropics and subtropics) reduces surface biodiversity, whereas an augmentation in temperature over temperate and polar regions increases biodiversity [27].

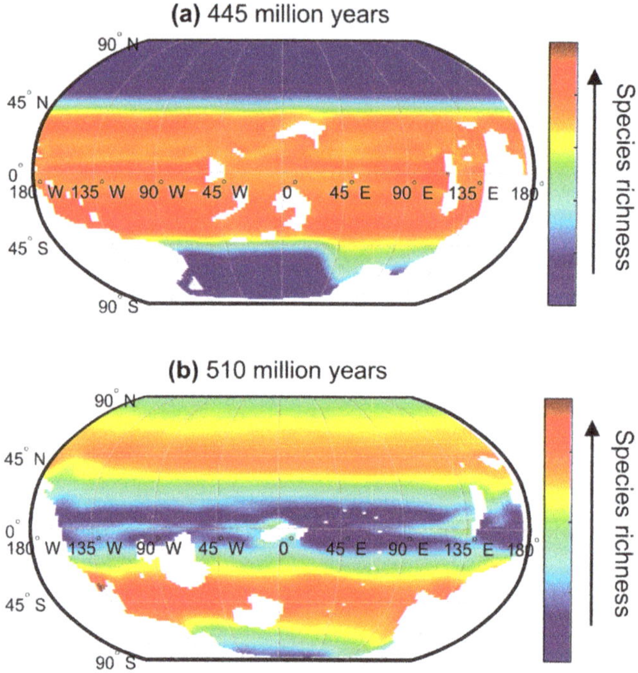

Figure 10. Expected marine surface biodiversity patterns using METAL for (**a**) the end of the Ordovician (445 million years) and (**b**) a warm phase of the Cambrian (510 million years). The position of the continents is indicated in white. Species richness is in relative unit with high richness values in red and low in blue. The conquest of the continents by the first terrestrial plants probably began around 500 million years ago. The carbon dioxide concentration was 5 times the pre-industrial concentration for the late Ordovician and 32 times the pre-industrial concentration for the Cambrian Stage 4. Modified, from Zacaï and colleagues [56].

8. Why Are There More Terrestrial Than Marine Species?

At a high taxonomic level, the number of phyla of metazoans is higher in the ocean than on land, and this number is greater for the benthic than the pelagic realm [173]. In sum, 32 phyla are found in the sea and 21 are exclusively marine, whereas 12 are found on land, with only one being endemic to this realm (Onychophora) [173], while 27 phyla inhabit the benthos (with 10 endemic) and only 11 the pelagos (with one endemic—Loricifera). These estimates strongly suggest that diversification began in the sea, and probably in or close to the benthic realm [174]. However, at a species level, there are more terrestrial than marine species. Robert May assessed that ~85% of all species are terrestrial [173,175], Michael Benton ~75% [176], and more recent estimates suggest they may be closer to ~80% [76]. About 77% of animal species live on land, the remaining being found in freshwater (11%) and in the marine environment (12%), and 93% of plants live on land, 5% being freshwater and 2% marine [177]. Why are there more terrestrial than marine species? This conundrum is all the more incomprehensible, since marine biodiversity appeared long before terrestrial species [176,178]. Since more time has passed since the emergence of life in the oceans, the higher number of terrestrial species over marine species seems to be a counterintuitive observation.

METAL reproduces the difference in biodiversity observed between the marine and terrestrial domains well [29]. Although results depended upon the choice of the total number of niches, modelled biodiversity scaled to catalogued (and estimated) species gave 1,111,186 (8,825,091) for the terrestrial domain and 316,069 (2,242,908) for the marine domain. These estimates are close to those given in Section 4 [76]. Moreover, an estimate of the deep-sea benthic biodiversity (894,881 benthic species in areas below 2000 m and 256,278 in areas between 2000 m and 200 m) is close to what has been calculated in some studies [179,180]. Species density is expected to be higher over the shelf (200–2000 m) than deep sea, but because the latter realm is larger (301 vs. 36 million km^2), there are more species in the deep-sea benthic realm [29].

Two mechanisms may explain why METAL reconstructs the difference between land and sea biodiversity well [29]. Firstly, Beaugrand and colleagues [29] suggested that the difference between the number of climatic dimensions among terrestrial and marine environments is fundamental. Water is evidently present everywhere in the ocean, which is not the case in the terrestrial environment. The additional discriminating climatic dimension in the terrestrial environment (water availability) arithmetically increases the number of climatic niches and thus the number of species that can establish in the terrestrial environment [29]. The addition of one climatic dimension increases the number of potential niches by ~100 [29].

Secondly, but to a lesser extent, Beaugrand and colleagues [29] suggested that the addition of a supplementary climatic dimension, combined with more pronounced geographical variations in the terrestrial environment (i.e., an increase in habitat heterogeneity), further fragment the geographical range of a species [181], increasing the possibility of allopatric speciation, i.e., the creation of species by prolonged or permanent geographic isolation of populations [29,41,182].

To conclude on this part, there are more terrestrial than marine species because there are more available climatic niches on land. This additional dimension inflates considerably the maximum number of niches and thereby species on the great chessboard of life (i.e., parameter L in Figure 8). This second fundamental dimension inflates considerably the number of species that the terrestrial realm may contain. The addition of the water availability dimension in the terrestrial realm (in addition to temperature) also morcellates species spatial distribution and increases the effect of landscape heterogeneity and the possibility for allopatric speciation [29]. In the ocean, the seascape is more uniform because there is only a single climate dimension (temperature). See, however, Ref. [181] for other important environmental dimensions. This influence is more prominent in the pelagic than the benthic environment, so it probably explains why there are more benthic than pelagic species [29]. The influence of seascape heterogeneity strongly affects local biodiversity over seamounts and shelves [183]. Therefore, local biodiversity should be higher over these areas, including heterogeneous shallow ones [29].

9. Conclusions

A central objective of biology and its sub-disciplines (e.g., biogeography, ecology) is to reveal the laws or general principles that govern the arrangement of life, but the sources of variations and exceptions seem inexhaustible. However, simple laws have been discovered in other areas of science, such as physics. Galileo Galilei wrote "*Philosophy [nature] is written in that great book which ever is before our eyes—I mean the universe—but we cannot understand it if we do not first learn the language and grasp the symbols in which it is written. The book is written in mathematical language, and the symbols are triangles, circles and other geometrical figures, without whose help it is impossible to comprehend a single word of it; without which one wanders in vain through a dark labyrinth*". We show here that the niche–environment interaction is fundamental because it controls a large number of phenomena, patterns of variability, and biological events. In this review, I only show how METAL can help in understanding the arrangement of biodiversity, but the theory also explains other phenomena, such as spatial range, biogeographical shifts, phenology, annual plankton succession, long-term changes in species abundance, and community composition, gradual or abrupt [26].

Like the Italian scholar of the Renaissance Galileo Galilei, I propose that the great book of life is also written in mathematical language. In particular, the niche–environment interaction, controlled in part by the climatic regime, generates a mathematical constraint on the large-scale arrangement of biodiversity. We have named this constraint the great chessboard of life (Figure 8). The mathematical effect is probably considerable such that an inverted latitudinal gradient is impossible under present climatic conditions for most taxonomic groups that presently exhibit increasing biodiversity from the poles to the equator. Moreover, a similar mathematical effect explains why there are more terrestrial than marine species, even if the number of phyla is higher in the marine than terrestrial realm. The establishment of a global theory of biodiversity, however, requires taking into account a large number of biological processes that also influence biodiversity (e.g., diversification rate and origination place of a clade), and Theodosius Dobzhansky was greatly inspired when he wrote his article 'Nothing in biology makes sense except in the light of evolution' [12]. In addition to other key ecological factors discussed in Section 5, METAL should therefore consider more explicitly some key evolutionary processes in the future. In the process of developing such a global theory of biodiversity, considering all the complexity of biological systems (Box 1), it is important to recognize that mathematical constraints caused by (i) the number of key dimensions that the niches include in the terrestrial and marine realms and (ii) the niche–environment interaction also control the arrangement of biodiversity.

Funding: This work has been partially financially supported by CNRS, Université du Littoral Côte d'Opale, the IFSEA Graduate School, the regional CPER programme IDEAL and the ANR ECO-BOOST.

Data Availability Statement: The main data used in this paper are available from the corresponding author on reasonable request.

Acknowledgments: The author thanks the four reviewers for helpful comments on an initial version of the manuscript.

Conflicts of Interest: The author declares no competing interest.

References

1. Purves, W.K.; Sadava, D.; Orians, G.H.; Heller, H.C. *Life: The Science of Biology*, 7th ed.; W. H. Freeman & Co.: New York, NY, USA, 2004; p. 1121.
2. Calenbuhr, V. Complexity Science in the Context of Policymaking. In *Science for Policy Handbook*; Šucha, V., Sienkiewicz, M., Eds.; Elsevier: Amsterdam, The Netherlands, 2020; pp. 118–127.
3. Wolfa, Y.I.; Katsnelsonb, M.I.; Koonin, E.V. Physical foundations of biological complexity. *Proc. Natl. Acad. Sci. USA* **2018**, *115*, E8678–E8687. [CrossRef]

4. Wickramasinghe, N.C.; Wickramasinghe, D.T.; Tout, C.A.; Lattanzio, J.C.; Steele, E.J. Cosmic biology in perspective. *Astrophys. Space Sci.* **2019**, *364*, 205. [CrossRef]
5. Kump, L.R.; Kasting, J.F.; Crane, R.G. *The Earth System*; Pearson Prentice Hall: Upper Saddle River, NJ, USA, 2004.
6. Gould, S.J. *Punctuated Equilibrium*; Harvard University Press: Cambridge, MA, USA, 2007.
7. Stearns, S.C. Life history evolution: Successes, limitations, and prospects. *Naturwissenschaften* **2000**, *87*, 476–486. [CrossRef]
8. Gould, S.J. *Ever Since Darwin (Reflexions in Natural History)*; W.W. Norton and Company: New York, NY, USA, 1977.
9. Jacob, F. Evolution and tinkering. *Science* **1977**, *196*, 1161–1166. [CrossRef]
10. Monod, J. *Chance and Necessity: An Essay on the Natural Philosophy of Modern Biology*; Vintage Books: New York, NY, USA, 1971.
11. Darwin, C. *The Origin of Species by Means of Natural Selection*; John Murray: London, UK, 1859.
12. Dobzhansky, T. Nothing in biology makes sense except in the light of evolution. *Am. Biol. Teach.* **1973**, *35*, 125–129. [CrossRef]
13. Kimura, M. *The Neutral Theory of Molecular Evolution*; Cambridge University Press: Cambridge, UK, 1983; p. 272.
14. Mayr, E. *Systematics and the Origin of Species from the Viewpoint of a Zoologist*; Columbia University Press: New York, NY, USA, 1942.
15. Müller, G.B. Evo–devo: Extending the evolutionary synthesis. *Nat. Rev. Genet.* **2007**, *8*, 943–949. [CrossRef]
16. Pennisi, E. What determines species diversity? *Science* **2005**, *309*, 90. [CrossRef]
17. Beaugrand, G. *Marine Biodiversity, Climatic Variability and Global Change*; Routledge: London, UK, 2015; p. 474.
18. Naeem, S.; Loreau, M.; Inchausti, P. Biodiversity and ecosystem functioning: The emergence of a synthetic ecological framework. In *Biodiversity and Ecosystem Functioning—Synthesis and Perspectives*; loreau, M., Naeem, S., Inchausti, P., Eds.; Oxford University Press: Oxford, UK, 2002; pp. 3–11.
19. Tilman, D. The ecological consequences of changes in biodiversity: A search for general principles. *Ecology* **1999**, *80*, 1455–1474. [CrossRef]
20. Worm, B.; Barbier, E.B.; Beaumont, N.; Duffy, J.E.; Folke, C.; Halpern, B.S.; Jackson, J.B.C.; Lotze, H.K.; Micheli, F.; Palumbi, S.R.; et al. Impacts of biodiversity loss on ocean ecosystem services. *Science* **2006**, *314*, 787–790. [CrossRef]
21. Beaugrand, G. Theoretical basis for predicting climate-induced abrupt shifts in the oceans. *Philos. Trans. R. Soc. B Biol. Sci.* **2015**, *370*, 20130264. [CrossRef]
22. Beaugrand, G.; Mackas, D.; Goberville, E. Applying the concept of the ecological niche and a macroecological approach to understand how climate influences zooplankton: Advantages, assumptions, limitations and requirements. *Prog. Oceanogr.* **2013**, *111*, 75–90. [CrossRef]
23. Beaugrand, G.; Goberville, E.; Luczak, C.; Kirby, R.R. Marine biological shifts and climate. *Proc. R. Soc. B Biol. Sci.* **2014**, *281*, 20133350. [CrossRef]
24. Beaugrand, G.; Kirby, R.R. Quasi-deterministic responses of marine species to climate change. *Clim. Res.* **2016**, *69*, 117–128. [CrossRef]
25. Beaugrand, G.; Balembois, A.; Kléparski, L.; Kirby, R.R. Addressing the dichotomy of fishing and climate in fishery management with the FishClim model. *Commun. Biol.* **2022**, *5*, 1146. [CrossRef]
26. Beaugrand, G.; Kirby, R.R. How do marine pelagic species respond to climate change? Theories and observations. *Annu. Rev. Mar. Sci.* **2018**, *10*, 169–197. [CrossRef]
27. Beaugrand, G.; Edwards, M.; Raybaud, V.; Goberville, E.; Kirby, R.R. Future vulnerability of marine biodiversity compared with contemporary and past changes. *Nat. Clim. Change* **2015**, *5*, 695–701. [CrossRef]
28. Beaugrand, G.; Conversi, A.; Atkinson, A.; Cloern, J.; Chiba, S.; Fonda-Unami, S.; Kirby, R.R.; Greene, C.G.; Goberville, E.; Otto, S.A.; et al. Prediction of unprecedented biological shifts in the global ocean. *Nat. Clim. Change* **2019**, *9*, 237–243. [CrossRef]
29. Beaugrand, G.; Kirby, R.R.; Goberville, E. The mathematical influence on global patterns of biodiversity. *Ecol. Evol.* **2020**, *10*, 6494–6511. [CrossRef]
30. Beaugrand, G.; Luczak, C.; Goberville, E.; Kirby, R.R. Marine biodiversity and the chessboard of life. *PLoS ONE* **2018**, *13*, e0194006. [CrossRef]
31. Hellweger, F.L.; Van Sebille, E.; Fredrick, N.D. Biogeographic patterns in ocean microbes emerge in a neutral agent-based model. *Science* **2014**, *345*, 1346–1349. [CrossRef]
32. Lladó Fernández, S.; Větrovský, T.; Baldrian, P. The concept of operational taxonomic units revisited: Genomes of bacteria that are regarded as closely related are often highly dissimilar. *Folia Microbiol.* **2018**, *64*, 19–23. [CrossRef] [PubMed]
33. Cohan, F.M.; Koeppel, A.F. The origins of ecological diversity in prokaryotes. *Curr. Biol.* **2008**, *18*, R1024–R1034. [CrossRef] [PubMed]
34. Suen, G.; Goldman, B.S.; Welch, R.D. Predicting prokaryotic eological niches using genome sequence analysis. *PLoS ONE* **2007**, *2*, e743. [CrossRef] [PubMed]
35. Hughes Martiny, J.B.; Bohannan, B.J.M.; Brown, J.H.; Colwell, R.K.; Fuhrman, J.A.; Green, J.L.; Horner-Devine, M.C.; Kane, M.; Krumins, J.A.; Kuske, C.R.; et al. Microbial biogeography: Putting microorganisms on the map. *Nat. Rev. Microbiol.* **2006**, *4*, 102–112. [CrossRef]
36. Lundberg, P.; Ranta, E.; Ripa, J.; Kaitala, V. Population variability in space and time. *Trends Ecol. Evol.* **2000**, *15*, 460–464. [CrossRef]
37. Russell, F.S. A summary of the observations of the occurence of planktonic stages of fish off Plymouth 1924–1972. *J. Mar. Biol. Assoc. U. K.* **1973**, *53*, 347–355. [CrossRef]
38. Edwards, M.; Beaugrand, G.; Helaouët, P.; Alheit, J.; Coombs, S.H. Marine ecosystem response to the Atlantic Multidecadal Oscillation. *PLoS ONE* **2013**, *8*, e57212. [CrossRef]

39. Alheit, J.; Hagen, E. Long-term forcing of European herring and sardine populations. *Fish. Oceanogr.* **1997**, *6*, 130–139. [CrossRef]
40. Poole, K.G. A review of the Canada Lynx, Lynx canadensis, in Canada. *Can. Field-Nat.* **2003**, *117*, 360–376. [CrossRef]
41. Lomolino, M.V.; Riddle, B.R.; Brown, J.H. *Biogeography*, 3rd ed.; Sinauer Associates, Inc.: Sunderland, MA, USA, 2006; p. 845.
42. Ricklefs, R.E. Evolutionary diversification, coevolution between populations and their antagonists, and the filling of niche space. *Proc. Natl. Acad. Sci. USA* **2010**, *107*, 1265–1272. [CrossRef]
43. van der Spoel, S. A biosystematic basis for pelagic biodiversity. *Bijdr. Tot De Dierkd.* **1994**, *64*, 3–31. [CrossRef]
44. Angel, M.V. Pelagic biodiversity. In *Marine Biodiversity: Patterns and Processes*; Ormond, R.F.G., Gage, J.D., Angel, M.V., Eds.; Cambridge University Press: Cambridge, UK, 1997; pp. 35–68.
45. Rex, M.A.; Etter, R.J.; Stuart, C.T. Large-scale patterns of species diversity in the deep-sea benthos. In *Marine Biodiversity: Patterns and Processes*; Ormond, R.F.G., Gage, J.D., Angel, M.V., Eds.; Cambridge University Press: Cambridge, UK, 1997; pp. 94–121.
46. Whittaker, R.H. *Communities and Ecosystems*, 2nd ed.; Macmillan: New York, NY, USA, 1975.
47. Beaugrand, G.; Ibañez, F.; Lindley, J.A. An overview of statistical methods applied to the CPR data. *Prog. Oceanogr.* **2003**, *58*, 235–262. [CrossRef]
48. Faillettaz, R.; Beaugrand, G.; Goberville, E.; Kirby, R.R. Atlantic Multidecadal Oscillations drive the basin-scale distribution of Atlantic bluefin tuna. *Sci. Adv.* **2019**, *5*, eaar6993. [CrossRef] [PubMed]
49. Caracciolo, M.; Beaugrand, G.; Hélaouët, P.; Gevaert, F.; Edwards, M.; Lizon, F.; Kléparski, L.; Goberville, E. Annual phytoplankton succession results from niche-environment interaction. *J. Plankton Res.* **2021**, *43*, 85–102. [CrossRef]
50. Kléparski, L.; Beaugrand, G.; Edwards, M.; Schmitt, F.G.; Kirby, R.R.; Breton, E.; Gevaert, F.; Maniez, E. Morphological traits, niche-environment interaction and temporal changes in diatoms. *Prog. Oceanogr.* **2022**, *201*, 102747. [CrossRef]
51. Levin, S.A. The problem of pattern and scale in ecology. *Ecology* **1992**, *73*, 1943–1967. [CrossRef]
52. Haury, L.R.; McGowan, J.A.; Wiebe, P.H. Patterns and processes in the time-space scales of plankton distributions. In *Spatial Pattern in Plankton Communities*; Steele, J.H., Ed.; Plenun Press: New York, NY, USA, 1978; pp. 277–327.
53. Wiens, J.A. Spatial scaling in ecology. *Funct. Ecol.* **1989**, *1989*, 385–397. [CrossRef]
54. Hutchinson, G.E. *The Ecological Theater and the Evolutionary Play*; Yale University Press: New Haven, CT, USA, 1965; p. 172.
55. Bachelard, G. *L'activité Rationaliste de la Physique Contemporaine*, 2nd ed.; Presses Universitaires de France: Paris, France, 1965; p. 223.
56. Zacaï, A.; Monnet, C.; Pohl, A.; Beaugrand, G.; Mullins, G.; Kröck, D.; Servais, T. Truncated bimodal latitudinal diversity gradient in early Paleozoic phytoplankton. *Sci. Adv.* **2021**, *7*, eabd6709. [CrossRef]
57. Hutchinson, G.E. Concluding remarks. *Cold Spring Harb. Symp. Quant. Biol.* **1957**, *22*, 415–427. [CrossRef]
58. Hutchinson, G.E. *An Introduction to Population Ecology*; Yale University Press: New Haven, CT, USA, 1978; p. 260.
59. Frontier, S.; Pichot-Viale, D.; Leprêtre, A.; Davoult, D.; Luczak, C. *Ecosystèmes. Structure, Fonctionnement et Évolution*, 3rd ed.; Dunod: Paris, France, 2004; p. 549.
60. Morin, E. *Introduction à la Pensée Complexe*; Editions du Seuil: Paris, France, 2005; p. 158.
61. Araújo, M.B.; Peterson, A.T. Uses and misuses of bioclimatic envelope modelling. *Ecology* **2012**, *93*, 1527–1539. [CrossRef]
62. Thuiller, W.; Lafourcade, B.; Engler, R.; Araújo, M.B. BIOMOD—A platform for ensemble forecasting of species distributions. *Ecography* **2009**, *32*, 369–373. [CrossRef]
63. Sillero, N. What does ecological modelling model? A proposed classification of ecological niche models based on their underlying methods. *Ecol. Model.* **2011**, *222*, 1343–1346. [CrossRef]
64. Goberville, E.; Beaugrand, G.; Hautekeete, N.-C.; Piquot, Y.; Luczak, C. Uncertainties in species distribution projections and general circulation models. *Ecol. Evol.* **2015**, *5*, 1100–1116. [CrossRef] [PubMed]
65. Raybaud, V.; Beaugrand, G.; Goberville, E.; Delebecq, G.; Destombe, C.; Valero, M.; Davoult, D.; Morin, P.; Gevaert, F. Decline in Kelp in West Europe and Climate. *PLoS ONE* **2013**, *8*, e66044. [CrossRef]
66. Rombouts, I.; Beaugrand, G.; Dauvin, J.-C. Potential changes in benthic macrofaunal distributions from the English Channel simulated under climate change scenarios. *Estuar. Coast. Shelf Sci.* **2012**, *99*, 153–161. [CrossRef]
67. Schickele, A.; Goberville, E.; Leroy, B.; Beaugrand, G.; Hattab, T.; Francour, P.; Raybaud, V. Redistribution of small pelagic fish in Europe and Climate Change. *Fish Fish.* **2021**, *22*, 212–225. [CrossRef]
68. Peterson, A.T.; Cobos, M.E.; Jiménez-García, D. Major challenges for correlational ecological niche model projections to future climate conditions: Climate change, ecological niche models, and uncertainty. *Ann. N. Y. Acad. Sci.* **2018**, *1429*, 66–77. [CrossRef]
69. Ehrlén, J.; Morris, W.F. Predicting changes in the distribution and abundance of species under environmental change. *Ecol. Lett.* **2015**, *18*, 303–314. [CrossRef]
70. Helaouët, P.; Beaugrand, G. Physiology, ecological niches and species distribution. *Ecosystems* **2009**, *12*, 1235–1245. [CrossRef]
71. Pauly, D. Ecosystem impacts of the world's marine fisheries. *Glob. Chang. Newsl.* **2003**, *55*, 21–23.
72. Jennings, S.; Kaiser, M.J.; Reynolds, J.D. *Marine Fisheries Ecology*; Blackwell Science Ltd.: Oxford, UK, 2001; p. 417.
73. Koch, G.W.; Vitousek, P.M.; Steffen, W.L.; Walker, B.H. Terrestrial transects for global change research. *Vegetatio* **1992**, *121*, 53–65. [CrossRef]
74. Raven, P.H.; Johnson, G.B.; Mason, K.A.; Losos, J.B.; Singer, S.R. *Biologie*; Deboeck Supérieur: Louvain-la-Neuve, Belgium, 2017; p. 1282.
75. Costello, M.J.; May, R.M.; Stork, N.E. Can we name Earth's species before they go extinct? *Science* **2013**, *339*, 413–416. [CrossRef] [PubMed]

76. Mora, C.; Tittensor, D.P.; Adl, S.; Simpson, A.G.B.; Worm, B. How Many Species Are There on Earth and in the Ocean? *PLoS Biol.* **2011**, *9*, e1001127. [CrossRef] [PubMed]
77. Appeltans, W.; Ahyong, S.T.; Anderson, G.; Angel, M.V.; Artois, T.; Bailly, N.; Bamber, R.; Barber, A.; Bartsch, I.; Berta, A.; et al. The magnitude of global marine species biodiversity. *Curr. Biol.* **2012**, *22*, 2189–2202. [CrossRef] [PubMed]
78. Erwin, T.L. Tropical forests: Their richness in Coleoptera and other arthropod species. *Coleopt. Bull.* **1982**, *36*, 74–75.
79. May, R.M. How Many Species are There on Earth? *Science* **1988**, *241*, 1441–1449. [CrossRef] [PubMed]
80. Thackeray, S.J.; Henrys, P.A.; Hemming, D.; Bell, J.R.; Botham, M.S.; Burthe, S.; Helaouet, P.; Johns, D.G.; Jones, I.D.; Leech, D.I.; et al. Phenological sensitivity to climate across taxa and trophic levels. *Nature* **2016**, *535*, 241–245. [CrossRef]
81. Parmesan, C.; Ryrholm, N.; Stefanescu, C.; Hill, J.K.; Thomas, C.D.; Descimon, H.; Huntley, B.; Kaila, L.; Kullberg, J.; Tammaru, T.; et al. Poleward shifts in geographical ranges of butterfly species associated with regional warming. *Nature* **1999**, *399*, 579–583. [CrossRef]
82. Parmesan, C.; Matthews, J. Biological impacts of climate change. In *Principles of Concervation Biology*; Groom, M.J., Meffe, G.K., Carroll, C.R., Eds.; Sinauer Associates, Inc.: Sunderland, MA, USA, 2006; pp. 333–360.
83. Beaugrand, G.; Brander, K.M.; Lindley, J.A.; Souissi, S.; Reid, P.C. Plankton effect on cod recruitment in the North Sea. *Nature* **2003**, *426*, 661–664. [CrossRef]
84. Beaugrand, G.; Reid, P.C.; Ibañez, F.; Lindley, J.A.; Edwards, M. Reorganisation of North Atlantic marine copepod biodiversity and climate. *Science* **2002**, *296*, 1692–1694. [CrossRef]
85. Post, E.; Peterson, R.O.; Stenseth, N.C.; McLaren, B.E. Ecosystem consequences of wolf behavioural response to climate. *Nature* **1999**, *1999*, 905–907. [CrossRef]
86. Poloczanska, E.S.; Brown, C.J.; Sydeman, W.J.; Kiessling, W.; Schoeman, D.S.; Moore, P.J.; Brander, K.; Bruno, J.F.; Buckley, L.B.; Burrows, M.T.; et al. Global imprint of climate change on marine life. *Nat. Clim. Change* **2013**, *3*, 919–925. [CrossRef]
87. Kennedy, D.; Norman, C. What don't we know? *Science* **2005**, *309*, 75. [CrossRef] [PubMed]
88. Tittensor, D.T.; Mora, C.; Jetz, W.; Lotze, H.K.; Ricard, D.; Berghe, E.V.; Worm, B. Global patterns and predictors of marine biodiversity across taxa. *Nature* **2010**, *466*, 1098–1101. [CrossRef]
89. Rex, M.A.; Stuart, C.T.; Hessler, R.R.; Allen, R.R.; Sanders, H.L.; Wilson, G.D.F. Global-scale latitudinal patterns of species diversity in the deep-sea benthos. *Nature* **1993**, *365*, 636–639. [CrossRef]
90. Pianka, E.R. Latitudinal gradients in species diversity: A review of concepts. *Am. Nat.* **1966**, *100*, 33–46. [CrossRef]
91. Roy, K.; Jablonski, D.; Valentine, J.W.; Rosenberg, G. Marine latitudinal diversty gradients: Tests of causal hypotheses. *Proc. Natl. Acad. Sci. USA* **1998**, *95*, 3699–3702. [CrossRef]
92. Gaston, K.J. Global patterns in biodiversity. *Nature* **2000**, *405*, 220–227. [CrossRef]
93. Hawkins, B.A.; Diniz-Filho, J.A.F.; Bini, L.M.; Araujo, M.B.; Field, R.; Hortal, J.; Kerr, J.T.; Rahbek, C.; Rodriguez, M.A.; Sanders, N.J. Metabolic theory and diversity gradients: Where do we go from here? *Ecology* **2007**, *88*, 1898–1902. [CrossRef]
94. Currie, D.J. Energy and large-scale patterns of animal- and plant- species richness. *Am. Nat.* **1991**, *137*, 27–49. [CrossRef]
95. Rohde, K. Latitudinal gradients in species diversity: The search for the primary cause. *Oikos* **1992**, *65*, 514–527. [CrossRef]
96. Rosenzweig, M.L.; Sandlin, E.A. Species diversity and latitudes: Listening to area's signal. *Oikos* **1997**, *80*, 172–176. [CrossRef]
97. Colwell, R.K.; Hurtt, G.C. Nonbiological gradients in species richness and a spurious rapoport effect. *Am. Nat.* **1994**, *144*, 570–595. [CrossRef]
98. Allen, A.P.; Brown, J.H.; Gillooly, J.F. Global biodiversity, biochemical kinetics, and the enrgetic-equivalence rule. *Science* **2002**, *297*, 1545–1548. [CrossRef] [PubMed]
99. Hubbell, S.P. *The Unified Neutral Theoy of Biodiversity and Biogeography*; Princeton University Press: Princeton, NJ, USA, 2001.
100. Beaugrand, G.; Rombouts, I.; Kirby, R.R. Towards an understanding of the pattern of biodiversity in the oceans. *Glob. Ecol. Biogeogr.* **2013**, *22*, 440–449. [CrossRef]
101. Zhang, Y.; Song, Y.-G.; Zhang, C.-Y.; Wang, T.-R.; Su, T.-H.; Huang, P.-H.; Meng, H.-H.; Li, J. Latitudinal Diversity Gradient in the Changing World: Retrospectives and Perspectives. *Diversity* **2022**, *14*, 334. [CrossRef]
102. Willig, M.R.; Presley, S.J. *Latitudinal Gradients of Biodiversity: Theory and Empirical Patterns*; Elsevier: Oxford, UK, 2018; Volume 3.
103. Muenchow, J.; Dieker, P.; Kluge, J.; Kessler, M.; von Wehrden, H. A review of ecological gradient research in the Tropics: Identifying research gaps, future directions, and conservation priorities. *Biodivers. Conserv.* **2018**, *27*, 273–285. [CrossRef]
104. Hawkins, B.A. Ecology's oldest pattern? *Trends Ecol. Evol.* **2001**, *16*, 470. [CrossRef]
105. Brown, J.H. Why are there so many species in the tropics? *J. Biogeogr.* **2014**, *41*, 8–22. [CrossRef]
106. Brown, J.H. Two decades of homage to Santa Rosalia: Toward a general theory of diversity. *Am. Zool.* **1981**, *21*, 877–888. [CrossRef]
107. Hutchinson, G.E. Homage to Santa Rosalia, or why are there so many kinds of animals? *Am. Nat.* **1959**, *93*, 145–159. [CrossRef]
108. Ricklefs, R.E. A comprehensive framework for global patterns in biodiversity. *Ecol. Lett.* **2004**, *7*, 1–15. [CrossRef]
109. Benton, M.J. Origins of Biodiversity. *PLoS Biol.* **2016**, *14*, e2000724. [CrossRef] [PubMed]
110. Rosenzweig, M.L. *Species Diversity in Space and Time*; Cambridge University Press: Cambridge, UK, 1995; p. 436.
111. Colwell, R.K.; Lees, D.C. The mid-domain effect: Geometric constraints on the geography of species richness. *Trends Ecol. Evol.* **2000**, *15*, 70–76. [CrossRef]
112. Colwell, R.K.; Gotelli, N.J.; Rahbek, C.; Entsminger, G.L.; Farrell, C.; Graves, G.R. Peaks, plateaus, canyons, and craters: The complex geometry of simple mid-domain effect models. *Evol. Ecol. Res.* **2009**, *11*, 355–370.
113. Simpson, G.G. Species density of North American recent mammals. *Syst. Zool.* **1964**, *13*, 57–73. [CrossRef]

114. Jablonski, D.; Roy, K.; Valentine, J.W. Out of the Tropics: Evolutionary dynamics of the latitudinal diversity gradient. *Science* **2006**, *314*, 102–106. [CrossRef]
115. Dowle, E.J.; Morgan-Richards, M.; Trewick, S.A. Molecular evolution and the latitudinal biodiversity gradient. *Heredity* **2013**, *110*, 501–510. [CrossRef]
116. Crame, J.A. Pattern and proceses in marine biogeography: A view from the poles. In *Fontiers of Biogeography I: New Directions in the Geography of Nature*; Lomolino, M.V., Heaney, L.R., Eds.; Sinauer Associates: Sunderland, MA, USA, 2004; pp. 272–292.
117. Economo, E.P.; Narula, N.; Friedman, N.R.; Weiser, M.D.; Guénard, B. Macroecology and macroevolution of the latitudinal diversity gradient in ants. *Nat. Commun.* **2018**, *9*, 1778. [CrossRef]
118. Moreau, C.S.; Bell, C.D. Testing the museum versus cradle tropical biological diversity hypothesis: Phylogeny, diversification, and ancestral biogeographic range evolution of the ants. *Evolution* **2013**, *67*, 2240–2257. [CrossRef]
119. Fine, P.V.A. Ecological and evolutionary drivers of geographic variation in species diversity. *Annu. Rev. Ecol. Evol. Syst.* **2015**, *46*, 369–392. [CrossRef]
120. Vasconcelos, T.; O'Meara, B.C.; Beaulieu, J.M. Retiring "Cradles" and "Museums" of Biodiversity. *Am. Nat.* **2022**, *199*, 194–205. [CrossRef] [PubMed]
121. Scholl, J.P.; Wiens, J.J. Diversification rates and species richness across the tree of life. *Proc. R. Soc. B* **2016**, *283*, 20161334. [CrossRef] [PubMed]
122. Gillooly, J.; Allen, A.P.; West, G.B.; Brown, J.H. The rate of DNA evolution: Effects of body size and temperature on the molecular clock. *Proc. Natl. Acad. Sci. USA* **2005**, *102*, 140–145. [CrossRef]
123. Hawkins, B.A.; Field, R.; Cornell, H.V.; Currie, D.J.; Guégan, J.-F.; Kaufman, D.M.; Kerr, J.T.; Mittelbach, G.G.; Oberdorff, T.; O'Brien, E.M.; et al. Energy, water, and broad-scale geographic patterns of species richness. *Ecology* **2003**, *84*, 3105–3117. [CrossRef]
124. Davies, T.J.; Barraclough, T.G.; Savolainen, V.; Chase, M.W. Environmental causes for plant biodiversity gradients. *Philos. Trans. R. Soc. B* **2004**, *359*, 1645–1656. [CrossRef]
125. Connell, J.H.; Orias, E. The ecological regulation of species diversity. *Am. Nat.* **1964**, *98*, 399–414. [CrossRef]
126. Schonbeck, M.W.; Norton, T.A. Factors controlling the upper limits of fucoid algae on the shore. *J. Exp. Mar. Biol. Ecol.* **1978**, *31*, 303–313. [CrossRef]
127. Sousa, W.P. The role of disturbance in natural communities. *Annu. Rev. Ecol. Evol. Syst.* **1984**, *15*, 353–391. [CrossRef]
128. Stein, A.; Kreft, H. Terminology and quantification of environmental heterogeneity in species-richness research. *Biol. Rev.* **2015**, *90*, 815–836. [CrossRef]
129. Hortal, J.; Triantis, K.A.; Meiri, S.; Thébault, E.; Sfenthourakis, S. Island species richness increases with habitat diversity. *Am. Nat.* **2009**, *174*, E205–E217. [CrossRef] [PubMed]
130. Turner, J.R.G.; Hawkins, B.A. The global biodiversity gradient. In *Frontiers of Biogeography: New Directions in the Geography of Nature*; Lomolino, M.V., Heaney, L.R., Eds.; Sinauer Associates, Inc.: Sunderland, MA, USA, 2004; pp. 171–190.
131. Emerson, B.C.; Kolm, N. Species diversity can drive speciation. *Nature* **2005**, *434*, 1015–1017. [CrossRef]
132. Erwin, D.H. Macroevolution of ecosystem engineering, niche construction and diversity. *Trends Ecol. Evol.* **2008**, *23*, 304–310. [CrossRef] [PubMed]
133. Erwin, D.H. Seeds of diversity. *Science* **2005**, *308*, 1752–1753. [CrossRef] [PubMed]
134. Cadena, C.D.; Ricklefs, R.E.; Jiménez, I.; Bermingham, E. Is speciation driven by species diversity? *Nature* **2005**, *438*, E1–E2. [CrossRef] [PubMed]
135. Emerson, B.C.; Kolm, N. Emerson & Kolm reply. *Nature* **2005**, *438*, E2.
136. Stehli, F.G.; Douglas, R.G.; Newell, R.G. Generation and maintenance of gradients in taxonomic diversity. *Science* **1969**, *164*, 947–949. [CrossRef]
137. Rombouts, I.; Beaugrand, G.; Ibañez, F.; Gasparini, S.; Chiba, S.; Legendre, L. Global latitudinal variations in marine copepod diversity and environmental factors. *Proc. R. Soc. B* **2009**, *276*, 3053–3062. [CrossRef]
138. Romdal, T.S.; Araújo, M.B.; Rahbek, C. Life on a tropical planet: Niche conservatism and the global diversity gradient. *Glob. Ecol. Biogeogr.* **2013**, *22*, 344–350. [CrossRef]
139. Crisp, M.D.; Cook, L.G. Phylogenetic niche conservatism: What are the underlying evolutionary and ecological causes? *New Phytol.* **2012**, *196*, 681–694. [CrossRef]
140. Sunday, J.M.; Bates, A.E.; Dulvy, N.K. Global analysis of thermal tolerance and latitude in ecthoterms. *Proc. R. Soc. B* **2011**, *278*, 1823–1830. [CrossRef] [PubMed]
141. Beaugrand, G.; Ibañez, F.; Lindley, J.A. Geographical distribution and seasonal and diel changes of the diversity of calanoid copepods in the North Atlantic and North Sea. *Mar. Ecol. Prog. Ser.* **2001**, *219*, 205–219. [CrossRef]
142. Jablonski, D.; Huang, S.; Roy, K.; Valentine, J.W. Shaping the latitudinal diversity gradient: New perspectives from a synthesis of paleobiology and biogeography. *Am. Nat.* **2017**, *189*, 1–12. [CrossRef] [PubMed]
143. Brayard, A.; Escarguel, G.; Bucher, H. Latitudinal gradient of taxonomic richness: Combined outcome of temperature and geographic mid-domains effects? *J. Zool. Syst. Evol. Res.* **2005**, *43*, 178–188. [CrossRef]
144. Kléparski, L.; Beaugrand, G. The species chromatogram, a new graphical method to represent, characterise and compare the ecological niches of different species. *Ecol. Evol.* **2022**, *12*, e8830. [CrossRef]

145. Speck, L. *Caractérisation De La Biodiversité Marine à L'aide De La Théorie METAL*; Université du Littoral Côte d'Opale: Wimereux, France, 2022; p. 22.
146. Goberville, E.; Hautekèete, N.-C.; Kirby, R.R.; Piquot, Y.; Luczak, C.; Beaugrand, G. Climate change and the ash dieback crisis. *Sci. Rep.* **2016**, *6*, 35303. [CrossRef]
147. Peters, R.H. *The Ecological Implications of Body Size*; Cambridge University Press: Cambridge, UK, 1983; p. 329.
148. Damuth, J. Population density and body size in mammals. *Nature* **1981**, *290*, 699–700. [CrossRef]
149. Damuth, J. Of size and abundance. *Nature* **1991**, *351*, 268–269. [CrossRef]
150. Brown, J.H. On the relationship between abundance and distribution of species. *Am. Nat.* **1984**, *124*, 255–279. [CrossRef]
151. Norris, R.D.; Hull, P.M. The temporal dimension of marine speciation. *Evol. Ecol.* **2012**, *26*, 393–415. [CrossRef]
152. Briggs, J.C. Modes of speciation: Marine indo-west Pacific. *Bull. Mar. Sci.* **1999**, *65*, 645–656.
153. Pierrot-Bults, A.C.; Van der Spoel, S. Speciation in macrozooplankton. In *Zoogeography and Diversity of Plankton*; Van der Spoel, S., Pierrot-Bults, A.C., Eds.; John Wiley: New York, NY, USA, 1979; pp. 144–167.
154. Momigliano, P.; Jokinen, H.; Fraimout, A.; Florin, A.-B.; Norkko, A.; Merilä, A. Extraordinarily rapid speciation in a marine fish. *Proc. Natl. Acad. Sci. USA* **2017**, *114*, 6074–6079. [CrossRef] [PubMed]
155. De Visser, J. Transition zones and salp speciation. In *Pelagic Biogeography*; Pierrot-Bults, A.C., Van der Spoel, S., Zahuranec, B.J., Johnson, R.K., Eds.; Unesco: Paris, France, 1985; pp. 266–269.
156. Knowlton, N. Sibling species in the sea. *Annu. Rev. Ecol. Evol. Syst.* **1993**, *24*, 189–216. [CrossRef]
157. Gause, G.F. *The Struggle for Coexistence*; Williams and Wilkins: Baltimore, MD, USA, 1934.
158. Rabosky, D.L.; Hurlbert, A.H. Species richness at continental sclaes is dominated by ecological limits. *Am. Nat.* **2015**, *185*, 572–583. [CrossRef]
159. McManus, M.A.; Woodson, C.B. Plankton distribution and ocean dispersal. *J. Exp. Biol.* **2012**, *215*, 1008–1015. [CrossRef]
160. Sexton, P.F.; Norris, R.D. Dispersal and biogeography of marine plankton: Long-distance dispersal of the foraminifer *Truncorotalia truncatulinoides*. *Geology* **2008**, *36*, 899–902. [CrossRef]
161. Machac, A.; Zrzavy, J.; Smrckova, J.; Storch, D. Temperature dependence of evolutionary diversification: Differences between two contrasting model taxa support the metabolic theory of ecology. *J. Evol. Biol.* **2012**, *25*, 2449–2456. [CrossRef]
162. Rohde, K. *Nonequilibrium Ecology*; Cambridge University Press: Cambridge, UK, 2005.
163. Benton, M.J. Diversification and extinction in the history of life. *Science* **1995**, *268*, 52–58. [CrossRef]
164. Morlon, H. Phylogenetic approaches for studying diversification. *Ecol. Lett.* **2014**, *17*, 508–525. [CrossRef]
165. Berta, A.; Adam, P. The evolutionary biology of pinnipeds. In *Secondary Adaptation of Tetrapods to Life in the Water*; de Buffrenil, V., Mazin, J.-M., Eds.; Verlag Dr Frederich Pfeil: Munchen, Germany, 2001; pp. 235–260.
166. Wiens, J.J.; Donoghue, M.J. Historical biogeography, ecology and species richness. *Trends Ecol. Evol.* **2004**, *19*, 639–644. [CrossRef]
167. Hawkins, B.A.; Porter, E.E. Does herbivore diversity depend on plant diversity? The case of California butterflies. *Am. Nat.* **2003**, *161*, 40–49. [CrossRef] [PubMed]
168. MacArthur, R.H.; Wilson, E.O. *The Theory of Island Biogeography*; Princeton University Press: Princeton, NJ, USA, 1967.
169. Brown, J.H.; Gillooly, J.F.; Allen, A.P.; Savage, V.M.; West, G.B. Toward a metabolic theory of ecology. *Ecology* **2004**, *85*, 1771–1789. [CrossRef]
170. Robinson, P.J.; Henderson-Sellers, A. *Contemporary Climatology*; Pearson Education Limited: Edinburgh, UK, 1999; p. 315.
171. Bonan, G. *Ecological Climatology*, 2nd ed.; Cambridge University Press: Cambridge, UK, 2008; p. 550.
172. Mannion, P.D.; Upchurch, P.; Benson, R.B.J.; Goswami, A. The latitudinal biodiversity gradient through deep time. *Trends Ecol. Evol.* **2014**, *29*, 42–50. [CrossRef] [PubMed]
173. May, R.M. Biological diversity: Differences between land and sea. *Philos. Trans. R. Soc. Lond. B* **1994**, *343*, 105–111.
174. Rozhnov, S.V. Appearance and evolution of marine benthic communities in the early Palaeozoic. *Paleontol. J.* **2005**, *40*, S444–S452. [CrossRef]
175. May, R.M. Conceptual aspects of the quantification of the extent of biological diversity. *Philos. Trans. R. Soc. Lond. B* **1994**, *345*, 13–20.
176. Benton, M.J. Biodiversity on land and in the sea. *Geol. J.* **2001**, *36*, 211–230. [CrossRef]
177. Román-Palacios, C.; Moraga-López, D.; Wiens, J.J. The origins of global biodiversity on land, sea and freshwater. *Ecol. Lett.* **2022**, *25*, 1376–1386. [CrossRef]
178. Pisani, D.; Poling, L.L.; Lyons-Weiler, M.; Hedges, S.B. The colonization of land by animals: Molecular phylogeny and divergence times among arthropods. *BMC Biol.* **2004**, *2*, 1. [CrossRef]
179. Grassle, C.F.; Maciolek, N.J. Deep-sea species richness: Regional and local diversity estimates from quantitative bottom samples. *Am. Nat.* **1992**, *139*, 313–341. [CrossRef]
180. Snelgrove, P.V.R. Getting to the bottom of marine biodiversity: Sedimentary habitats. *Bioscience* **1999**, *49*, 129–138. [CrossRef]
181. Kléparski, L.; Beaugrand, G.; Kirby, R.R. How do plankton species coexist in an apparently unstructured environment? *Biol. Lett.* **2022**, *18*, 20220207. [CrossRef] [PubMed]

182. Briggs, J.C. *Global Biogeography*; Elsevier: Amsterdam, The Netherland, 1995; Volume 14.
183. Swanborn, D.J.B.; Huvenne, V.A.I.; Pittman, S.J.; Rogers, A.D.; Taylor, M.L.; Woodall, L.C. Mapping, quantifying and comparing seascape heterogeneity of Southwest Indian Ridge seamounts. *Landsc. Ecol.* **2023**, *38*, 185–203. [CrossRef]

biology

MDPI

Article

Nutrient Accumulation Pattern in Mixtures of Wheat and Faba Bean Is Strongly Influenced by Cultivar Choice and Co-Existing Weeds

James Ajal * and Martin Weih

Department of Crop Production Ecology, Swedish University of Agricultural Sciences, P.O. Box 7043, SE-75007 Uppsala, Sweden; martin.weih@slu.se
* Correspondence: james.ajal@slu.se; Tel.: +46-702157239

Simple Summary: Growing a mixture of two or more crop species, particularly cereals and legumes, can enhance resource use efficiency for growth-limiting resources, such as nutrients. We evaluated the patterns of nutrient accumulation efficiency in different cultivars of faba bean and wheat grown in mixtures with and without the presence of weeds in a growth container experiment. The cultivar used in the mixture determined the amount of nitrogen accumulated by the legumes, but cereals generally accumulated more nitrogen in the mixtures than when grown as sole crops. Competition from weeds resulted in lower nitrogen accumulation in the crop plants, and plant neighbor identity affected the accumulation of other nutrients relative to the accumulation of nitrogen and phosphorus in the plants. Cultivar choice is therefore important for resource limitation and thereby the growth performance of plants grown in mixtures.

Abstract: Cereal–legume mixtures are often associated with higher yields than the components grown as sole crops, but the underlying mechanisms are unclear. The study aims to evaluate how different cultivars in a two-species wheat–faba bean mixture influence above- and below-ground nitrogen (N) accumulation in the plant biomass, whether crop mixing affected the accumulation of other nutrients relative to the accumulation of N and phosphorus (P), and how the nutrient accumulation pattern in sole crops and mixtures is influenced by weed competition. Using a growth container experiment, we investigate nutrient accumulation patterns on specific wheat and faba bean cultivars grown as sole crops and mixtures, and with and without weed competition. We found that cereals in the mixture accumulated more N than in the sole crops, and the cultivar used influenced biomass accumulation in the legumes. Competition from weeds reduced the amount of plant N pools accumulated in the crop plant biomass. Based on stoichiometric scaling exponents, the plant neighbor affected the accumulation of other nutrients relative to the accumulation of N and P. These results are relevant for species and cultivar selection, all of which are important prerequisites for maximizing mixture performance.

Keywords: scaling exponent; cereal–legume; weed competition; species mixtures; nutrient accumulation; plant–plant interaction

Citation: Ajal, J.; Weih, M. Nutrient Accumulation Pattern in Mixtures of Wheat and Faba Bean Is Strongly Influenced by Cultivar Choice and Co-Existing Weeds. *Biology* **2022**, *11*, 630. https://doi.org/10.3390/biology11050630

Academic Editors: Daniel Puppe, Panayiotis Dimitrakopoulos and Baorong Lu

Received: 10 March 2022
Accepted: 19 April 2022
Published: 20 April 2022

1. Introduction

Increasing the diversity of crop systems could enhance and stabilize crop yields while increasing ecological sustainability. Legume-supported mixtures are often exploited to achieve increased nutrient uptake, and hence reduce the requirement for external fertilizer input [1,2]. While the realized benefits of growing mixtures have previously been associated with low input systems, recent studies have shown that intercropping could realize similar benefits also in high-input systems [2,3]. Species mixtures, especially those of cereal–legumes, make use of positive plant–plant interactions, which facilitate a complementary use of resources, such as light and nutrients [4,5]. In addition, mixtures can deliver services that may not be realized in sole crops. For example, legumes are weak competitors against weeds, but growing them as mixtures with cereals significantly improves the weed suppressive ability of the crop [6,7].

Based on ecological theory, it should be possible to realize the benefits of cereals and legumes grown as mixtures irrespective of the specific crop component used. In practice, specific cereal–legume combinations are commonly preferred, which could be related to cropping regimes that either result in the buildup of diseases [8], or beneficial fungal and bacterial communities in the soil [9]. Mixtures of wheat (*Triticum aestivum* L.) and faba bean (*Vicia faba* L.) are among the most commonly grown cereal–legume mixtures in Europe and offer great potential for use in fodder production. The selection of cultivars to include in these mixtures is crucial in designing productive mixtures, as specific cultivar traits may influence plant nitrogen (N) dynamics and thus the productivity of the mixture [10]. This is particularly important if the cultivars in the mixture differ in traits that influence competition between the species and hence increase spatial complementarity, for example, the shoot architecture, which may affect canopy structure and light interception [11], or the root length density, which is important for below-ground interactions and the uptake of nutrients that limit plant growth [12].

Nitrogen and phosphorus (P) are considered the most important for limiting plant growth [13,14]. For this reason, N and P are the center of attention for most studies that focus on nutrient use [15], and specifically in cereal–legume mixtures [1,16–18]. However, other nutrients play a crucial role in different plant processes and can co-limit plant growth. For example, [15] found evidence for co-limitation by magnesium (Mg) and P in field-grown wheat using scaling exponents. Scaling exponents determine the balance between different nutrients and how this balance is affected by different environmental factors. In this study, we use scaling exponents to quantify the stoichiometric relationship of N and P with respect to a set of other macronutrients, calcium (Ca), potassium (K), Mg, and sulphur (S). A scaling exponent >1 indicates that the relative accumulation of the other nutrients in the plant biomass occurs at a greater rate compared to the combined accumulation of N and P, and a scaling exponent <1 indicates that the relative accumulation of the other nutrients occurs at a lower rate compared to the accumulation of N and P [13]. In that context, resource complementarity is an important mechanism positively influencing plant interaction in mixtures [5]. Although complementarity in mixtures is mostly known to occur for N and P, we expect that the same may apply also for other (co-) limiting nutrients. Moreover, due to plant–plant interactions, the presence of some plants can alter the nutrient concentrations of other plants growing in the vicinity, for example, as shown for P availability in maize grown with mixed species of weeds [19].

Using cultivars of wheat and faba bean grown in sole crops and mixtures as our model system, we aim to evaluate how the different cultivars in the two-species wheat–faba-bean mixtures influence above- and below-ground N accumulation in the plant biomass, and whether crop mixing results in different scaling exponents (N and P vs. other macronutrients) than in the sole crops. In addition, we aim to investigate how the N accumulation and scaling exponents in sole crops and mixtures are influenced by weed competition. The aims are evaluated by means of a semi-controlled growth container experiment in Uppsala during the 2019 growing season. Specifically, we hypothesized that: *(H1)* the wheat grown in a mixture with faba bean accumulates more N and biomass in the shoots and roots than when grown in the sole crop, as a result of increased N fixation from the legumes and the high competitiveness of the cereals; *(H2)* above- and below-ground N accumulation depends on the specific composition (i.e., cultivar choice) of the mixture; *(H3)* resource competition between the crop and weeds results in lower accumulated plant N pool and grain yield of the crops; and *(H4)* scaling exponents differ between sole crops and the same crop grown in mixture, and are affected by the presence of weeds.

2. Materials and Methods

2.1. Plant Material

Two cultivars (cv.) of spring wheat (*Triticum aestivum* L.), cv. 'Diskett' and 'KWS Alderon', and faba bean (*Vicia faba* L.), cv. 'Boxer' and 'Fuego', were grown as sole crops and in wheat–faba-bean mixtures. The wheat cv. 'KWS Alderon' is hereafter referred to as 'Alderon'. All seed material was provided by Lantmännen Lantbruk, Svalöv, Sweden. The cultivars and species combinations were selected as they were previously grown under field conditions in Sweden and have contrasting characteristics. Specifically, the cereals 'Alderon' and 'Diskett' showed contrasting N accumulation, with 'Alderon' having higher N uptake and N accumulation efficiencies than 'Diskett' when grown in mixtures with faba bean in one of the years [10]. In addition to the cultivars used in the different species combinations, the various sole crops and mixtures were grown with and without a weed species (*Chenopodium album* L.), which commonly occurs in agricultural fields in the region, and has been characterized as highly responsive to the addition of nutrient fertilization [7].

2.2. Experimental Setup

The experiment was conducted in Ultuna, Uppsala, Central Sweden (59°49′01.3″ N 17°39′26.0″ E) under semi-controlled conditions with plants exposed to similar weather conditions to those experienced outdoor or in the field, between 13 May–28 August 2019 (Figure S1). The experiment was arranged in a split-plot design with weed treatment as the main plots and the species combinations in sole crops and mixtures as subplots. Each block of 17 pots comprised (a) 2 cultivars of faba bean and wheat in sole crops and 2-species mixtures of the different cultivars grown together with weeds (+weed) and (b) 2 cultivars of faba bean and wheat in sole crops and 2-species mixtures of the different cultivars grown without weeds (−weed). An additional pot containing only weeds was included in each block. The experiment was replicated 8 times resulting in a total number of 136 pots. The growth medium was composed of mineral soil of 45% sand, 25% gravel, 18% coarse silt, 5% fine silt, and 5% clay. The pH of the mineral soil was 7.6, cation exchange capacity (CEC) was approx. 5.3 meq/100 g, and base saturation was >80%. The mineral soil was placed in plastic pots measuring 26 cm in diameter, 21.1 cm in height, and 7.5 L in volume. Mineral soil was chosen as it facilitated root washing for below-ground assessments. Seeds of wheat and faba bean were placed in Petri dishes with dump filter paper and left at room temperature for 48 h. Only pre-germinated seeds were used to ensure 100% seedling survival at the initial growth stage. The pre-germinated seedlings of each plant species and cultivar were transplanted to the pots. Based on the different growth characteristics of the two plant species, which determine space occupancy in a pot, eight and four seedlings were transplanted for wheat and faba bean, respectively, in the sole-crop pots, while the mixtures contained half the plant population of each species grown as a sole crop (Figure 1). Seven days after the planting (DAP) of the crop plants, *C. album* seedlings were transplanted to the respective pots with weed treatment (+weed). The *C. album* seedlings used in the experiment were of uniform size and selected from a weed population growing naturally in the vicinity of the experiment. The weeds were added in an additive design to pots that were identical to the ones without weed treatment (−weed). The plants were supplied with a uniform concentration of a 2 mL L^{-1} nutrient solution with a pH of 3.0. At the seedling stage, 100 mL of the nutrient solution (of concentration 2 mL L^{-1}) was applied per pot once a week for the first four weeks. With increased crop demand during the growth period, 200 mL per pot was applied twice a week, and after crop flowering, 200 m of the nutrient solution was applied once a week until harvest. The nutrient solution was prepared from a balanced, complete fertilizer, 'Wallco' (Cederroth International, Upplands Väsby, Sweden), with NH_4^+ and NO_3^- N in the proportion 2:3 and containing (g L^{-1}): 51 N, 10 P, 43 K, 4 S, 3 Ca, 4 Mg, 0.17 Fe, 0.20 Mn, 0.10 B, 0.03 Zn, 0.015 Cu and 0.004 Mo. Watering was conducted when necessary to maintain a sufficient water supply for plant growth.

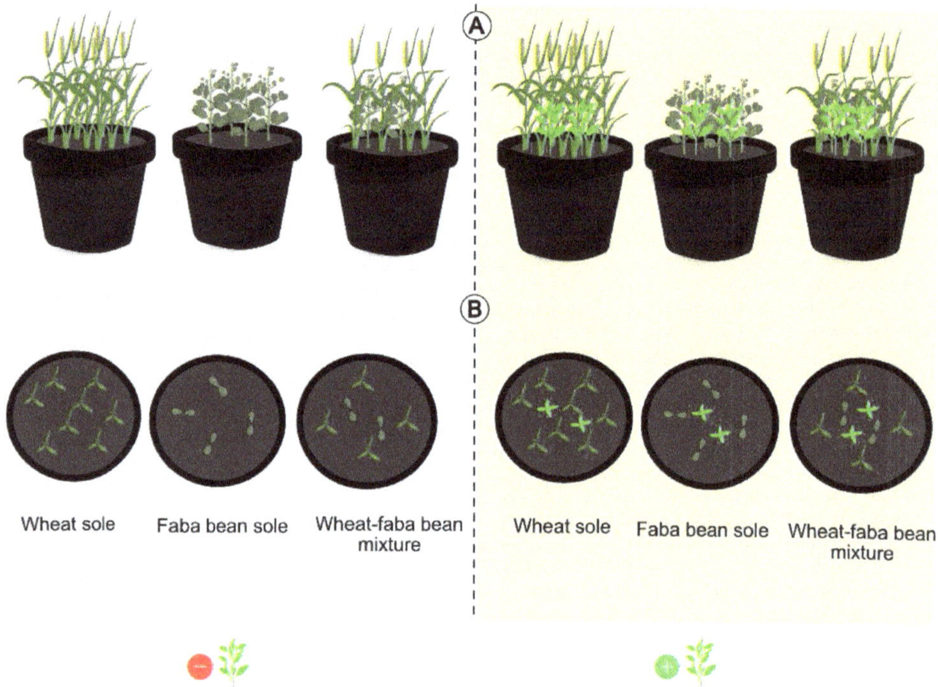

Figure 1. Illustration of a pot experimental treatment including wheat and faba bean with each crop grown as sole crop and in mixture. In half of the pots, the crop plants were grown with the weed *C. album* (+weed) and the other half without the weed treatment (−weed). Two cultivars of the cereal (cv. 'Diskett' and 'Alderon') and legume (cv. 'Boxer' and 'Fuego') were used in each case. (**A**) is a two-dimensional representation of the treatments and (**B**) represents the aerial view of the pots, showing the planting design during the early growth stages.

2.3. Measurements and Sampling

Phenological measurements were taken throughout the growing period, and periodic destructive samplings were conducted at crop flowering and maturity of the cereals. Whole-plant harvests were conducted at 78 DAP (wheat flowering) and at 107 DAP (wheat maturity) to determine above- and below-ground biomass accumulation. Half of the total number of blocks (4 blocks) were randomly assigned to destructive harvests at each vegetative sampling point. All biomass samples were dried at 70 °C for 48 h. For the plants harvested at crop flowering, the relative concentrations of N and the macronutrients Ca, K, Mg, P, and S were analyzed in the biomass of the different species. We focused on the dynamics of both N and macronutrients at the flowering stage because flowering marks an important transition from vegetative to reproductive phases in flowering plants. At maturity, only N was analyzed, but separately for the vegetative plant parts and grains. The nutrient contents in the biomass were used to calculate the nutrient concentrations and pools in the different plant parts grown in the sole crops and different mixture combinations, with and without weeds. We were unable to separate the roots of the different species in the mixtures, and the biomass and N accumulation in the roots of each component in the mixtures were therefore calculated based on the proportions of shoot biomass and N accumulation observed in the corresponding sole crops. This procedure is based on the assumption that the proportion of above- and below-ground accumulation of each species is similar in the sole crops and the same crop grown in a mixture.

2.4. Calculation of Nitrogen Accumulation Efficiency

Nitrogen accumulation efficiency (NAE) and its components were used to evaluate N acquisition and conversion of the cereals and legumes grown as sole crops and in mixtures, with and without the presence of weeds. The evaluations were based on the formula: $NAE = U_N \times E_{N,g} \times C_{N,g}$ [20,21]. U_N, the mean N uptake efficiency, is the component that represents the mean N uptake of the plant during the growth period and accounts for the N acquisition aspect of the plant per unit N in the sown seed. $E_{N,g}$ is the grain specific N efficiency, calculated as a ratio of grain biomass at harvest and mean plant N content during the entire growth period. The grain specific N efficiency represents the N conversion aspect of the plant. $C_{N,g}$ is the grain N concentration: the ratio of grain N content and grain biomass produced at harvest.

2.5. Calculations of Scaling Exponents for Nutrient Concentrations

A niche volume concept was applied to analyze the relationships between the vegetative tissue concentrations of N and P (expressed as their volume, VNP) and the concentrations of other macronutrients (VOth; Ca, K, Mg, S) [13]. Thus, based on the scaling relationship $VOth = \beta(VNP)^{\alpha}$, a faster increase in VOth in relation to VNP is reflected by scaling exponents (α) > 1, whilst a slower increase in VOth in proportion to VNP is reflected by scaling exponents < 1. Calculations of scaling exponents were based on data from the flowering stage only. All regressions for the stoichiometric volumes were calculated as reduced major axes (RMA; SPSS version 26) using ln-transformed values as is conventional in this type of allometric analyses [22].

2.6. Statistical Analysis

Apart from the calculations of scaling exponents, all other statistical analyses were performed in R version 4.0.2 (R Core Team, 2020). Before analysis, data were checked to ensure that the assumptions for normal distribution and equal variances were met. Data that did not meet the assumption for normality were log-transformed ($\log_e$) before analysis. Linear mixed-effects models (nlme Package: *'lme'* by Pinheiro et al. [23]) were used, with each species analyzed separately for each sampling occasion. Mixture treatment (Mix.Sole), the cultivar in the mixture (Cultivar) and the weed treatment (Weed) were treated as fixed effects, while Weed nested within block as random effects. The calculation for NAE was based on data from both flowering and maturity, and instead of evaluating the different species depending on the mixture and weed treatments within each block, species were grouped based on the treatment for the whole experiment. Therefore, the model structure was similar as above (i.e., Cultivar, Mix.Sole and Weed as fixed effects), but the assigned treatment (Treatment) was regarded as random effect.

3. Results

3.1. Differences between Sole and Mixed Culture Were Mostly Related to N Uptake Efficiency, Not N Conversion Efficiency

Wheat grown as mixtures accumulated greater amounts of N in the shoots compared to the sole-cropped wheat. For example, at maturity, wheat cv. 'Alderon' accumulated 41.5% more N in the shoots when grown as a mixture than as a sole crop together with weeds (Figure 2, Table 1). In contrast, the plant N pools of the legumes were higher in the sole crops than in the mixtures at both flowering and maturity (Table 1). At flowering, this translated to 55.1% and 6.4% greater N pool in sole crops of Faba bean cv. 'Boxer' and cv. 'Fuego', respectively, compared to growth in the mixed setup without weed competition. Based on the calculated NAE components (Figure 3), N uptake efficiency (U_N) followed a similar trend as the accumulated N in the biomass. The high U_N of wheat in the mixture was associated with more biomass and N accumulation in the shoots than in the sole crop, while in the legumes, high U_N in the sole crop was associated with more biomass and N accumulation in the sole crop than in the mixtures. The grain specific N efficiency ($E_{N,g}$), which represents the N conversion aspect of the plant, and grain N concentration ($C_{N,g}$) were similar in the sole crops and mixtures of both wheat and faba bean.

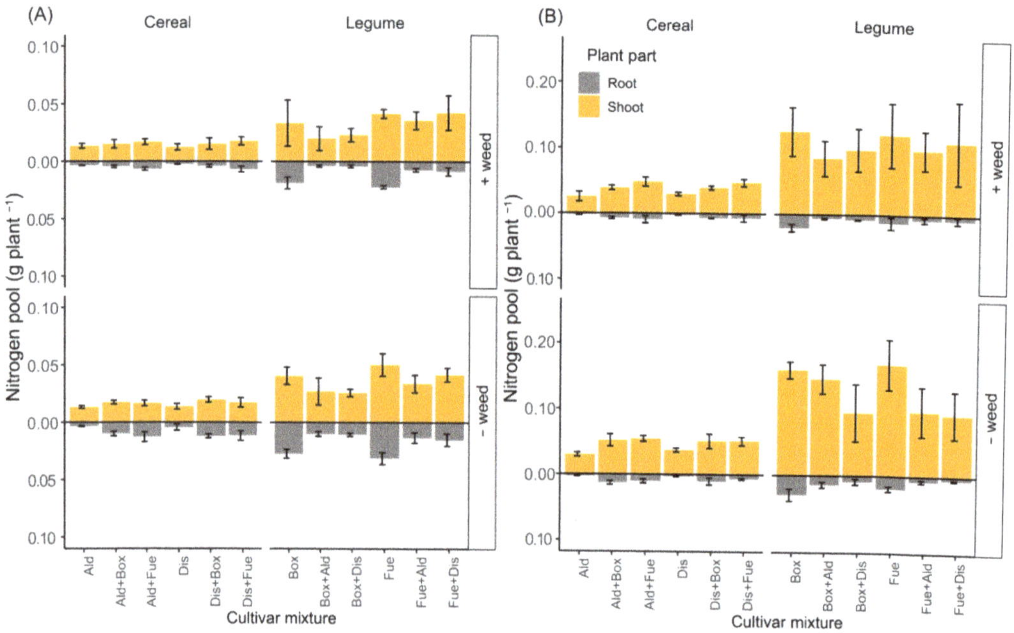

Figure 2. Above and below ground mean plant N pools for different wheat and faba bean cultivars pot-grown as sole crops and in mixtures, and sampled at flowering (**A**) and maturity (**B**). Each cultivar (Ald, 'Alderon'; Dis, 'Diskett'; Box, 'Boxer'; and Fue, 'Fuego') was grown as a sole crop and in mixture, with (+) and without (−) the weed *C. album*. Above-ground N pools were calculated from leaves, stems and heads/pods (grains at maturity). Note the different scales in (**A**,**B**). The error bars represent ± 1SD.

Table 1. Analysis of variance for plant nitrogen pools in the shoot and roots of different cultivars of wheat and faba bean grown as sole crops and mixtures sampled at flowering and maturity. Symbols show results with significant levels: *** = $p \leq 0.001$; ** = $p \leq 0.01$; * = $p \leq 0.05$.

Factor	Flowering		Maturity	
	Cereal	**Legume**	**Cereal**	**Legume**
Shoot				
Weed	0.2028	0.4512	0.0246 *	0.1430
Cultivar	0.4714	<0.0001 ***	0.5768	0.7630
Mix.Sole	<0.0001 ***	0.0006 ***	<0.0001 ***	0.0008 ***
Weed × Cultivar	0.4912	0.4661	0.9034	0.3167
Weed × Mix.Sole	0.5915	0.2649	0.4683	0.1726
Cultivar × Mix.Sole	0.4418	0.3256	0.0893	0.7012
Weed × Cultivar × Mix.Sole	0.8027	0.4774	0.5317	0.2482
Root				
Weed	0.0080 **	0.0190 *	0.1781	0.0645
Cultivar	0.9514	<0.0001 ***	0.0870	0.0017 **
Mix.Sole	<0.0001 ***	<0.0001 ***	<0.0001 ***	<0.0001 ***
Weed × Cultivar	0.3825	0.9960	0.5227	0.1756
Weed × Mix.Sole	0.0012 **	0.1450	0.3454	0.0513
Cultivar × Mix.Sole	0.8327	0.8606	0.2458	0.0297*
Weed × Cultivar × Mix.Sole	0.6327	0.8662	0.6260	0.5824

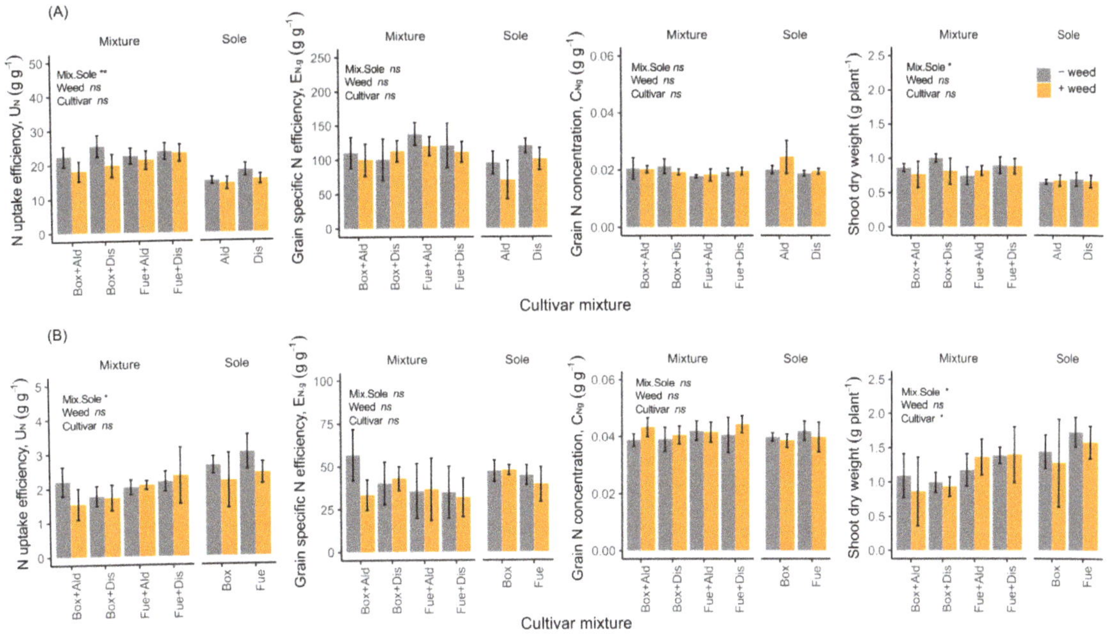

Figure 3. Nitrogen accumulation efficiency (NAE) components for cereals (**A**) and legumes (**B**), showing N uptake efficiency (U_N), grain specific N efficiency, $E_{N,g}$), grain N concentration ($C_{N,g}$) and shoot dry weight at flowering. The different cereal and legume cultivars were pot-grown as sole crops and mixtures, with (+weed) and without (−weed) *C. album*. The full cultivar names are the same as in Figure 2. Symbols show ANOVA results with significant levels: ** = $p \leq 0.01$; * = $p \leq 0.05$; and ns= non-significant. Error bars represent ± 1SD.

The specific cultivar of the legume used influenced biomass and shoot N accumulation at flowering (Table 1), while the presence of the weed *C. album* reduced the N pool only for the wheat at maturity. There was no effect of weeds on any of the NAE components, as similar U_N, $E_{N,g}$ and $C_{N,g}$ were found in all cases (Figure 3, Table S1). At both crop flowering and maturity, wheat accumulated more root N when grown in a mixture than as a sole crop ($p < 0.000$). In contrast, the legumes had higher root N in the sole crop than when grown as a mixture at both sampling periods ($p < 0.000$). The cultivar variable only had marginal effects on root N accumulation in the mixtures ($p = 0.057$ and $p = 0.054$ at flowering and maturity, respectively).

3.2. The Reduction in Plant N Pools in the Mixtures Is Correlated with Weed Biomass Accumulation

The total plant N pools in the cereals grown in the mixtures decreased with increasing amounts of biomass accumulated by the weeds at both flowering and maturity. (Figure 4A,B; $p = 0.001$ and $p < 0.009$ for flowering and maturity, respectively.) The crop cultivars used in the mixtures did not influence the relationships between the plant N pools and accumulated shoot biomass in *C. album*. Similarly, growing weeds in the same pots with the legumes did not significantly reduce the plant N pools for the legumes at both flowering and maturity (flowering; $R^2 = 0.02$, $p = 0.567$ and maturity; $R^2 = 0.18$, $p = 0.108$). Despite the reduction in plant N pools of the cereals, the cereal yield, assessed as head weight, was not affected by weed growth as no difference was found between crops grown with and without the weed (*C. album*). All cereal cultivars grown in the mixtures had higher yields than the same cultivars grown as sole crops (Figure 5A, Table S2). The greatest yield difference was observed in the wheat cv. 'Alderon', where 50.7% greater head weight was observed when intercropped with faba bean cv. 'Fuego' in the absence of weed competition. For legumes, 'Boxer' in a 'Boxer'–'Alderon' mixture grown without *C. album* produced a higher yield (pod weight) than when grown with *C. album* (Figure 5B, Table S2). The yield-related loss attributed to weed competition was 1.59 g plant^{-1}, equivalent to a 49.8% lower yield under weed competition. Similarly, only 'Boxer' both in mixtures with 'Alderon' and 'Diskett' had yields higher in the mixture than the sole crop.

3.3. Plant Neighbor Affects the Accumulation of Other Nutrients (Ca, K, Mg and S) Relative to the Accumulation of N and P

We found a higher scaling exponent for nutrient concentrations in the legumes than the cereals, implying the relative accumulation of nutrients other than N and P (i.e., Ca, K, Mg, S) was greater in the legumes than the cereals (Table 2). Crop mixing significantly decreased the scaling exponent of the cereals compared to growing them as sole crops, implying less accumulation of other nutrients (Ca, K, Mg and S) in relation to the accumulation of N and P (p = 0.016, Table 3). Growing the weed *C. album* in the same pots significantly increased the scaling exponents of the legumes compared to growing the same legumes in the absence of the weed (p = 0.005). The scaling exponents did not significantly differ between the cultivars.

Table 2. Scaling exponents (α) for cereals and legumes grown as sole crops and in mixtures, with and without the weed *C. album*, based on the nutrient volumes for N and P (VNP) vs. the volumes of other nutrients (VOth; K, Ca, Mg and S). The scaling exponents were based on data from the flowering stage.

Group	Scaling Exponent (α)	N	SE
Cereals sole	1.349	8	0.026
Legumes sole	2.001	8	0.012
Cereals mix	1.284	16	0.012
Legumes mix	2.033	16	0.025
Cereals + Weed	1.329	8	0.034
Legumes + Weed	2.077	8	0.019

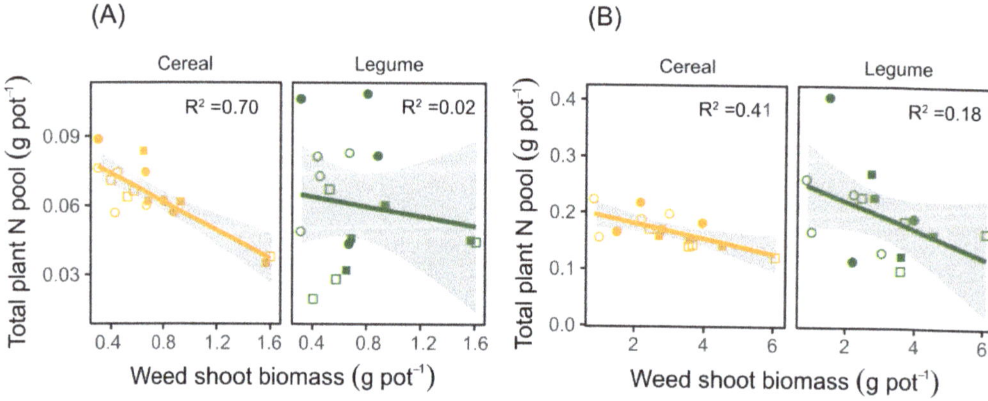

Figure 4. Relationships between total plant N pools and weed shoot biomasses for the cultivars of cereals and legumes pot-grown in mixtures and with the weed *C. album*. The plants were sampled at (**A**) flowering and (**B**) maturity. Both cultivars of cereals and legumes were considered in the analysis. Open and closed square symbols represent 'Boxer'–'Alderon' and 'Boxer'–'Diskett' cultivar mixtures, respectively, while open and closed circles represent 'Fuego'–'Alderon' and 'Fuego'–'Diskett' cultivar mixtures, respectively. Lines indicate linear regressions at crop flowering: cereals y= 0.087–0.031x and legumes y = 0.068–0.011x with $p \leq 0.001$ and $p = 0.567$, respectively; and at maturity: cereals y = 0.21–0.014x and legumes y = 0.27–0.025x with $p < 0.0098$ and $p = 0.108$, respectively.

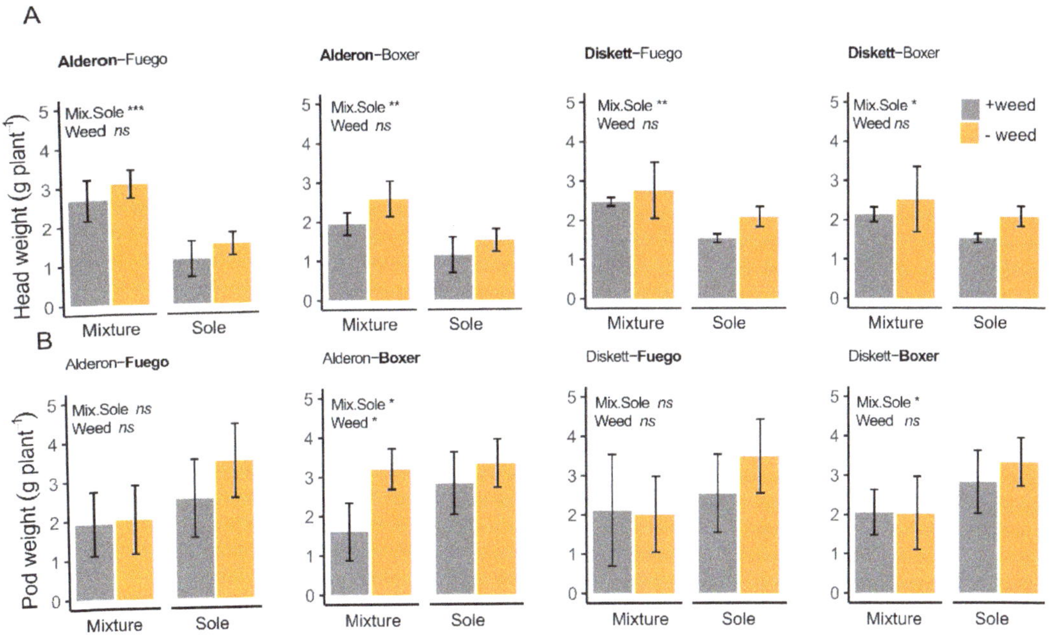

Figure 5. The mean (**A**) head height for wheat and (**B**) pod weight for two faba bean cultivars pot-grown as sole crops and in mixture, with and without the weed *C. album* at maturity. The error bars represent ±1 SD of the different cultivar combinations. The cultivars whose names appear in bold correspond to the ones presented in the graph. Symbols show ANOVA results with significance levels: *** = $p \leq 0.001$; ** = $p \leq 0.01$; * = $p \leq 0.05$; and ns= non-significant.

Table 3. Pairwise comparison of scaling exponents following analysis of variance for cereals and legumes grown as sole crops and mixtures, with and without the *C. album* weed. The scaling exponents were based on data from the flowering stage. Symbols show results with significant levels: *** = $p \leq 0.001$; ** = $p \leq 0.01$; * = $p \leq 0.05$.

Comparison	F-Value	*p*-Value
Cereals sole vs. Legumes sole	518.42	0.000 ***
Cereals sole vs. Cereals mix	6.85	0.016 *
Cereals sole vs. Cereals + Weed	0.22	0.648
Legumes sole vs. Legumes mix	0.76	0.393
Legumes sole vs. Legumes + Weed	11.44	0.005 **
Cereals mix vs. Legumes mix	729.52	0.000 ***
Cereals + Weed vs. Legumes + Weed	368.82	0.000 ***

4. Discussion

The point of departure for this study was the set of results from a field experiment in which the same wheat and faba bean cultivars were grown as sole crops and in mixtures for two years [10], but with the difficulties associated with field studies, such as the limited possibilities to control growth conditions (e.g., plant neighbors, including weeds) and the consideration of below-ground plant parts. This investigation therefore complements the field study by Ajal et al. [10] and provides additional insights into the mechanisms underlying plant–plant interactions, especially with regard to nutrient accumulation patterns, including nutrients other than N. Furthermore, by linking plant functional traits to plant diversity and ecosystem processes, our research contributes to the further understanding of the mechanisms underpinning plant community productivity, stability and nutrient dynamics, and thus to biodiversity–ecosystem function (BEF) theory development, for which a clear mechanistic understanding is still lacking [24,25].

4.1. Effect of Diversity on N Accumulation Efficiency and Plant Biomass: Do Cultivars Matter?

We observed the contrasting effects of diversity on shoot N accumulation by wheat and faba bean, with the wheat and faba bean having higher shoot N pools in the mixtures and sole crops, respectively. Plant–plant interactions between crop species with different functional trait values facilitate complementarity, which ensures a more complete resource use for the components in the mixture [26]. Furthermore, the spatial distribution of plants or temporal availability of resources ensures that a more complete resource capture is achieved by the plants in the community [27]. In line with the above, our results were consistent with hypothesis *H1*. More biomass and N accumulation of the cereals in the mixture than sole crop were associated with a higher N uptake efficiency (U_N), which suggests that the mixtures had accumulated more plant N than the sole crops. Although contrary to the cereals, the legume performance supports the results from several other studies that have shown the competitive disadvantage of legumes, especially when in interaction with cereals [6,10,28]. Compared to the legumes, cereals can achieve a more developed root system at a faster rate, especially in the early stages of crop development, which enables the exploration of larger soil volumes for the uptake of resources such as water and nutrients [16,29,30]. Ideally, a high N acquisition rate, as we observed in cereals that were grown in mixtures, would need to be complemented by a high conversion efficiency of the acquired N in order to maximize grain yield. However, the yield increase in the mixture was exclusively explained by N acquisition since we found no evidence for enhanced N conversion efficiency (represented by grain specific N efficiency, $E_{N,g}$, in our study) in the mixtures. The high accumulation of shoot N pool in the legume sole crops may be explained by a low competitive pressure due to the absence of a highly competitive plant neighbor (wheat), and the reliance on N fixation to meet part of its N requirement. The faba bean cultivar used, either in the sole crop or mixture, influenced the amount of N pools accumulated both above- and below-ground at flowering and maturity. Although we do not have any empirical evidence for the cultivar differences, a plausible explanation could be related to the differences in biological N fixation, which ensured that a different amount of N was utilized by the crop during the growth period as was, for example, illustrated in the pea by Hauggaard-Nielsen et al. [31]. The differences between the cultivars of faba bean may have implications for management, especially in relation to cultivar selection, when designing a mixed crop system. Although we here focused on cultivar differences for plant N pools, the same trend may also apply for N uptake or leaf area index (LAI), because N is known to enhance plant leaf and canopy development [32]. Incidentally, canopy structure is one of the traits that can be manipulated to allow maximum light interception in crop mixtures [33]. Therefore, farming practitioners could consider more functionally different canopy characteristics when choosing components to include in mixtures because they, together with other factors, influence radiation use efficiency (RUE). As expected, we found that the shoot N pool increased between flowering and maturity in both cereals and legumes. However, in a few cases, the N pools of the cereals were similar or even decreased between flowering and maturity when the same cultivars were grown in the field [10]. From the onset of crop flowering to maturity, many physiological changes occur that relate to N remobilization from the leaves or N re-allocation to the grains, which may be influenced by various factors. For example, the time of leaf senescence (i.e., delayed senescence), which favors prolonged photosynthesis, and the high N stored at flowering can favor more grain N accumulation at maturity [34].

4.2. Weed Influences N Accumulation Efficiency in Wheat–Faba-Bean Mixtures

The diversity in crop mixtures, especially when accomplished by species or cultivars with great differences in functional traits, may facilitate more efficient resource use, which is associated with the complementary sharing of resources from a common pool [5,35]. In such cases, complementarity or facilitation supersedes interspecific competition between the components of the mixture. Competition from weeds is undesirable and associated with the reduction in the quality or quantity of yield [36,37]. In the case of this study, competition from the weed *C. album* resulted in less accumulated plant N pools in the crops at both flowering and maturity, which is consistent with our hypothesis *H3*. It is likely that most of the N that the wheat and faba bean were deprived of, or did not take up, constituted the portion accumulated by the weeds for biomass production. This is based on the assumption that the crop and weeds draw resources from the same resource pool given that resource pool similarity or diversity can have different influence on crop–weed competition intensity [38]. The lower plant N accumulation by the crop plants when grown together with *C. album* is consistent with the results by Jäck et al. [7], in which accumulated N by the weeds (*C. album*) was positively correlated and increased with increasing N available to the weeds, although the study was based on a pea–barley mixture. It should also be noted that *C. album* is a highly responsive weed in relation to N availability [7]. These two factors taken together may explain the observed relationship between the total plant N pools of the crop plants and the biomass accumulation of the weeds. However, in general, the competition from *C. album* was not strong enough to compromise the crop yield, except in the faba bean cv. 'Boxer'. We think this was associated with the weak competition between the crop and the weed for above-ground resources. Under controlled conditions (e.g., pot experiments), where interaction is usually among few plants, individual species in the interaction may efficiently utilize light (hence light becomes non-limiting for growth), and stronger competition for resources, such as water and nutrients, is likely to occur below- than above-ground [39]. In the field, changes in the canopy structure make competition for light between the crop and weed an important element of the interaction, which is significant in the suppression of weeds. Therefore, the better performance of all cereal cultivars in the mixtures compared to the sole crops, irrespective of whether in interaction with the weed or not, is due to the higher competitiveness of the cereals than the legumes. Cereals generally have faster initial growth than the legumes under similar growth conditions and possess more developed and deeper root systems that favor greater resource uptake.

4.3. How Does N and P Uptake Relate to Other Major Nutrients? Does Mixing Improve the Accumulation of the Other Nutrients?

Compared to other nutrients, N and P have been extensively studied, which is attributed to their role as being the most limiting for plant growth [13]. The relative accumulation of nutrients other than N and P, especially in cereals and legumes grown as sole crops and mixtures, provided new insights for nutrient balance as a result of species interactions. Through this study, we showed that the specific cultivar or plant species used significantly affects the nutrient accumulation and stoichiometry of the neighboring plants grown together in the same pot. We expected the scaling exponents for wheat and faba bean to differ when grown in sole crop or mixture (hypothesis *H4*). Indeed, the cereals in the mixture (with legumes) significantly decreased their scaling exponent, implying that the interaction with the legumes reduced the relative accumulation of nutrients other than N and P. The cereals rely on N and P accumulation for important growth processes, such as tillering (N) [40] and root development and energy metabolism (P) [41]. Therefore, it is not surprising that, when the accumulation of other nutrients increased at a lower rate, more N and P were accumulated. Although we found no significant variation in the scaling exponents of faba bean grown in sole crop and mixture, the scaling exponent of faba bean (legume) was generally higher than that of wheat (cereal). The difference in the crop species scaling exponents implies that the relative accumulation of nutrients other than N and P was greater in the legumes than the cereals. We attribute this to the N-fixing nature of the legumes, which promotes the reliance on fixed N to meet part of its own N requirement. Thus, the increase in demand (and accumulation) of other nutrients is more important for the legumes than the cereals [17]. In the presence of weed competition, weeds pose a strong competition for N with the legumes, e.g., as earlier illustrated for pea in competition with *C. album* [7]. Since the legumes often and to a high degree rely on N from atmospheric N fixation, the need to accumulate other nutrients might become more important than accumulating N. In addition, it has been shown in another study that weed mixtures (including *C. album*) can facilitate the greater uptake of P due to the mycorrhizal colonization of the roots [19]. This may partly explain why in our study the scaling exponents of the legumes increased in the presence of weeds. It is worth highlighting the magnitude of scaling exponents (scaling exponent >2) of the legumes in the mixture or when grown together in a mixture with the cereals and weed, which resulted in a strong and significant effect of the neighboring plant on nutrient stoichiometry. This observation is particularly important as it may have implications for nutrient (co-) limitation as shown earlier by Weih et al. [15]. The co-limitation may be applicable in this study since having a legume or a weed as a neighboring plant may result in more accumulation of one or more nutrients other than N and P, but may also be applicable in the field, especially when crop mixtures are grown. However, the observed effects of plant neighbor on nutrient stoichiometry must be validated in field experiments.

4.4. Implications for Field Conditions and Management

This study was based on species combinations that have previously, under a different set of objectives, been tested under field conditions [7,10]. We conducted this study under semi-controlled conditions, specifically to enable us to sample both above- and below-ground plant parts. Interestingly, some trends were consistent with the findings of the field studies by Ajal et al. [10] and Jäck et al. [7], for example, the relationship between crop and weed N accumulation. However, as in the case of many controlled studies, concerns of their applicability under field conditions often arise. Nevertheless, the findings from this study are important in making informed decisions when designing crop mixtures. Specifically, the difference in N accumulation and biomass in the shoots of different faba bean cultivars suggests the need to focus on specific cultivars with traits that enhance nutrient accumulation efficiency. The increased N pools between flowering and maturity may also have implications for nutrient management. This suggests that split application should be considered, with one application performed relatively late in the season (for example, around booting of the cereals) to facilitate more N accumulation at a later stage during the growth period, and translocation to the grains for higher quality products.

4.5. Study Limitations

The number of cereal and legume cultivars in this study was limited to only two cultivars per species, which limited the potential to explore the scaling exponent variability compared to if more cultivars had been used. We also estimated the dry weight and plant N pools of the roots based on the assumption that their proportions above- and below-ground are similar when grown as sole crop and in mixture. We acknowledge that, although this enabled us to estimate root characteristics, the plasticity of traits may be influenced by the different growth conditions in the mixtures compared to the sole crops. The mentioned limitations notwithstanding, we think this study generated valuable information for nutrient accumulation and species interaction in crop mixtures.

5. Conclusions

This study demonstrated some mechanistic links explaining why species and cultivar choice is important for resource limitation, plant-internal resource allocation and use and, ultimately, the growth performance of plants grown in mixtures with and without the presence of co-existing weeds. Crop interaction with weeds (*C. album*) results in reduced amounts of N pools accumulated by the crop plants, especially wheat, a mechanism similar to the effect of cereals on legumes when the two are grown together. These interactions are strongly modulated by the characteristics of the crop cultivars used in the mixture. Results from the analysis of stoichiometric scaling exponents show that the relative accumulation of other nutrients (Ca, K, Mg and S) by the crops compared to the accumulation of N and P depends on which plant neighbors the focal plant has. Since this study involved different crop species, cultivars and crop interaction with weeds, it provides new insights necessary for evidence-based species selection when designing crop mixtures. Future plant physiological studies on the mechanisms underlying the differential accumulation of nutrients when grown with different neighbors are important for our understanding of plant–plant interactions. Furthermore, studies involving specialized techniques for quantifying below-ground interaction in mixtures, for example, DNA-based methods or Infrared Spectroscopy, could provide more precise measures for quantifying root interactions between different species in the mixture than was possible with the method used in our study.

Supplementary Materials: The following supporting information can be downloaded at: https://www.mdpi.com/article/10.3390/biology11050630/s1, Figure S1: Daily precipitation, average temperature and relative humidity for Ultuna near Uppsala, Central Sweden, during the period of experimentation in 2019. The arrows show the time points for planting, and destructive samplings at flowering and maturity. Relative humidity and air temperature data were recorded using a weather logger installed in net yard. Rainfall data were accessed from Ultuna Climate Station, Uppsala (Sweden); Table S1: Analysis of variance for Nitrogen Accumulation Efficiency (NAE) components and shoot dry weight at flowering for different cultivars of cereals and legumes. The different cultivars (Cultivars) were grown as sole crops and in mixtures (Mix.Sole), with and without *C. album* (Weed); Table S2: Analysis of variance for yield assessed as head weight and pod weight in the cereals and legumes, respectively. Analysis was performed for different cultivar combinations pot-grown in sole crops and mixtures (Mix.Sole), with and without the weed C. album (Weed).

Author Contributions: Conceptualization, J.A. and M.W.; methodology, J.A. and M.W.; formal analysis, J.A.; investigation, J.A.; data curation, J.A.; writing—original draft preparation, J.A.; writing—review and editing, J.A. and M.W.; supervision, M.W.; funding acquisition, M.W. and J.A. All authors have read and agreed to the published version of the manuscript.

Funding: This research was funded through the DIVERSify project, a grant from the European Union's Horizon 2020 research and innovation program under grant agreement No. 727284, and C L Behm's fund for legume cultivation.

Data Availability Statement: All data supporting results in this paper are openly accessible at http://doi.org/10.6084/m9.figshare.19607040.

Acknowledgments: We are grateful to Dewi Berniot, Hedvig Säll, Varwi Jacob Tavaziva, Ewa Magnuski and Nils-Erik Nordh for assistance with plant sampling.

Conflicts of Interest: The authors declare no conflict of interest.

References

1. Jensen, E.S.; Carlsson, G.; Hauggaard-Nielsen, H. Intercropping of grain legumes and cereals improves the use of soil N resources and reduces the requirement for synthetic fertilizer N: A global-scale analysis. *Agron Sustain. Dev.* **2020**, *40*, 5. [CrossRef]
2. Li, C.; Hoffland, E.; Kuyper, T.W.; Yu, Y.; Zhang, C.; Li, H.; Zhang, F.; van der Werf, W. Syndromes of production in intercropping impact yield gains. *Nat. Plants* **2020**, *6*, 653–660. [CrossRef] [PubMed]
3. Tilman, D. Benefits of intensive agricultural intercropping. *Nat. Plants* **2020**, *6*, 604–605. [CrossRef] [PubMed]
4. Stomph, T.J.; Dordas, C.; Baranger, A.; de Rijk, J.; Dong, B.; Evers, J.; Gu, C.; Li, L.; Simon, J.; Jensen, E.S. Designing intercrops for high yield, yield stability and efficient use of resources: Are there principles? In *Advances in Agronomy*; Academic Press Inc.: Cambridge, MA, USA, 2020.
5. Brooker, R.W.; Bennett, A.E.; Cong, W.F.; Daniell, T.J.; George, T.S.; Hallett, P.D.; Hawes, C.; Iannetta, P.P.; Jones, H.G.; Karley, A.J.; et al. Improving intercropping: A synthesis of research in agronomy, plant physiology and ecology. *New Phytol.* **2015**, *206*, 107–117. [CrossRef]
6. Corre-Hellou, G.; Dibet, A.; Hauggaard-Nielsen, H.; Crozat, Y.; Gooding, M.; Ambus, P.; Dahlmann, C.; von Fragstein, P.; Pristeri, A.; Monti, M.; et al. The competitive ability of pea–barley intercrops against weeds and the interactions with crop productivity and soil N availability. *Field Crops Res.* **2011**, *122*, 264–272. [CrossRef]
7. Jäck, O.; Ajal, J.; Weih, M. Altered Nitrogen Availability in Pea–Barley Sole- and Intercrops Changes Dominance of Two Nitrophilic Weed Species. *Agronomy* **2021**, *11*, 679. [CrossRef]
8. Lv, J.; Dong, Y.; Dong, K.; Zhao, Q.; Yang, Z.; Chen, L. Intercropping with wheat suppressed Fusarium wilt in faba bean and modulated the composition of root exudates. *Plant. Soil* **2020**, *448*, 153–164. [CrossRef]

9. Granzow, S.; Kaiser, K.; Wemheuer, B.; Pfeiffer, B.; Daniel, R.; Vidal, S.; Wemheuer, F. The Effects of Cropping Regimes on Fungal and Bacterial Communities of Wheat and Faba Bean in a Greenhouse Pot Experiment Differ between Plant Species and Compartment. *Front. Microbiol* **2017**, *8*, 902. [CrossRef]
10. Ajal, J.; Jäck, O.; Vico, G.; Weih, M. Functional trait space in cereals and legumes grown in pure and mixed cultures is influenced more by cultivar identity than crop mixing. *Perspect. Plant. Ecol. Evol. Syst.* **2021**, *50*, 125612. [CrossRef]
11. Gou, F.; van Ittersum, M.K.; Simon, E.; Leffelaar, P.A.; van der Putten, P.E.L.; Zhang, L.; van der Werf, W. Intercropping wheat and maize increases total radiation interception and wheat RUE but lowers maize RUE. *Eur. J. Agron.* **2017**, *84*, 125–139. [CrossRef]
12. Wang, Y.F.; Qin, Y.Z.; Chai, Q.; Feng, F.X.; Zhao, C.; Yu, A.Z. Interspecies Interactions in Relation to Root Distribution Across the Rooting Profile in Wheat-Maize Intercropping Under Different Plant Densities. *Front. Plant. Sci.* **2018**, *9*, 483. [CrossRef] [PubMed]
13. Agren, G.I.; Weih, M. Multi-Dimensional Plant Element Stoichiometry-Looking Beyond Carbon, Nitrogen, and Phosphorus. *Front. Plant. Sci.* **2020**, *11*, 23. [CrossRef] [PubMed]
14. Elser, J.J.; Bracken, M.E.; Cleland, E.E.; Gruner, D.S.; Harpole, W.S.; Hillebrand, H.; Ngai, J.T.; Seabloom, E.W.; Shurin, J.B.; Smith, J.E. Global analysis of nitrogen and phosphorus limitation of primary producers in freshwater, marine and terrestrial ecosystems. *Ecol. Lett.* **2007**, *10*, 1135–1142. [CrossRef] [PubMed]
15. Weih, M.; Liu, H.; Colombi, T.; Keller, T.; Jack, O.; Vallenback, P.; Westerbergh, A. Evidence for magnesium-phosphorus synergism and co-limitation of grain yield in wheat agriculture. *Sci. Rep.* **2021**, *11*, 9012. [CrossRef]
16. Corre-Hellou, G.; Fustec, J.; Crozat, Y. Interspecific Competition for Soil N and its Interaction with N2 Fixation, Leaf Expansion and Crop Growth in Pea–Barley Intercrops. *Plant. Soil* **2006**, *282*, 195–208. [CrossRef]
17. Tang, X.; Zhang, C.; Yu, Y.; Shen, J.; van der Werf, W.; Zhang, F. Intercropping legumes and cereals increases phosphorus use efficiency; a meta-analysis. *Plant. Soil* **2021**, *460*, 89–104. [CrossRef]
18. Ramirez-Garcia, J.; Martens, H.J.; Quemada, M.; Thorup-Kristensen, K. Intercropping effect on root growth and nitrogen uptake at different nitrogen levels. *J. Plant. Ecol.* **2015**, *8*, 380–389. [CrossRef]
19. Zacher, A.; Baum, C.; de Mol, F.; Dehmer, K.J.; Gerowitt, B. Mixed Growth with Weeds Promotes Mycorrhizal Colonization and Increases the Plant-Availability of Phosphorus under Maize (*Zea mays* L.). *Agronomy* **2021**, *11*, 1304. [CrossRef]
20. Weih, M.; Asplund, L.; Bergkvist, G. Assessment of nutrient use in annual and perennial crops: A functional concept for analyzing nitrogen use efficiency. *Plant. Soil* **2011**, *339*, 513–520. [CrossRef]
21. Weih, M.; Hamnér, K.; Pourazari, F. Analyzing plant nutrient uptake and utilization efficiencies: Comparison between crops and approaches. *Plant. Soil* **2018**, *430*, 7–21. [CrossRef]
22. Niklas, K.J. Plant allometry, leaf nitrogen and phosphorus stoichiometry, and interspecific trends in annual growth rates. *Ann. Bot.* **2006**, *97*, 155–163. [CrossRef] [PubMed]
23. Pinheiro, J.; Bates, D.; DebRoy, S.; Sarkar, D.; Heisterkamp, S.; Van Willigen, B.; Maintainer, R. Package 'nlme'. Linear and Nonlinear Mixed Effects Models, Version. 2017. Available online: http://cran.rapporter.net/web/packages/nlme/nlme.pdf (accessed on 26 April 2020).
24. Turnbull, L.A.; Levine, J.M.; Loreau, M.; Hector, A. Coexistence, niches and biodiversity effects on ecosystem functioning. *Ecol. Lett.* **2013**, *16*, 116–127. [CrossRef] [PubMed]
25. O'Connor, M.I.; Gonzalez, A.; Byrnes, J.E.; Cardinale, B.J.; Duffy, J.E.; Gamfeldt, L.; Griffin, J.N.; Hooper, D.; Hungate, B.A.; Paquette, A. A general biodiversity–function relationship is mediated by trophic level. *Oikos* **2017**, *126*, 18–31. [CrossRef]
26. Bedoussac, L.; Journet, E.-P.; Hauggaard-Nielsen, H.; Naudin, C.; Corre-Hellou, G.; Jensen, E.S.; Prieur, L.; Justes, E. Ecological principles underlying the increase of productivity achieved by cereal-grain legume intercrops in organic farming. A review. *Agron Sustain. Dev.* **2015**, *35*, 911–935. [CrossRef]
27. Finney, D.M.; White, C.M.; Kaye, J.P. Biomass Production and Carbon/Nitrogen Ratio Influence Ecosystem Services from Cover Crop Mixtures. *Agron. J.* **2016**, *108*, 39–52. [CrossRef]
28. Hauggaard-Nielsen, H.; Gooding, M.; Ambus, P.; Corre-Hellou, G.; Crozat, Y.; Dahlmann, C.; Dibet, A.; von Fragstein, P.; Pristeri, A.; Monti, M.; et al. Pea–barley intercropping for efficient symbiotic N2-fixation, soil N acquisition and use of other nutrients in European organic cropping systems. *Field Crops Res.* **2009**, *113*, 64–71. [CrossRef]
29. Hauggaard-Nielsen, H.; Ambus, P.; Jensen, E.S. Temporal and spatial distribution of roots and competition for nitrogen in pea-barley intercrops–a field study employing 32P technique. *Plant. Soil* **2001**, *236*, 63–74. [CrossRef]
30. Li, L.; Sun, J.; Zhang, F.; Guo, T.; Bao, X.; Smith, F.A.; Smith, S.E. Root distribution and interactions between intercropped species. *Oecologia* **2006**, *147*, 280–290. [CrossRef]
31. Hauggaard-Nielsen, H.; Jensen, E.S. Evaluating pea and barley cultivars for complementarity in intercropping at different levels of soil N availability. *Field Crops Res.* **2001**, *72*, 185–196. [CrossRef]
32. Gastal, F.; Lemaire, G. N uptake and distribution in crops: An agronomical and ecophysiological perspective. *J. Exp. Bot.* **2002**, *53*, 789–799. [CrossRef]
33. Burgess, A.J.; Retkute, R.; Pound, M.P.; Mayes, S.; Murchie, E.H. Image-based 3D canopy reconstruction to determine potential productivity in complex multi-species crop systems. *Ann. Bot.* **2017**, *119*, 517–532. [CrossRef] [PubMed]
34. Gaju, O.; Allard, V.; Martre, P.; Le Gouis, J.; Moreau, D.; Bogard, M.; Hubbart, S.; Foulkes, M.J. Nitrogen partitioning and remobilization in relation to leaf senescence, grain yield and grain nitrogen concentration in wheat cultivars. *Field Crops Res.* **2014**, *155*, 213–223. [CrossRef]

35. Barry, K.E.; Mommer, L.; van Ruijven, J.; Wirth, C.; Wright, A.J.; Bai, Y.; Connolly, J.; De Deyn, G.B.; de Kroon, H.; Isbell, F.; et al. The Future of Complementarity: Disentangling Causes from Consequences. *Trends Ecol. Evol.* **2019**, *34*, 167–180. [CrossRef] [PubMed]
36. Gibson, D.J.; Young, B.G.; Wood, A.J. Can weeds enhance profitability? Integrating ecological concepts to address crop-weed competition and yield quality. *J. Ecol.* **2017**, *105*, 900–904. [CrossRef]
37. Stephens, R. Effects of weed infestation on crop yield and quality. In *Theory and Practice of Weed Control*; Springer: Berlin/Heidelberg, Germany, 1982; pp. 1–14.
38. Smith, R.G.; Mortensen, D.A.; Ryan, M.R. A new hypothesis for the functional role of diversity in mediating resource pools and weed-crop competition in agroecosystems (vol 50, pg 37, 2009). *Weed Res.* **2010**, *50*, 185. [CrossRef]
39. Thorsted, M.D.; Weiner, J.; Olesen, J.E. Above- and below-ground competition between intercropped winter wheat Triticum aestivum and white clover Trifolium repens. *J. Appl. Ecol.* **2006**, *43*, 237–245. [CrossRef]
40. Wang, Y.; Lu, J.; Ren, T.; Hussain, S.; Guo, C.; Wang, S.; Cong, R.; Li, X. Effects of nitrogen and tiller type on grain yield and physiological responses in rice. *AoB Plants* **2017**, *9*, plx012. [CrossRef]
41. Campos, P.; Borie, F.; Cornejo, P.; López-Ráez, J.A.; López-García, Á.; Seguel, A. Phosphorus Acquisition Efficiency Related to Root Traits: Is Mycorrhizal Symbiosis a Key Factor to Wheat and Barley Cropping? *Front. Plant. Sci.* **2018**, *9*, 752. [CrossRef]

biology

MDPI

Article

Effects of Trace Metals and Municipal Wastewater on the Ephemeroptera, Plecoptera, and Trichoptera of a Stream Community

Marek Let *, Jan Černý, Petra Nováková, Filip Ložek and Martin Bláha *

Faculty of Fisheries and Protection of Waters, South Bohemian Research Center of Aquaculture and Biodiversity of Hydrocenoses, University of South Bohemia in České Budějovice, Zátiší 728/II, 389 25 Vodňany, Czech Republic; cernyj18@frov.jcu.cz (J.Č.); novakovapetra@frov.jcu.cz (P.N.); lozekf@frov.jcu.cz (F.L.)

* Correspondence: mlet@frov.jcu.cz (M.L.); blaha@frov.jcu.cz (M.B.)

Simple Summary: Mayflies (Ephemeroptera), stoneflies (Plecoptera), and caddisflies (Trichoptera) (EPT) are aquatic insects that are well known to the general public and are commonly used as indicators of environmental quality in water management. Knowledge of how EPT communities react to human-induced gradients in real environments can be important, for example, during the assessment of the implications of newly planned or currently active human disturbances for natural or cultural landscapes. We sampled a stream ecosystem affected by mining and smelting industries and communal wastewaters with pronounced concentrations of cadmium, lead, and zinc, as well as high levels of pesticides, pharmaceuticals, illegal drugs, and sewage-derived organic matter. Changes in other environmental factors such as increases in temperature were also studied at the affected sites. The abundance and species richness of stoneflies fell rapidly at the study sites. The richness of mayfly families also declined, from four to one, even though overall mayfly abundance was not affected. Conversely, the abundances of caddisflies were higher at the affected sites, and their richness did not decrease. This study will provide feedback for ecotoxicologists who perform better controlled and manipulated tests in laboratories, although any such test results are limited by simple artificial environments.

Abstract: Abundances of EPT larvae sampled in a Central European locality affected by mining and smelting, as well as by the continual inflow of treated communal wastewaters (WWs), were recorded. High concentrations of trace metals in water (maximum 1200 $\mu g \cdot L^{-1}$ for zinc) and sediments (maximum 140,000 $mg \cdot kg^{-1}$ in dry weight for lead) were found at the most contaminated sites. The highest loads of pesticides, pharmaceuticals, and illegal drugs were found under the WW effluent. Other associated factors such as the physicochemical parameters of the water and alterations to microhabitats were also evaluated and taken into account. Although EPT richness was lower at affected sites, abundances did not fall. Stoneflies were dominant at unaffected sites, while caddisflies dominated at affected sites. Only baetid mayflies were detected at the sites contaminated by trace metals and WWs; ephemerellid, heptageniid, and leptophlebiid mayflies were absent from these sites. The site contaminated by trace metals was also inhabited by numerous limnephilid caddisflies, in which limb malformations were detected in up to 11.8% of all specimens of a single taxon. Downstream from the entrance of the WWs, the locality was dominated by hydropsychid caddisflies. The increasing prevalence of predator or passive filter-feeding strategies in these EPT communities was significantly related to increasing water conductivity and acute ecosystemic exposure to 'poorly treated' WWs.

Keywords: anthropogenic disturbances; aquatic insect; environmental gradients; heavy metals; industrial pollution; wastewater treatment plant

Citation: Let, M.; Černý, J.; Nováková, P.; Ložek, F.; Bláha, M. Effects of Trace Metals and Municipal Wastewater on the Ephemeroptera, Plecoptera, and Trichoptera of a Stream Community. *Biology* **2022**, *11*, 648. https://doi.org/10.3390/biology11050648

Academic Editors: Daniel Puppe, Panayiotis Dimitrakopoulos and Baorong Lu

Received: 8 April 2022
Accepted: 21 April 2022
Published: 24 April 2022

1. Introduction

The aquatic insect orders of mayflies (Ephemeroptera), stoneflies (Plecoptera), and caddisflies (Trichoptera) (henceforth, EPT) are frequently used indicators of environmental quality, particularly in running waters [1,2]. Their larvae dominate in unpolluted headwaters where proportions of fine sediment deposits are low and are essential for correct nutrient flow cycling in these environments [3–5]. The contribution of EPT taxa adapted to conditions in downstream ecosystems (lower rhithral and potamal zones) is likewise important, although populations of these specialists, namely certain species of mayflies and stoneflies, are prone to be decimated by the cumulative effects of intensive human activities in lowland areas [6–8].

Given their rapid expansion through the natural environment, the ecological impact of human-induced gradients needs to be monitored, and possible threats are objectively predicted. The contamination of stream ecosystems by toxicants is a global issue; nevertheless, engineering work severely modifies the morphology of whole catchment areas and causes substantial changes in hydrological regimes, habitat structure, and water quality [8–10]. Hence, communities in running waters in cultural landscapes are habitually subjected to multiple anthropogenic factors [11]. Therefore, the manipulative testing of stressors within controlled conditions is now seen as increasingly important. Despite this, natural experiments (observational studies) are still needed to reflect the validity of derived results and vice versa [12].

The experiment described in this work explored an EPT community in a temperate zone in the European Central Highlands ecoregion [13], specifically, in a catchment area severely affected by industrial pollution (trace metals) and treated municipal wastewaters (WWs) whose levels are relatively high in terms of current European standards [14,15]. Measured confounding environmental variables (EVs)—concentrations of pesticides, physicochemical parameters of water, choriotopic composition, and current velocity—were included in the analyses of the changes in community composition. The shift in artificially/hierarchically assigned values for feeding strategies in the EPT community along with measured gradients of the EVs was also tested.

The contamination of streams by trace metals (often generalised as 'heavy metals') is a global phenomenon, and many studies tackling the responses of macroinvertebrate communities have identified mining as the cause of this type of pollution [16–19]. Increased concentrations of cadmium, copper, lead, and zinc—the so-called 'bivalent metals' that bind to both sulphur/nitrogen and oxygen groups—are most often reported as the cause of shifts in stream macroinvertebrate compositions in real environments or as the potential origin of impaired bottom-up control in ecosystems [20,21]. It is easy to misinterpret an oversimple comparison of concentrations measured in a real environment with the lethal values estimated for several EPT taxa under controlled conditions since many contradictions between observational studies and laboratory or mesocosm experiments exist in assessments of the toxicity of trace metals for aquatic insects [22].

Effluent from wastewater treatment plants (WWTPs) often represents an 'entrance gate' for many pollutants, including sewage-derived particulate organic matter (SDPOM), nutrients, pesticides, active pharmaceutical compounds (PhACs), synthetic organic compounds, phthalates, and trace metals into streams [14,23–27]. Over the past two decades, there has been a general decrease in organic matter and nutrient content in outputs from modernised WWTPs in the European Union [28]. Nonetheless, the continual release of compounds that are resistant to conventional purification technologies from WWTPs still occurs [29,30]. In addition, uncontrolled outflows of raw sewage from WWTPs and sewer infrastructure, especially during storm events, persist and represent a further challenge to the protection of natural water resources [31].

This article attempts to decipher the relationships between these human-induced gradients and EPT taxa and their feeding strategies which could be beneficial and permeable in such an impaired environment. It provides a list of taxa that can be considered tolerant to particular types of pollution, and the possible susceptibility of missing taxa is compared and discussed with previous reports. Since EPT taxa usually represent the essential part of macroinvertebrates in streams [3–5] and the shift in feeding strategies may reflect the change in available food resources, the article also provides partial evidence about the human-induced changes in the functioning of lotic ecosystems, namely, in nutrient cycling and energy flow [32].

2. Materials and Methods

2.1. Locality Description

The two surveyed localities—(i) Obecnický brook and (ii) Litavka river—are situated in central Bohemia (Czech Republic, Central Europe; spring to confluence: (i) N 49.7186939, E 13.8732253–N 49.7083778, E 13.9831206 and (ii) N 49.6563228, E 13.8557881–N 49.9602636, E 14.0847414; Figure 1). EPT larvae were sampled at four sites:

(i) Site 1 was a stretch of the Obecnický brook in the Brdy highlands above the Obecnice reservoir (third-order watercourse according to Strahler's system; N 49.7181961, E 13.9110308–N 49.7176411, E 13.9214808; Figures 1 and S1). This site was assumed to be the least affected by human activities, although forestry management has taken place in the area for many years, and there was once here a military firing range. Although increased acidity in this catchment area due to high emissions of sulphur and nitrogen compounds has been reported in the past, this area seems to have partially recovered [33]. The catchment area consists of spruce plantations, forest-free artificial moorlands (the former artillery range), and peat bogs. The brook's water has very low turbidity, and its rusty brown colour is likely to be caused by humic substances. By comparison, a few spring-fed tributaries had transparent water (an abundant population of acid-sensitive gammarids was observed in one unnamed tributary). Mature spruces grow along the brook edges, along with a few young alders. Particulate organic matter consists mainly of spruce needles, cones, and twigs, as well as leaf litter from the alders, birches, and old unharvested beeches. The brook bottom is densely covered in places by mosses (e.g., *Fontinalis antipyretica*) and species of Marchantiophyta. Despite not monitoring fish populations by electrofishing, brown trout (*Salmo trutta*) were observed during the macrozoobenthic sampling (Table S9). Fišer et al. [34] reported the presence of the brook lamprey (*Lampetra planeri*), brook trout (*Salvelinus fontinalis*), and European perch (*Perca fluviatilis*) at this site four years before our study began. Insects were prevalent in the macrozoobenthic samples, although only a few small oligochaetes, crustaceans, and molluscs were found;

(ii) Site 2 consists of a stretch of the Obecnický brook below the Obecnice reservoir (third-order watercourse according to Strahler's system; N 49.7161703, E 13.9312433–N 49.7161931, E 13.9346444; Figures 1 and S1). Discharge here was usually lower than at Site 1 because of continuous water abstraction from the reservoir. Water retention and discharge manipulation also influence the colour and turbidity, which were both different from those observed at Site 1 (from dark brown with greater turbidity to crystal clear). Site 2 is surrounded by forest where little management occurs, and the brook edges are lined by old alders in the lower part, which increases the site's heterogeneity by forming wide pools or dividing the brook into several branches (the gradient was also approximately one quarter less). Proportionally more leaf litter from deciduous trees was observed here than at Site 1. The brook bottom is covered in places by mosses (*F. antipyretica*). Fish stocks consist of brown trout, brook lamprey, and stone loach (*Barbatula barbatula*) (in decreasing order of abundance; Table S9). Insects prevailed in the macrozoobenthic samples, although small oligochaetes, crustaceans (*Gammarus* sp.), and molluscs (*Pisidium* sp. and *Ancylus fluviatilis*) were relatively abundant;

(iii) Site 3 was on a stretch of the Litavka river below its confluence with the Obecnický brook (fourth-order watercourse according to Strahler's system; from N 49.7100606, E 13.9884817–N 49.7112883, E 13.9961850; Figures 1 and S2). This site has been heavily affected by the local industrial activity (mining, smelting and processing of silver, iron, lead, zinc, and uranium) that has been active for several centuries [15]. An active industrial complex that recycles, above all, lead waste occupies part of the river's alluvial plain. Heaps of waste material (especially sodium slag) are still stored several meters from the riverbank. Nevertheless, all mining activity has ceased. Here, the effects of agricultural disturbance are also likely to occur as arable fields cover approximately half of the deforested catchment area (excluding human settlements). The rest of the catchment is covered by pastures and meadows. The effect of the municipal WWs is presumably low because of a relatively small human population upstream. The riparian vegetation has been substantially reduced upstream along the Litavka river (third-order watercourse according to Strahler's system), which has been channelled in places between artificial banks. In addition, several shallow reservoirs (up to 7 ha), including some small sludge deposits, are situated upstream. We sampled the upper channelled parts with old reinforced banks and little riparian vegetation, as well as a lower unchannelled, naturally heterogenic stretch (morphologically similar to those described for Site 2) with banks lined with old alders and willows (Figure 1). The fish stocks here consist of common minnow (*Phoxinus phoxinus*), brown trout, and European perch (in descending order of abundance; Table S9). Insects prevailed in the macrozoobenthic samples, and small oligochaetes were relatively abundant, but crustaceans and molluscs were almost totally absent from the macrozoobenthic samples;

(iv) Site 4 was a stretch of the Litavka river downstream from its confluence with the Příbramský brook (fourth-order watercourse according to Strahler's system; from N 49.7113508, E 14.0093011–N 49.7198211, E 14.0133511; Figures 1 and S2). This site was expected to be the most affected by anthropogenic activities because of the combination of industrial pollution from mining and smelting and contamination by treated municipal WWs from the town of Příbram. Treated WWs are continually pumped into the Příbramský brook approximately 900 m upstream from its confluence with the Litavka river. According to Grabicova et al. [14], Site 4 has the greatest concentrations of psychoactive PhACs of all the inspected localities, probably because of the low dilution possibilities (i.e., a small watercourse receiving relatively large amounts of municipal WWs). As at Site 3, we sampled an upper channelled stretch with artificial banks and little riparian vegetation, as well as a lower unchanneled section. Nonetheless, this morphologically heterogenic stretch only had a narrow and irregular fringe of riparian trees (alders and willows), possibly because of the dynamic migration of the river's course across its alluvial plain [15]. The fish stock consists of common minnow, common roach (*Rutilus rutilus*), chub (*Squalius cephalus*), gudgeon (*Gobio gobio*), brown trout, European perch, common bream (*Abramis brama*), common rudd (*Scardinius erythrophthalmus*), stone loach, and rainbow trout (*Oncorhynchus mykiss*) (in descending order of abundance; Table S9). The occurrence of limnophilic fish was due to the presence of aquaculture ponds upstream. There was no clear dominance by insects in the macrozoobenthic samples since small oligochaetes, leeches (e.g., *Erpobdella octoculata*, *Helobdella stagnalis*), and the water louse (*Asellus aquaticus*) were very abundant. Molluscs were detected in low numbers.

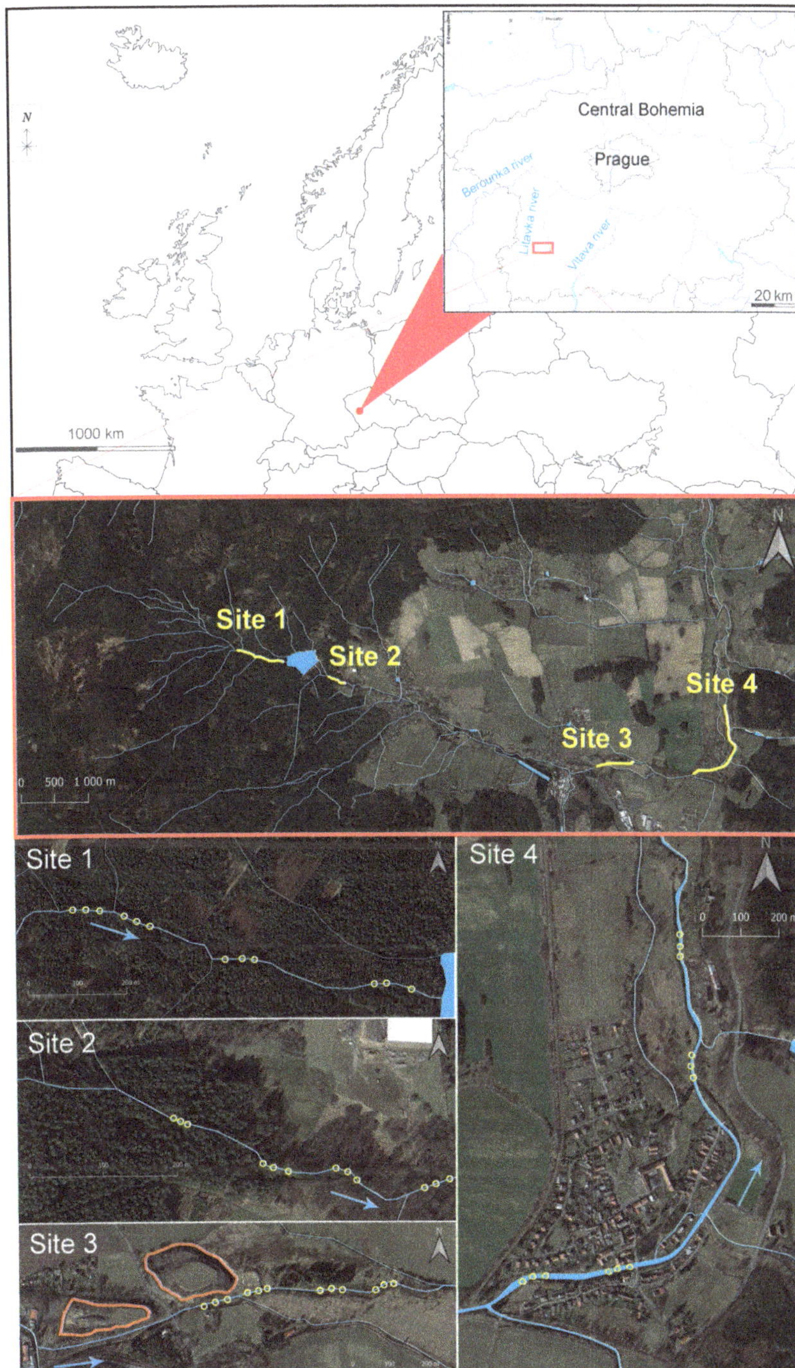

Figure 1. Map of the study area. The yellow circles indicate the sampling points. A set of three sampling points represents a whole plot. The contoured orange-coloured areas in the close-up view of Site 3 (lower left-hand corner) are slag heaps and waste material generated by the smelting industry. The river's flow direction is indicated by blue arrows.

2.2. Sampling and Analysis of Samples

The whole sampling campaign was conducted in 2020, specifically on 11 May, 24 June, 11 August, and 24 September. The macrozoobenthos were sampled using a Surber sampler (30 × 30-cm frame, 500-μm mesh) at four sampling sites and three points (triplicates) at each site (see part 2.1. 'Locality description' and Figure 1). Each triplicate consisted of three subplots within a whole plot or site. The location of the whole plot was chosen at random during each sampling campaign. A diversification of samples within each whole plot was made by sampling riffles (shallow and with high water velocity), inflows into pools (deep and with high water velocity) and the backs of pools (deep and with low water velocity). Samples were processed using a round steel sieve (40 cm diameter, 500 μm mesh) and preserved in 70% technical ethanol. Organisms were sorted in the laboratory from the material taken from the stream bottom. Mayfly, stonefly, and caddisfly larvae were determined to the deepest possible taxonomic level using a binocular microscope and stereomicroscope (Olympus SZX16) and determination keys; their frequencies (abundances) in the samples were recorded. The methodology used to analyse the water samples for pesticides, PhACs and drugs, and the water and sediment samples for trace metals is described in detail in the Supplementary Materials (Section Materials and Methods).

Environmental data were taken for both the whole and subplot levels. The environmental variables (EVs) differing only at the whole-plot level (hereafter referred to as 'whole-plot EVs') consisted of (i) physicochemical parameters—conductivity, pH, oxygen concentrations, and water temperature—measured using a HACH® multi-meter, (ii) pesticide and PhAC concentrations in water (measured from samples taken on 11 August 2020), (iii) cadmium, lead, and zinc concentrations measured in sediments and water (measured from samples taken on 24 September 2020), and (iv) daily volumes of sewage (untreated or 'poorly treated' municipal WWs) recorded up to 15 days prior to each sampling day at the Příbram WWTP (data provided by 1. SčV JSC, Příbram, Czech Republic). Detailed monthly data of water and sediment analyses from Sites 1, 2, and 3 (2009–2020) and data of actual water discharge from the same sites were provided by the Czech Hydrometeorological Institute (Prague, Czech Republic) and Povodí Vltavy, State Enterprise (Prague, Czech Republic). In addition, the water temperature was continuously monitored during the sampling campaign by four data loggers (TFA) placed approximately in the middle of each site.

The EVs also differed at the subplot level (hereafter referred to as 'subplot EVs') in terms of their (i) choriotopic compositions, which were empirically estimated for each Surber sampling spot; (ii) current velocities [$m \cdot s^{-1}$] (minimum, maximum and average), measured continuously for 0.5 min above the Surber sampling spot with a flowmeter (MiniController MC20 with the Flowprobe for MiniWater20, Schiltknecht Messtechnik AG, Gossau, Switzerland) placed close to the bottom, in the water column and close to the water surface; (iii) widths and depths at each subplot; and (iv) classification as riffles, pools, inflows into pools, or backs of pools.

2.3. Statistical Analyses

Data were analysed using Canoco 5 version 5.12 (written by Cajo J.F. ter Braak and Petr Šmilauer; Wageningen, the Netherlands, and P. Šmilauer, Czech Republic) [35] and R-studio version 4.1.1 (written by RStudio team; Boston, MA, United States) [36] software. The analysis of shifts in feeding strategies was based on the preferences of individual species according to Graf et al. [37], Buffagni et al. [38], and Graf et al. [39]. The analyses with species community composition as response variables were conducted using unimodal Canonical correspondence analysis (CCA), while analyses with feeding strategies as response variables were performed using linear Redundancy analysis (RDA). Species abundances were log + 1 transformed before the CCA. Detrending was not used because no dependency between positions of samples on the first or second ordination axes was observed. Feeding strategies were averaged by species composition-weighted standardisation (hereafter referred to as 'averaged feeding strategies'). Only centring by species was applied using RDA.

The significances of axes constrained by (i) site identity, (ii) the identity of sampling times, (iii) whole-plot EVs, and (iv) subplot EVs (see part 2.2. 'Sampling and sample analysis') were tested using Monte-Carlo permutation tests (999 permutations used). Permutations were carried out according to the experiment's hierarchical design: (i) when the whole-plot EVs or the identity of sites were used as explanatory variables, both whole-plots and subplots in the four blocks defined by the four sampling times used as covariates were permuted using cyclic shifts; (ii) when the subplot EVs were used as explanatory variables, only the subplots in the four blocks defined by sampling times were permuted using cyclic shifts; and (iii) when the identity of sampling times was used as an explanatory variable, both whole plots and subplots in the four blocks defined by the four sites were permuted using cyclic shifts. A false discovery rate was used for adjusting p values. Forward selection (FS) was used for selecting significant whole- and subplot EVs. 'Hybrid analysis' was chosen when tests on visualised constrained axes were not significant ($p > 0.05$). The significance of differences in choriotopic compositions between sites was tested in a similar way; individual values were transformed by the angular (inverse sine) function and the transformed values were used as response variables. This transformation was also applied when the choriotopic compositions were used as explanatory variables.

Generalised linear models (GLMs) were employed to model trends in shifts of multiple averaged feeding strategies along environmental gradients or along the resulting ordination axes. The significance of differences between sites and sampling times in EPT abundances and richness was tested using GLMs with negative binomial and Poisson distributions, respectively (the quasi-likelihood estimation method was applied in the case of the Poisson distributions). The significance of improvements using Generalised linear mixed-effects models (GLMMs) rather than GLMs was tested by likelihood ratio tests. A post hoc test with Tukey's method for p value adjustment was applied.

3. Results

3.1. Environmental Conditions

The differences between sites are summarised in Table 1. Several environmental gradients, including increasing temperature, pH, and conductivity and decreasing oxygen concentrations, changed along with the longitudinal downstream profile (from Site 1 to Site 4). For more information about the basic physicochemical parameters at other sites in the longitudinal profile, see Supplementary Materials (Section Materials and Methods; Table S1). The greatest total concentration of cadmium, zinc, and lead in the waters was detected at Site 3, followed by Site 4, while the highest loads of pesticides and PhACs were found at Site 4 below the discharge of effluent from the WWTP (Table 1). Treated WWs constituted 1/13–1/3 of all discharges at Site 4 (Table S2). The periodical releasing of untreated or 'poorly treated' WWs containing organic material and coarse communal waste (mainly hygienic products) also occurred. The monthly released volumes of this sewage ranged from 299 to 130,016 m^3 in 2020 (Table 2); however, considerable volumes also leaked from sewers upstream from the WWTP but were not included in the analyses. For more information about contamination at other sites along the longitudinal profile, see Supplementary Materials (Section Materials and Methods; Tables S3–S6).

The average current velocities at individual sites were as follows: Site 4 >Site 3 > Site 1 > Site 2 (in descending order from the fastest to the slowest; Table 3). Differences in the choriotopic compositions at sites tested using a RDA were marginal (test on first axis: pseudo-F = 1.2, p = 0.085; test on all axes: pseudo-F = 2.8, p = 0.003; Figure S4); only Site 4 and Site 2 were significantly different from the others when tested with the RDA (explained variation = 7.5%, pseudo-F = 3.8, p (adj.) = 0.025 and explained variation = 6.0%, pseudo-F = 2.9, p (adj.) = 0.049, respectively). Samples from Site 1 compared with those from Site 2 had a much greater proportion of smaller mineral particles (size < 6.3 cm—'microlithal', 'akal', and 'psammal') and 'mosses' (Marchantiophyta only at the Site 1), and a lower proportion of light deposits ('coarse particulate organic matter'—CPOM and 'fine particulate organic matter'—FPOM), debris, and xylal (Table 4). Samples from Site 3 resembled most those from Site 1, with similar proportions of smaller mineral particles, deposits, debris, and xylal (Table 4). Finally, Site 4 differed most from the others because of the highest proportions of deposits and debris and the occurrence of filamentous algae and anthropogenic trash (especially wet wipes; Table 4).

Table 1. Resulted mean physicochemical parameters measured at the given sites and concentrations of metals, pesticides, and active pharmaceutical compounds (PhACs) in water samples taken at these sites.

	Site 1	Site 2	Site 3	Site 4
	n = 4 Mean ± SEM	n = 4 Mean ± SEM	n = 4 Mean ± SEM	n = 4 Mean ± SEM
Temperature (°C)	12.05 ± 0.34	13.33 ± 0.61	17.05 ± 1.11	16.80 ± 0.68
pH	7.09 ± 0.24	7.35 ± 0.17	7.45 ± 0.17	7.45 ± 0.09
Oxygen concentration (mg·L^{-1})	9.90 ± 0.10	9.02 ± 0.18	8.91 ± 0.16	7.61 ± 0.27
Conductivity (μS·cm^{-1})	71.40 ± 5.80	84.35 ± 5.98	390.60 ± 39.18	571.00 ± 56.74
	Concentrations in Water Samples			
Cadmium (μg·L^{-1})	0.26	0.23	7.90	3.80
Lead (μg·L^{-1})	5.80	9.80	64.00	53.00
Zinc (μg·L^{-1})	15.00	13.00	1200.00	530.00
Sum pesticides (ng·L^{-1})	132.00	70.32	462.90	877.87 ± 125.71 (n = 3)
Sum PhACs (ng·L^{-1})	15.17	13.40	874.58	4636.73 ± 254.79 (n = 3)

Table 2. Monthly volumes of untreated or 'poorly treated' wastewater released from the wastewater treatment plant into Site 4 in 2019 and 2020.

	Volume [m^3]	
	2019	**2020**
January	164,799	1522
February	80,750	55,224
March	58,862	35,861
April	183	3239
May	26,930	24,749
June	34,483	130,016
July	13,340	11,305
August	26,333	22,382
September	817	12,064
October	4967	67,559
November	3270	5941
December	71	299
Total	**414,805**	**370,161**

Table 3. Resulted mean current velocities measured for the macrozoobenthic samples from each site.

	Site 1	Site 2	Site 3	Site 4
Velocity (m·s^{-1})	n = 12 Mean ± SEM	n = 12 Mean ± SEM	n = 12 Mean ± SEM	n = 9 Mean ± SEM
Average	0.19 ± 0.04	0.14 ± 0.03	0.27 ± 0.04	0.29 ± 0.05
Maximum	0.44 ± 0.09	0.32 ± 0.06	0.70 ± 0.14	0.62 ± 0.10
Minimum	0.04 ± 0.04	0.02 ± 0.03	0.04 ± 0.04	0.06 ± 0.05

Table 4. Resulted mean choriotopic percentual compositions empirically estimated for each macrozoobenthic sample at all given sites.

Choriotopic Parameter	Site 1 n = 12 Mean ± SEM (%)	Site 2 n = 12 Mean ± SEM (%)	Site 3 n = 12 Mean ± SEM (%)	Site 4 n = 12 Mean ± SEM (%)
Megalithal Large cobbles, boulders and blocks, bedrock; >40 cm	5.75 ± 2.63	18.29 ± 7.87	6.08 ± 2.16	18.50 ± 7.30
Macrolithal Coarse blocks, cobbles, gravel and sand; 20–40 cm	29.92 ± 6.88	17.38 ± 5.28	19.10 ± 4.62	12.60 ± 2.95
Mesolithal Fist to hand-sized cobbles; 6.3–20.0 cm	16.58 ± 3.75	22.29 ± 3.37	28.50 ± 2.41	19.70 ± 3.48
Microlithal Coarse gravel; 2.0–6.3 cm	11.08 ± 1.89	8.67 ± 2.23	18.30 ± 3.29	11.20 ± 0.99
Akal Fine to medium-sized gravel; 0.2–2.0 cm	15.50 ± 3.32	5.21 ± 0.93	11.60 ± 1.97	11.90 ± 2.06
Psammal Sand; <0.2 cm	3.88 ± 1.12	1.21 ± 0.58	3.00 ± 1.04	4.50 ± 2.00
Xylal Deadwood, cones	4.79 ± 1.33	10.38 ± 2.71	2.04 ± 0.82	1.63 ± 0.80
Coarse particulate organic matter CPOM; coarse deposits, e.g., leaves	1.83 ± 0.82	2.04 ± 0.81	0.88 ± 0.26	3.42 ± 1.38
Fine particulate organic matter FPOM; fine deposits	0.75 ± 0.41	3.96 ± 1.54	1.25 ± 0.62	4.25 ± 1.53
Debris Hard and coarse matter—organic or inorganic	2.67 ± 0.77	6.63 ± 2.04	3.13 ± 0.57	4.63 ± 1.10
Mosses *Fontinalis antipyretica* and Marchantiophyta at the Site 1	6.25 ± 3.27	2.71 ± 1.34	1.58 ± 1.21	2.13 ± 0.83
Filamentous algae	0	0	0	2.33 ± 1.37
Roots of riparian trees *Alnus* sp., *Salix* sp. or conifers	0.58 ± 0.42	1.25 ± 0.86	3.33 ± 1.70	0
Submerged riparian vegetation	0.42 ± 0.40	0	1.25 ± 0.63	0.83 ± 0.54
'Wet wipes' Anthropogenic trash	0	0	0	1.38 ± 0.61
'Metal sheets' Anthropogenic trash	0	0	0	1.17 ± 1.12

3.2. EPT Abundance and Richness

In total, 87 EPT taxa were detected in the samples: 11 mayflies, 29 stoneflies, and 47 caddisflies (Table S7). EPT abundances were not substantially different between sites (Figure 2A; Table 5) or sampling times (Table 5). Nevertheless, total EPT richness estimated for individual sites did show a gradual decreasing and significant trend from Site 1 to Site 4 (Figure 2B; Table 5). Sampling time had a nonsignificant effect on EPT richness (Table 5).

The greatest differences between sites in terms of abundance or richness were observed in stoneflies followed by mayflies (Table 5), both of which were impaired at disturbed Sites 3 and 4 (Figure 2C–E). Conversely, caddisflies were more abundant at disturbed Sites 3 and 4 (Figure 2C), although caddisfly richness and family richness did not significantly differ between sites (Figure 2D,E; Table 5). Significant differences between the four sampling times in abundance or richness were detected in mayflies and stoneflies but not in caddisflies (Table 5). The abundance of mayfly nymphs was lowest in August and highest in May, whilst the opposite was found in stoneflies: the highest stonefly abundance was detected in August and the lowest in May.

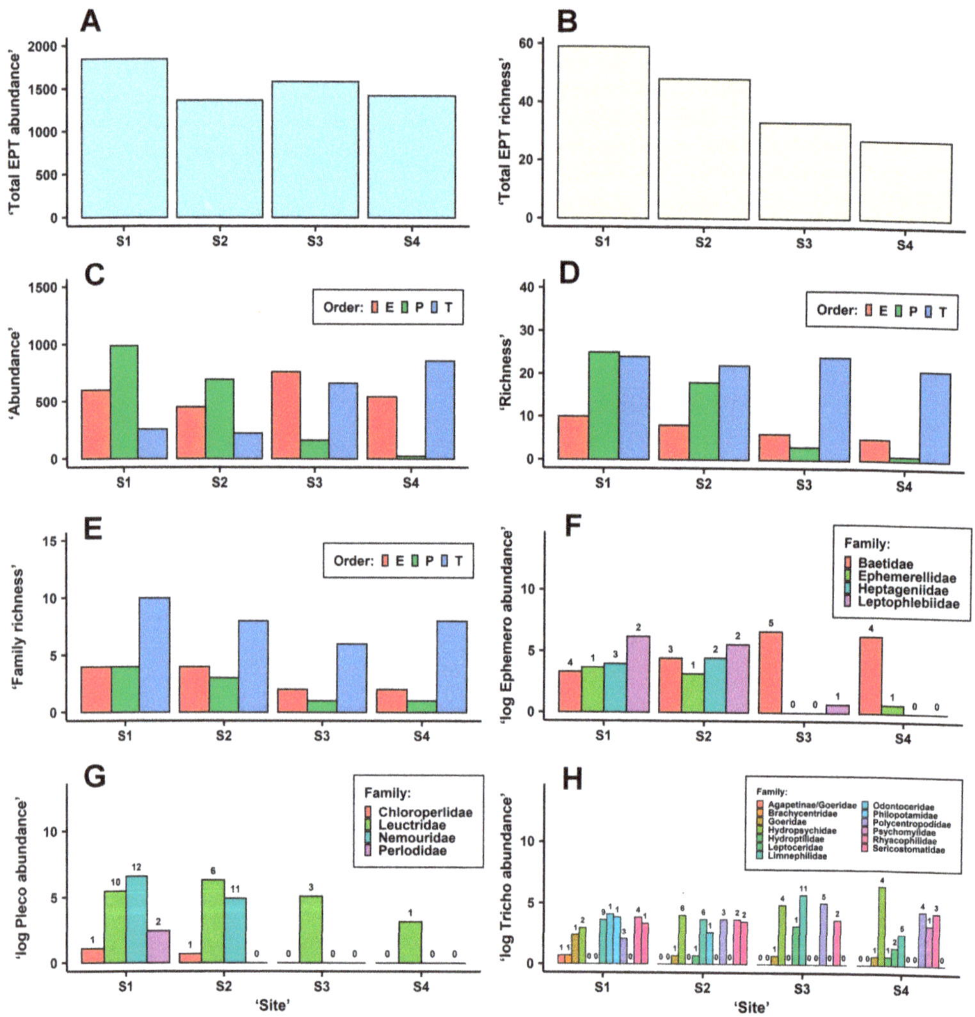

Figure 2. Bar graphs visualising (**A**,**C**) 'abundance' (number of individuals), (**B**,**D**) 'richness' (number of taxa), (**E**) 'family richness' (number of families), and (**F**–**H**) abundances and richness of individual detected families of Ephemeroptera, Plecoptera, and Trichoptera, respectively, summarised for each 'site' from the samples taken at all sampling times. Values in the bar graphs (**F**–**H**) are transformed by a natural logarithm, and the 'richness' is displayed above each column. E—Ephemeroptera, P—Plecoptera, T—Trichoptera.

The stonefly and caddisfly orders significantly differed in abundance between sites (Table 5). Stonefly abundances rapidly declined at disturbed Sites 3 and 4 compared with Sites 1 and 2, whilst caddisfly abundances were lower at Sites 1 and 2 (Figure 2C). Mayfly abundance did not significantly differ between sites (Figure 2C, Table 5) but marginally differed between sampling times (Table 5). No significant interaction was detected between the four-level factor site and sampling time.

Table 5. Results of testing the significance of differences within factors site and sampling time (both four-level factors) in given response variables using Generalised linear models (GLMs) or Generalised linear mixed-effects models (GLMMs). Significant ($p < 0.05$) results are in bold text. No significant interaction between site and sampling time was detected for any response variable ($p > 0.05$). E—Ephemeroptera, P—Plecoptera, and T—Trichoptera.

	Factors	
Response Variable	**Site**	**Sampling Time**
EPT abundance	$\chi^2_3 = 1.19, p = 0.76$	$\chi^2_3 = 3.96, p = 0.27$
EPT richness	$\mathbf{F_{3,44} = 2.82, p = 0.049}$	$F_{3,41} = 1.14, p = 0.34$
E abundance	$\chi^2_3 = 1.22, p = 0.75$	$\chi^2_3 = 7.56, p = 0.055$
E richness	$F_{3,44} = 0.14, p = 0.93$	$\mathbf{F_{3,44} = 3.19, p = 0.03}$
P abundance	$\mathbf{\chi^2_3 = 56.33, p < 0.001}$	$\mathbf{\chi^2_3 = 9.70, p = 0.02}$
P richness	$\mathbf{\chi^2_3 = 124.03, p < 0.001}$	$\mathbf{\chi^2_3 = 8.14, p = 0.04}$
T abundance	$\mathbf{\chi^2_3 = 12.59, p = 0.006}$	$\chi^2_3 = 6.11, p = 0.11$
T richness	$F_{3,44} = 1.03, p = 0.39$	$F_{3,44} = 0.43, p = 0.73$
E family richness	$\mathbf{F_{3,41} = 6.83, p < 0.001}$	$F_{3,41} = 2.48, p = 0.074$
P family richness	$\mathbf{F_{3,44} = 23.70, p < 0.001}$	$F_{3,41} = 1.02, p = 0.39$
T family richness	$F_{3,44} = 0.47, p = 0.70$	$F_{3,44} = 0.23, p = 0.88$

The total abundances and species richness shown in Figure 2F,G indicate that practically only members of the Baetidae family (namely *Baetis fuscatus*, *B. rhodani*, *B.* aff. *scambus*, and *B. vernus*) and the Leuctridae family (stoneflies) were detected at disturbed Sites 3 and 4. At these sites, *Habrophlebia lauta* (Leptophlebidae) and *Seratella ignita* (Ephemerellidae) all but completely disappeared, while *Paraleptophlebia submarginata* (Leptophlebiidae), *Ecdyonurus torrentis*, and members of the genus *Rhithrogena* (both Heptageniidae) were totally absent, although these latter taxa were detected upstream at the interconnected Sites 1 and 3 (Figure 2F, Table S7). Of the stoneflies, only *Leuctra albida*, *L. fusca*, and *L. geniculata* were detected at Site 3 (*L. geniculata* was only detected at this site, Table S7) and only *L. fusca* in low abundances and body sizes was detected at Site 4 (Figure 2G, Table S7). *Siphonoperla torrentium* (Chloroperlidae), several species belonging to the genera *Protonemura*, *Amphinemura*, and *Nemoura* (Nemouridae), all very abundant at Sites 1 and 2, along with *Diura bicaudata* and *Perlodes* aff. *microcephalus* (Perlodidae) were completely absent in the samples taken from disturbed Sites 3 and 4 (Figure 2G, Table S7). Nevertheless, compared with Site 1, the richness of Leuctridae decreased also at Site 2 below the water reservoir (Figure 2G), and, for instance, *Leuctra nigra*, frequent in samples from Site 1, was missing from samples from Site 2 (Table S7).

Of the caddisflies, Site 1 had a greater richness of trichopterans than Site 2; the abundant family Philopotamidae, for example, was no more detected at Sites 2, 3, and 4; finally, members of the Goeridae were detected only sporadically at particular sites (Figure 2H). On the other hand, the richness of Hydropsychidae was lowest at Site 1. *Odontocerum albicorne* (Odontoceridae) and *Sericostoma flavicorne/personatum* (Sericostomatidae), very abundant at both Sites 1 and 2, were completely missing from the samples from Sites 3 and 4 (Figure 2H, Table S7). One of the dominant families at Site 3 was Limnephilidae, represented above all by species from the genera *Potamophylax*, *Halesus*, and *Chaetopteryx* (in descending order of total abundance; Figure 2H, Table S7). One of the dominant families at Site 4 below the WWTP was Hydropsychidae, represented by *Hydropsyche siltalai*, *H. angustipennis*, and *H. incognita/pellucidula* (in descending order of total abundance; Figure 2H, Table S7); the population of limnephilids at Site 4 was substantially lower (Figure 2H). At both disturbed Sites 3 and 4, the family Polycentropodidae was much more abundant than at Sites 1 and 2 (Figure 2H). A substantially greater proportion in the total abundances of *Cyrnus trimaculatus* than *Polycentropus flavomaculatus* was observed at Site 4 than at Site 3 (2.54 and 0.03, respectively; Table S7). The Leptoceridae represented by *Athripsodes bilineatus* was most abundant at Site 3, while the Psychomyiidae, represented only by *Psychomyia pusilla*, were detected only at Site 4 (Figure 2H; Table S7). The total abundances of the family Rhyacophilidae were almost identical at all sites (Figure 2H). Even though the clear morphological determination of larvae from the group *Rhyacophila* sensu stricto is not always possible [40], *Rhyacophila* cf. *aurata* was dominant in samples from Sites 3 and 4 (especially Site 3), whilst *Rhyacophila nubila* gr. was commonest in samples from Sites 1 and 2.

3.3. Shift in EPT Community Composition along Environmental Gradients

The differences in species compositions between sites were significant (test on first axis: pseudo-F = 1.9, p = 0.001; test on all axes: pseudo-F = 4.0, p = 0.001; Figure S5), as were the differences in composition of species between sampling times (but only the composition of specimens > 0.5 mm in size; test on first axis: pseudo-F = 0.7, p = 0.024; test on all axes: pseudo-F = 1.8, p = 0.001; Figure S6). The whole- and subplot EVs significantly correlated with the shift in species composition (p < 0.05); those selected by FS are shown in Figures 3 and 4, respectively.

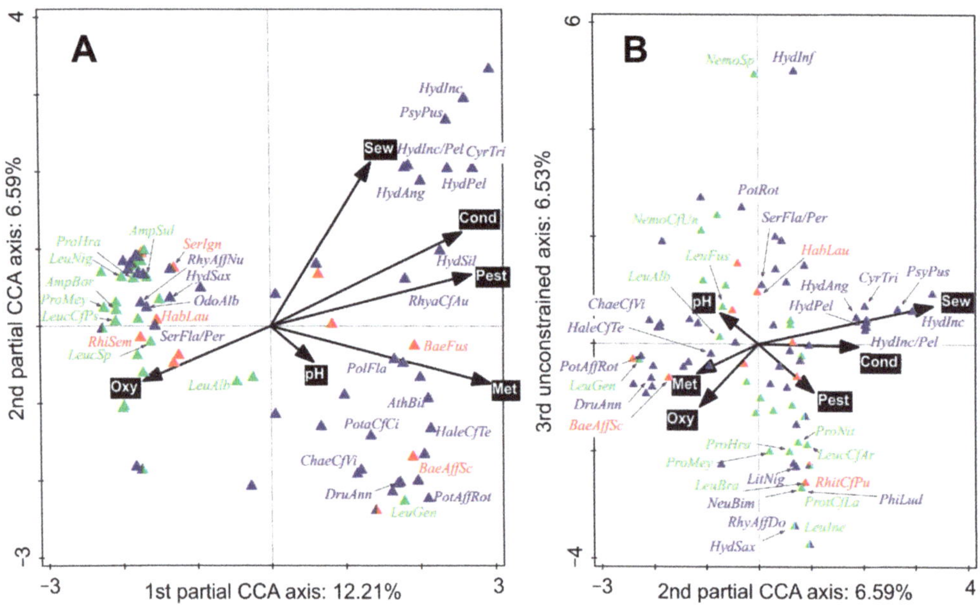

Figure 3. Species + whole-plot environmental variables (EVs) biplots: the first 33 best fitting species are labelled (Ephemeroptera, Plecoptera, and Trichoptera labelled in red, green, and blue, respectively). A partial canonical correspondence analysis (CCA) was used: (**A**) first and second ordination axes; (**B**) second and third ordination axes. Ordination axes were constrained by the EVs selected by forward selection, which accounted for 30.48% of the explained variation. Test on first axis: pseudo-F = 0.5, p = 0.001; test on all axes: pseudo-F = 2.3, p = 0.001. Conditional effects (from the greatest to the lowest explained variation): 'Log-transformed sum of cadmium, lead and zinc (Met)': explained variation = 11.7%, pseudo-F = 5.7, p (adj.) = 0.003; 'Conductivity (Cond)': explained variation = 6.1%, pseudo-F = 3.1, p (adj.) = 0.020; 'Oxygen concentration (Oxy)': explained variation = 4.0%, pseudo-F = 2.1, p (adj.) = 0.012; 'Log-transformed sum pesticides (Pest)': explained variation = 3.0%, pseudo-F = 1.6, p (adj.) = 0.045; 'pH' explained variation = 2.9%, pseudo-F = 1.6, p (adj.) = 0.047; 'Total volume of released sewage for three days prior to sampling time (Sew)': explained variation = 2.9%, pseudo-F = 1.6, p (adj.) = 0.036. Abbreviations: *AmpBor—Amphinemura borealis*, *AmpSul—A. sulcicollis*, *AthBil—Athripsodes bilineatus*, *BaeAffSc—Baetis* aff. *scambus*, *BaeFus—B. fuscatus*, *ChaeCfVi—Chaetopteryx* cf. *villosa*, *CyrTri—Cyrnus trimaculatus*, *DruAnn—Drusus annulatus*, *HabLau—Habrophlebia lauta*, *HaleCfTe—Halesus* cf. *tesselatus*, *HydInf—Hydatophylax infumatus*, *HydAng—Hydropsyche angustipennis*, *HydInc—H. incognita*, *HydInc/Pel—H. incognita/pellucidula*, *HydPel—H. pellucidula*, *HydSax—H. saxonica*, *HydSil—H. siltalai*, *LeuAlb—Leuctra albida*, *LeucCfAr—L.* cf. *armata*, *LeuBra—L. braueri*,

LeucCfPs—*L.* cf. *pseudocingulata*, *LeucSp*—*Leuctra* sp., *LeuFus*—*L. fusca*, *LeuGen*—*L. geniculata*, *LeuIne*—*L. inermis*, *LeuNig*—*L. nigra*, *NemoCfUn*—*Nemoura* cf. *uncinata*, *NemoSp*—*Nemoura* sp., *NeuBim*—*Neureclipsis bimaculata*, *OdoAlb*—*Odontocerum albicorne*, *PhiLud*—*Philopotamus ludificatus*, *PolFla*—*Polycentropus flavomaculatus*, *PotaCfCi*—*Potamophylax* cf. *cingulatus*, *PotAffRot*—*P.* aff. *rotundipennis*, *PotRot*—*P. rotundipennis*, *ProHra*—*Protonemura hrabei*, *ProMey*—*P. meyeri/nitida*, *ProNit*—*P. nitida*, *ProtCfLa*—*P.* cf. *lateralis*, *PsyPus*—*Psychomyia pusilla*, *RhiSem*—*Rhithrogena semicolorata*, *RhitCfPu*—*R.* cf. *puytoraci*, *RhyaCfAu*—*Rhyacophila* cf. *aurata*, *RhyAffDo*—*R.* aff. *dorsalis*, *RhyAffNu*—*R.* aff. *nubila*, *SerFla/Per*—*Sericostoma flavicorne/personatum*, *SerIgn*—*Serratella ignita*.

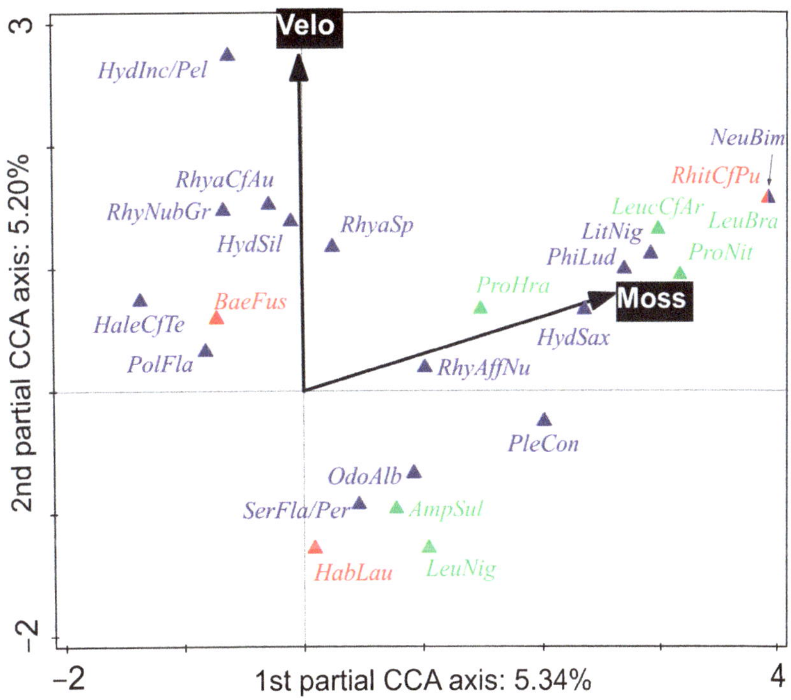

Figure 4. Species + subplot environmental variables (EVs) biplot: the first 24 best-fitting species are shown (Ephemeroptera, Plecoptera, and Trichoptera labelled in red, green, and blue, respectively). A partial Canonical correspondence analysis (CCA) was used. Ordination axes were constrained by the EVs selected by forward selection, which accounted for 10.54% of the explained variation. Test on first axis: pseudo-F = 0.1, p = 0.024; test on all axes: pseudo-F = 1.4, p = 0.001. Conditional effects (from the greatest to the lowest explained variation): 'Arcsine-transformed proportion of mosses in choriotope (Mosses)': explained variation = 5.3%, pseudo-F = 2.4, p (adj.) = 0.015 and 'Mean current velocity (Velo)': explained variation = 5.2%, pseudo-F = 2.5, p (adj.) = 0.023. Abbreviations: *AmpSul*—*Amphinemura sulcicollis*, *BaeFus*—*B. fuscatus*, *HabLau*—*Habrophlebia lauta*, *HaleCfTe*—*Halesus* cf. *tesselatus*, *HydInc/Pel*—*Hydropsyche incognita/pellucidula*, *HydSax*—*H. saxonica*, *HydSil*—*H. siltalai*, *LeuBra*—*L. braueri*, *LeucCfAr*—*L.* cf. *armata*, *LeuNig*—*L. nigra*, *LitNig*—*Lithax niger*, *NeuBim*—*Neureclipsis bimaculata*, *OdoAlb*—*Odontocerum albicorne*, *PhiLud*—*Philopotamus ludificatus*, *PleCon*—*Plectrocnemia conspersa*, *PolFla*—*Polycentropus flavomaculatus*, *ProHra*—*Protonemura hrabei*, *RhitCfPu*—*Rhithrogena* cf. *puytoraci*, *RhyaCfAu*—*Rhyacophila* cf. *aurata*, *RhyAffNu*—*R.* aff. *nubila*, *RhyaSp*—*Rhyacophila* sp. (early instar larvae), *RhyNubGr*—*R. nubila* gr., *SerFla/Per*—*Sericostoma flavicorne/personatum*.

3.4. Shift in Averaged Feeding Strategies along Environmental Gradients

The differences in compositions of the averaged feeding strategies between sites were significant in general; however, only the composition at Site 4 was significantly different from all other sites ($p < 0.05$; Figure 5A). There was also a significant shift in the whole-plot EVs ($p < 0.05$). Significant relationships with 'conductivity' (32.5% of explained variability), 'total volume of released sewage for three days prior to sampling time' (5.9% of explained variability), 'oxygen concentration' (2.8% of explained variability), and 'total concentration of PhACs' (2.3% of explained variability) were detected ($p < 0.05$). 'Passive filter feeder' and 'predator' feeding strategies were dominant in the WW-polluted environment; the 'grazer and scraper' feeding strategy seemed to be limited by the most polluted WW; and the 'gatherer/collector' and 'shredders' feeding strategies were generally suppressed in the most polluted environment (Figures 5B and S7B,C). The shift in the subplot EVs was not significant (test on first axis: pseudo-$F = 0.6$, $p = 0.379$; test on all axes: pseudo-$F = 1.9$, $p = 0.168$).

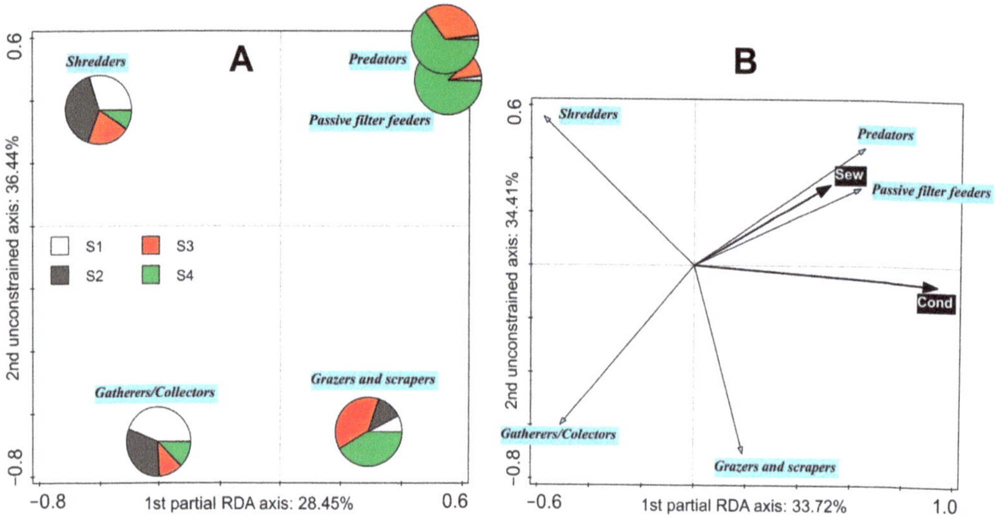

Figure 5. (**A**) Averaged feeding strategies pie chart and (**B**) Averaged feeding strategies + whole-plot environmental variables (EVs) biplot. Partial Redundancy analyses (RDA) were used. First ordination axes were constrained: (**A**) by the factor 'Site' that accounted for 30.98% of the explained variation and (**B**) by the EVs selected by forward selection, which accounted for 38.91% of the explained variation. (**A**) Test on first axis: pseudo-$F = 5.4$, $p = 0.003$; test on all axes: pseudo-$F = 6.1$, $p = 0.026$; (**B**) test on first axis: pseudo-$F = 2.5$, $p = 0.006$; test on all axes: pseudo-$F = 5.7$, $p = 0.009$. Conditional effects (from the greatest to the lowest explained variation): (**A**) 'Site 4 (S4)' explained variation = 22.0%, pseudo-$F = 12.1$, p (adj.) = 0.027; 'Site 3 (S3)' explained variation = 8.6%, pseudo-$F = 5.2$, p (adj.) = 0.090; and 'Site 2 (S2)' explained variation = 0.3%, pseudo-$F = 0.2$, p (adj.) = 0.943. (**B**) 'Conductivity (Cond)': explained variation = 32.5%, pseudo-$F = 20.7$, p (adj.) = 0.005 and 'Total volume of released sewage for three days prior to sampling time (Sew)': explained variation = 6.4%, pseudo-$F = 4.4$, p (adj.) = 0.045.

3.5. Malformations and Mortality of Caddisflies

A substantial proportion of final instar caddisfly larvae in *Halesus* cf. *tesselatus*, *Potampohylax* aff. *rotundipennis*, and *R.* cf. *aurata* sampled at Site 3 had malformed limbs (11.8%, 7.0%, and 7.7%, respectively). Peculiarly fused terminal parts of walking legs (between tibiae and tarsal claws) or overall shortened walking legs were observed in the two given limnephilid taxa; deformed and shortened anal claws were observed in *R.* cf. *aurata* (Figure 6). Malformed specimens of *H.* cf. *tesselatus* were also observed at Site 4 (28.6%), although at the latter site, the total number of sampled specimens was substantially lower than at Site 3 (7 vs. 76 specimens at Site 3). There were no malformed specimens in the abundant *R.* cf. *aurata* in Site 4; however, there was one malformed specimen out of a total of 20 specimens of *R. nubila* gr. at Site 4.

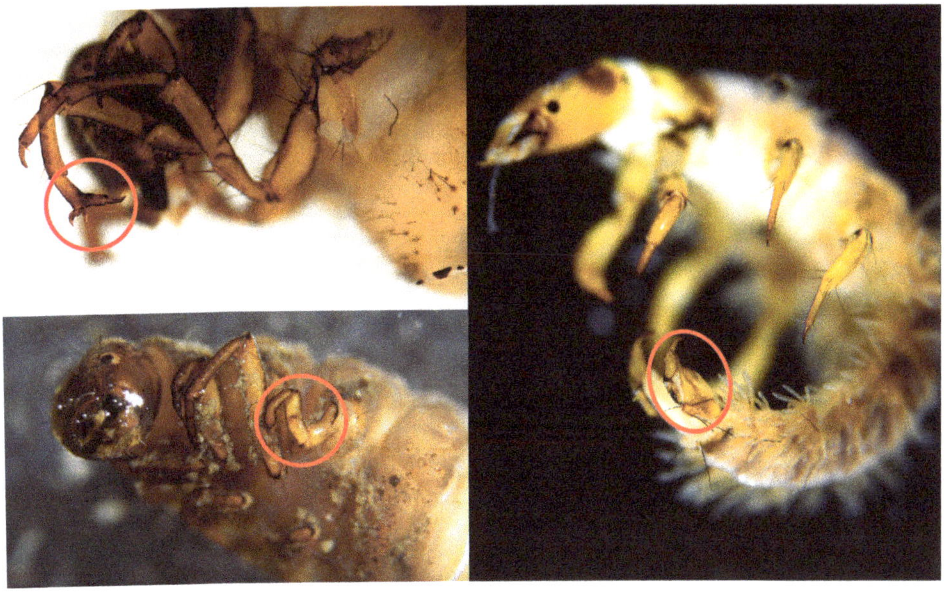

Figure 6. Malformed limbs of *Halesus* cf. *tesselatus* (two pictures on the left) and *Rhyacophila* cf. *aurata* (right side) sampled at Sites 3 and 4.

We also observed a drift of dead limnephilid pupae at Site 3 in September during the macrozoobenthic sampling. We found dead and decaying larvae and pupae of *Drusus annulatus*, *H.* cf. *tesselatus*, and Limnephilidae gen. sp. inc. in the given samples (14 dead limnephilid specimens and 7 living limnephilid specimens in total). Several dead and decaying pupae of *Hydropsyche* sp. were also found in the samples from Site 2 taken in June.

4. Discussion

4.1. Variability in the EPT Community along an Environmental Gradient

The EPT community can significantly reflect the quality of the environment (Figures 3–5). Even though our study only consisted of a small-scale natural experiment (i.e., we only sampled two mutually interlinked watercourses), the results match, in general, those of a study of whole macroinvertebrate communities carried out in 23 Swiss streams [41]. As in our work, in the Swiss study, about 30% of community variability was explained by water quality; conductivity (which correlates to total dissolved solids) together with oxygen concentrations explained 10.1% in our study, while concentrations of pesticides explained 3.0% of community variability in both studies (Figure 3).

The changes in EPT abundance and richness related to pollution by trace metals observed in our study generally resemble reports from other sites, the greatest similarity being in the susceptibility of mayflies—and especially heptageniids—to this type of pollution (Figure 2). Sites 3 and 4 were both warmer—maximums up to 23 °C—than unpolluted Sites 1 and 2 (Table 1; Figure S3), a relevant finding given that the temperature gradient, which is often related to the toxicity of chemical compounds and the solubility of oxygen in water, is considered a good predictor of change in the composition of EPT communities [42]. The increasing temperature gradient observed when moving downstream is undoubtedly natural; however, the difference in the temperature regime between the contaminated and uncontaminated sites is also likely to be the product of the canalisation, the removal of the riparian vegetation, and the presence of shallow reservoirs upstream on the Litavka river (see part 2.1 'Locality description' [43]). That many species of mayflies and stoneflies would avoid high temperatures, mine effluent and certain trace metals was predicted beforehand [17–19,44–46]. Our results show that baetids were the most resistant (or resilient) to the trace metal contamination, and within this group, the dominant taxa were the *Baetis fuscatus* group (i.e., *B. fuscatus* and *B.* aff. *scambus*, whose separation was not possible since the taxon identified and presented here as *B.* aff. *scambus* shared features of both species). The total abundance of *B. fuscatus* gr. compensated for the decrease in abundances of other mayfly taxa so that this small-sized baetid could act as a proxy for other mayflies in human-disturbed sites. The other most commonly detected baetids, *B. rhodani* and *B. vernus*, did not show any substantial changes in abundance between sites. Nevertheless, trace metal pollution can have an effect later on in the life cycle of baetids and other aquatic insects by rapidly causing increased mortality in emerging imagines (or subimagines) [47,48].

Stoneflies are generally the aquatic insects most sensitive to conditions modified by humans [49,50]. There was significantly decreased richness in stoneflies—even at Site 2 compared with Site 1—and their abundance only significantly fell at Sites 3 and 4 (Figure 2; Table 5). We interpret their decrease in richness at Site 2, situated below the reservoir, as the result of the discontinuity in the river network, i.e., a change in the deposition regime (a lower proportion of smaller mineral particles and a greater proportion of light deposits; Table 4), and the separation of Site 2 from the network of tributaries in the crenal zone situated upstream from the reservoir (Figure 1). However, there are several indications that stoneflies are less sensitive to metals than mayflies. Many stoneflies (as well as caddisflies) are more tolerant to increased acidity than many mayflies [51]. Lower pH leads to natural access to metals (e.g., aluminium) originally bound to substrates [52,53]. Observational work studying the effects of contamination from mining has reported relatively high resistance in some stoneflies, including *Amphinemura sulcicollis*, *Leuctra fusca*, and *L. inermis*, taxa that are similar to those found in our study [16,19]. Little evidence exists regarding the levels of trace metals concentrations that are lethal for stoneflies [54], although certain North American stoneflies have been found to withstand (in significantly reduced numbers) concentrations of 11, 120, and 1100 $\mu g \cdot L^{-1}$ of cadmium, copper, and zinc, respectively, for 10 days in microcosm conditions. By comparison, numbers of heptageniids and *Baetis tricaudatus* fell significantly in only half of these concentrations [46].

Therefore, the significant reduction in stonefly abundance and richness at both contaminated sites cannot be conclusively attributed to pollution by trace metals. The absence there of most of the stoneflies detected upstream could also be due to water warming since most stoneflies prefer lower temperatures than those recorded at Sites 3 and 4 [39,42]. However, the interaction of multiple effects, including habitat degradation and several types of pollution, are assumed to take place.

According to our data, the less sensitive stoneflies were able to withstand the heavy industrial trace metal pollution but were then eliminated by the municipal WW, despite the advanced technology used in the WW treatment [14]. A synergic effect of multiple stressors is likely to occur. We can demonstrate that specimens of all sizes of *Leuctra fusca* gr. (species identified as *L. albida* and *L. fusca* in this study [55]) and *L. geniculata* occurred in the water contaminated by trace metals at Site 3 (Table S7) but not in the water contaminated by municipal WWs. Only *L. fusca* gr. (the only species identified as *L. fusca*) of small sizes were detected at Site 4, contaminated by both trace metals and municipal WWs. It is important to note that the organic pollution produced by the WWs sometimes reaches high levels. According to unpublished data provided by the Czech Hydrometeorological Institute and Povodí Vltavy, State Enterprise, the average estimated five-day biochemical oxygen demand (BOD_5) was equal to 2.39 $mg \cdot L^{-1}$ at Site 4; however, the estimated maximum was 17.00 $mg \cdot L^{-1}$ just four months before the beginning of our sampling campaign. In the Příbramský brook (Auxiliary Site 4 downstream from the effluent from the WWTP, see Supplementary Materials), the estimated BOD_5 at the same time was

24.00 mg·L^{-1} (the average and maximum of the estimated BOD_5 were 4.76 and 60.00 mg·L^{-1} in 2013–2020, respectively).

Caddisflies were the least harmed insects in the EPT community. Their richness was not significantly different between sites, and their abundance was significantly higher at Site 4 than at Sites 1 and 2 (Figure 2; Table 5). Many caddisflies are acid-tolerant or acid environment specialists (e.g., *Chaetopterygopsis maclachlani* and *Rhyacophila polonica* detected during our study [51,56]), and some are reported to be tolerant to trace metal pollution (e.g., *Hydropsyche* sp., leptocerids, *Rhyacophila* sp. [17,18,46]). Ubiquitous caddisfly taxa can occupy warmer sites polluted by organic materials with low oxygen concentrations (e.g., potamal hydropsychids and *Psychomyia pusilla* [56,57]). Nevertheless, caddisflies are generally very sensitive to pesticides and avoid sedimentation and droughts, as does the EPT group in general [11,58,59].

The vast majority of EPT taxa situated on the right-hand side of the CCA biplot (Figure 3A) were caddisflies. The relative abundances of these caddisflies were positively correlated to the human-induced gradients. The caddisfly communities at Sites 1 and 2 were quite similar; however, net-spinning philopotamids were lacking from Site 2 and were replaced by net-spinning hydropsychids and polycentropodids (Figure 2H). Changing conditions along the longitudinal profile (e.g., food source and current velocity) probably played an important role in shaping the net-spinning caddisfly community. Philopotamids prefer diatoms and fine detritus and have been reported to be quite sensitive to organic pollution [60,61]). According to our data, philopotamids were found at the subplots with the greatest maximum water velocity (0.73–0.97 m·s^{-1}). Slower water speeds caused by water abstraction [60] combined with the natural longitudinal stream gradient [37,60] favoured hydropsychids and polycentropodids. Unlike philopotamids, hydropsychids and polycentropodids are either facultatively or obligatory carnivorous [37,62].

The altered conditions at Sites 3 and 4 probably prevent the occurrence of the eruciform caddisflies *O. albicorne* and *S. flavicorne/personatum*, which were dominant in the samples from Sites 1 and 2. The warmer temperatures (and poorer oxygen conditions) are expected to have a negative effect. Haidekker and Hering [42] have reported that *O. albicorne* tolerates well mean summer temperatures up to 16 °C. However, this figure was exceeded at Site 4 by 1.24 °C (Figure S2), and so, based on individual measurements (Table 1), we assume that a similar temperature regime exists at Site 3. Furthermore, both these taxa are partially endobenthic and are more exposed to the toxic compounds that accumulate in sedimented particles [63]; the long larval development time in *S. flavicorne/personatum* [64] will also prolong the exposure of its larvae to toxicants. On the other hand, high abundances of large eruciform caddisflies—limnephilids of the genus *Halesus* and *Potamophylax*—were found at Site 3. These taxa are probably tolerant to trace metal pollution. Nonetheless, we found malformed walking legs in up to 11.8% of individuals that, in light of previous studies, could be a sign of trace metal contamination [65] or other persistent and hazardous compounds generated by the industrial complex upstream from Site 3 [66].

The lower abundances and richness of limnephilids observed at Site 4 downstream from the WWTP effluent could be a consequence of the interaction between trace metals and WW pollution (or the WW pollution alone). Hydropsychids, polycentropodids, rhyacophilids, and psychomyiids (*P. pusilla*) were most abundant downstream from the WW effluent. The taxa are shown in the upper right-hand corner of Figure 3A were positively related to the 'total volume of released sewage three days prior to the sampling' (hereafter referred to as '3-day sewage volumes'), which probably contained a high proportion of SDPOM—a mixture of organic detritus and microorganisms such as bacteria and algae that is a high-quality food resource downstream from WW effluents [23]. On the other hand, the released sewage could trigger a drift of benthic prey organisms with greater oxygen demands (e.g., mayfly nymphs), and SDPOM could be used by periphyton and macrophytes (aquatic and riparian), which can serve as microhabitats or food. For example, the carnivorous *C. trimaculatus* with a more pronounced microhabitat preference for macrophytes [37] was more common than *P. flavomaculatus* at Site 4 than Site 3. The most abundant hydropsychids can use multiple food resources and, aside from predating on small benthic organisms and consuming detritus, they also graze on algae [37,67]. Besides crustaceans, insects (including EPT), and terrestrial prey, detritus also constitute a substantial portion of gut content in polycentropodids [68]. It cannot be rejected that hydropsychids and polycentropodids could have a detrimental impact on the populations of other EPT taxa. Nevertheless, their net spinning increases heterogeneity of current velocities and nets represent valuable microhabitats, e.g., positive relationships to ephemerellid mayfly and chironomids were reported [69,70]. We associate the high abundance of early instars of *H. siltalai* in September at Sites 3 and 4 with the presence of aquatic mosses (Figure 4).

The proportion of mosses in the choriotope (explained community variation = 5.3%) and mean current velocity (explained variation community variation = 5.2%) had the highest parsimonious and significant power to explain EPT community composition at the subplot level (Figure 4). These two variables were positively correlated because mosses were detected mainly in shallow, fast-flowing habitats. Mosses (as well as roots) provide better vertical zonation possibilities for aquatic organisms, as well as food and shelter [71]. The information provided by Buffagni et al. [38], Graf et al. [37], and Graf et al. [39] does not correspond particularly well to our results: of the species best corresponding to the presence of aquatic mosses only *Hydropsyche saxonica, Plectrocnemia conspersa*, and *Protonemura hrabei* were reported to have a certain preference for macrophytes (mosses included). Nevertheless, Krno [72] suggests that the goerid *Lithax niger* is a bryophyte dweller and reports that *Rhyacophila obliterata* can only live on emergent mosses, whilst *R. nubila*, *R. polonica* can also live on submerged mosses. Similar findings have been reported by Glime [71]. Conversely, Graf et al. [37] consider all goerids and rhyacophilids to inhabit micro-/mesolithal and macro-/megalithal and suggest that they are 'specialised for lithal'.

A high mean velocity best corresponded to the highest relative abundance of *H. pellucidula/incognita* (the larvae of these two species are often not separable [40]): fitted optimum over 0.60 m·s^{-1}; *R.* cf. *aurata* and *R. nubila* gr.: fitted optimum around 0.50 m·s^{-1}; and *H. siltalai*: fitted optimum 0.47 m·s^{-1} (Figure 4). The mayfly *B. fuscatus* had a fitted optimum around 0.35 m·s^{-1} (Figure 4). By contrast, according to our data, *O. albicorne* had a fitted optimal mean current velocity around 0.14 m·s^{-1}, *S. flavicorne/personatum* 0.10 m·s^{-1}, and the mayfly *H. lauta* less than 0.10 m·s^{-1}.

4.2. Shift in Averaged Feeding Strategies in the EPT Community along Environmental Gradients

The analysis of shifts in ecological traits and, above all, their relationship to environmental characteristics has recently become a key topic in community ecology [73]. Substantial changes in the composition of feeding strategies (functional feeding groups) due to disturbances are assumed to occur in the macrozoobenthos [74,75]. Significant differences in the composition of average EPT feeding strategies were detected between localities (Figure 5A). Conductivity and 3-day sewage volumes were found to be the most parsimonious significant predictors in our dataset (Figure 5B). Both these EVs were positively related to the input of various materials and were positively related to both passive filter feeder and predator feeding strategies (represented almost only by caddisflies; Figure S7B). It suggests in higher susceptibility of EPT shredders to the pollution or unavailability of the CPOM (main food resource utilised by shredders) edible for these organisms (e.g., crystalline deposits containing trace metals can precipitate on the external surface of alder leaves because of fungal grown [21]). Even though shredder abundance will be compensated by numerous *Asellus aquaticus* and gatherers/collectors by chironomids and oligochaetes (see part 2.1. 'Locality description'), the abundance of winged adults of the aquatic insect directly utilising CPOM may be substantially lowered and could cause an alteration in nutrient and energy flows through the environment.

Despite that, 3-day sewage volumes positively correlated to the responses of passive filter feeders, predators, and shredders and negatively to the responses of gatherers/collectors (Figure S8A), the greatest 15-day sewage volumes (Figure S8B) positively correlated to the responses of gatherers/collectors and grazers and scrapers. Nevertheless, this trend was not strongly associated with the '15-day sewage volumes' because the dominance of gatherers/collectors and grazers and scrapers was caused by a high abundance of nymphs of the mayfly *B. fuscatus* gr., which probably colonised this site after drifting from an upstream stretch after a high discharge event (up to 8.5 times the average discharge). This hydrological situation occurred at the same time as the release of a high volume of sewage. The sewage release ended four days before the sampling time. The greatest similarity in the overall EPT community of samples to the samples from Site 3 is shown in the overlap of the ellipses in Figure S4. This similarity was subsequentially lost when the heavy rains stopped: the greatest dissimilarity between samples from individual sites was observed in August, followed by September. Nonzero values of high 3-day sewage volumes were linked to EPT compositions sampled in both these months, as well as in May. The significant linkage between 3-day sewage volumes and the composition of feeding strategies, as a result of FS in RDA (Figure 5B), supports the presumption that the population of campodeiform net-spinning caddisflies is directly or indirectly dependent on the supply of 'poorly treated' WWs.

5. Conclusions

This study highlights the impacts on the EPT community of trace metal pollution originating from mining and smelting and effluent from a municipal WWTP, which releases continuously treated WW and, periodical, 'poorly' treated WW. According to our data and previously published studies, trace metal pollution can have quite different consequences for the EPT community than, for example, pesticides and several caddisflies, especially limnephilids, were found to be very abundant at the affected site. Additionally, malformations in caddisfly larvae and mortality of caddisfly pupae were observed. Even though the affected sites were warmer, the absence of ephemerellid, heptageniid, and leptophlebiid mayflies is attributable to trace metal pollution, given the results from previous studies. The factors responsible for the decline in the stonefly community are less clear, and, except for higher water temperature, other kinds of pollutants and habitat degradation can be assumed to have a negative impact. Nevertheless, *L. fusca* gr. and *L. geniculata* can be considered tolerant to trace metal pollution but sensitive to WW pollution or its combination with trace metals. The municipal WW combined with trace metal contamination led to the greatest changes in the EPT community, including changes in the composition of feeding strategies. Caddisfly passive filter feeders and predators—mainly hydropsychids, polycentropodids, and rhyacophilids—dominated the most polluted environment. EPT shredders and collectors/gatherers dominated in unaffected sites. Our results demonstrate how the EPT community reacts to human-induced gradients in natural environments, which is important knowledge for assessing the implication of planned or current human disturbances in natural or cultural landscapes.

Supplementary Materials: The following supporting information can be downloaded at: https://www.mdpi.com/article/10.3390/biology11050648/s1, Materials and Methods; Results, Figure S1: photographic documentation of Site 1 and Site 2, Figure S2: photographic documentation of Site 3, Site 4 and Auxiliary Site 4, Figure S3: summer temperature regime monitored by data loggers (TFA), Table S1: physicochemical parameters of water at longitudinal profile, Table S2: parameters of discharge at monitored sites, Table S3: contamination of water and sediments by trace metals at longitudinal profile, Table S4: contamination of water by trace metals at Auxiliary Site 3 and Site 4 during 2018–2020, Table S5: pesticides and pharmaceutical active compounds in water samples taken at the longitudinal profile, Table S6: list of pesticide and pharmaceutical active compounds analysed in water samples with particular limits of quantification (LOQs), Figure S4: classified sample diagram resulted from choriotopic compositions using the RDA (sampling sites), Table S7: list of EPT taxa detected at given sites and times, Figure S5: classified CCA sample diagram resulted from species compositions and CCA species pie chart (sampling sites), Figure S6: CCA Species pie chart (sampling times), Figure S7: plotted significant averaged feeding strategies response curves, Figure S8: plotted significant averaged feeding strategies response curves against '3-d Sew' and '15-d Sew', Table S8: the statistics of relationship strengths for plots in Figures S7 and S8, Table S9: abundance and biomass of fishes detected at the particulate sites during 20 May and 8 October in 2020, Figure S9: plotted significant relationship between EPT family richness and Metal index. References [14,76–78] are cited in the supplementary materials.

Author Contributions: Conceptualization, M.L. and M.B.; methodology, M.L.; formal analysis, M.L.; investigation, M.L., J.Č., P.N., F.L., and M.B.; writing—original draft preparation, M.L.; writing—review and editing, M.B. and P.N.; supervision, M.B. All authors have read and agreed to the published version of the manuscript.

Funding: Project was supported by the Czech Science Foundation (20-16111S) and by the Ministry of Education, Youth and Sports of the Czech Republic (project CENAKVA no. LM2018099).

Institutional Review Board Statement: Not applicable.

Informed Consent Statement: Not applicable.

Data Availability Statement: Not applicable.

Acknowledgments: We would like to thank Michael Lockwood for English language editing, Wei Guo, Pavel Franta, and Jiří Jakš for their contribution during field works, Olga Valentová for analysis of water hardness, the Czech Hydrometeorological Institute (Prague, Czech Republic), Povodí Vltavy, State Enterprise (Prague, Czech Republic), and 1. SčV JSC (Příbram, Czech Republic) for providing environmental data, and finally to the Local Organisations of the Czech Anglers Union Hořovice and Příbram for enabling the fish stock monitoring.

Conflicts of Interest: The authors declare no conflict of interest.

References

1. Eriksen, T.E.; Brittain, J.E.; Soli, G.; Jacobsen, D.; Goethals, P.; Friberg, N. A global perspective on the application of riverine macroinvertebrates as biological indicators in Africa, South-Central America, Mexico and Southern Asia. *Ecol. Indic.* **2021**, *126*, 107609. [CrossRef]
2. Carter, J.L.; Resh, V.H.; Hannaford, M.J. Macroinvertebrates as biotic indicators of environmental quality. In *Methods in Stream Ecology*; Lamberti, G.A., Hauer, F.R., Eds.; Elsevier: London, UK, 2017; Volume 2: Ecosystem Function, pp. 293–318, ISBN 978-0-12-813047-6.
3. Johnson, R.C.; Jin, H.S.; Carreiro, M.M.; Jack, J.D. Macroinvertebrate community structure, secondary production and trophic-level dynamics in urban streams affected by non-point-source pollution. *Freshw. Biol.* **2013**, *58*, 843–857. [CrossRef]
4. Wagenhoff, A.; Townsend, C.R.; Matthaei, C.D. Macroinvertebrate responses along broad stressor gradients of deposited fine sediment and dissolved nutrients: A stream mesocosm experiment. *J. Appl. Ecol.* **2012**, *49*, 892–902. [CrossRef]
5. Johnson, R.C.; Carreiro, M.M.; Jin, H.S.; Jack, J.D. Within-year temporal variation and life-cycle seasonality affect stream macroinvertebrate community structure and biotic metrics. *Ecol. Indic.* **2012**, *13*, 206–214. [CrossRef]
6. Zwick, P. Historische Dokumente zur Fauna der Elbe bei Dresden vor hundert Jahren. *Lauterbornia* **1999**, *37*, 97–112.
7. Marten, M. Environmental monitoring in Baden-Württemberg with special reference to biocoenotic trend-monitoring of macrozoobenthos in rivers and methodical requirements for evaluation of long-term biocoenotic changes. *Aquat. Ecol.* **2001**, *35*, 159–171. [CrossRef]
8. Štěrba, O.; Měkotová, J.; Benář, V.; Šarapatka, B.; Rychnovská, M.; Kubíček, F.; Řehořek, V. *River Landscape and Its Ecosystems*, 1st ed.; Univerzita Palackého: Olomouc, Czech Republic, 2008; pp. 295–335. ISBN 978-80-244-2203-9.
9. Blann, K.L.; Anderson, J.L.; Sands, G.R.; Vondracek, B. Effects of agricultural drainage on aquatic ecosystems: A review. *Crit. Rev. Environ. Sci. Technol.* **2009**, *39*, 909–1001. [CrossRef]
10. Stanford, J.A.; Ward, J.; Liss, W.J.; Frissell, C.A.; Williams, R.N.; Lichatowich, J.A.; Coutant, C.C. A general protocol for restoration of regulated rivers. *Regul. Rivers Res. Manag.* **1996**, *12*, 391–413. [CrossRef]
11. Beermann, A.J.; Elbrecht, V.; Karnatz, S.; Ma, L.; Matthaei, C.D.; Piggott, J.J.; Leese, F. Multiple-stressor effects on stream macroinvertebrate communities: A mesocosm experiment manipulating salinity, fine sediment and flow velocity. *Sci. Total Environ.* **2018**, *610–611*, 961–971. [CrossRef]
12. Diamond, J.M. Ecology—Laboratory, Field and Natural Experiments. *Nature* **1983**, *304*, 586–587. [CrossRef]
13. *Directive 2000/60/EC, PE-CONS 3639/1/100 Rev 1*; Establishing a Framework for Community Action in the Field of Water Policy; European Commission: Luxemburg, 2000.
14. Grabicova, K.; Grabic, R.; Fedorova, G.; Kolarova, J.; Turek, J.; Brooks, B.W.; Randak, T. Psychoactive pharmaceuticals in aquatic systems: A comparative assessment of environmental monitoring approaches for water and fish. *Environ. Pollut.* **2020**, *261*, 114150. [CrossRef] [PubMed]
15. Kotková, K.; Nováková, T.; Tůmová, Š.; Kiss, T.; Popelka, J.; Faměra, M. Migration of risk elements within the floodplain of the Litavka River, the Czech Republic. *Geomorphology* **2019**, *329*, 46–57. [CrossRef]

16. Clements, W.H.; Carlisle, D.M.; Lazorchak, J.M.; Johnson, P.C. Heavy metals structure benthic communities in Colorado mountain streams. *Ecol. Appl.* **2000**, *10*, 626–638. [CrossRef]
17. Qu, X.; Wu, N.; Tang, T.; Cai, Q.; Park, Y.-S. Effects of heavy metals on benthic macroinvertebrate communities in high mountain streams. *Ann. De Limnol.-Int. J. Limnol.* **2010**, *46*, 291–302. [CrossRef]
18. Norris, R.; Lake, P.; Swain, R. Ecological effects of mine effluents on the South Esk River, North-eastren Tasmania. III. Benthic Macroinvertebrates. *Mar. Freshw. Res.* **1982**, *33*, 789–809. [CrossRef]
19. Malmqvist, B.; Hoffsten, P.O. Influence of drainage from old mine deposits on benthic macroinvertebrate communities in central Swedish streams. *Water Res.* **1999**, *33*, 2415–2423. [CrossRef]
20. Hedtke, S.F. Structure and Function of Copper-Stressed Aquatic Microcosms. *Aquat. Toxicol.* **1984**, *5*, 227–244. [CrossRef]
21. Ehrman, J.M.; Barlocher, F.; Wennrich, R.; Krauss, G.J.; Krauss, G. Fungi in a heavy metal precipitating stream in the Mansfeld mining district, Germany. *Sci. Total Environ.* **2008**, *389*, 486–496. [CrossRef]
22. Brix, K.V.; DeForest, D.K.; Adams, W.J. The sensitivity of aquatic insects to divalent metals: A comparative analysis of laboratory and field data. *Sci. Total Environ.* **2011**, *409*, 4187–4197. [CrossRef]
23. DeBruyn, A.M.; Rasmussen, J.B. Quantifying assimilation of sewage-derived organic matter by riverine benthos. *Ecol. Appl.* **2002**, *12*, 511–520. [CrossRef]
24. Zhang, X.Y.; Yu, T.; Li, X.; Yao, J.Q.; Liu, W.G.; Chang, S.L.; Chen, Y.G. The fate and enhanced removal of polycyclic aromatic hydrocarbons in wastewater and sludge treatment system: A review. *Crit. Rev. Environ. Sci. Technol.* **2019**, *49*, 1425–1475. [CrossRef]
25. Franco, A.A.; Arellano, J.M.; Albendin, G.; Rodriguez-Barroso, R.; Zahedi, S.; Quiroga, M.; Coello, M.D. Mapping microplastics in Cadiz (Spain): Occurrence of microplastics in municipal and industrial wastewaters. *J. Water Process Eng.* **2020**, *38*, 101596. [CrossRef]
26. Gani, K.M.; Tyagi, V.K.; Kazmi, A.A. Occurrence of phthalates in aquatic environment and their removal during wastewater treatment processes: A review. *Environ. Sci. Pollut. Res.* **2017**, *24*, 17267–17284. [CrossRef] [PubMed]
27. Matheri, A.N.; Eloko, N.S.; Ntuli, F.; Ngila, J.C. Influence of pyrolyzed sludge use as an adsorbent in removal of selected trace metals from wastewater treatment. *Case Stud. Chem. Environ. Eng.* **2020**, *2*, 100018. [CrossRef]
28. Soares, A. Wastewater treatment in 2050: Challenges ahead and future vision in a European context. *Environ. Sci. Ecotechnol.* **2020**, *2*, 100030. [CrossRef]
29. Fedorova, G.; Grabic, R.; Grabicova, K.; Turek, J.; Van Nguyen, T.; Randak, T.; Brooks, B.W.; Zlabek, V. Water reuse for aquaculture: Comparative removal efficacy and aquatic hazard reduction of pharmaceuticals by a pond treatment system during a one year study. *J. Hazard. Mater.* **2022**, *421*, 126712. [CrossRef] [PubMed]
30. Grabicova, K.; Grabic, R.; Blaha, M.; Kumar, V.; Cerveny, D.; Fedorova, G.; Randak, T. Presence of pharmaceuticals in benthic fauna living in a small stream affected by effluent from a municipal sewage treatment plant. *Water Res.* **2015**, *72*, 145–153. [CrossRef]
31. Suchowska-Kisielewicz, M.; Nowogonski, I. Influence of storms on the emission of pollutants from sewage into waters. *Sci. Rep.* **2021**, *11*, 18788. [CrossRef]
32. Covich, A.P.; Palmer, M.A.; Crowl, T.A. The role of benthic invertebrate species in freshwater ecosystems: Zoobenthic species influence energy flows and nutrient cycling. *BioScience* **1999**, *49*, 119–127. [CrossRef]
33. Beneš, F.; Horecký, J.; Senoo, T.; Kamasová, L., Lamačová, A.; Tátosová, J.; Hardekopf, D.W.; Stuchlík, E. Evidence for responses in water chemistry and macroinvertebrates in a strongly acidified mountain stream. *Biologia* **2017**, *72*, 1049–1058. [CrossRef]
34. Fišer, D.; Muška, M.; Vlach, P.; Dort, H.; Ťuláková, A.; Blabolil, P.; Vašek, M.; Kočvara, L. Fish communities of the Brdy Protected Landscape Area, current threats and management suggestions. *Bohemia Cent.* **2018**, *34*, 231–272.
35. Ter Braak, C.; Šmilauer, P. *Windows Release*; Version 5.12; Software for Mutivariate Data Exploration, Testing, and Summarization; Biometris, Plant Research International: Wageningen, The Netherlands; Germany University & Research: Berlin, Germany; Petr Šmilauer: České Budějovice, Czech Republic, 2012–2019.
36. R Core Team. *R: A Language and Environment for Statistical Computing*; R Foundation for Statistical Computing: Vienna, Austria, 2021.
37. Graf, W.; Murphy, J.; Dahl, J.; Zamora-Munoz, C.; López-Rodríguez, M.J. Trichoptera. In *Distribution and Ecological Preferences of European Freshwater Organisms*; Schmidt-Kloiber, A., Hering, D., Eds.; Pensoft Publishing: Sofia, Bulgaria, 2008; Volume 1, pp. 1–388, ISBN 978-954-642-441-9.
38. Buffagni, A.; Cazzola, M.; López-Rodríguez, M.J.; Alba-Tercedor, J.; Armanini, D.G. Ephemeroptera. In *Distribution and Ecological Preferences of European Freshwater Organisms*; Schmidt-Kloiber, A., Hering, D., Eds.; Pensoft Publishing: Sofia, Bulgaria, 2009; Volume 3, pp. 1–254. ISBN 978-954-642-508-9.
39. Graf, W.; Lorenz, A.W.; Tierno de Figueroa, J.M.; Lücke, S.; López-Rodríguez, M.J.; Davies, C. Plecoptera. In *Distribution and Ecological Preferences of European Freshwater Organisms*; Schmidt-Kloiber, A., Hering, D., Eds.; Pensoft Publishing: Sofia, Bulgaria, 2009; Volume 2, pp. 1–262. ISBN 978-954-642-479-2.
40. Waringer, J.; Graf, W. *Atlas of Central European Trichoptera Larvae/Atlas der Mitteleuropäischen Köcherfliegenlarven*; Erik Mauch Verlag: Dinkelscherben, Germany, 2011; pp. 1–470. ISBN 978-3-00-032177-1.
41. Burdon, F.J.; Munz, N.A.; Reyes, M.; Focks, A.; Joss, A.; Rasanen, K.; Altermatt, F.; Eggen, R.I.L.; Stamm, C. Agriculture versus wastewater pollution as drivers of macroinvertebrate community structure in streams. *Sci. Total Environ.* **2019**, *659*, 1256–1265. [CrossRef] [PubMed]

42. Haidekker, A.; Hering, D. Relationship between benthic insects (Ephemeroptera, Plecoptera, Coleoptera, Trichoptera) and temperature in small and medium-sized streams in Germany: A multivariate study. *Aquat. Ecol.* **2008**, *42*, 463–481. [CrossRef]
43. Moore, R.D.; Spittlehouse, D.; Story, A. Riparian Microclimate and Stream Temperature Response to Forest Harvesting: A Review. *J. Am. Water Resour. Assoc.* **2005**, *41*, 813–834. [CrossRef]
44. Marten, M. Interspecific variation in temperature dependence of egg development of five congeneric stonefly species (Protonemura Kempny, 1898, Nemouridae, Plecoptera). *Hydrobiologia* **1990**, *199*, 157–172. [CrossRef]
45. Humpesch, U.; Elliott, J. Effect of temperature on the hatching time of eggs of three Rhithrogena spp. (Ephemeroptera) from Austrian streams and an English stream and river. *J. Anim. Ecol.* **1980**, *49*, 643–661. [CrossRef]
46. Kiffney, P.M.; Clements, W.H. Effects of Heavy-Metals on a Macroinvertebrate Assemblage from a Rocky-Mountain Stream in Experimental Microcosms. *J. N. Am. Benthol. Soc.* **1994**, *13*, 511–523. [CrossRef]
47. Wesner, J.S.; Kraus, J.M.; Schmidt, T.S.; Walters, D.M.; Clements, W.H. Metamorphosis enhances the effects of metal exposure on the mayfly. *Environ. Sci. Technol.* **2014**, *48*, 10415–10422. [CrossRef]
48. Schmidt, T.S.; Kraus, J.M.; Walters, D.M.; Wanty, R.B. Emergence flux declines disproportionately to larval density along a stream metals gradient. *Environ. Sci. Technol.* **2013**, *47*, 8784–8792. [CrossRef]
49. Bojková, J.; Komprdova, K.; Soldán, T.; Zahrádková, S. Species loss of stoneflies (Plecoptera) in the Czech Republic during the 20th century. *Freshw. Biol.* **2012**, *57*, 2550–2567. [CrossRef]
50. Master, L.L.; Stein, B.A.; Kutner, G.A.; Hammerson, G.A.K. Vanishing assets, conservation status of US species. In *Precious Heritage, the Status of Biodiversity in the United States. The Nature Conservancy & Association for Biodiversity Information*; Stein, B.A., Kutner, L.S., Adams, J.S., Eds.; Oxford University Press: Oxford, UK, 2000; pp. 93–118. ISBN 978195125191.
51. Hardekopf, D.W.; Horecky, J.; Kopacek, J.; Stuchlik, E. Predicting long-term recovery of a strongly acidified stream using MAGIC and climate models (Litavka, Czech Republic). *Hydrol. Earth Syst. Sci.* **2008**, *12*, 479–490. [CrossRef]
52. Vuori, K.-M. Direct and indirect effects of iron on river ecosystems. In Proceedings of the Annales Zoologici Fennici, Helsinki, Finland, 1 November 1995; pp. 317–329.
53. Dangles, O. Functional plasticity of benthic macroinvertebrates: Implications for trophic dynamics in acid streams. *Can. J. Fish. Aquat. Sci.* **2002**, *59*, 1563–1573. [CrossRef]
54. Malaj, E.; Grote, M.; Schafer, R.B.; Brack, W.; von der Ohe, P.C. Physiological sensitivity of freshwater macroinvertebrates to heavy metals. *Environ. Toxicol. Chem.* **2012**, *31*, 1754–1764. [CrossRef] [PubMed]
55. Gattolliat, J.L.; Vincon, G.; Wyler, S.; Pawlowski, J.; Sartori, M. Toward a comprehensive COI DNA barcode library for Swiss Stoneflies (Insecta: Plecoptera) with special emphasis on the genus Leuctra. *Zoosymposia* **2016**, *11*, 135–155. [CrossRef]
56. Skála, I.; Lapšanská, N.; Špaček, J. Macrozoobenthos of brooks in the Brdy Highlands Protected Landscape Area (Czech Republic). *Bohemia Cent.* **2019**, *35*, 291–358.
57. Roux, C.; Tachet, H.; Bournaud, M.; Cellot, B. Stream continuum and metabolic rate in the larvae of five species of Hydropsyche (Trichoptera). *Ecography* **1992**, *15*, 70–76. [CrossRef]
58. Liess, M.; Von Der Ohe, P.C. Analyzing effects of pesticides on invertebrate communities in streams. *Environ. Toxicol. Chem.* **2005**, *24*, 954–965. [CrossRef]
59. Aspin, T.W.H.; Hart, K.; Khamis, K.; Milner, A.M.; O'Callaghan, M.J.; Trimmer, M.; Wang, Z.N.; Williams, G.M.D.; Woodward, G.; Ledger, M.E. Drought intensification alters the composition, body size, and trophic structure of invertebrate assemblages in a stream mesocosm experiment. *Freshw. Biol.* **2019**, *64*, 750–760. [CrossRef]
60. Pitsch, T. Contribution to larval taxonomy, ecology and distribution of the central European species of the genus Philopotamus (Trichoptera: Philopotamidae). In Proceedings of the Fifth International Symposium on Trichoptera, Lyon, France, 21–26 July 1986; pp. 331–335. [CrossRef]
61. Edington, J. Habitat preferences in net-spinning caddis larvae with special reference to the influence of water velocity. *J. Anim. Ecol.* **1968**, *37*, 675–692. [CrossRef]
62. Reiso, S.; Brittain, J.E. Life cycle, diet and habitat of Polycentropus flavomaculatus, Plectrocnemia conspersa and Rhyacophila nubila (Trichoptera) in Øvre Heimdalen, Jotunheimen Mountains, Norway. *Nor. J. Entomol.* **2000**, *47*, 113–124.
63. Stehle, S.; Schulz, R. Agricultural insecticides threaten surface waters at the global scale. *Proc. Natl. Acad. Sci. USA* **2015**, *112*, 5750–5755. [CrossRef] [PubMed]
64. Let, M.; Spacek, J.; Ferencik, M.; Kouba, A.; Blaha, M. Insecticides and Drought as a Fatal Combination for a Stream Macroinvertebrate Assemblage in a Catchment Area Exploited by Large-Scale Agriculture. *Water* **2021**, *13*, 1352. [CrossRef]
65. Di Veroli, A.; Santoro, F.; Pallottini, M.; Selvaggi, R.; Scardazza, F.; Cappelletti, D.; Goretti, E. Deformities of chironomid larvae and heavy metal pollution: From laboratory to field studies. *Chemosphere* **2014**, *112*, 9–17. [CrossRef] [PubMed]
66. Ratia, H.; Vuori, K.M.; Oikari, A. Caddis larvae (Trichoptera, Hydropsychidae) indicate delaying recovery of a watercourse polluted by pulp and paper industry. *Ecol. Indic.* **2012**, *15*, 217–226. [CrossRef]
67. Buntha, P.; Traichaiyaporn, S.; Thapanya, D. Food Source for Hydropsychid Larvae during an Algae Bloom in Nan River, Nan Province, Thailand (Trichoptera: Hydropsychidae). *Zoosymposia* **2020**, *18*, 9–16. [CrossRef]
68. Otto, C. Prey size and predation as factors governing the distribution of lotic polycentropodid caddisfly larvae. *Oikos* **1985**, *44*, 439–447. [CrossRef]
69. Englund, G.; Evander, D. Interactions between sculpins, net-spinning caddis larvae and midge larvae. *Oikos* **1999**, *85*, 117–126. [CrossRef]

70. Nakano, D.; Yamamoto, M.; Okino, T. Ecosystem engineering by larvae of net-spinning stream caddisflies creates a habitat on the upper surface of stones for mayfly nymphs with a low resistance to flows. *Freshw. Biol.* **2005**, *50*, 1492–1498. [CrossRef]
71. Glime, J.M. Bryophyte Ecology, Chapter 11—Aquatic Insects. Available online: http://digitalcommons.mtu.edu/bryophyte-ecology2 (accessed on 27 March 2022).
72. Krno, I. Longitudinal changes in the structure of macrozoobenthos and its microdistribution in natural and moderately eutrophicated waters of the River Rajcianka (Strázovské vrchy). *Acta Fac. Rerum Nat. Univ. Comen. Zool.* **1990**, *33*, 31–48.
73. McGill, B.J.; Enquist, B.J.; Weiher, E.; Westoby, M. Rebuilding community ecology from functional traits. *Trends Ecol. Evol.* **2006**, *21*, 178–185. [CrossRef]
74. Boulton, A.J.; Peterson, C.G.; Grimm, N.B.; Fisher, S.G. Stability of an Aquatic Macroinvertebrate Community in a Multiyear Hydrologic Disturbance Regime. *Ecology* **1992**, *73*, 2192–2207. [CrossRef]
75. Lima, M.; Firmino, V.C.; de Paiva, C.K.S.; Juen, L.; Brasil, L.S. Land use changes disrupt streams and affect the functional feeding groups of aquatic insects in the Amazon. *J. Insect Conserv.* **2022**, *26*, 137–148. [CrossRef]
76. Grabic, R.; Fick, J.; Lindberg, R.H.; Fedorova, G.; Tysklind, M. Multi-residue method for trace level determination of pharmaceuticals in environmental samples using liquid chromatography coupled to triple quadrupole mass spectrometry. *Talanta* **2012**, *100*, 183–195. [CrossRef]
77. Lindberg, R.H.; Ostman, M.; Olofsson, U.; Grabic, R.; Fick, J. Occurrence and behaviour of 105 active pharmaceutical ingredients in sewage waters of a municipal sewer collection system. *Water Res.* **2014**, *58*, 221–229. [CrossRef]
78. Giddings, E.M.; Hornberger, M.I.; Hadley, H.K. Trace-metal concentrations in sediment and water and health of aquatic macroinvertebrate communities of streams near Park City, Summit County, Utah. In *Utah*; Science for a Changing World: Salt Lake City, UT, USA, 2001; pp. 9–10. [CrossRef]

biology

MDPI

Article

Plant Diversity and Fungal Richness Regulate the Changes in Soil Multifunctionality in a Semi-Arid Grassland

Zhuo Li [1,2], Xiaowei Liu [1,2], Minghui Zhang [1,2] and Fu Xing [1,2,*]

1 Key Laboratory of Vegetation Ecology, Institute of Grassland Science, Northeast Normal University, Ministry of Education, Changchun 130024, China; liz347@nenu.edu.cn (Z.L.); liuxw101@nenu.edu.cn (X.L.); zhangmh740@nenu.edu.cn (M.Z.)

2 Jilin Songnen Grassland Ecosystem National Observation and Research Station, Changchun 130024, China

* Correspondence: xingf@nenu.edu.cn

Simple Summary: Understanding relationships between biodiversity and ecosystem functions is important in the context of global plant diversity loss. We evaluated the relationships between soil bacterial and fungal diversity, rare microbial taxa, and soil multifunctionality in a semi-arid grassland with varied plant diversity levels. The fungal richness rather than bacterial richness was positively related to soil multifunctionality. The relative abundance of saprotrophs was positively correlated with soil multifunctionality, and the relative abundance of pathogens was negatively correlated with soil multifunctionality. Furthermore, the rare fungal taxa played a disproportionate role in regulating soil multifunctionality. The shift of plant biomass allocation patterns increased plant below-ground biomass in the high diversity plant assemblages, which can alleviate soil microbial carbon limitations and enhance the fungal richness, thus promoting soil multifunctionality. This study provides a new perspective for evaluating the relative roles of fungal and bacterial diversity in maintaining multiple soil functions under global plant diversity loss scenarios.

Abstract: Loss in plant diversity is expected to impact biodiversity and ecosystem functioning (BEF) in terrestrial ecosystems. Soil microbes play essential roles in regulating ecosystem functions. However, the important roles and differences in bacterial and fungal diversity and rare microbial taxa in driving soil multifunctionality based on plant diversity remain poorly understood in grassland ecosystems. Here, we carried out an experiment in six study sites with varied plant diversity levels to evaluate the relationships between soil bacterial and fungal diversity, rare taxa, and soil multifunctionality in a semi-arid grassland. We used Illumina HiSeq sequencing to determine soil bacterial and fungal diversity and evaluated soil functions associated with the nutrient cycle. We found that high diversity plant assemblages had a higher ratio of below-ground biomass to above ground biomass, soil multifunctionality, and lower microbial carbon limitation than those with low diversity. Moreover, the fungal richness was negatively and significantly associated with microbial carbon limitations. The fungal richness was positively related to soil multifunctionality, but the bacterial richness was not. We also found that the relative abundance of saprotrophs was positively correlated with soil multifunctionality, and the relative abundance of pathogens was negatively correlated with soil multifunctionality. In addition, the rare fungal taxa played a disproportionate role in regulating soil multifunctionality. Structural equation modeling showed that the shift of plant biomass allocation patterns increased plant below-ground biomass in the highly diverse plant plots, which can alleviate soil microbial carbon limitations and enhance the fungal richness, thus promoting soil multifunctionality. Overall, these findings expand our comprehensive understanding of the critical role of soil fungal diversity and rare taxa in regulating soil multifunctionality under global plant diversity loss scenarios.

Keywords: plant diversity loss; soil multifunctionality; fungal diversity; saprotrophic fungi; rare microbial taxa

Citation: Li, Z.; Liu, X.; Zhang, M.; Xing, F. Plant Diversity and Fungal Richness Regulate the Changes in Soil Multifunctionality in a Semi-Arid Grassland. *Biology* **2022**, *11*, 870. https://doi.org/10.3390/biology11060870

Academic Editors: Daniel Puppe, Panayiotis Dimitrakopoulos and Baorong Lu

Received: 7 May 2022
Accepted: 3 June 2022
Published: 6 June 2022

1. Introduction

Soil supports a wide range of ecological functions and services that are important to human welfare, and soil multifunctionality can be used as a comprehensive index to evaluate soil quality [1–4]. Multifunctionality is an essential biological and management concept that describes the ability of an ecosystem to maintain multiple ecological functions simultaneously [1,2,5]. As significant drivers of ecosystem functions, soil microbial communities can be regulated by plant diversity via changing the nutrient availability and microenvironmental conditions [6–8]. However, global climate changes and intensive anthropogenic activities have led to the loss of plant diversity in grassland ecosystems [2,9–11]. Therefore, there remains an urgent need to understand the mechanisms by which soil microorganisms drive soil multifunctionality in the context of global plant diversity loss.

Below-ground microbes comprise a large portion of global terrestrial ecosystem life's genetic diversity and support important ecosystem functions and services, such as biomass production, nutrient cycling, and decomposition of organic matter [10,12,13]. Additionality, soil microorganisms are an essential link between the above-ground and below-ground components of terrestrial ecosystems [14–17]. Bacteria and fungi constitute vital parts of the soil microbiome [18]. Nevertheless, bacteria and fungi play different roles in regulating ecosystem processes [14]. In detail, bacteria are the essential drivers of soil N cycle processes, such as N_2 fixation, nitrification, and denitrification [19–21]. However, other studies have suggested that soil fungal communities might dominate in driving soil functions [14]. Soil fungi can decompose recalcitrant plant litter efficiently and form links with roots to capture and transport subsurface carbon (C) [22]. For instance, saprophytic fungi are primary mediators for the decomposition of plant litter, and their mycelium networks across the soil litter interface and networks are a highly dynamic channel through which nutrients can be easily distributed [23]. Despite the above studies, most research has concentrated on how soil microbes affect individual or specific nutrient cycling functions. However, few studies have evaluated the relative significance of bacterial and fungal diversity in regulating soil multifunctionality in the context of plant diversity in a semi-arid grassland.

Furthermore, the effects of rare microbial taxa on soil functions have often been ignored in most previous soil multiple functions studies because they have mainly concentrated on predominant taxa; a majority of rare microbial taxa were frequently eliminated from original datasets. However, rare microbial taxa have important ecological roles because these may be activated to perform critical functions under environmental changes [24,25]. For instance, rare taxa were found to play essential roles in nutrient cycling after disturbances in aquatic ecosystems [26]. Rare microbial taxa have also been used as indicators of changes in ecosystem functions under long-term greenhouse cultivation conditions in subtropical agricultural soils [27]. Therefore, rare microbial taxa can represent a microbial "seed bank" that may be activated when the environment is disturbed [24,28]. The intensity and frequency of plant diversity loss due to global climate change are expected to increase in the future [9,29,30]; therefore, it is of great significance to link the ecological functions of rare microbial taxa with soil multifunctionality. Unfortunately, relevant knowledge is still scarce, especially in semi-arid grassland ecosystems. Such knowledge is critical for developing a management framework to maintain rare taxa involved in functionality and decrease the impact of future climate change on a semi-arid grassland.

Loss in plant diversity of terrestrial ecosystems might aggravate C and other nutrient limitations [30–32]. Heterotrophic organisms depend on plants to obtain C substrates; therefore, there is ample evidence that soil microbes are usually limited by C [33]. Microorganisms play an essential role in ecosystem functions [6,18,34]; thus, more research is undoubtedly required to evaluate how the shift in microbial nutrient limitations affects microbes under changes in plant diversity. Soil enzymes produced by microbes transform substrates in soil C and nutrient cycles [35]. Thus, enzymatic stoichiometry provides a method for this study [35]. Several studies have shown that less diverse plant communities allocate fewer photosynthates to below-ground ecosystems by decreasing the exudation of root exudates [32]. Root exudates are an important C source for microbial growth [36];

therefore, a less diverse plant community might intensify microbial C limitation [37]. C limitation could affect microbial biosynthesis processes [35]; thus, lower below-ground biomass allocation would reduce soil functions such as microbial growth and respiration by increasing microbial C limitation [36,38]. The above results show that plant diversity affects soil microbial communities by trophic interactions and altering soil biochemical processes [12,13]. Thus, the loss of plant diversity may alter the relationships between plant and soil microbial communities. Considering that soil microbial communities mediate ecosystem functions, it is essential to elucidate the mechanisms by which plant-diversity-induced changes in microbial C limitation regulate soil multifunctionality through impacting soil microorganisms in a meadow grassland.

In addition, ecosystem multifunctionality is different from ecosystem services multifunctionality in that the former represents the overall performance of an ecosystem without considering stakeholders; whereas the latter is defined as the ability of ecosystems to synergistically provide various ecosystem functions and services that translate into a variety of social benefits and welfare [1]. Therefore, investigations of ecosystem service multifunctionality have significant importance in future semi-arid grassland ecosystems management and sustainable development. In this study, a meadow grassland in northeast China was selected to discuss the soil multifunctionality and microbial driving mechanisms under a plant diversity loss scenario. The objectives of this study were to 1) evaluate the important roles of soil bacterial and fungal diversity and rare taxa in driving soil multifunctionality under different plant diversity conditions; 2) reveal the underlying mechanisms by which plant-diversity-induced alterations of microbial C limitation regulate soil multifunctionality through soil microorganisms; 3) provides a scientific reference for the evaluation of semi-arid grassland ecosystem services multifunctionality.

2. Materials and Methods

2.1. Study Site

The experiment was conducted at the Tongyu Observatory in a semi-arid climate and environment (44° 25′ N, 122° 52′ E). This site is located within the Songnen grassland ecosystem of northeastern China (Figure 1a). The vegetation is a meadow steppe, situated at the eastern end of the Eurasian grassland belt with *Leymus chinensis* (Trin.) Tzvelev (https://www.ipni.org/) (accessed on 1 June 2022) is the dominant species. The soil texture is clay loam, according to the International Society of Soil Science Standard. The study area has a temperate continental monsoon climate. The mean annual precipitation is ~404 mm, and more than 80% of the rainfall is concentrated during the growing season (from May to September). The mean annual temperature is approximately 5.7 °C.

2.2. Experimental Design and Sampling

Within approximately 1 km^2 of the study area, we randomly selected three single plant species patches i.e., *Carex duriuscula* C.A.Mey., *Lespedeza hedysaroides* (Pall.) Kitag., and *Calamagrostis rigidula* A.I.Baranov and Skvortsov assemblages and three multiple species coexisting plant assemblages with different dominant *Lespedeza daurica* (Laxm.) Schindl., *L. chinensis*, and *Hierochloe glabra* Trin. naturally existing in the grassland. Single species assemblages and multiple species assemblages represent low and high diversity levels, respectively. The plant species composition is shown in Table S1. A vegetation survey was conducted in mid-August 2018 when the standing biomass reached its maximum. For the six study sites selected above, the interval between every two sites was greater than 100 m. For plant and soil sampling, five plots (1 m × 1 m) with an interval of more than 10 m were randomly selected at each study site (Figure 1b). In this study, soil characteristics of low diversity plant assemblages and high diversity plant assemblages were similar ($p > 0.05$) (Figure S1); thus, soils were comparable across different plant diversity plots. We selected six sites with similar soil characteristics but different levels of plant diversity. It is reasonable to obtain data based on one-time sampling to explore the impact of different levels of plant diversity on soil, microorganisms, and soil functions. Additionally, this method has been widely adopted by many studies in the field of ecology [32,39].

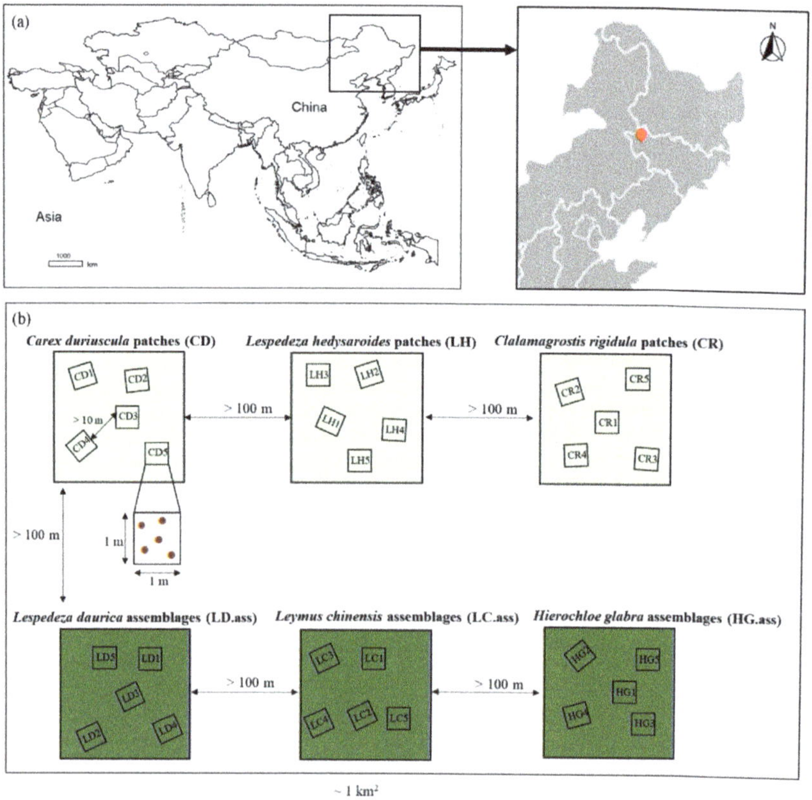

Figure 1. Locations of the study site (**a**) and experimental design for plant and soil sampling (**b**) in the Songnen grassland, China. Three single plant species patches, i.e., *Carex duriuscula* C.A.Mey. (CD), *Lespedeza hedysaroides* (Pall.) Kitag. (LH), and *Calamagrostis rigidula* A.I.Baranov and Skvortsov (CR) assemblages and three multiple species coexisting plant assemblages with different dominant *Lespedeza daurica* (Laxm.) Schindl. (LD.ass), *Leymus chinensis* (Trin.) Tzvelev (LC.ass), and *Hierochloe glabra* Trin. (HG.ass) naturally existing in the grassland. For the six study sites selected above, the interval between every two sites was greater than 100 m. For plant and soil sampling, five plots (1 m × 1 m) with an interval of more than 10 m were randomly selected at each study site. Five soil cores were randomly selected in each plot by a soil auger.

All the plant species were identified and recorded in each plot; then, soil and plant samples were collected simultaneously from each plot. Five soil cores (0–10 cm) were randomly selected in each plot by a soil auger (5 cm diameter), then mixed to form a composite sample. Next, the soil samples were sieved (2 mm) to remove any roots and stones. We harvested above-ground living plants to evaluate each plot's above-ground biomass (AGB). The below-ground biomass (BGB) of each plot was measured by root biomass from three soil cores with a diameter of 10 cm and a depth of 30 cm. All the soil samples were placed in well-sealed zippered bags and transported to the laboratory within 24 h in cooling boxes.

2.3. Plant and Soil Samples Analysis

The plant AGB and BGB were determined by plant materials heated at 105 °C for 30 min to rapidly cease metabolic activities [40] and then oven-dried at 65 °C for 48 h until the weight remained constant [41]. The soil samples were divided into three subsamples for physiochemical analyses and the assessment of microbial communities. One aliquot of soil was stored at 4 °C and used to determine the soil water content (SWC), soil available N (AN) content (NH_4^+-N and NO_3^--N), soil net nitrification rate (Nn), and the net N mineralization rate (Nm). Another aliquot of soil was stored at −80 °C and used for the analysis of high-throughput sequencing and to assess activities of α-1,4-glucosidase (αG), β-1,4-glucosidase (βG), β-1,4-xylosidase (βX), β-D-cellobiohydrolase (CBH), leucine aminopeptidase (LAP), β-1,4-N-acetylglucosaminidase (NAG), and alkaline phosphatase (ALP). Finally, the third aliquot of soil was air-dried at room temperature for soil pH, electrical conductivity (EC), soil total organic carbon (TOC), total soil nitrogen (TN), total soil phosphorus (TP), and available soil phosphorus (AP) analyses.

The soil pH was detected in deionized water (soil: water = 1:5, w:v) with a portable pH meter (Leichi PHBJ-260, Shanghai, China), and the soil EC was measured with an electronic conductivity meter (Leici DDS307, Shanghai, China) [42]. The SWC was measured as the weight loss recorded after the fresh soils had been oven-dried to a constant weight at 105 °C [10]. The soil TN content was measured with an elemental analyzer (vario EL cube, Elementar, Langenselbold, Germany). Briefly, 30 mg of the air-dried soil was put into a tinfoil cup and tightly wrapped in soil and then put into an automatic sampling tray; the results were analyzed on the machine. Soil TOC was determined with an elemental analyzer (Isoprime 100, Isoprime Ltd., Manchester, UK). The NH_4^+-N and NO_3^--N concentrations were measured by extraction with 2 M KCl (soil: water = 1: 5, w:v) and analyzed with a continuous flow analyzer (Futura, Alliance-AMS) [41]. The Nn and Nm were calculated according to alterations in the concentrations of NH_4^+-N and NO_3^--N before and after incubation [43]. For soil TP, 0.5 g air-dried soil and ball milling were digested with $HClO_4$ (7.7 mL, 75%) at 203 °C for 75 min [44]. Soil AP was extracted with 0.5 M $NaHCO_3$, and the molybdenum blue colorimetric method was used to analyze [45]. All enzyme activities were measured using a 4-methylumbelliferyl (MUB) substrate, except for 7-amino-4-methyl-coumarin (7-AMC) for the LAP [46]. Seven enzyme activities were determined in black 96-well microplates. Four replicate wells were set up for each enzyme test sample (200 μL slurry + 50 μL substrate) and corresponding substrate control (200 μL buffer + 50 μL substrate). Four replicate wells were set up for standard fluorescence (200 μL buffer + 50 μL standard), slurry control (200 μL slurry + 50 μL buffer), and quench standards (200 μL slurry + 50 μL standard). The assay plate was incubated in a dark environment at 25 °C for 3 h. Fluorescence was measured using a microplate fluorometer (TECAN infinite F200, Tecan Group, Switzerland) with excitation and emission filters of 360 nm and 460 nm, respectively.

2.4. Assessment of Microbial Communities

According to the manufacturer's protocol, total genomic DNA was extracted from 0.3 g of soil with the PowerSoil DNA Isolation Kit (MO BIO Laboratories, Inc., Carlsbad, CA, USA). The concentration and purity of the extracted DNA were measured with a NanoDrop 2000 spectrophotometer (Thermo Fisher Scientific, Waltham, MA, USA). DNA quality was evaluated by 1% agarose gel electrophoresis. We acquired information on the diversity and composition of soil bacteria and fungi by performing 16S rRNA and ITS genes amplicon sequencing and the 338F/806R (5′-ACTCCTACGGGAGGCAGCA-3′, 5′-GGACTACHVGGGTWTCTAAT-3′) and ITS1F/ITS2R (5′-CTTGGTCATTTAGAGGAAGTAA-3′, 5′-GCTGCGTTCTTCATCGATGC-3′) primer pairs, respectively. The total volume of the PCR amplification for bacteria and fungi was 50 μL, including 10 μL buffer, 0.2 μL Q5 High-Fidelity DNA Polymerase, 10 μL High GC Enhancer, 1 μL dNTP, a 10 μM concentration of each primer, and 60 ng genome DNA. The thermal cycling conditions were 95 °C denaturation for 5 min, 15 cycles at 95 °C for 1 min, 50 °C for 1 min, 72 °C for 1 min, and 72 °C final extension for 7 min. Ultimately, we quantified all PCR products by Quant-iT™ dsDNA HS Reagent and pooled them together. Furthermore, we conducted a high-throughput sequencing analysis of 16S rRNA and ITS genes by the Illumina

Hiseq 2500 platform (2 × 250 paired ends) at Biomarker Technologies Corporation (Beijing, China). Bacterial and fungal sequences were quality-filtered with the QIIME software package and merged using the FLASH software package [47]. The operational taxonomic units (OTUs) were defined by 97% similarity. Representative sequences of bacteria and fungi were annotated with the SILVA and UNITE databases [48]. The final total data set retained 1852 and 1325 OTU numbers and 2,057,790 and 2,071,174 clean reads for bacteria and fungi, respectively. The raw bacterial and fungal reads were deposited into the National Center for Biotechnology Information (NCBI) Sequence Read Archive (SRA) database under accession numbers PRJNA810930 and PRJNA810946, respectively.

2.5. Definition of Abundant and Rare Taxa

Previous studies have widely used relative abundance as a metric to describe microbial taxa or OTUs in their environment. Thus, relative abundance is useful for classifying abundant or rare taxa in microbial communities [24,49]. We identified relative abundance thresholds as 1% for abundant taxa and 0.01% for rare taxa [26,27,50]. These classifications ignored intermediate taxa (relative abundance between 0.01 and 1%) and oscillatory taxa (rare and abundant under different environment conditions) [27]. All OTUs were classified into six categories based on the criteria used in recent studies [26,27,50]: always abundant taxa (AAT, relative abundance $\geq$ 1%); conditionally abundant taxa (CAT, relative abundance $\geq$ 0.01% in all samples and $\geq$ 1% in some samples); always rare taxa (ART, relative abundance < 0.01% in all samples); conditionally rare taxa (CRT, relative abundance < 0.01% in some samples but never >1% in any samples); moderate taxa (MT, relative abundance between 0.01% and 1% in all samples); and conditionally rare and abundant taxa (CRAT, relative abundance < 0.01% in some samples and $\geq$ 1% in some samples). Finally, AAT and CAT were classified as abundant taxa, and ART and CRT were classified as rare taxa [26,27,50].

2.6. Assessment of Multifunctionality

Soil multifunctionality was quantified based on ten soil functions widely adopted in previous studies [51–55]. These parameters are related to organic matter decomposition, climate regulation, and nutrient cycling, including αG, βG, βX, CBH, LAP, NAG, Nn, Nm, ALP, and AP. Of them, αG, βG, βX, and CBH represent the C cycle, LAP, NAG, Nn, and Nm represent the N cycle, and ALP and AP represent the P cycle (Table S2). The averaging and threshold approaches have commonly been used to estimate the relationship between biodiversity and multifunctionality [2,34,53,56–58]. In this study, we evaluated the soil multifunctionality index using three methods, which contained averaging approach [56], single-threshold, and multiple-threshold approach [57,58]. To obtain an average soil multifunctionality index, each function was standardized by Z-score transformation and then averaged [2,56]. We used a single-threshold approach to calculate the number of soil functions exceeding a given threshold (25%, 50%, 75%, and 90%) [2,54]. However, the multiple-threshold method can demonstrate the effect of bacterial and fungal richness on soil multifunctionality in the full threshold range.

2.7. Enzymatic Stoichiometry

We constructed the vector analysis of soil enzymatic stoichiometry to evaluate soil microbial C limitation levels [59]. A longer vector length represents a stronger C limitation.

$$\text{Vector length (unitless)} = \sqrt{(\ln \beta\text{G}/\ln(\text{NAG}+\text{LAP}))^2 + (\ln \beta\text{G}/\ln\text{ALP})^2}$$

2.8. Statistical Analysis

We used an independent sample *t*-test to evaluate the differences in plant and soil properties and soil multifunctionality between different levels of plant diversity. When the data did not meet the requirements for normality (Shapiro Wilk test) and homogeneity of variance (Bartlett test), log transformations were performed on the data. We calculated the microbial richness with the "*vegan*" package [60]. Ordinary least squares (OLS) linear regression was conducted to explore the relationships between microbial C limitation and microbial richness; OLS linear regression also was conducted to explore the relationships between bacterial richness, fungal richness, relative pathogen abundance, saprotrophic relative abundance, relative symbiont abundance, and soil multifunctionality [14,15,34,61]. Three methods of soil multifunctionality were evaluated with the "multifunc" package [58]. According to recent studies [14,47,61,62], functions were obtained based on fungal taxa by the FUNGuild database (http://www.stbates.org/guilds/app.php) (accessed on 6 January 2019). Random forest modeling analysis was applied with the "randomForest" package [63] to identify the major statistically significant microbial taxa (OTU level) in impacting soil multifunctionality. Then, seven microbial taxa, including four bacterial taxa and three fungal taxa, were selected in the random forest modeling. The "A3" package [64] was used to assess the importance of each predictor on soil multifunctionality. In addition, we used piecewise structural equation modeling (SEM) with the "piecewiseSEM" [65], "nlme" and "lme4" packages [66] to investigate the potential direct and indirect effects of plant diversity on soil multifunctionality. The prior model was constructed based on current ecological knowledge of the grassland ecosystems [7,36,53,67] to evaluate the link between soil biodiversity and multifunctionality (averaging), assuming that plant diversity altered plant biomass allocation patterns and microbial C limitation, thus promoting soil microbial richness and soil multifunctionality (Figure S2a). Fisher's C test ($0 \leq$ Fisher's $C \leq 2\ df$ and $0.05 < p \leq 1.00$) was used to verify the rationality of the modeling results. Among significant models, the one with the lowest Akaike information criterion (AIC) value was selected for the final SEM analysis [68] (Figure S2b–d). All statistical analyses and visualizations of this study were performed in R v.4.1.2 software.

3. Results

3.1. Plant and Soil Properties and Soil Multifunctionality

The ratio of BGB to AGB was significantly higher with high plant diversity than with low plant diversity (Figure 2a; $p < 0.001$). Microbial C limitation was higher with low diversity than with high diversity (Figure 2b; $p < 0.01$). The results showed that plant diversity significantly altered soil multifunctionality (Figure 2c), i.e., high diversity significantly increased soil multifunctionality compared with low plant diversity (Figure 2c; $p < 0.001$). SWC, soil pH, EC, AN, and TOC showed no significant differences between the two plant diversities (Figure S3a–d,g). Soil TN and TP were higher with low diversity than with high diversity (Figure S3e,f; $p < 0.05$ and $p < 0.01$). In addition, the soil C/N was much higher with high diversity than with low diversity (Figure S3h; $p < 0.001$).

3.2. Soil Microbial Community Composition

The rarefaction curves of OTU richness of bacterial and fungal communities almost approached saturation (Figure S4a,b), showing that the amount of data of sequenced reads were reasonable. The bacterial OTUs were assigned to 23 phyla, 76 classes, 110 orders, 188 families, 271 genera, and 226 species; similarly, the fungal OTUs were assigned to 10 phyla, 23 classes, 51 orders, 103 families, 177 genera, and 152 species. The bacterial communities were dominated by Acidobacteria, Proteobacteria, and Actinobacteria (Figure S5). A significantly higher relative abundance of Acidobacteria was observed with low plant diversity than with high plant diversity (Figure S5a; $p < 0.001$). The relative abundance of Actinobacteria and Chloroflexi was higher with high diversity than with low diversity (Figure S5b,d; $p < 0.001$). The fungal communities were dominated by Ascomycota and Basidiomycota (Figure S5). The bacterial communities were dominated by the genera *RB41*, *uncultured_bacterium_c_Subgroup_6*, and *uncultured_bacterium_f_Gemmatimonadaceae* (Figure 3a). However, the fungal communities were dominated by the genera *Ceratobasidium*, *Mortierella*, and *Cladosporium* (Figure 3b). The richness (OTU richness) of fungal and bacterial were significantly greater with high diversity than with low diversity (Figure S6; $p < 0.001$, $p < 0.05$).

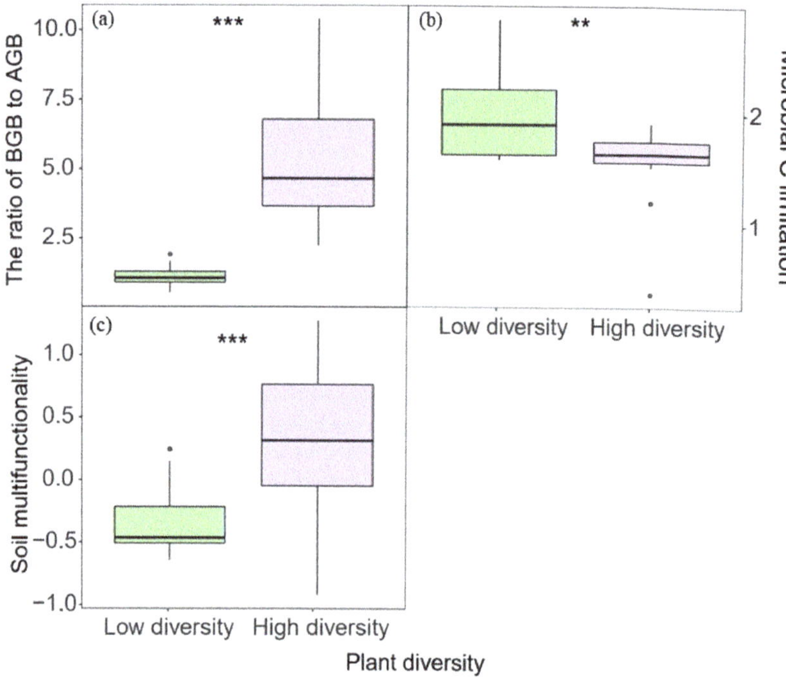

Figure 2. The ratio of BGB to AGB (**a**), microbial C limitation (**b**), and soil multifunctionality (**c**) in response to plant diversity. ** $p < 0.01$ and *** $p < 0.001$ (t-test). AGB: above-ground biomass, BGB: below-ground biomass.

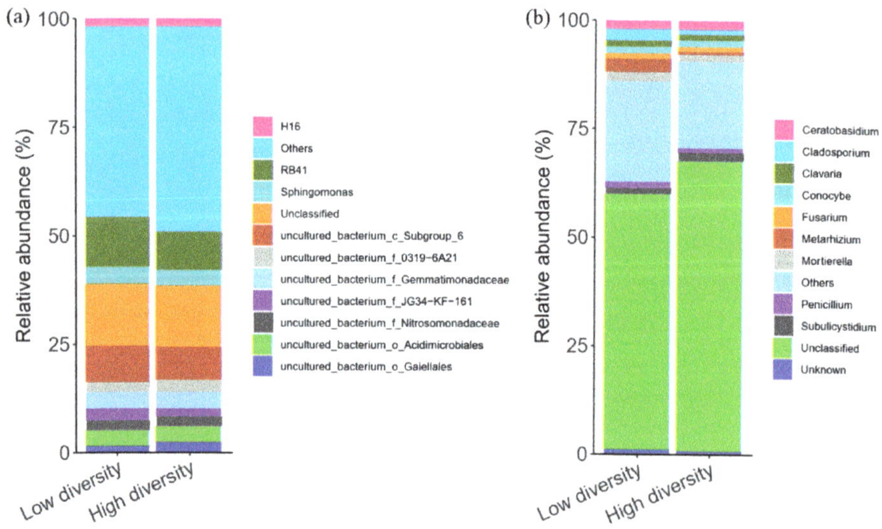

Figure 3. Relative abundance of bacterial (**a**) and fungal (**b**) dominant genera in different plant diversity levels, "Others" represents all genera with relative abundance < 1%.

3.3. Microbial Taxa in Relation to Soil Multifunctionality

The OLS regression models showed that microbial C limitation was significantly and negatively related to fungal richness (Figure 4b; $R^2 = 0.309$, $p < 0.001$). Furthermore, the results showed that fungal richness was significantly and positively associated with average soil multifunctionality (Figure 5b; $R^2 = 0.233$, $p = 0.003$). However, bacterial richness showed no correlation with soil multifunctionality (Figure 5a). Based on the positive correlation between fungal richness and soil multifunctionality, we aimed to identify the relationships between fungal guilds and soil multifunctionality. A negative linear correlation was found between the relative abundance of pathogens and soil multifunctionality (Figure 5c; $R^2 = 0.140$, $p = 0.023$). However, we found a significantly positive relationship between the relative abundance of saprotrophs and soil multifunctionality (Figure 5d; $R^2 = 0.165$, $p = 0.014$).

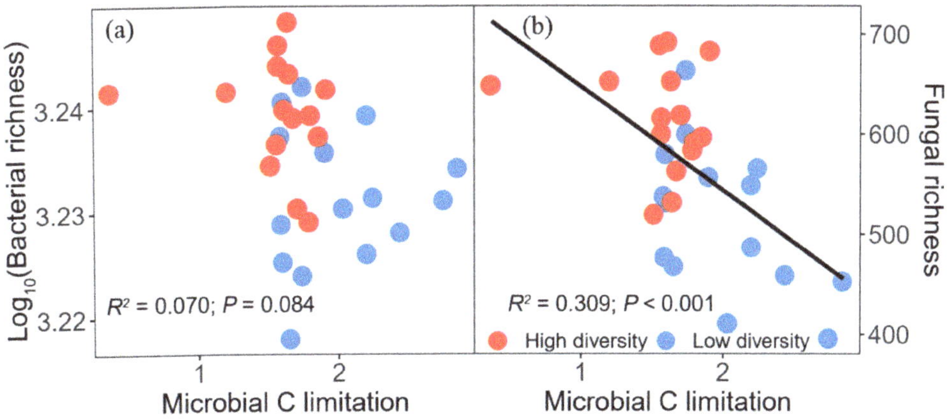

Figure 4. Relationships between soil microbial carbon (C) limitation and bacterial richness (**a**) and fungal richness (**b**). The black lines represent the fitted ordinary least squares (OLS) linear regressions. Red circles represent high plant diversity, while blue ones indicate low plant diversity.

Soil multifunctionality was positively and significantly related to fungal richness rather than bacterial richness. This result was confirmed using the single-threshold method (Figure S7) and the multiple-threshold method (Figure 6). Threshold methods were also used to assess whether multiple functions were performed at high levels simultaneously. Specifically, we performed single-threshold analysis; positive and significant correlations were found between the bacterial richness and soil multifunctionality at given thresholds of 75% and 90% (Figure S7g,h; $R^2 = 0.108$, $p = 0.042$, $R^2 = 0.105$, $p = 0.044$). However, there were significant and positive correlations between the fungal richness and soil multifunctionality at given thresholds of 25%, 50%, 75%, and 90% (Figure S7a–d; $R^2 = 0.220$, $p = 0.005$, $R^2 = 0.111$, $p = 0.039$, $R^2 = 0.313$, $p < 0.001$, $R^2 = 0.292$, $p = 0.001$). However, the multiple-threshold method does not require a threshold to be set and studies a continuous threshold gradient (Figure 6a,b). The minimum threshold (T_{min}) for fungi was 16%, the lowest threshold at which fungal richness began to have a positive effect on soil multifunctionality. When the threshold was 36%, the maximum richness effect (R_{mde}) was 0.016, i.e., the relationship strength of fungal richness with the strongest positive effect, indicating that the addition of one species of fungus could increase the function by 0.016 (Figure 6d).

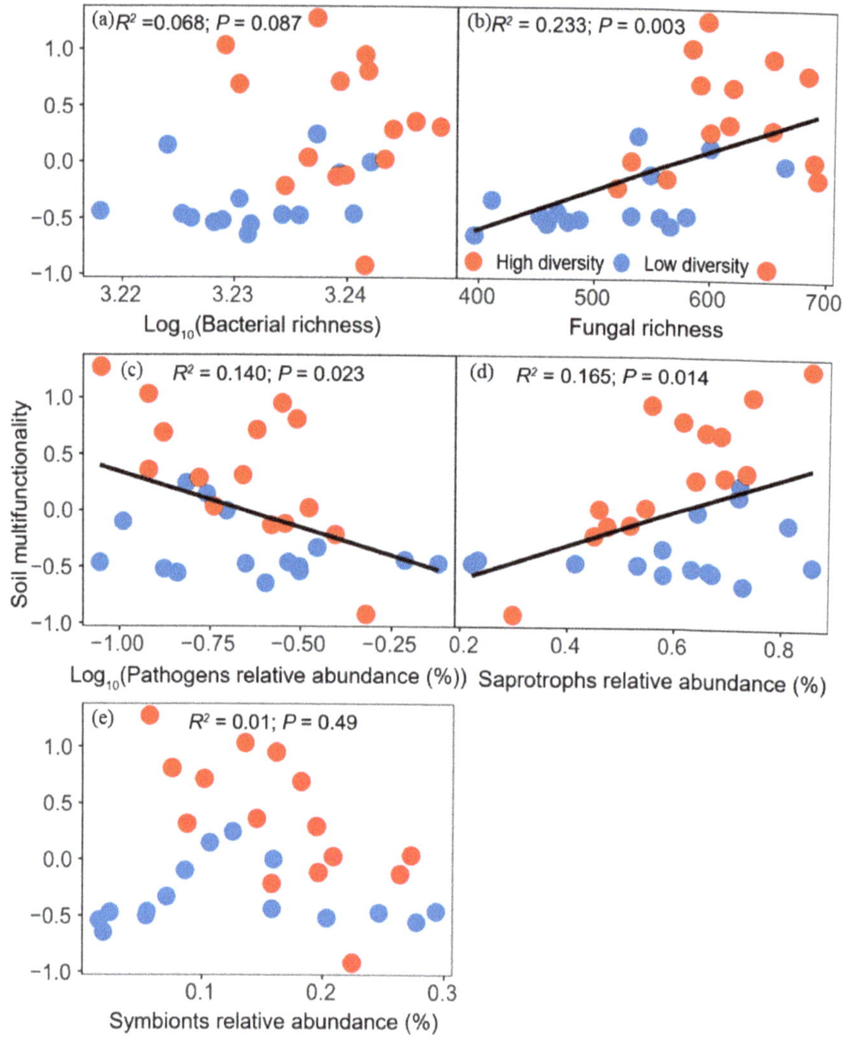

Figure 5. Relationships between soil multifunctionality and bacterial richness (**a**), fungal richness (**b**), pathogens relative abundance (**c**), saprotrophs relative abundance (**d**), and symbionts relative abundance (**e**). The black lines represent the fitted ordinary least squares (OLS) linear regressions. Red circles represent high plant diversity, while blue ones indicate low plant diversity.

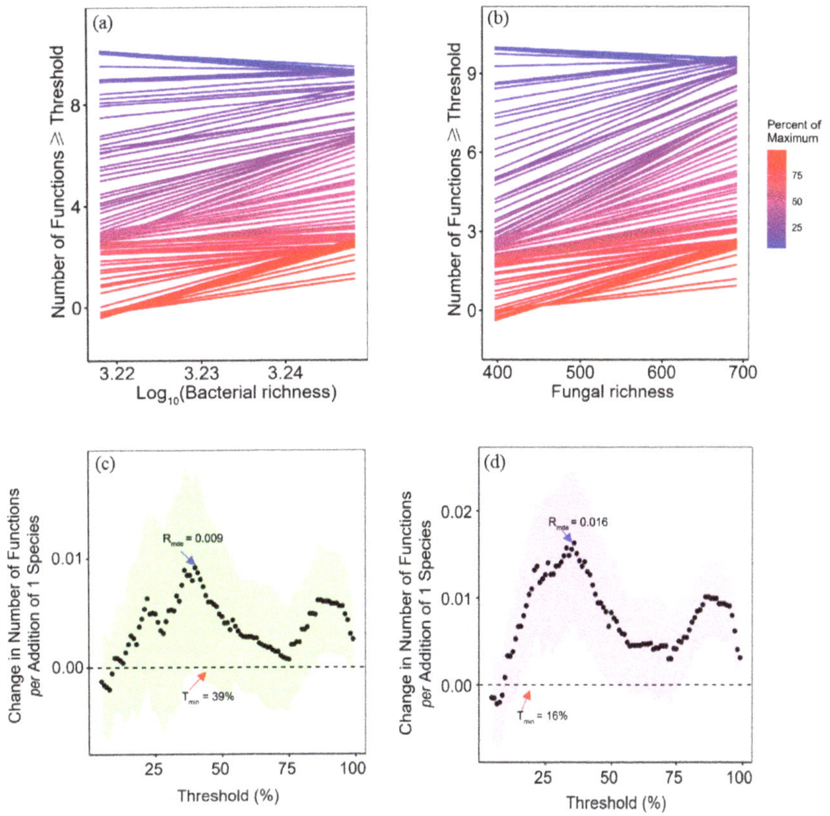

Figure 6. Effects of bacterial (**a**) and fungal (**b**) richness on the number of functions above thresholds. Lines represent the slope between soil microbial richness and the number of functions greater than or equal to a threshold value ranging from 5% to 99% of the maximum for each function. The dotted curves indicate the changes in the number of functions per unit increment of the richness of bacteria (**c**) and fungi (**d**). T_{min} is the minimum threshold that soil multifunctionality becomes influenced by changes in microbial richness, and R_{med} is the realized maximum effect of richness on soil multifunctionality.

3.4. Microbial Taxa Predicting Soil Multifunctionality

Bacterial rare taxa accounted for 81.64% of total OTUs and 40.90% of the relative abundance, respectively (Figure 7a). In comparison, abundant bacterial taxa accounted for 1.13% of total OTUs and 17.55% of the relative abundance, respectively (Figure 7a). In addition, bacterial conditionally rare and abundant taxa accounted for 0.11% of total OTUs and 0.67% of the relative abundance, respectively (Figure 7a). Bacterial moderate taxa accounted for 17.12% of total OTUs and 40.86% of the relative abundance, respectively (Figure 7a). Fungal rare taxa accounted for 81.66% of total OTUs and 23.93% of the relative abundance, respectively (Figure 7a). Fungal abundant taxa accounted for 0.91% of total OTUs and 11.52% of the relative abundance, respectively (Figure 7a). Finally, fungal conditionally rare and abundant taxa accounted for 17.28% of total OTUs and 64.02% of the relative abundance, respectively (Figure 7a). Fungal moderate taxa accounted for 0.15% of total OTUs and 0.54% of the relative abundance, respectively (Figure 7a). Random forest modeling results indicated that fungal ART (p = 0.012) was the most important microbial taxa in predicting soil multifunctionality (Figure 7b).

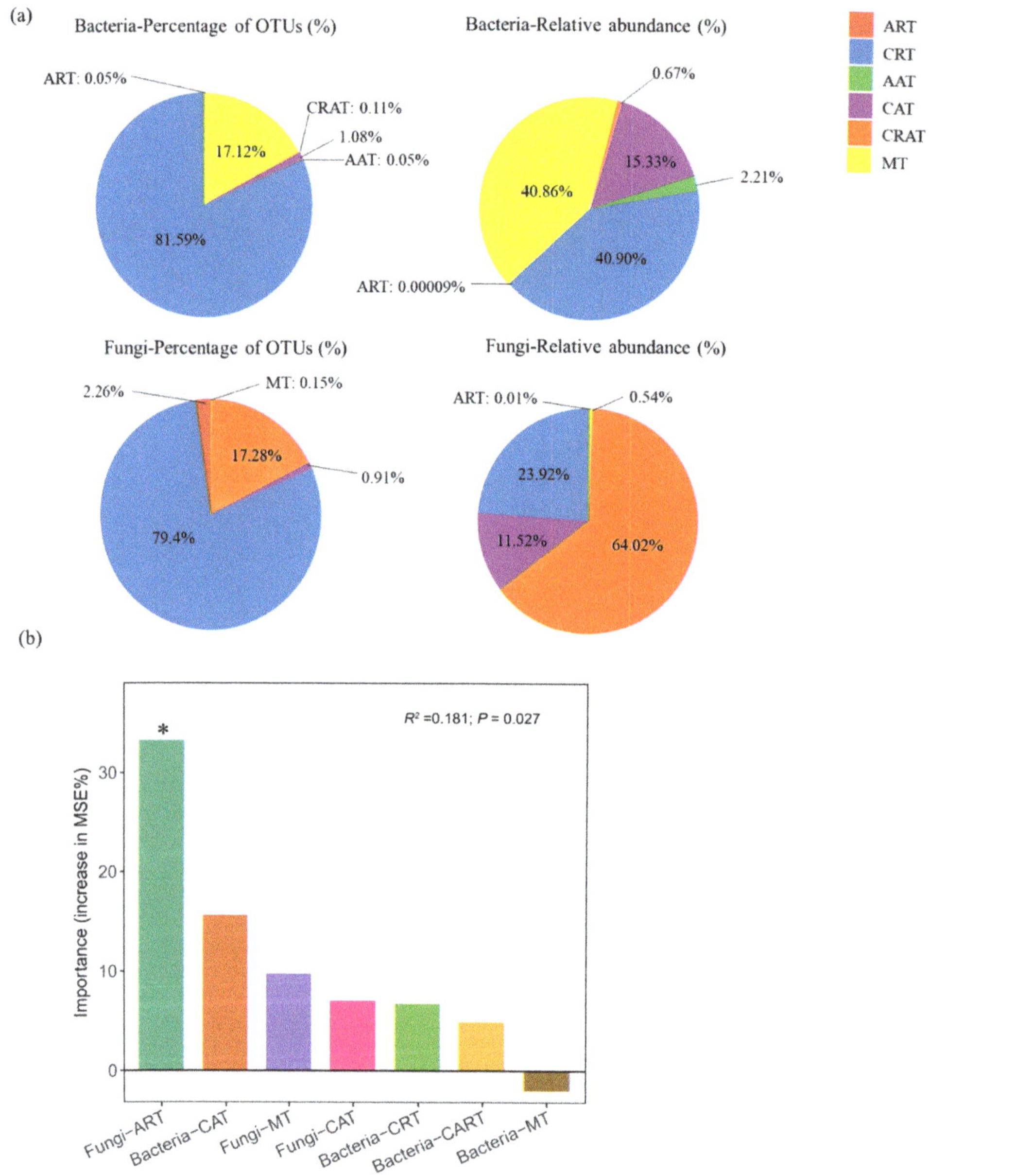

Figure 7. The percentage of OTUs and relative abundance of six categories of microbial taxa (**a**). The random forest regression model indicates the main microbial drivers of soil multifunctionality (**b**). MSE is the mean square error. This accuracy importance measure is computed for each tree and averaged over the forest (5000 trees) and determined by the increase in the MSE. * $p < 0.05$ on the bar shows that the associated microbial taxa have a significant effect on soil multifunctionality. ART: always rare taxa; CRT: conditionally rare taxa; AAT: always abundant taxa; CAT: conditionally abundant taxa; CRAT: conditionally rare and abundant taxa; MT: moderate taxa.

3.5. Direct and Indirect Effects of Plant Diversity on Soil Multifunctionality

The SEM analysis explained 53% of the total variation in soil multifunctionality (Figure 8a). Plant diversity indirectly affected soil multifunctionality by influencing the ratio of BGB to AGB, microbial C limitation, and fungal richness but not bacterial richness (Figure 8a and Figure S3). Plant diversity, the ratio of BGB to AGB, and fungal richness displayed positive effects on soil multifunctionality (Figure 8b). However, microbial C limitation exhibited a negative effect on soil multifunctionality (Figure 8b).

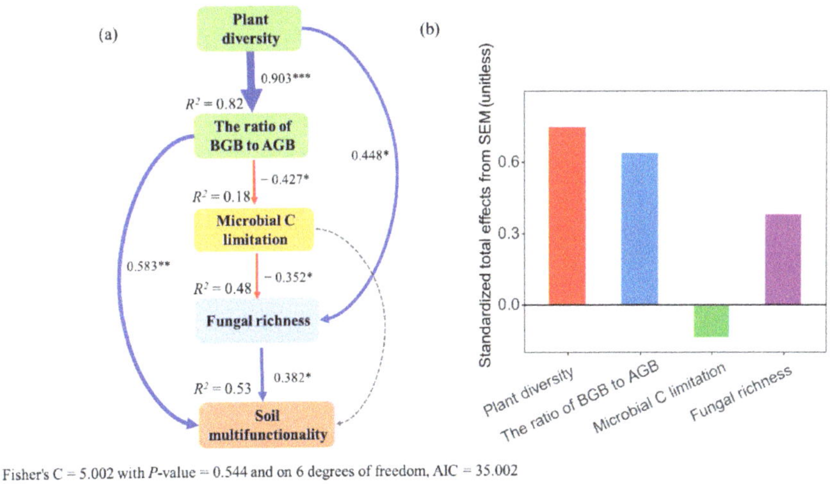

Figure 8. The piecewise structural equation model was used to test the direct and indirect relationships between plant diversity and soil multifunctionality (**a**). The corresponding values of the solid line arrows are the width of normalized path system arrows that reflect the size of the normalized path coefficient. The blue and red arrows show significant positive and negative correlations, respectively (* $p < 0.05$, ** $p < 0.01$, and *** $p < 0.001$). Dashed lines show non-significant relationships. The values above each variable represent the explanatory degree (R^2) of each variable in the model. Standardized total effects (direct plus indirect effects) derived from the piecewise structural equation model (**b**).

4. Discussion

Our study provides evidence that fungal richness rather than bacterial richness was significantly related to soil multifunctionality in a semi-arid grassland in northeast China. Saprotrophic fungi and rare fungal taxa were essential for maintaining soil functions. Furthermore, under high diversity plant assemblages, the changes in plant biomass allocation patterns increased plant below-ground biomass, which can alleviate microbial C limitation and thus enhance the fungal richness, ultimately promoting soil multifunctionality. Such results suggest that the above-ground and below-ground biodiversity, as well as rare fungal taxa, are vital to maintaining ecosystem functions in a semi-arid grassland ecosystem.

Soil microorganisms are some of the most sensitive components to precipitation change in semi-arid grasslands [69]. Therefore, biodiversity in semi-arid zones can be seriously compromised if rainfall changes. For example, reducing precipitation significantly increased fungal diversity [70]. However, one study showed a significant positive correlation between fungal diversity and precipitation [71]. Previous studies have suggested that bacterial communities are more sensitive to changes in precipitation than fungal communities in a semi-arid grassland [72]. However, we found that fungal richness and saprotrophic fungi were principal biotic factors in regulating soil multifunctionality in a semi-arid grassland. Fungi can create an environment around themselves by secreting polysaccharides from their hyphae to prevent dehydration [73]. Moreover, substrate diffusion restrictions might force the soil fungal mycelium network to expand, which is conducive to absorbing water and nutrients [74]. Therefore, fungi are more resistant to drought than bacteria, which may play an important role in arid and semi-arid grasslands [75]. For instance, soil fungi are crucial to organic matter decomposition, and root-associated fungi are important regulators of ecosystem C dynamics [22]. Moreover, fungal richness was significantly positively correlated with denitrification activity, indicating that fungi could promote soil N cycling [14]. Our results also showed that saprotrophic fungi significantly affected soil multifunctionality (Figure 5d). Saprotrophic fungi mainly grow throughout the soil litter interface and are involved in plant litter decomposition [76]. Previous studies have also suggested that saprotrophic fungi might affect soil C storage through interactions with ectomycorrhizal fungi [77]. Moreover, free-living saprotrophs generally play an essential ecological role in dead plant material because they can derive C by propagules or hyphae from dead organic material [78]. Saprotrophic fungi can also obtain fresh nutrients from recalcitrant organic polymers using extracellular enzymes and the nonenzymatic Fenton reaction [79,80]. For example, *Podospora anserina* is a potent plant biomass degrader and efficiently utilizes lignocellulose as a C source through dedicated lignin degradation enzymes [81,82]. Therefore, the ability of saprophytic fungi to translocate C resources implies that saprophytic fungi are essential agents of nutrient redistribution in soils [83]. Furthermore, the hyphal network formed by saprotrophic fungi is involved in the formation of soil aggregates, which is of great significance for soil water retention and erosion resistance [84]. Collectively, the present study suggested the non-negligible roles played by the fungal richness and saprophytic fungi in regulating soil functions.

An interesting result was that fungal ART was identified as the main driver of soil multifunctionality (Figure 7b). Previous studies have suggested that rare fungal community composition and functional guilds are more stable than those of the abundant taxa under certain conditions [85]. In other words, rare taxa contribute to maintaining the microbiome's function under environmental stress because some may be highly resistant to stress [86]. For example, a recent study found that rare microbial taxa might modulate the adverse effects of semi-arid grassland degradation drivers such as vegetation loss and eutrophication on soil organic matter decomposition [24]. Furthermore, rare taxa contributed more to soil C and N cycling and crop yield than the abundant taxa [25]. Thus, rare microbial taxa play a unique role in maintaining ecosystem functions [28,87]. However, our understanding of rare microbial taxa is still preliminary, and more attention should be given to rare microbial taxa in future studies of biodiversity and ecosystem multifunctionality.

As expected, in line with previous studies, high plant diversity increased soil multifunctionality as compared with low plant diversity [4,56,88]. However, our study elucidated the underlying mechanisms by which high plant diversity could enhance soil multifunctionality through fungal richness. According to current knowledge [89], the possible reason was that the competition for below-ground resources was less than that for above-ground resources in low diversity plant assemblages. Therefore, low diversity plant assemblages might increase the allocation of above-ground biomass to compete for light [89]. Although above-ground biomass was expected to be associated with litter production, below-ground C from rhizodeposition was vital for soil microbial communities [90]. C decomposes from above-ground plant litter to the soil surface; thus, it is not readily available to microorganisms inside the soil [36]. Therefore, a low diversity plant assemblage might intensify microbial C limitation [91], mainly by changes in plant-root-derived substrates [37]. Our results indicated that microbial C limitation significantly affected soil fungal richness but not bacterial richness (Figure 4a,b). A reasonable explanation might be that members of the fungal community are the first consumers of below-ground plant-derived C inputs to the soil [24,92]. Thus, intensifying microbial C limitation reduced soil fungal richness and ultimately decreased soil multifunctionality in the low diversity plant assemblage (Figure 9). However, high diversity plant assemblages increased below-ground biomass allocation (Figure 2a). A possible explanation for the increase in below-ground biomass may be that the interactions between different plant species roots may affect biomass allocation patterns, i.e., interspecific competition can increase plant below-ground biomass [93]. The total soil organic carbon content under high plant diversity was higher than that under low plant diversity in this study area (Figure S3g). Although this study did not test the contents of soil organic matter mineralization, previous literature has shown that the quantity and quality of soil organic matter are the main driving factors affecting microorganisms [94]. The increasing plant below-ground biomass may improve the quantity and quality of soil organic matter in semi-arid zones [32], and the increased soil C resources provide rich substrates and energy for microorganism growth and basal metabolism [7], thus increasing fungal diversity. In this study, the decreased microbial C limitation favored fungal richness under high plant diversity, thus promoting soil multifunctionality (Figure 9). Furthermore, our results indicate that semi-arid grassland provides essential ecosystem services such as carbon sequestration, climate regulation, and biodiversity protection under high diversity plant assemblages. Our findings suggest that high diversity plant assemblages make semi-arid grassland ecosystems more efficient in regulating ecological processes, which local stakeholders or grassland conservation agencies want. Therefore, ecosystem services through the TESSA methodology (http://tessa.tools/) (accessed on 1 June 2021) [95] should contain future multifunctionality studies. The study of ecosystem services multifunctionality can provide important enlightenment for ecological protection and sustainable development under the increasing pressure of human activities and climate changes. In summary, this study revealed the mechanism of changes in soil multifunctionality in which plant-diversity-induced alterations of microbial C limitation regulate soil multifunctionality by affecting soil fungal richness in a semi-arid grassland.

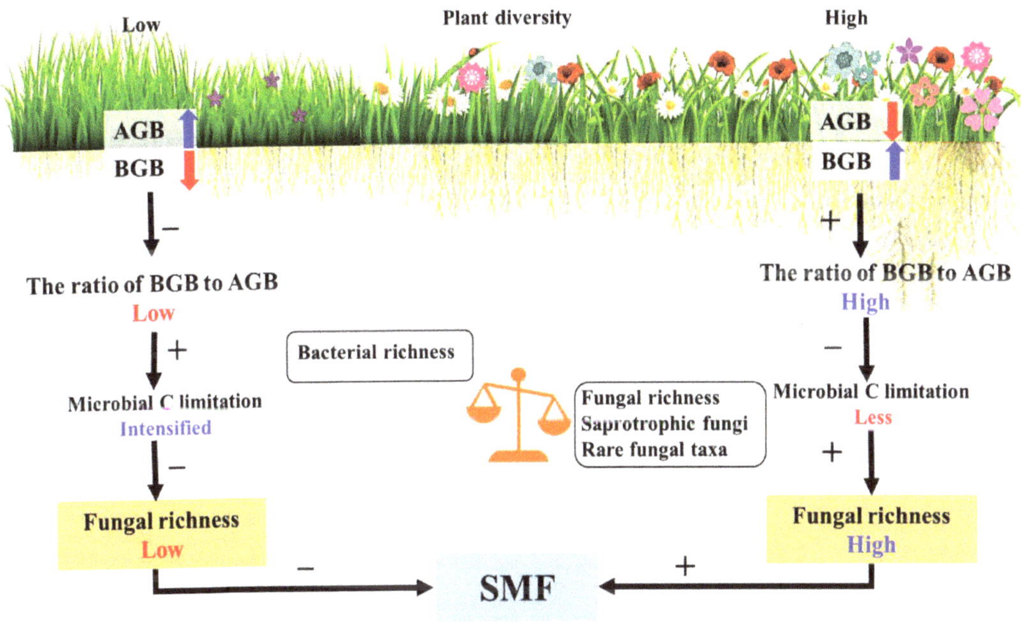

Figure 9. A conceptual framework for understanding the effects of plant and fungal richness on soil multifunctionality in a semi-arid grassland. Changing plant biomass allocation patterns increased the ratio of plant below-ground biomass to above-ground biomass under high diversity plant assemblages, which can alleviate microbial carbon (C) limitation and thus enhance the fungal richness, finally promoting soil multifunctionality. The fungal richness was positively related to soil multifunctionality, but the bacterial richness was not. Saprotrophic fungi and rare fungal taxa were essential for maintaining the soil functions. The blue arrow represents an increase, and the red arrow represents a decrease. + and − describe promotion and inhibition effects. AGB: plant above-ground biomass; BGB: plant below-ground biomass; SMF: soil multifunctionality.

5. Conclusions

Our results demonstrated that the positive effect of plant diversity on soil multifunctionality was mainly due to the high plant diversity increasing plant below-ground biomass allocation, which alleviated microbial C limitation and favored fungal richness, finally promoting soil multifunctionality. Additionality, the high diversity of plant assemblages enhanced ecosystem services for semi-arid grasslands protection, and plant-fungus relationships were important to improve the assessment of ecosystem services. Instead of bacterial richness, the fungal richness was the crucial biotic predictor of soil multifunctionality in a semi-arid grassland. Furthermore, saprotrophic fungi and rare fungal taxa were major drivers of soil multifunctionality. In conclusion, this study provides a new perspective for evaluating the relative roles of fungal and bacterial diversity and biomass allocation patterns in maintaining soil functions in the context of global plant diversity loss and has important implications for biodiversity conservation and sustainable development in semi-arid grasslands. Ecosystem multifunctionality and ecosystem services multifunctionality should be popularized and applied in future multifunctionality studies.

Supplementary Materials: The following supporting information can be downloaded at: https://www.mdpi.com/article/10.3390/biology11060870/s1. Table S1: The plant species composition of the selected low diversity with a single species and high diversity with multiple species assemblages in the Songnen grassland, China; Table S2: Variables used to estimate soil multifunctionality and their importance; Figure S1: Principal component analysis (PCA) to determine the difference of soil characteristics between low diversity plant assemblages and high diversity plant assemblages. The soil characteristics included soil water content, soil pH, soil electrical conductivity, soil available nitrogen, total soil nitrogen, total soil phosphorus, soil total organic carbon, and soil C/N; Figure S2: *Prior* structural equation models (SEM) of hypothetical relationships between plant diversity and soil multifunctionality (a). Models were accepted for the significant Fisher's C test ($p > 0.05$). Among significant models, the one with the lowest Akaike information criterion (AIC) was selected for the final SEM analysis (b–d); Figure S3: Soil water content (SWC) (a), soil pH (b), soil electrical conductivity (EC) (c), soil available nitrogen (AN) (d), soil total nitrogen (TN) (e), soil total phosphorus (TP) (f), soil total organic carbon (TOC) (g), and soil C/N (h) in response to low and high plant diversities. * $p < 0.05$, ** $p < 0.01$, and *** $p < 0.001$ (t-test); Figure S4: Rarefaction curves for the soil bacterial (a) and fungal (b) communities at 97% sequence similarity in different plant diversity assemblages. Low diversity assemblages including single species *Carex duriuscula* C.A.Mey. (CD), *Lespedeza hedysaroides* (Pall.) Kitag. (LH), and *Calamagrostis rigidula* A.I.Baranov and Skvortsov (CR) assemblages. High diversity assemblages including multiple species assemblages with the dominant *Lespedeza daurica* (Laxm.) Schindl. (LD.ass), *Leymus chinensis* (Trin.) Tzvelev (LC.ass), and *Hierochloe glabra* Trin. (HG.ass); Figure S5: The relative abundance of bacterial and fungal communities at the phylum level. "Others" represents all phyla with relative abundance < 1%. ** $p < 0.01$, and *** $p < 0.001$ (t-test); Figure S6: Fungal (a) and bacterial richness (b) in response to plant diversity. *** $p < 0.001$ (t-test); Figure S7: The relationship between fungal (a–d) and bacterial richness (e–h) and soil multifunctionality, at four different thresholds 25%, 50%, 75%, and 90% of maximum. The black lines represent the fitted ordinary least squares (OLS) linear regressions. Red circles represent high plant diversity, while blue ones indicate low plant diversity.

Author Contributions: Conceptualization, Z.L. and F.X.; Data curation, Z.L.; Formal analysis, F.X.; Funding acquisition, F.X.; Investigation, Z.L. and F.X.; Methodology, Z.L.; Project administration, F.X.; Resources, Z.L.; Software, Z.L.; Supervision, Z.L. and M.Z.; Validation, Z.L.; Visualization, Z.L.; Writing—original draft, Z.L.; Writing—review and editing, Z.L., X.L. and F.X. All authors have read and agreed to the published version of the manuscript.

Funding: This research was financially supported by the National Natural Science Foundation of China (31670524).

Institutional Review Board Statement: Not applicable.

Informed Consent Statement: Not applicable.

Data Availability Statement: The data supporting this study's findings are available from the corresponding author upon reasonable request.

Acknowledgments: We gratefully acknowledge the staff of the Tongyu Observatory of semi-arid climate and environment for help with the logistics and permission to access the study site. Furthermore, we thank Yu Bi for her field and lab work support.

Conflicts of Interest: The authors declare no conflict of interest.

References

1. Manning, P.; van der Plas, F.; Soliveres, S.; Allan, E.; Maestre, F.T.; Mace, G.; Whittingham, M.J.; Fischer, M. Redefining ecosystem multifunctionality. *Nat. Ecol. Evol.* **2018**, *2*, 1515. [CrossRef]
2. Wang, L.; Delgado-Baquerizo, M.; Wang, D.L.; Isbell, F.; Liu, J.; Feng, C.; Liu, J.S.; Zhong, Z.W.; Zhu, H.; Yuan, X.; et al. Diversifying livestock promotes multidiversity and multifunctionality in managed grasslands. *Proc. Natl. Acad. Sci. USA* **2019**, *116*, 6187–6192. [CrossRef]
3. Turco, R.; Kennedy, A.; Jawson, M. Microbial indicators of soil quality. *Defin. Soil Qual. A Sustain. Environ.* **1994**, *35*, 73–90. [CrossRef]
4. Xu, Y.J.; Dong, K.; Jiang, M.; Liu, Y.L.; He, L.Y.; Wang, J.L.; Zhao, N.X.; Gao, Y.B. Soil moisture and species richness interactively affect multiple ecosystem functions in a microcosm experiment of simulated shrub encroached grasslands. *Sci. Total Environ.* **2022**, *803*, 149950. [CrossRef] [PubMed]
5. Troudet, J.; Grandcolas, P.; Blin, A.; Vignes-Lebbe, R.; Legendre, F. Taxonomic bias in biodiversity data and societal preferences. *Sci. Rep.* **2017**, *7*, 1–14. [CrossRef] [PubMed]
6. Zhang, R.; Wang, Z.; Niu, S.; Tian, D.; Wu, Q.; Gao, X.; Schellenberg, M.P.; Han, G. Diversity of plant and soil microbes mediates the response of ecosystem multifunctionality to grazing disturbance. *Sci. Total Environ.* **2021**, *776*, 145730. [CrossRef] [PubMed]
7. Shen, C.; Wang, J.; Jing, Z.; Qiao, N.-H.; Xiong, C.; Ge, Y. Plant diversity enhances soil fungal network stability indirectly through the increase of soil carbon and fungal keystone taxa richness. *Sci. Total Environ.* **2021**, *818*, 151737. [CrossRef]
8. Aira, M.; Gomez-Brandon, M.; Lazcano, C.; Baath, E.; Dominguez, J. Plant genotype strongly modifies the structure and growth of maize rhizosphere microbial communities. *Soil Biol. Biochem.* **2010**, *42*, 2276–2281. [CrossRef]
9. Cardinale, B.J. Biodiversity loss and its impact on humanity. *Nature* **2012**, *489*, 326. [CrossRef]
10. Shi, Y.J.; Wang, J.F.; Ao, Y.N.; Han, J.Y.; Guo, Z.H.; Liu, X.Y.; Zhang, J.W.; Mu, C.S.; Le Roux, X. Responses of soil N_2O emissions and their abiotic and biotic drivers to altered rainfall regimes and co-occurring wet N deposition in a semi-arid grassland. *Glob. Chang. Biol.* **2021**, *27*, 4894–4908. [CrossRef] [PubMed]
11. Wang, X.X.; Dong, S.K.; Yang, B.; Li, Y.Y.; Su, X.K. The effects of grassland degradation on plant diversity, primary productivity, and soil fertility in the alpine region of Asia's headwaters. *Environ. Monit. Assess.* **2014**, *186*, 6903–6917. [CrossRef] [PubMed]
12. Bardgett, R.D.; van der Putten, W.H. Belowground biodiversity and ecosystem functioning. *Nature* **2014**, *515*, 505–511. [CrossRef]
13. van der Heijden, M.G.A.; Bardgett, R.D.; van Straalen, N.M. The unseen majority: Soil microbes as drivers of plant diversity and productivity in terrestrial ecosystems. *Ecol. Lett.* **2008**, *11*, 296–310. [CrossRef] [PubMed]
14. Li, J.; Delgado-Baquerizo, M.; Wang, J.T.; Hu, H.W.; Cai, Z.J.; Zhu, Y.N.; Singh, B.K. Fungal richness contributes to multifunctionality in boreal forest soil. *Soil Biol. Biochem.* **2019**, *136*, 107526. [CrossRef]
15. Luo, G.W.; Rensing, C.; Chen, H.; Liu, M.Q.; Wang, M.; Guo, S.W.; Ling, N.; Shen, Q.R. Deciphering the associations between soil microbial diversity and ecosystem multifunctionality driven by long-term fertilization management. *Funct. Ecol.* **2018**, *32*, 1103–1116. [CrossRef]
16. Bona, E.; Massa, N.; Toumatia, O.; Novello, G.; Cesaro, P.; Todeschini, V.; Boatti, L.; Mignone, F.; Titouah, H.; Zitouni, A.; et al. Climatic zone and soil properties determine the biodiversity of the soil bacterial communities associated to native plants from desert areas of north-central algeria. *Microorganisms* **2021**, *9*, 1359. [CrossRef]
17. Zeng, Q.C.; An, S.S. Identifying the biogeographic patterns of rare and abundant bacterial communities using different primer sets on the loess plateau. *Microorganisms* **2021**, *9*, 139. [CrossRef]
18. Six, J.; Frey, S.D.; Thiet, R.K.; Batten, K.M. Bacterial and fungal contributions to carbon sequestration in agroecosystems. *Soil Ence Soc. Am. J.* **2006**, *70*, 555–569. [CrossRef]
19. Ma, W.B.; Jiang, S.J.; Assemien, F.; Qin, M.S.; Ma, B.B.; Xie, Z.; Liu, Y.J.; Feng, H.Y.; Du, G.Z.; Ma, X.J.; et al. Response of microbial functional groups involved in soil N cycle to N, P and NP fertilization in Tibetan alpine meadows. *Soil Biol. Biochem.* **2016**, *101*, 195–206. [CrossRef]
20. Li, H.; Yang, S.; Xu, Z.W.; Yan, Q.Y.; Li, X.B.; van Nostrand, J.D.; He, Z.L.; Yao, F.; Han, X.G.; Zhou, J.Z.; et al. Responses of soil microbial functional genes to global changes are indirectly influenced by aboveground plant biomass variation. *Soil Biol. Biochem.* **2017**, *104*, 18–29. [CrossRef]
21. Yang, K.N.; Luo, S.W.; Hu, L.G.; Chen, B.B.; Xie, Z.; Ma, B.B.; Ma, W.B.; Du, G.Z.; Ma, X.J.; Le Roux, X. Responses of soil ammonia-oxidizing bacteria and archaea diversity to N, P and NP fertilization: Relationships with soil environmental variables and plant community diversity. *Soil Biol. Biochem.* **2020**, *145*, 107795. [CrossRef]
22. Clemmensen, K.E.; Bahr, A.; Ovaskainen, O.; Dahlberg, A.; Ekblad, A.; Wallander, H.; Stenlid, J.; Finlay, R.D.; Wardle, D.A.; Lindahl, B.D. Roots and associated fungi drive long-term carbon sequestration in boreal forest. *Science* **2013**, *339*, 1615–1618. [CrossRef] [PubMed]

23. Crowther, T.W.; Boddy, L.; Jones, T.H. Functional and ecological consequences of saprotrophic fungus–grazer interactions. *ISME J.* **2012**, *6*, 1992–2001. [CrossRef] [PubMed]
24. Wu, Y.; Chen, D.M.; Saleem, M.; Wang, B.; Hu, S.J.; Delgado-Baquerizo, M.; Bai, Y.F. Rare soil microbial taxa regulate the negative effects of land degradation drivers on soil organic matter decomposition. *J. Appl. Ecol.* **2021**, *58*, 1658–1669. [CrossRef]
25. Xiong, C.; He, J.Z.; Singh, B.K.; Zhu, Y.G.; Wang, J.T.; Li, P.P.; Zhang, Q.B.; Han, L.L.; Shen, J.P.; Ge, A.H.; et al. Rare taxa maintain the stability of crop mycobiomes and ecosystem functions. *Environ. Microbiol.* **2021**, *23*, 1907–1924. [CrossRef]
26. Xue, Y.Y.; Chen, H.H.; Yang, J.R.; Liu, M.; Huang, B.Q.; Yang, J. Distinct patterns and processes of abundant and rare eukaryotic plankton communities following a reservoir cyanobacterial bloom. *ISME J.* **2018**, *12*, 2263–2277. [CrossRef] [PubMed]
27. Xue, M.D.; Guo, Z.Y.; Gu, X.Y.; Gao, H.L.; Weng, S.M.; Zhou, J.; Gu, D.G.; Lu, H.X.; Zhou, X.Q. Rare rather than abundant microbial communities drive the effects of long-term greenhouse cultivation on ecosystem functions in subtropical agricultural soils. *Sci. Total Environ.* **2020**, *706*, 136004. [CrossRef]
28. Shade, A.; Jones, S.E.; Caporaso, J.G.; Handelsman, J.; Knight, R.; Fierer, N.; Gilbert, J.A. Conditionally rare taxa disproportionately contribute to temporal changes in microbial diversity. *mBio* **2014**, *5*, e01371-14. [CrossRef]
29. Wardle, D.A.; Bardgett, R.D.; Callaway, R.M.; Van der Putten, W.H. Terrestrial ecosystem responses to species gains and losses. *Science* **2011**, *332*, 1273–1277. [CrossRef]
30. Peng, S.S.; Chen, A.P.; Xu, L.; Cao, C.X.; Fang, J.Y.; Myneni, R.B.; Pinzon, J.E.; Tucker, C.J.; Piao, S.L. Recent change of vegetation growth trend in China. *Environ. Res. Lett.* **2011**, *6*, 044027. [CrossRef]
31. Chen, H.; Tang, J.J.; Sun, X.B.; Ma, K.Y.; Chen, H.H.; Li, D.J. Topography modulates effects of nitrogen deposition on microbial resource limitation in a nitrogen-saturated subtropical forest. *For. Ecosyst.* **2021**, *8*, 1–19. [CrossRef]
32. Li, G.Y.; Cai, J.T.; Song, X.X.; Pan, X.B.; Pan, D.F.; Jiang, S.C.; Sun, J.Y.; Zhang, M.N.; Wang, L. Herbivore grazing mitigates the negative effects of nitrogen deposition on soil organic carbon in low-diversity grassland. *J. Appl. Ecol.* **2021**, *59*, 483–491. [CrossRef]
33. Demoling, F.; Figueroa, D.; Bååth, E. Comparison of factors limiting bacterial growth in different soils. *Soil Biol. Biochem.* **2007**, *39*, 2485–2495. [CrossRef]
34. Wang, J.; Wang, X.T.; Liu, G.B.; Zhang, C.; Wang, G.L. Bacterial richness is negatively related to potential soil multifunctionality in a degraded alpine meadow. *Ecol. Indic.* **2021**, *121*, 106996. [CrossRef]
35. Jing, X.; Chen, X.; Fang, J.Y.; Ji, C.J.; Shen, H.H.; Zheng, C.Y.; Zhu, B. Soil microbial carbon and nutrient constraints are driven more by climate and soil physicochemical properties than by nutrient addition in forest ecosystems. *Soil Biol. Biochem.* **2020**, *141*, 107657. [CrossRef]
36. Ning, Q.S.; Hattenschwiler, S.; Lu, X.T.; Kardol, P.; Zhang, Y.H.; Wei, C.Z.; Xu, C.Y.; Huang, J.H.; Li, A.; Yang, J.; et al. Carbon limitation overrides acidification in mediating soil microbial activity to nitrogen enrichment in a temperate grassland. *Glob. Chang. Biol.* **2021**, *27*, 5976–5988. [CrossRef]
37. Geisseler, D.; Lazicki, P.A.; Scow, K.M. Mineral nitrogen input decreases microbial biomass in soils under grasslands but not annual crops. *Appl. Soil Ecol.* **2016**, *106*, 1–10. [CrossRef]
38. Schleuss, P.M.; Widdig, M.; Heintz-Buschart, A.; Guhr, A.; Martin, S.; Kirkman, K.; Spohn, M. Stoichiometric controls of soil carbon and nitrogen cycling after long-term nitrogen and phosphorus addition in a mesic grassland in South Africa. *Soil Biol. Biochem.* **2019**, *135*, 294–303. [CrossRef]
39. Song, X.X.; Wang, L.; Zhao, X.; Liu, C.; Chang, Q.; Wang, Y.; Xu, T.T.; Wang, D.L. Sheep grazing and local community diversity interact to control litter decomposition of dominant species in grassland ecosystem. *Soil Biol. Biochem.* **2017**, *115*, 364–370. [CrossRef]
40. Rillig, M.C.; Wright, S.F.; Eviner, V.T. The role of arbuscular mycorrhizal fungi and glomalin in soil aggregation: Comparing effects of five plant species. *Plant. Soil* **2002**, *238*, 325–333. [CrossRef]
41. Cui, H.Y.; Sun, W.; Delgado-Baquerizo, M.; Song, W.Z.; Ma, J.Y.; Wang, K.Y.; Ling, X.L. Cascading effects of N fertilization activate biologically driven mechanisms promoting P availability in a semi-arid grassland ecosystem. *Funct. Ecol.* **2021**, *35*, 1001–1011. [CrossRef]
42. Yang, Y.R.; Chen, S.Y.; Wu, X.F.; Syed, S.I.; Syed, I.U.S.; Huang, B.T.; Guan, P.T.; Wang, D.L. Grazing affects bacterial and fungal diversities and communities in the rhizosphere and endosphere compartments of leymus chinensis through regulating nutrient and ion distribution. *Microorganisms* **2021**, *9*, 476. [CrossRef] [PubMed]
43. Han, Z.Q.; Xu, P.S.; Li, Z.T.; Lin, H.Y.; Zhu, C.; Wang, J.Y.; Zou, J.W. Microbial diversity and the abundance of keystone species drive the response of soil multifunctionality to organic substitution and biochar amendment in a tea plantation. *Glob. Chang. Biol. Bioenergy* **2022**, *14*, 481–495. [CrossRef]
44. Sommers, L.; Nelson, D. Determination of total phosphorus in soils: A rapid perchloric acid digestion procedure. *Soil Sci. Soc. Am. J.* **1972**, *36*, 902–904. [CrossRef]
45. Olsen, S.R.; Watanabe, F.S.; Cosper, H.R.; Larson, W.; Nelson, L. Residual phosphorus availability in long-time rotations on calcareous soils. *Soil Sci.* **1954**, *78*, 141–152. [CrossRef]
46. German, D.P.; Weintraub, M.N.; Grandy, A.S.; Lauber, C.L.; Rinkes, Z.L.; Allison, S.D. Optimization of hydrolytic and oxidative enzyme methods for ecosystem studies. *Soil Biol. Biochem.* **2011**, *43*, 1387–1397. [CrossRef]

47. Cui, H.Y.; Sun, W.; Delgado-Baquerizo, M.; Song, W.Z.; Ma, J.Y.; Wang, K.Y.; Ling, X.L. The effects of mowing and multi-level N fertilization on soil bacterial and fungal communities in a semiarid grassland are year-dependent. *Soil Biol. Biochem.* **2020**, *151*, 108040. [CrossRef]
48. Amato, K.R.; Yeoman, C.J.; Kent, A.; Righini, N.; Carbonero, F.; Estrada, A.; Gaskins, H.R.; Stumpf, R.M.; Yildirim, S.; Torralba, M. Habitat degradation impacts black howler monkey (Alouatta pigra) gastrointestinal microbiomes. *ISME J.* **2013**, *7*, 1344–1353. [CrossRef]
49. Fuhrman, J.A. Microbial community structure and its functional implications. *Nature* **2009**, *459*, 193–199. [CrossRef] [PubMed]
50. Li, H.Q.; Li, H.; Zhou, X.Y.; Shen, Y.J.; Su, J.Q. Distinct patterns of abundant and rare subcommunities in paddy soil during wetting-drying cycles. *Sci. Total Environ.* **2021**, *785*, 147298. [CrossRef] [PubMed]
51. Cui, H.Y.; Sun, W.; Delgado-Baquerizo, M.; Song, W.Z.; Ma, J.Y.; Wang, K.Y.; Ling, X.L. Phosphorus addition regulates the responses of soil multifunctionality to nitrogen over-fertilization in a temperate grassland. *Plant Soil* **2020**, *473*, 73–87. [CrossRef]
52. Delgado-Baquerizo, M.; Eldridge, D.J.; Ochoa, V.; Gozalo, B.; Singh, B.K.; Maestre, F.T. Soil microbial communities drive the resistance of ecosystem multifunctionality to global change in drylands across the globe. *Ecol. Lett.* **2017**, *20*, 1295–1305. [CrossRef]
53. Delgado-Baquerizo, M.; Maestre, F.T.; Reich, P.B.; Jeffries, T.C.; Gaitan, J.J.; Encinar, D.; Berdugo, M.; Campbell, C.D.; Singh, B.K. Microbial diversity drives multifunctionality in terrestrial ecosystems. *Nat. Commun.* **2016**, *7*, 10541. [CrossRef]
54. Delgado-Baquerizo, M.; Reich, P.B.; Trivedi, C.; Eldridge, D.J.; Abades, S.; Alfaro, F.D.; Bastida, F.; Berhe, A.A.; Cutler, N.A.; Gallardo, A.; et al. Multiple elements of soil biodiversity drive ecosystem functions across biomes. *Nat. Ecol. Evol.* **2020**, *4*, 210–220. [CrossRef] [PubMed]
55. Shi, X.Z.; Wang, J.Q.; Lucas-Borja, M.E.; Wang, Z.Y.; Li, X.; Huang, Z.Q. Microbial diversity regulates ecosystem multifunctionality during natural secondary succession. *J. Appl. Ecol.* **2021**, *58*, 2833–2842. [CrossRef]
56. Maestre, F.T.; Quero, J.L.; Gotelli, N.J.; Escudero, A.; Ochoa, V.; Delgado-Baquerizo, M.; Garcia-Gomez, M.; Bowker, M.A.; Soliveres, S.; Escolar, C.; et al. Plant species richness and ecosystem multifunctionality in global drylands. *Science* **2012**, *335*, 214–218. [CrossRef]
57. Li, J.; Cui, L.; Delgado-Baquerizo, M.; Wang, J.; Zhu, Y.; Wang, R.; Li, W.; Lei, Y.; Zhai, X.; Zhao, X. Fungi drive soil multifunctionality in the coastal salt marsh ecosystem. *Sci. Total Environ.* **2022**, *818*, 151673. [CrossRef]
58. Byrnes, J.E.K.; Gamfeldt, L.; Isbell, F.; Lefcheck, J.S.; Griffin, J.N.; Hector, A.; Cardinale, B.J.; Hooper, D.U.; Dee, L.E.; Duffy, J.E. Investigating the relationship between biodiversity and ecosystem multifunctionality: Challenges and solutions. *Methods Ecol. Evol.* **2014**, *5*, 111–124. [CrossRef]
59. Moorhead, D.L.; Rinkes, Z.L.; Sinsabaugh, R.L.; Weintraub, M.N. Dynamic relationships between microbial biomass, respiration, inorganic nutrients and enzyme activities: Informing enzyme-based decomposition models. *Front. Microbiol.* **2013**, *4*, 223. [CrossRef] [PubMed]
60. Oksanen, J.; Blanchet, F.G.; Kindt, R.; Legendre, P.; Wagner, H. vegan: Community ecology package version 2.0–10 [Software]. *J. Stat. Softw.* **2013**, *2*, 1–295.
61. Jiao, S.; Lu, Y.; Wei, G. Soil multitrophic network complexity enhances the link between biodiversity and multifunctionality in agricultural systems. *Glob. Change Biol.* **2022**, *28*, 140–153. [CrossRef] [PubMed]
62. Nguyen, N.H.; Song, Z.W.; Bates, S.T.; Branco, S.; Tedersoo, L.; Menke, J.; Schilling, J.S.; Kennedy, P.G. FUNGuild: An open annotation tool for parsing fungal community datasets by ecological guild. *Fungal Ecol.* **2016**, *20*, 241–248. [CrossRef]
63. Liaw, A.; Wiener, M. Classification and Regression by randomForest. *R News* **2002**, *2*, 18–22.
64. Fortmann-Roe, S. Consistent and clear reporting of results from diverse modeling techniques: The A3 method. *J. Stat. Softw.* **2015**, *66*, 1–23. [CrossRef]
65. Lefcheck, J.S. piecewiseSEM: Piecewise structural equation modeling in R for ecology, evolution, and systematics. *Methods Ecol. Evol.* **2016**, *7*, 573–579. [CrossRef]
66. Bates, D.M.; Maechler, M.; Bolker, B.M.; Walker, S.C. Package lme4: Linear mixed-effects models using Eigen and S4. *J. Stat. Softw.* **2014**, *67*. Available online: https://cran.r-project.org/web/packages/lme4/lme4.pdf (accessed on 1 May 2022).
67. Delgado-Baquerizo, M.; Trivedi, P.; Trivedi, C.; Eldridge, D.J.; Reich, P.B.; Jeffries, T.C.; Singh, B.K. Microbial richness and composition independently drive soil multifunctionality. *Funct. Ecol.* **2017**, *31*, 2330–2343. [CrossRef]
68. Burnham, K.P.; Anderson, D.R. A practical information-theoretic approach. *Model. Sel. Multimodel Inference* **2002**, *2*, 70–71.
69. Chen, H.; Zhao, X.R.; Lin, Q.M.; Li, G.T.; Kong, W.D. Using a combination of PLFA and DNA-based sequencing analyses to detect shifts in the soil microbial community composition after a simulated spring precipitation in a semi-arid grassland in China. *Sci. Total Environ.* **2019**, *657*, 1237–1245. [CrossRef]
70. Wang, H.; Ta, N.; Jin, K.; Ji, B.M.; Schellenberg, M.P.; Wei, Z.J.; Wang, Z. Interactive effects of nitrogen fertilizer and altered precipitation on fungal communities in arid grasslands of northern China. *J. Soils Sediments* **2020**, *20*, 1344–1356. [CrossRef]
71. Wang, D.; Rui, Y.C.; Ding, K.; Cui, X.Y.; Hao, Y.B.; Tang, L.; Pang, Z.; Zhang, B.; Zhou, S.T.; Wang, K.; et al. Precipitation drives the biogeographic distribution of soil fungal community in Inner Mongolian temperate grasslands. *J. Soils Sediments* **2018**, *18*, 222–228. [CrossRef]
72. Yang, X.C.; Zhu, K.; Loik, M.E.; Sun, W. Differential responses of soil bacteria and fungi to altered precipitation in a meadow steppe. *Geoderma* **2021**, *384*, 114812. [CrossRef]
73. Borken, W.; Matzner, E. Reappraisal of drying and wetting effects on C and N mineralization and fluxes in soils. *Glob. Chang. Biol.* **2009**, *15*, 808–824. [CrossRef]

74. Barnard, R.L.; Osborne, C.A.; Firestone, M.K. Responses of soil bacterial and fungal communities to extreme desiccation and rewetting. *Isme J.* **2013**, *7*, 2229–2241. [CrossRef]
75. Austin, A.T.; Yahdjian, L.; Stark, J.M.; Belnap, J.; Porporato, A.; Norton, U.; Ravetta, D.A.; Schaeffer, S.M. Water pulses and biogeochemical cycles in arid and semiarid ecosystems. *Oecologia* **2004**, *141*, 221–235. [CrossRef]
76. Snajdr, J.; Valaskova, V.; Merhautova, V.; Herinkova, J.; Cajthaml, T.; Baldrian, P. Spatial variability of enzyme activities and microbial biomass in the upper layers of Quercus petraea forest soil. *Soil Biol. Biochem.* **2008**, *40*, 2068–2075. [CrossRef]
77. Averill, C.; Turner, B.L.; Finzi, A.C. Mycorrhiza-mediated competition between plants and decomposers drives soil carbon storage. *Nature* **2014**, *505*, 543–545. [CrossRef] [PubMed]
78. Promputtha, I.; Lumyong, S.; Dhanasekaran, V.; McKenzie, E.H.C.; Hyde, K.D.; Jeewon, R. A phylogenetic evaluation of whether endophytes become saprotrophs at host senescence. *Microb. Ecol.* **2007**, *53*, 579–590. [CrossRef]
79. Eastwood, D.C.; Floudas, D.; Binder, M.; Majcherczyk, A.; Schneider, P. The plant cell wall-decomposing machinery underlies the functional diversity of forest fungi. *Science* **2011**, *333*, 762–765. [CrossRef]
80. Smith, G.R.; Finlay, R.D.; Stenlid, J.; Vasaitis, R.; Menkis, A. Growing evidence for facultative biotrophy in saprotrophic fungi: Data from microcosm tests with 201 species of wood-decay basidiomycetes. *New Phytol.* **2017**, *215*, 747–755. [CrossRef]
81. Demoor, A.; Silar, P.; Brun, S. Appressorium: The Breakthrough in Dikarya. *J. Fungi* **2019**, *5*, 72. [CrossRef] [PubMed]
82. Espagne, E.; Lespinet, O.; Malagnac, F.; Silva, C. The genome sequence of the model ascomycete fungus Podospora anserina. *Genome Biol.* **2008**, *9*, R77. [CrossRef]
83. Cairney, J. Basidiomycete mycelia in forest soils: Dimensions, dynamics and roles in nutrient distribution. *Mycol. Res.* **2005**, *109*, 7–20. [CrossRef]
84. Beare, M.H.; Hu, S.; Coleman, D.C.; Hendrix, P.F. Influences of mycelial fungi on soil aggregation and organic matter storage in conventional and no-tillage soils. Appl. *Soil Ecol.* **1997**, *5*, 211–219. [CrossRef]
85. Ziegler, M.; Eguiluz, V.M.; Duarte, C.M.; Voolstra, C.R. Rare symbionts may contribute to the resilience of coral-algal assemblages. *ISME J.* **2018**, *12*, 161–172. [CrossRef] [PubMed]
86. Jousset, A.; Bienhold, C.; Chatzinotas, A.; Gallien, L.; Gobet, A.; Kurm, V.; Kusel, K.; Rillig, M.C.; Rivett, D.W.; Salles, J.F.; et al. Where less may be more: How the rare biosphere pulls ecosystems strings. *ISME J.* **2017**, *11*, 853–862. [CrossRef]
87. Shade, A.; Gilbert, J.A. Temporal patterns of rarity provide a more complete view of microbial diversity. *Trends Microbiol.* **2015**, *23*, 335–340. [CrossRef]
88. Valencia, E.; Gross, N.; Quero, J.L.; Carmona, C.P.; Ochoa, V.; Gozalo, B.; Delgado-Baquerizo, M.; Dumack, K.; Hamonts, K.; Singh, B.K.; et al. Cascading effects from plants to soil microorganisms explain how plant species richness and simulated climate change affect soil multifunctionality. *Glob. Chang. Biol.* **2018**, *24*, 5642–5654. [CrossRef] [PubMed]
89. Poorter, H.; Niklas, K.J.; Reich, P.B.; Oleksyn, J.; Poot, P.; Mommer, L. Biomass allocation to leaves, stems and roots: Meta-analyses of interspecific variation and environmental control. *New Phytol.* **2012**, *193*, 30–50. [CrossRef]
90. Sokol, N.W.; Kuebbing, S.E.; Karlsen-Ayala, E.; Bradford, M.A. Evidence for the primacy of living root inputs, not root or shoot litter, in forming soil organic carbon. *New Phytol.* **2019**, *221*, 233–246. [CrossRef]
91. Zhang, N.L.; Wan, S.Q.; Li, L.H.; Bi, J.; Zhao, M.M.; Ma, K.P. Impacts of urea N addition on soil microbial community in a semi-arid temperate steppe in northern China. *Plant Soil* **2008**, *311*, 19–28. [CrossRef]
92. Fanin, N.; Kardol, P.; Farrell, M.; Kempel, A.; Ciobanu, M.; Nilsson, M.C.; Gundale, M.J.; Wardle, D.A. Effects of plant functional group removal on structure and function of soil communities across contrasting ecosystems. *Ecol. Lett.* **2019**, *22*, 1095–1103. [CrossRef] [PubMed]
93. Mommer, L.; van Ruijven, J.; de Caluwe, H.; Smit-Tiekstra, A.E.; Wagemaker, C.A.M.; Ouborg, N.J.; Bogemann, G.M.; van der Weerden, G.M.; Berendse, F.; de Kroon, H. Unveiling below-ground species abundance in a biodiversity experiment: A test of vertical niche differentiation among grassland species. *J. Ecol.* **2010**, *98*, 1117–1127. [CrossRef]
94. Ding, J.J.; Zhang, Y.G.; Wang, M.M.; Sun, X.; Cong, J.; Deng, Y.; Lu, H.; Yuan, T.; Van Nostrand, J.D.; Li, D.Q.; et al. Soil organic matter quantity and quality shape microbial community compositions of subtropical broadleaved forests. *Mol. Ecol.* **2015**, *24*, 5175–5185. [CrossRef] [PubMed]
95. Peh, K.S.H.; Balmford, A.; Bradbury, R.B.; Brown, C.; Butchart, S.H.M.; Hughes, F.M.R.; Stattersfield, A.; Thomas, D.H.L.; Walpole, M.; Bayliss, J.; et al. TESSA: A toolkit for rapid assessment of ecosystem services at sites of biodiversity conservation importance. *Ecosyst. Serv.* **2013**, *5*, E51–E57. [CrossRef]

biology

MDPI

Article

Soil Microbial Diversity and Community Composition in Rice–Fish Co-Culture and Rice Monoculture Farming System

Noppol Arunrat [1,*], Chakriya Sansupa [2], Praeploy Kongsurakan [3], Sukanya Sereenonchai [1] and Ryusuke Hatano [4]

1 Faculty of Environment and Resource Studies, Mahidol University, Nakhon Pathom 73170, Thailand
2 Department of Biology, Faculty of Science, Chiang Mai University, Chiang Mai 50200, Thailand
3 Graduate School of Fisheries Science and Environmental Studies, Nagasaki University, 1–14 Bunkyo-machi, Nagasaki 852–8521, Japan
4 Laboratory of Soil Science, Research Faculty of Agriculture, Hokkaido University, Sapporo 060–8589, Japan
* Correspondence: noppol.aru@mahidol.ac.th

Simple Summary: The integration of fish in rice fields can influence the diversity and structural composition of soil microbial communities. Therefore, soil microorganisms between rice–fish co-culture (RF) and rice monoculture (MC) were compared. The key findings revealed that Actinobacteria, Chloroflexi, Proteobacteria, Acidobacteria, and Planctomycetes were the most dominant taxa across both paddy fields. The most abundant genus in MC belonged to *Anaeromyxobacter*, whereas that in RF was *Bacillus.* Nitrogen fixation, aromatic compound degradation, and hydrocarbon degradation were more abundant in RF. Phosphatase, β-glucosidase, cellulase, and urease enzymes were detected in both paddy fields. However, a 2-year conversion from organic rice to rice–fish co-culture may not be long enough to significantly alter alpha diversity indices.

Abstract: Soil microorganisms play an important role in determining nutrient cycling. The integration of fish into rice fields can influence the diversity and structural composition of soil microbial communities. However, regarding the rice–fish co-culture (RF) farming system in Thailand, the study of the diversity and composition of soil microbes is still limited. Here, we aim to compare the microbial diversity, community composition, and functional structure of the bacterial communities between RF and rice monoculture (MC) farming systems and identify the environmental factors shaping bacterial community composition. Bacterial taxonomy was observed using 16s rRNA gene amplicon sequencing, and the functional structures of the bacterial communities were predicted based on their taxonomy and sequences. The results showed that soil organic carbon, total nitrogen (TN), organic matter, available phosphorous, and clay content were significantly higher in RF than in MC. The most dominant taxa across both paddy rice fields belonged to Actinobacteria, Chloroflexi, Proteobacteria, Acidobacteria, and Planctomycetes. The taxa Nitrosporae, Rokubacteria, GAL15, and Elusimicrobia were significantly different between both rice fields. At the genus level, *Bacillus*, *Anaeromyxobacter*, and HSB OF53-F07 were the predominant genera in both rice fields. The most abundant genus in MC was *Anaeromyxobacter*, whereas RF belonged to *Bacillus*. The community composition in MC was positively correlated with magnesium and sand content, while in RF was positively correlated with pH, TN, and clay content. Nitrogen fixation, aromatic compound degradation, and hydrocarbon degradation were more abundant in RF, while cellulolysis, nitrification, ureolysis, and phototrophy functional groups were more abundant in MC. The enzymes involved in paddy soil ecosystems included phosphatase, β-glucosidase, cellulase, and urease. These results provide novel insights into integrated fish in the paddy field as an efficient agricultural development strategy for enhancing soil microorganisms that increase soil fertility.

Keywords: microbial diversity; microbial community composition; 16s rRNA gene; rice–fish co-culture; rice monoculture

Citation: Arunrat, N.; Sansupa, C.; Kongsurakan, P.; Sereenonchai, S.; Hatano, R. Soil Microbial Diversity and Community Composition in Rice–Fish Co-Culture and Rice Monoculture Farming System. *Biology* **2022**, *11*, 1242. https://doi.org/10.3390/biology11081242

Academic Editors: Daniel Puppe, Panayiotis Dimitrakopoulos and Baorong Lu

Received: 3 August 2022
Accepted: 18 August 2022
Published: 20 August 2022

1. Introduction

Microorganisms play important roles in soil and agricultural ecosystems [1,2]. They are responsible for several processes, such as biomass decomposition, nutrient circulation, and soil formation, which subsequently affect plant growth and production [1–3]. In recent years, soil microbial communities have been extensively investigated, as they may reflect soil fertility and ecosystem function [4,5]. Furthermore, the soil microbial community can be used as an indicator to track changes in various land management methods, such as tracking changes in restoration outcomes [6,7] or evaluating agricultural management methods [8].

Integrated rice and fish farming has been practiced in Thailand for more than 200 years by capturing wild seed fish in the rice fields [9]. The rice–fish co-culture (RF) is eulogized for improving ecosystems and alleviating poverty [10] and promoted as increasing biodiversity, reducing fertilizer and pesticide utilization, and contributing to system stability and sustainability [11,12]. Several studies (i.e., [13–15]) reported that RF generated the extra production of aquaculture, which increased farmers' income. Due to fish eating insects, pests, and weeds, the use of pesticides and herbicides can be reduced [16] while organic fertilizers and organic amendments are more applied. These promote the suitable condition for the abundant and diversified population of soil microorganisms, especially bacteria that play a crucial role in soil carbon and nitrogen mineralization. Proteobacteria, Bacteroidetes, Acidobacteria, and Chloroflexi were generally dominant phyla in the paddy soil [17–19], which play important roles in soil nutrient cycles [20,21]. Thus, soil microbial communities can be used as an indicator to explain soil health [22].

To date, the scientific knowledge on soil microbial taxonomic and functional composition, and their interactions with environmental factors of integrating fish in paddy fields remain unclear. Therefore, our study was carried out to fill this gap, aiming to (i) compare microbial diversity and community composition between rice monoculture (MC) and RF fields, (ii) identify the environmental factors shaping the bacterial community composition, and (iii) compare the functional structure of the bacterial communities between both study sites. This study can provide scientific knowledge for the development of a rice–fish co-culture farming system.

2. Materials and Methods

2.1. Description of Study Sites

The study sites were located in the Samnak Khun Nen Subdistrict, Dong Charoen District, Phichit Province, Lower North of Thailand. The maximum and minimum temperatures were 32.9 and 23.3 °C, respectively, while the average precipitation was 1264.8 mm year^{-1}. A rice–fish co-culture farm (16°04′04.1″ N, 100°32′31.1″ E, Figure 1a) which has been producing organic rice for more than 10 years was selected. The International Federation of Organic Agriculture Movements (IFOAM) Standard was first certified in 2016, while the EU/USDA Organic Standard was approved in 2018. The "Riceberry" rice variety was usually grown once per year from August to December. Since 2019, this farm has been raising fish in the paddy field. The main species of fish were common snakehead (*Channa striata*), walking catfish (*Clarias batrachus* (L.)), and Nile tilapia (*Oreochromis niloticus*). The rice bran, vegetable, and fruit residues and cattle manure were applied in the paddy field as the food for the fish and nutrients for rice. The weeds were removed by hand, while biofermented juice was produced from lemongrass, neem leaves, fruits, and vegetables and then mixed with molasses and animal dung (poultry and cattle) to dispose of the insects. The type of rice–fish co-culture field was the canal refuge (Figure 1b). The transplanting method was used for rice planting, which was performed by hand. One-month-old fish were released into the paddy field 30 days after rice planting. The water level in the field was maintained at around 20–30 cm during rice growing. Rice harvesting was performed by hand, and all rice residues were left in the paddy field. Rice yield was approximately 3.6 tons ha^{-1}, while the yield of fish was around 300 kg ha^{-1}.

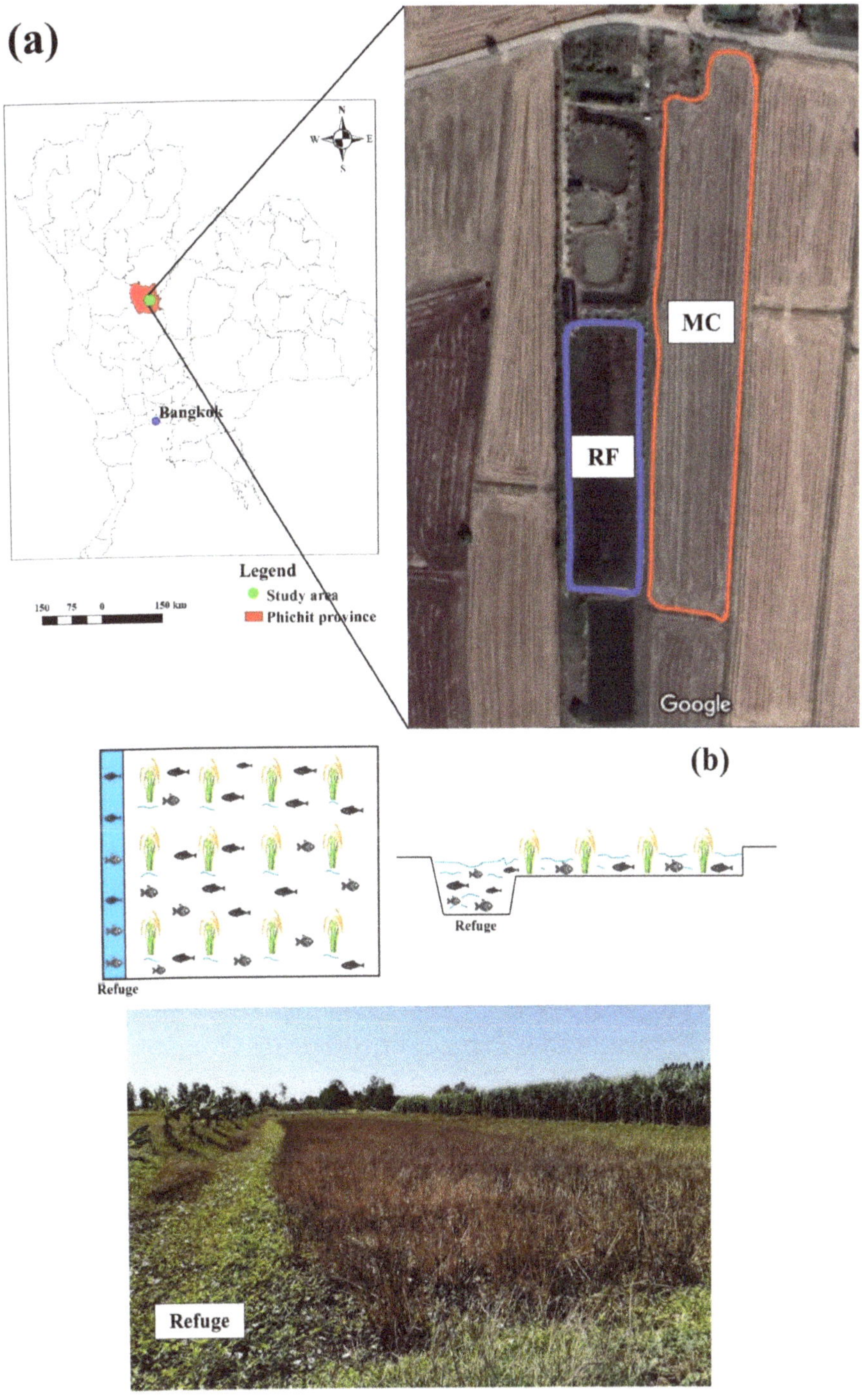

Figure 1. Study areas. (**a**) study sites, (**b**) canal refuge. The aerial image was taken from Google maps on 20 April 2022. The photo was taken on 27 December 2021 by Noppol Arunrat.

For a fair comparison, an adjacent conventional rice farm (16°04′04.6″ N, 100°32′31.9″ E) was selected as the comparison site (Figure 1a). The "RD41" (105 days) or "RD57" (110 days) rice varieties were chosen for planting once a year (August to November). The pregerminated rice seeds were sown by hand (broadcasting method). Then, N, P_2O_5, and K_2O chemical fertilizers were applied, namely 46-0-0 (125 kg ha^{-1}) and 16-20-0 (156.3 kg ha^{-1}). Glyphosate (48% *w*/*v* SL) and alachlor (48% *w*/*v* EC) were used to kill the weeds, while acephate (75% S) and chlorpyrifos (40% EC) were applied to eliminate the diseases and insects. A harvesting machine was usually used for rice harvesting, then all rice residues were left in the paddy field. The rice yield was approximately 4.7 ton ha^{-1}.

2.2. Soil Sampling and Measurements

In December 2021, soil samples were collected from the top layer (0–10 cm) of the rice–fish co-culture field and conventional rice field. Five duplicates of soil samples were collected for each field, and 10 soil samples were collected in total. The 50 g soil samples were kept in a cold storage box and brought to the laboratory for the extraction of soil microbial DNA.

Moreover, the soil cores (5.0 cm width × 5.5 cm length) were used to collect the soil samples for soil bulk density measurement and were then measured after 24 h of drying in an oven at 105 °C. The 1 kg soil samples were taken for soil physical and chemical properties analysis.

Soil texture was measured using a hydrometer. Electrical conductivity (ECe) was measured using an EC meter in saturation paste extracts (1:5) [23]. Soil pH was determined using a pH meter in a 1:1 soil-to-water mixture [24]. Available phosphorus (Avail. P) was determined following the molybdate blue method (Bray II extraction) [25]. Available potassium (Avail. K), available calcium (Avail. Ca), and available magnesium (Avail. Mg) were measured using atomic absorption spectrometry (NH4OAc extraction pH 7.0) [26]. Total nitrogen (TN) was measured using the micro-Kjeldahl method. Organic carbon (OC) was analyzed using the Walkley and Black [27] method and converted to organic matter (OM) by multiplying by 1.724. The SOC stock was calculated using the following equation:

$$SOCstock = OC \times BD \times L, \quad (1)$$

where *SOCstock* is the soil organic carbon stock (Mg C ha^{-1}), *OC* is organic carbon (%), *BD* is soil bulk density (Mg m^{-3}), and *L* is soil depth (cm).

2.3. DNA Extraction and Amplicon Sequencing of 16s rRNA Gene

DNA was extracted from 0.25 g of soil using DNeasy PowerSoil Pro DNA Kit (Qiagen, Germantown, MD, USA) following the manufacturer's instructions. The extracted DNA was subjected to amplicon library preparation and sequencing. Briefly, PCR amplification, targeting the V3–V4 variable of the 16s rRNA gene, was performed using the universal primers 341F (5′-CCTAYGGGDBGCWSCAG) and 805R (5′-GGA CTACNVGGGTHTC-TAAT) (Klindworth et al., 2013). The amplicons were then sequenced on the Illumina Miseq platform (2 × 250 bp). The amplification and sequencing steps were run by Omics Sciences and Bioinformatics Center (Chulalongkorn University, Bangkok, Thailand).

2.4. Sequencing Analysis and Microbial Taxonomic Identification

The raw sequence dataset was analyzed with QIIME 2 v. 2021.8 [28]. The 16s rRNA primers were trimmed from forward and reverse reads using cutadapt [29]. The trimmed sequences were quality-filtered (MaxEE = 2; no ambiguous nucleotide) and merged (minimum overlap = 12 nucleotides), and chimeras were removed using the DADA2 plugin [30]. The high-quality sequence was clustered at 97% sequence identity into operational taxonomic units (OTUs) using the VSEARCH plugin [31,32]. Representative sequences of each OTU were taxonomically identified against the Silva v.138 database [33,34]. To eliminate potential sequencing error, rare OTUs (singletons, doubletons, and tripletons) were removed from the dataset. After that, the number of reads that remained in each sample was randomly subsampled and normalized to the smallest number of reads per sample (24,676 reads/sample), to avoid sequencing depth bias, using rarefy was implemented in QIIME 2. These normalized datasets were used for further analysis.

2.5. Functional Prediction

Microbial associated functions were predicted using FAPROTAX [35,36] and PICRUSt2 [37]. The FAPROTAX predicted the ecologically relevant function of each taxon based on data of the cultured taxa. For example, if all cultured taxa of a bacterial genus were identified as nitrogen-fixing bacteria, all uncultured members of that genus will also be identified as nitrogen-fixing bacteria. On the other hand, PICRUSt2 predicted potential functions based on gene sequences presented in each taxon. In this study, the PICRUSt function was emphasized as enzyme activities that were potentially performed by the detected taxon. These functional analyses were performed following the instructions on the FAPROTAX (http://www.loucalab.com/archive/FAPROTAX/lib/php/index.php?section=Home accessed on 27 March 2022) and PICRUSt (https://github.com/picrust/picrust2/wiki accessed on 27 March 2022) webpages.

2.6. Statistical Analysis

Statistical analysis was performed on PAST [38] and R statistical software [39]. Independent *t*-tests were employed for comparison of soil physicochemical properties between monoculture rice fields and rice–fish co-culture fields. The correlations among the physicochemical variables were observed via Pearson's correlation matrixes. Differences in the relative abundance of microbial taxa detected in the two study sites were indicated using the linear discriminant analysis (LDA) effect size (LEfSe) [40]. Taxa with significant *p*-values ($p < 0.05$) and LDA score ≥ 2 were considered differentially abundant taxa. The LEfSe analysis was performed on an online interface of the Huttenhower lab Galaxy server (http://huttenhower.sph.harvard.edu/galaxy accessed on 27 March 2022). Alpha diversity, including observed OTU richness, Shannon, and Simpson indices were estimated using the diversity indices function in PAST. Differences in the alpha diversity indices between the two study sites were tested via t-test. Beta diversity, representing community composition, was analyzed using non-metric multidimensional scaling (NMDS) ordination based on Bray–Curtis dissimilarity, which was computed using the metaMDS function in the vegan R package. Permutational MANOVA (PERMANOVA) was used to the calculated difference between the two community compositions using the adonis function. The influence of soil properties on soil bacterial community composition was estimated by correlation analysis. The correlations were calculated using the envfit function in the vegan R package, and the *p*-values were corrected by Bonferroni's correction using the p.adjust function in the stat R package. The NMDS ordination with significantly correlated soil parameters was plotted using the ggplot2 R package. Bacterial-associated functions, predicted by both FAPROTAX and PICRUSt2, were visualized as extended bar plots in STAMP software. Statistical differences between each function were tested via *t*-test, and all *p*-values were corrected using Bonferroni's correction. Functional compositions were analyzed following the community composition (beta diversity) analysis as described above.

3. Results

3.1. Soil Physicochemical Properties in Rice Monoculture and Rice–Fish Co-Culture Fields

The soil samples, both from the rice monoculture (MC) and the rice–fish co-culture fields (RF), were silty clay. However, significant variances in soil physiochemical properties were found (Table 1). Lower acidity (6.0 ± 0.2, $p < 0.01$), bulk density (1.4 ± 0.02 Mg m^{-3}, $p < 0.05$), ECe (0.4 ± 0.01 dS m^{-1}, $p < 0.01$), available Ca (2279.0 ± 90.0 mg kg^{-1}, $p < 0.01$), available Mg (175.1 ± 3.6 mg kg^{-1}, $p < 0.01$), and %Sand (10.1 ± 0.8, $p < 0.01$) were easily observed in the RF field. Meanwhile, the RF soils also contained significantly higher contents of OM fraction (3.4% ± 0.2, $p < 0.01$), SOC (80.9 ± 3.5 Mg C ha^{-1}, $p < 0.01$), TN (0.5% ± 0.02, $p < 0.01$), available P (20.0 ± 0.9 mg kg^{-1}, $p < 0.01$), and %Clay content

(46.3 ± 0.9 mg kg^{-1}, $p < 0.01$). However, there was no significant difference in available content or %Silt content between the two sampling sites. Negative correlations betwee SOC stock and %Silt content were found in MC (r = -0.887, $p < 0.05$). In RF, negativ correlations were found between SOC and ECe (r = –0.904, $p < 0.05$) as well as betwee available K and %Silt content (r = –0.992, $p < 0.05$). Moreover, total nitrogen positivel correlated with %Sand content (r = 0.917, $p < 0.05$) (Table 2).

Table 1. Comparison of soil physicochemical properties of rice monoculture and rice–fis co-culture fields.

Soil Properties	Rice Monoculture	Rice–fish Co-Culture	T	Sig.
pH (1:2.5)	4.7 ± 0.3	6.0 ± 0.2	8.031	**
BD (Mg m^{-3})	1.4 ± 0.02	1.4 ± 0.02	−2.414	*
OM (%)	2.1 ± 0.4	3.4 ± 0.2	6.997	**
SOC (Mg C ha^{-1})	51.0 ± 9.2	80.9 ± 3.5	13.878	**
TN (%)	0.3 ± 0.01	0.5 ± 0.2	15.637	**
ECe (dS m^{-1})	1.0 ± 0.4	0.4 ± 0.01	−3.636	**
Avail. P (mg kg^{-1})	13.6 ± 1.9	20.0 ± 0.9	6.928	**
Avail. K (mg kg^{-1})	162.8 ± 6.2	170.0 ± 4.1	2.182	NS
Avail. Ca (mg kg^{-1})	2554.4 ± 85.2	2279.0 ± 90.0	−4.967	**
Avail. Mg (mg kg^{-1})	225.0 ± 5.6	175.1 ± 3.6	−16.709	**
Sand (%)	17.4 ± 0.9	10.1 ± 0.8	−14.141	**
Silt (%)	42.0 ± 1.3	43.6 ± 1.0	2.123	NS
Clay (%)	40.6 ± 0.9	46.3 ± 0.9	10.486	**
Soil texture	Silty Clay	Silty Clay	-	-

*, ** indicate statistically significant with p-value < 0.05 and < 0.01, respectively. NS: No significant, BD = bulk density, OM = organic matter, TN = total nitrogen, ECe = electrical conductivity, CEC = cation exchange capacity, Avail. P = available P, Avail. K = available K, Avail. Ca = available Ca, Avail. Mg = available Mg.

Table 2. Pearson's correlation matrixes of soil properties in rice–fish co-culture (n = 5, white area) and rice monoculture fields (n = 5, grey area).

Soil Properties	pH	BD	SOC	TN	EC$_e$	P	K	Ca	Mg	%Sand	%Silt	%Clay
pH		−0.718	0.391	−0.387	−0.086	−0.302	−0.613	0.275	−0.401	0.061	−0.056	0.010
BD	−0.562		0.106	−0.015	−0.160	0.558	0.216	−0.662	0.204	−0.197	−0.313	0.537
SOC	−0.316	−0.187		−0.787	−0.485	0.744	−0.367	−0.293	0.328	0.247	**−0.887**	0.796
TN	−0.373	0.775	−0.454		−0.157	−0.434	0.773	−0.226	−0.516	−0.709	0.871	−0.361
ECe	−0.087	0.524	**−0.904**	0.741		−0.565	−0.493	0.803	0.231	0.626	0.180	−0.770
P	−0.101	−0.515	−0.243	−0.316	0.149		0.194	−0.549	0.587	0.121	−0.819	0.831
K	0.258	0.618	−0.266	0.512	0.307	−0.868		−0.452	−0.026	−0.543	0.377	0.056
Ca	0.700	−0.745	0.046	−0.277	−0.323	0.049	−0.120		0.292	0.733	0.128	−0.808
Mg	0.474	0.198	−0.551	0.618	0.498	−0.435	0.668	0.445		0.791	−0.681	0.069
%Sand	−0.126	0.811	−0.618	**0.917**	0.808	−0.446	0.744	−0.333	0.691		−0.546	−0.274
%Silt	−0.333	−0.559	0.341	−0.535	−0.355	0.827	**−0.992**	0.022	−0.756	−0.762		−0.656
%Clay	0.631	0.051	0.090	−0.090	−0.258	−0.818	0.774	0.295	0.468	0.173	−0.770	

All values in bold print are significant ($p < 0.05$). BD = bulk density, OM = organic matter, TN = total nitrogen, ECe = electrical conductivity, CEC = cation exchange capacity, P = available P, K = available K, Ca = available Ca, Mg = available Mg.

3.2. General Overview of the Sequencing Analysis

A total of 337,778 high quality and abundance readings, representing 4597 OTUs, were derived from this study. After normalization, 4582 OTUs remained. Rarefaction curves of the detected OTUs derived from both MC and RF samples were flattened at the analysis sequencing depth (24,676 reads/sample), indicating that the detected OTUs were sufficient to represent the microbial community in each sample (Figure 2).

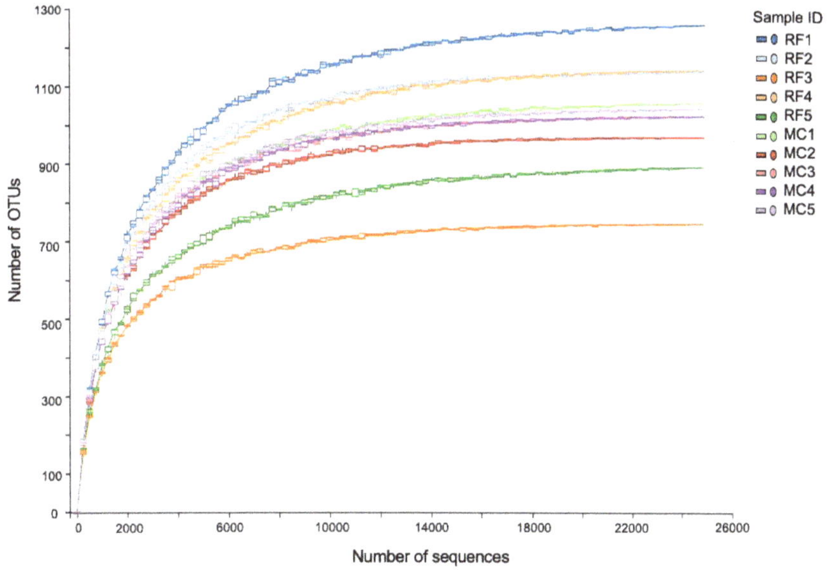

Figure 2. Rarefaction curve of observed microbial OTUs detected in monoculture (MC) and rice–fish (RC) fields.

3.3. Taxonomic Distribution and Differential Abundance of Soil Bacteria in Rice Monoculture and Rice–Fish Co-Culture Fields

According to microbial analysis based on the 16s rRNA gene, microbial taxa detected in this study were classified into 47 phyla, 128 classes, 235 orders, 318 families, 479 genera and 4582 OTUs. Taxonomic distribution of the abundant bacteria (total relative abundance > 0.1%) is presented in Figure 2. The most dominant taxa across all samples belonged to Actinobacteria (MC = 22.78%, RF= 24.17%, on average), followed by Chloroflexi (MC = 18.44%, RF= 17.77%), Proteobacteria (MC = 18.25%, RF= 17.28%), Acidobacteria (MC = 11.16%, RF= 11.88%), and Planctomycetes (MC = 10.44%, RF= 8.76%) (Figure 3a). Taxa that were significantly different between the two sites were Nitrosporae, Rokubacteria, GAL15, and Elusimicrobia. The latter was enriched in RF, whereas the other three were enriched in MC (Figure 3b).

The difference in the microbial community was more noticeable at a deeper taxonomic level, as shown in Figure 4. More than 50% of all detected OTUs were found uniquely in one of the two study sites (Figure 4a). According to the LEfSe analysis at the phylum to genus level, a total of 135 differentially abundant taxa ($p < 0.05$; LDA score > 2) were detected, 111 of which were more abundant in MC than RF and 24 of which were more abundant in RF than MC (Figure 4b, Figure S1). At the genus level, *Bacillus*, *Anaeromyxobacter*, and HSB OF53-F07 were the predominant genera in both rice fields. The most abundant genus in MC was *Anaeromyxobacter*, whereas that in RF was *Bacillus* (Figure 4c). Seven out of the top 30 most prevalent genera, for example, *Bradyrhizobium*, *Bryobacter*, *Conexibacter*, *Nocadiodides*, and *Solirubrobacter*, were significantly more abundant in MC than in RF (Figure 4c). However, some low-abundance genera, such as *Chlorobaculum*, *Niastella*, and *Vicinamibacter*, were more abundant in RF than in MC (Figure S1).

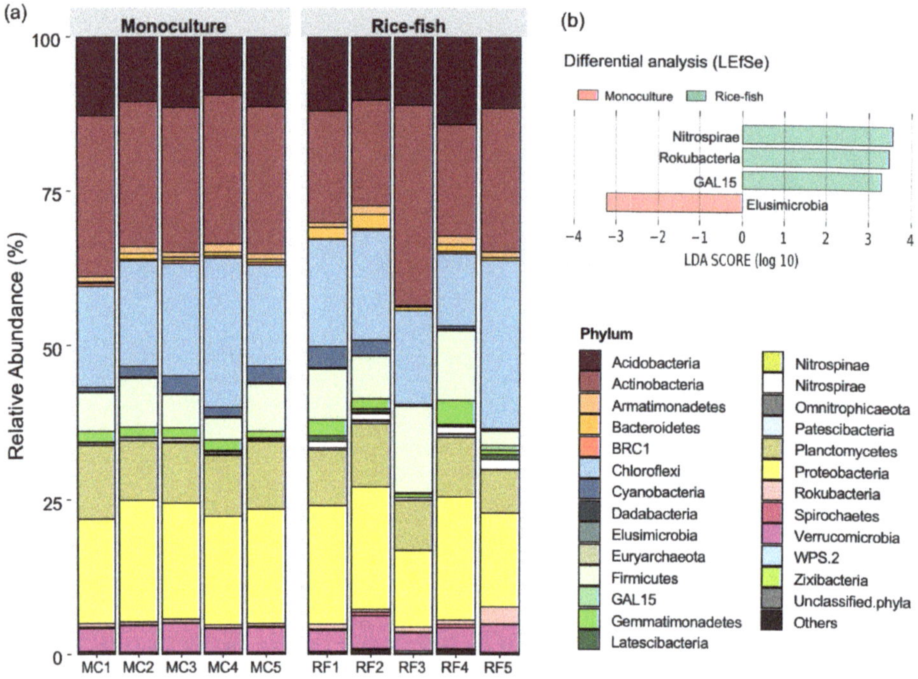

Figure 3. Taxonomic distribution at phylum level. (**a**) The relative abundance of microbial phyla in each sample. (**b**) LEfSe analysis of differential taxonomy abundance at phylum level between the two study sites. The orange and blue horizontal bars represent the taxa enriched in monoculture and rice–fish fields, respectively.

3.4. Richness, Diversity, Community Composition, and Their Correlation to Soil Properties

Alpha diversity, reflecting richness and diversity, was indicated by observed OTU richness, Shannon, and Simpson indices. As shown in Figure 5, all alpha diversity indices were slightly higher in MC than in RF, but no significant difference was found between the two sites ($p > 0.05$, Figure 5a–c). On the contrary, beta diversity, presented by NMDS ordination with Bray–Curtis distance, showed a separated community between MC and RF (Figure 5d). This indicated that the community composition of bacteria in MC was different from that in RF. The difference was confirmed by PERMANOVA test ($F = 0.251$, $p = 0.008$).

The correlations between soil properties and bacterial community composition are shown in Table 3; 5 out of 12 measured parameters were significantly correlated with bacterial community composition in both study sites. Whilst the community composition in MC was positively correlated with Mg and sand, the same in RF was positively correlated with pH, TN, and clay (Figure 5). Mg was the most correlated factor ($r^2 = 0.880$), following closely by Sand ($r^2 = 0.866$), pH ($r^2 = 0.857$) and TN ($r^2 = 0.834$) (Table 3).

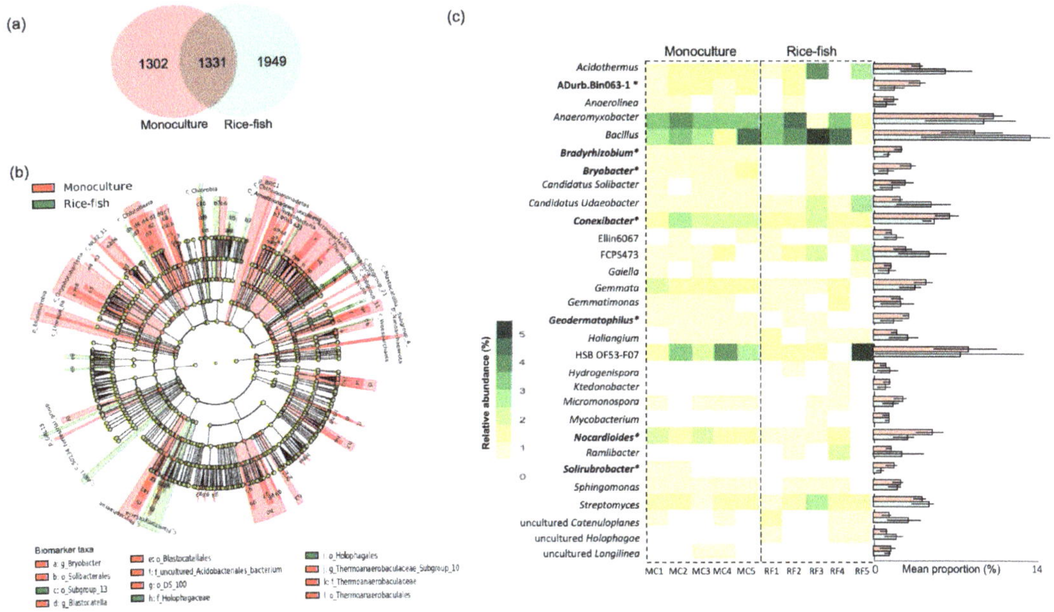

Figure 4. Taxonomic differences of bacteria in monoculture and rice–fish co-culture fields. (**a**) The Venn diagram shows the number of OTUs found in monoculture (red) and rice–fish (green) fields. (**b**) The cladogram shows differential abundance taxa among the two rice fields. More information on the differential taxa is provided in Supplementary Figures S1 and S2. (**c**) Heatmap shows the relative abundance of the thirty most abundant microbial genera detected in each sample. The bar plot presents the mean relative abundance of the microbial genera detected in monoculture (orange) and rice–fish fields (green). Genus names with an asterisk are statistically different between the two sites ($p < 0.05$). MC: Monoculture, RF: rice–fish.

Table 3. Pearson's correlation between microbial community composition and soil properties.

Parameter	NMDS1	NMDS2	*r*	*p*-Value
pH	0.284	−0.959	0.857	0.024 *
BD	−0.523	0.853	0.697	0.108
OM	0.286	−0.958	0.762	0.084
TN	0.211	−0.978	0.834	0.012 *
ECe	−0.238	0.971	0.563	0.516
Avail. P	0.266	−0.964	0.662	0.240
Avail. K	0.176	−0.984	0.261	1.000
Avail. Ca	−0.088	0.996	0.649	0.240
Avail. Mg	−0.214	0.977	0.880	0.024 *
Sand	−0.190	0.982	0.866	0.036 *
Silt	0.104	−0.995	0.460	1.000
Clay	0.220	−0.976	0.764	0.048 *

* Indicate significant parameters ($p < 0.05$). BD = bulk density, OM = organic matter, TN = total nitrogen, ECe = electrical conductivity, CEC = cation exchange capacity, Avail. P = available P, Avail. K = available K, Avail. Ca = available Ca, Avail. Mg = available M.

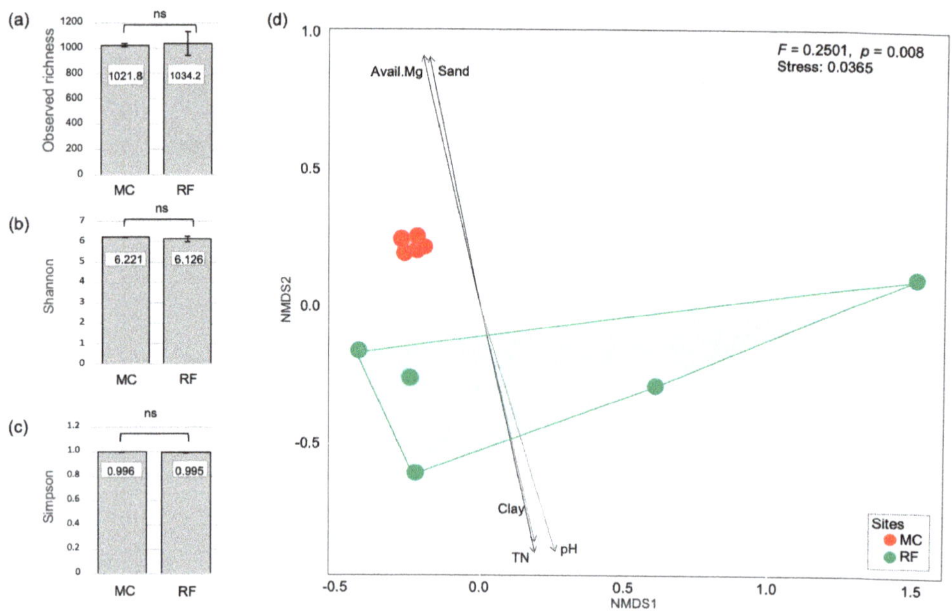

Figure 5. Bacterial diversity and community composition. Bar plot shows (**a**) OTU richness, (**b**) Shannon's, and (**c**) Simpson's indices measured in monoculture and rice–fish fields. (**d**) NMDS ordination, based on the Bray–Curtis dissimilarity measure, presents the community composition of soil microbes in the two study sites. MC: Monoculture, RF: rice–fish.

3.5. Predictive Function

Totals of 807 (17.61%) and 4532 (98.90%) OTUs were assigned to at least one function of the 63 ecologically relevant functions and 2238 enzymes, respectively. For FAPROTAX analysis, Cellulolysis, nitrification, ureolysis, phototrophy, nitrogen fixation, aromatic compound degradation, and hydrocarbon degradation, were found among the top 20 most abundant functions (Figure 6a). While the first four functional groups were more abundant in MC, the last three groups were more abundant in RF. However, no significant change was detected in any of those functions ($p > 0.05$), which is consistent with the results from the PICRUSt analysis. Enzymes involved in soil systems, such as phosphatase, β-glucosidase, cellulase, and urease, were presented in this study. As shown in Figure 6b, no significant difference was found between the enzymes detected in MC and RF ($p > 0.05$). The functional potential structures, created from all detected ecologically relevant functions and enzymes, were shown in Figures 6c and 6d, respectively. Based on NMDS ordination and PERMANOVA analysis, no significant difference was detected ($p > 0.05$) between the functional structures in MC and RF.

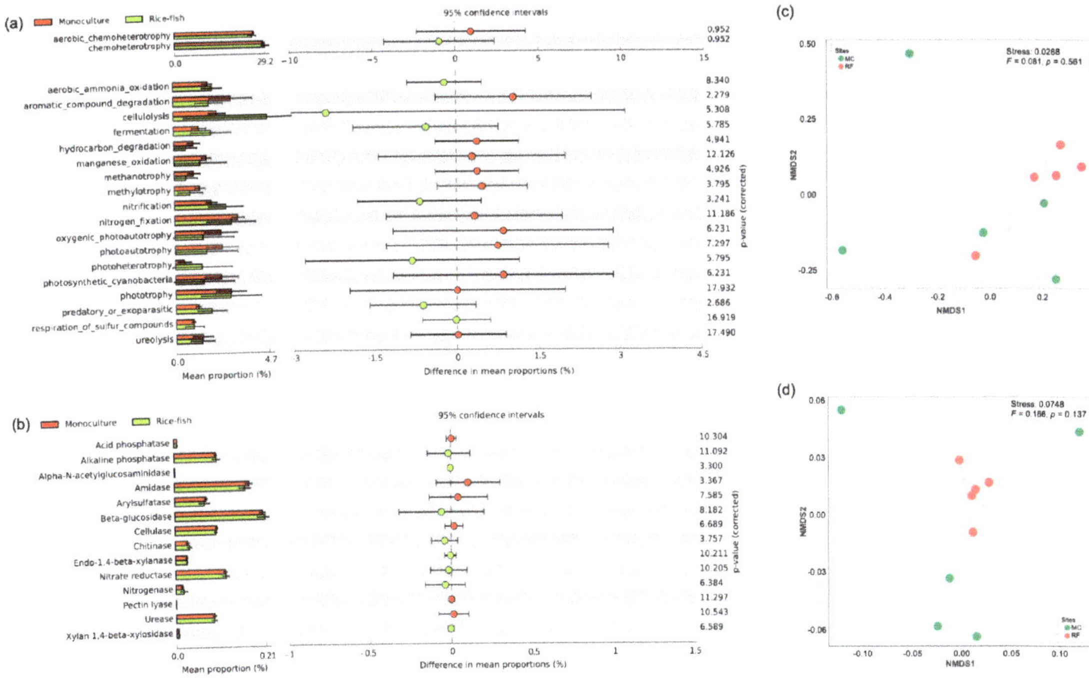

Figure 6. Predictive functions. Extended bar plots show mean and difference in mean proportions of microbial function predicted by (**a**) FAPROTAX and (**b**) PICRUST2. Functions overrepresented in the monoculture (red) have a positive difference in the proportion, whereas functions overrepresented in the rice–fish (green) have a negative difference. NMDS ordinations based on the Bray–Curtis dissimilarity measure show the composition of bacterial functions, predicted by (**c**) FAPROTAX and (**d**) PICRUST2, in monoculture (red) and rice–fish field (green). MC: Monoculture, RF: rice–fish.

4. Discussion

4.1. RF Farming System Can Increase Soil Nutrients

Our results (Table 1) reveal that the rice–fish co-culture system significantly increased SOC stock and TN similarly to other studies [41–44], which found that the rice–fish co-culture system can potentially increase the content of organic carbon and nitrogen in the soil through its mineralization of organic matter. This is due to the uneaten and excess feed as well as excreta produced during fish growth increasing SOC content and TN, which is consistent with the findings in the rice–crayfish–turtle co-culture by Li et al. [45] and rice–crayfish farming system by Si et al. [46]. They also revealed that crayfish and turtle feeding, molting, and excretion could increase the SOC and TN, while these aquatic animals help increase soil permeability by penetrating the soil surface in the paddy field. The supplies of some available elements (i.e., Ca and Mg) in the RF farming system were significantly lower than in MC (Table 1), which were possibly shaped by its divalent charge fraction on the ECe [47], following the lower ECe value in RF compared to MC (Table 1). The rice–fish co-culture system, however, shows a higher content of available P (Table 1), while in rice monoculture, it generally decreased due to the long-term cultivation. Slowly cycling P and labile P in soil increasing with time was also previously reported in the rice-frog-fish culture soil [48]. Higher clay content and organic matter were found in RF compared with MC due to a higher amount of organic amendments being added. Hassink [49] reported that clay particles could absorb organic matter and protect it from microbial decomposition.

4.2. Microbial Diversity and Community Composition under Rice–Fish Co-Culture and Rice Monoculture

To our knowledge, this study was among a few studies investigating the differentiation of both community and functional potential structures of soil bacteria in rice monoculture fields, compared with rice–fish co-culture fields. Here, we found that although the alpha diversity of bacteria did not differ significantly between MC and RF, the community composition did. Even though the alpha diversity indices, which present observed richness, Shannon's, and Simpson's indices, differed between the two sites, the differences were minor and did not reveal a significant difference. This result is consistent with previous studies on other rice co-culture fields, including rice–frog (Yi et al., 2019), rice–crab [19], and rice–fish fields [7,45]. Whilst Zhao et al. [7] reported that microbial diversity in rice–fish co-culture fields was similar to those in monoculture rice fields in the first year of cultivation and substantially changed after 5 years, Li et al. [45] found that it took up to 12 years to see the differences in rice–fish–turtle fields. Here, we showed that changes were still not found after three years of rice–fish cultivation. However, when it comes to community composition, all studies [7,45,50], including this one, found significant differences between co-culture and monoculture rice fields. A total of 135 differentially abundant taxa were found in both the MC and RF sites, four of which were among the abundant phyla and seven of which were among the top 30 most abundant genera, respectively. These results implied that three years of rice–fish co-culture agriculture, which was implemented after long-term monoculture rice cultivation, had a significant impact on microbial community composition, while diversity remained unchanged.

As shown in Figure 3a, the most dominant taxa in paddy fields (both RF and MC farming systems) belonged to Actinobacteria, Chloroflexi, Proteobacteria, Acidobacteria, and Planctomycetes. Meanwhile, all alpha diversity indices showed no significant difference between the two sites. This is consistent with the study of Jiang et al. [19], who found that the most abundant phyla in paddy soil and ditch sediment under the rice–crab co-culture system were Proteobacteria, Bacteroidetes, and Chloroflexi, while alpha diversity of bacterial diversity in paddy soil and ditch sediment was similar. The phylum Actinobacteria, which contributes to the turnover of the complex biopolymers and plant residue decomposition [51,52] by producing various carbon cycling enzymes [53], was the most dominant taxa in both RF and MC sites. The phylum Chloroflexi, which involves nitrification was the second-most dominant that was detected in RF and MC sites [54]. The phylum Proteobacteria is usually classified as "copiotrophs" (R-strategists), which indicates that its members are more abundant and have high growth rates under nutrient-rich conditions [55]. It plays a key role in OM decomposition, produces many types of glycosyl hydrolases, and is involved in nitrogen fixation, which promotes plant growth [1]. The phylum Acidobacteria is involved in the carbon cycle of humus decomposition [56] and adjusts soil pH [57]. Members of the bacterial phylum Planctomycetes can act as slow-acting degraders of various biopolymers, cellulose, and chitin. They can degrade exopolysaccharides produced by other bacteria [58]. García-Orenes et al. [59] reported that some members of the phylum Planctomycete, such as *Blastopirellula*, are good indicators of soil fertility because they respond to the applications of manure or fertilizers faster than they react to soil physicochemical properties.

Nitrosporae, Rokubacteria, GAL15, and Elusimicrobia taxa were significantly different between the RF and MC fields (Figure 3b). This result indicated that the structure of the soil bacterial community was significantly changed after integrating fish into the paddy field. The phylum Nitrospirae plays a crucial role of decomposing soil mineral nitrogen and improving nutrient availability [60], and its functions can enhance crop productivity [61]. At the genus level, *Bacillus*, *Anaeromyxobacter*, and HSB OF53-F07 were found in both rice fields. The most abundant genus in MC was *Anaeromyxobacter*, whereas *Bacillus* was the most dominant in RF (Figure 4c). *Bacillus* plays multiple functions in the soil ecosystem for nutrient cycling, which is involved in nitrogen fixation, phosphorus nutrition, and potassium solubilization that promotes plant growth [62]. Similarly, *Anaeromyxobacter* can fix and assimilate N_2 gas via its nitrogenase [63]. This is why soil nutrients (TN, Avail. P, and Avail. K) were higher in RF than in MC soil (Table 1).

Soil physicochemical properties have a great influence on bacterial community composition [7,50,64,65]. Our findings indicated that the different soil management practices of RF and MC farming systems cause a significant difference in soil physicochemical properties, resulting in the different composition of soil bacterial communities and their metabolic functions (Figure 5 and Table 3). This is in line with the study of Viruel et al. [66], who reported that changes in SOC stock, TN, and pH cause the changes in soil bacterial communities and metabolic functions in farming systems of the semi-arid Chaco region, Argentina. Hartmann et al. [67] found that soil pH, SOC, and TN were the main predictors of bacterial community structure in long-term organic and conventional farming. As presented in Figure 5 and Table 3, we found that Mg, TN, pH, and soil texture (percent of sand and clay) were the main factors determining the community composition of bacteria in MC and RF. TN and pH have previously been identified as the key factors that influence the microbial community in rice co-culture fields [7,50]. While Mg was rarely measured in earlier works, this study found that it was one of the most important elements affecting community composition in rice fields. Hou et al. [68] found a strong effect of Mg on bacterial nitrification, which is in line with the study of Zhang et al. [69], who reported that the appropriate concentrations of Mg^{2+} could promote nitrification activity in the soil.

In addition, this study also showed the functional potential structure of the bacterial community, based on FAPROTAX and PICRUSt2, in MC and RF (Figure 6). Interestingly, despite differences in community composition between MC and RF, no difference in functional potential structure was found. Furthermore, there was no difference in the abundance of the individual function predicted by either tool. Recently, Chen et al. [70] demonstrated on global soil metagenomic data that microbial functional structure could remain stable while taxonomic diversity and composition changed, which is consistent with our findings. We reveal that, whilst bacterial community composition changed during the transition from monoculture to rice–fish co-culture, the functional structure could be maintained, and this was the first study to report such results on monoculture and rice–fish fields. However, it should be noted that the limitation of the prediction tools was that only taxa or genes from the databases were functionally assigned [35–37,71]. Regardless, using both tools to estimate the availability and abundance of genes or functions within the community could provide more insight into information on the complex community in soil.

5. Conclusions

This study shows that soil physicochemical properties (SOC, TN, OM, Avail. P, and clay content) were significantly higher in RF than in MC sites. The most dominant taxa across both paddy rice-farming systems belonged to Actinobacteria, Chloroflexi, Proteobacteria, Acidobacteria, and Planctomycetes. The taxa Nitrosporae, Rokubacteria, GAL15, and Elusimicrobia were significantly different between the RF and MC sites. At the genus level, *Bacillus*, *Anaeromyxobacter*, and HSB OF53-F07 were the predominant genera in both rice fields. The most abundant genus in MC was *Anaeromyxobacter*, that in RF was *Bacillus*. All alpha diversity indices between the RF and MC sites were not significantly different. The community composition in MC was positively correlated with Mg and sand, while in RF was positively correlated with pH, TN, and clay. Nitrogen fixation, aromatic compound degradation, and hydrocarbon degradation were more abundant in RF, while cellulolysis, nitrification, ureolysis, and phototrophy functional groups were more abundant in MC. The enzymes involved in paddy soil ecosystems included phosphatase, β-glucosidase, cellulase, and urease, but no significant difference was detected between MC and RF fields. However, more observation fields and long-term monitoring are necessary to conclusively confirm our findings in this study.

Supplementary Materials: The following supporting information can be downloaded at: https://www.mdpi.com/article/10.3390/biology11081242/s1, Figure S1: LEfSE analysis from phylum to genus level. The cladogram shows differential abundance in taxa ($p > 0.05$, LDA score ≥ 2) between monoculture rice fields and rice–fish co-culture fields. Taxa in red are significantly enriched in the monoculture, whereas taxa in green are significantly enriched in the rice–fish field. Figure S2: Bar plot shows differential abundance taxa with effect size (LDA score). Taxa in red are significantly enriched in the monoculture, whereas taxa in green are significantly enriched in the rice–fish field.

Author Contributions: Conceptualization, N.A.; methodology, N.A., C.S. and P.K.; investigation, N.A. and S.S.; writing—original draft preparation, N.A., C.S., P.K., S.S. and R.H.; writing—review and editing, N.A., C.S., P.K., S.S. and R.H.; supervision, R.H. All authors have read and agreed to the published version of the manuscript.

Funding: This research project is supported by Mahidol University (contract number: A14/2564).

Institutional Review Board Statement: The study was conducted according to the guidelines of the Declaration of Helsinki and approved by the Institutional Review Board of Institute for Population and Social Research, Mahidol University (IPSR-IRB) (COA. No. 2021/01-001).

Informed Consent Statement: Not applicable.

Data Availability Statement: Raw sequence data generated for this study are available in the Sequence Read Archives (SRA) of the National Center for Biotechnology Information (NCBI) under BioProject accession number: PRJNA819169.

Acknowledgments: The authors extend their appreciation to Mahidol University (contract number: A14/2564).

Conflicts of Interest: The authors declare no conflict of interest.

References

1. Aislabie, J.; Deslippe, J.; Dymond, J. *Soil Microbes and Their Contribution to Soil Services. Ecosystem Services in New Zealand: Conditions and Trends*; Manaaki Whenua Press: Lincoln, New Zealand, 2013; pp. 143–161.
2. Kumari, A.; Kumar, R.; Rani, P.; Beniwal, V.; Kapoor, K.; Sharma, P. Role of Microbes in Sustainable Agriculture. In *Microbes in the Service of Mankind*; Nagpal, R., Kumar, A., Singh, R., Eds.; JBC Press: Delhi, India, 2014; pp. 178–200.
3. Jacoby, R.; Peukert, M.; Succurro, A.; Koprivova, A.; Kopriva, S. The Role of Soil Microorganisms in Plant Mineral Nutrition—Current Knowledge and Future Directions. *Front. Plant Sci.* **2017**, *8*, 1617. [CrossRef] [PubMed]
4. Falkowski, P.G.; Fenchel, T.; Delong, E.F. The Microbial Engines That Drive Earth's Biogeochemical Cycles. *Science* **2008**, *320*, 1034–1039. [CrossRef] [PubMed]
5. Wang, R.; Xiao, Y.; Lv, F.; Hu, L.; Wei, L.; Yuan, Z.; Lin, H. Bacterial Community Structure and Functional Potential of Rhizosphere Soils as Influenced by Nitrogen Addition and Bacterial Wilt Disease under Continuous Sesame Cropping. *Appl. Soil Ecol.* **2018**, *125*, 117–127. [CrossRef]
6. Gellie, N.J.C.; Mills, J.G.; Breed, M.F.; Lowe, A.J. Revegetation Rewilds the Soil Bacterial Microbiome of an Old Field. *Mol. Ecol.* **2017**, *26*, 2895–2904. [CrossRef] [PubMed]
7. Zhao, Z.; Chu, C.B.; Zhou, D.P.; Wang, Q.F.; Wu, S.H.; Zheng, X.Q.; Song, K.; Lv, W.G. Soil bacterial community composition in rice–fish integrated farming systems with different planting years. *Sci. Rep.* **2021**, *11*, 10855. [CrossRef] [PubMed]
8. Gautam, A.; Sekaran, U.; Guzman, J.; Kovács, P.; Hernandez, J.L.G.; Kumar, S. Responses of Soil Microbial Community Structure and Enzymatic Activities to Long-Term Application of Mineral Fertilizer and Beef Manure. *Environ. Sustain. Indic.* **2020**, *8*, 100073. [CrossRef]
9. MacKay, K.T.; Chapman, G.; Sollows, J.; Thongpan, N. Rice-fish culture in Northeast Thailand: Stability and sustainability. In *Proceedings of the IFOAM 6th International Scientific Conference*; University of California: Santa Cruz, CA, USA, 18–21 August 1986; pp. 355–365.
10. Halwart, M.; Gupta, M.V. *Culture of Fish in Rice Fields*; FAO: Rome, Italy; The WorldFish Center: Penang, Malaysia, 2004; pp. 1–83.
11. Heckman, C.W. *Rice Field Ecology in Northeastern Thailand*; Springer: Dordrecht, The Netherlands, 1979; Volume 34, pp. 1–228.
12. Arunrat, N.; Sereenonchai, S.; Chaowiwat, W.; Wang, C.; Hatano, R. Carbon, nitrogen and water footprints of organic rice and conventional rice production over 4 years of cultivation: A case study in the Lower North of Thailand. *Agronomy* **2022**, *12*, 380. [CrossRef]
13. Xie, J.; Hu, L.; Tang, J.; Wu, X.; Li, N.; Yuan, Y.; Yang, H.; Zhang, J.; Luo, S.; Chen, X. Ecological mechanisms underlying the sustainability of the agricultural heritage rice-fish co-culture system. *Proc. Natl. Acad. Sci. USA* **2011**, *8*, E1381–E1387.
14. Zhang, J.; Hu, L.; Ren, W.; Guo, L.; Tang, J.; Shu, M.; Chen, X. Rice-soft shell turtle coculture effects on yield and its environment. *Agric. Ecosyst. Environ.* **2016**, *224*, 116–122. [CrossRef]

15. Arunrat, N.; Sereenonchai, S. Assessing ecosystem services of rice–fish co-culture and rice monoculture in Thailand. *Agronomy* **2022**, *12*, 1241. [CrossRef]
16. Clasen, B.; Loro, V.L.; Murussi, C.R.; Tiecher, T.L.; Moraes, B.; Zanella, R. Bioaccumulation and oxidative stress caused by pesticides in Cyprinus carpio reared in a rice-fish system. *Sci. Total Environ.* **2018**, *626*, 737–743. [CrossRef] [PubMed]
17. Ahring, B.K. Perspectives for anaerobic digestion. *Adv. Biochem. Eng. Biotechnol.* **2003**, *81*, 1–30. [PubMed]
18. Janssen, P.H. Identifying the dominant soil bacterial taxa in libraries of 16S rRNA and 16S rRNA genes. *Appl. Environ. Microbiol.* **2006**, *72*, 1719–2172. [CrossRef]
19. Jiang, X.; Ma, H.; Zhao, Q.; Yang, J.; Xin, C.; Chen, B. Bacterial Communities in Paddy Soil and Ditch Sediment under Rice-Crab Co-Culture System. *AMB Express* **2011**, *11*, 163. [CrossRef] [PubMed]
20. Thomsen, T.R.; Kong, Y.; Nielsen, P.H. Ecophysiology of abundant denitrifying bacteria in activated sludge. *FEMS Microbiol. Ecol.* **2007**, *60*, 370–382. [CrossRef] [PubMed]
21. Chen, L.; Xu, J.; Wan, W.; Xu, Z.; Hu, R.; Zhang, Y.; Zheng, J.; Gu, Z. The Microbiome Structure of a Rice-Cray fish Integrated Breeding Model and Its Association with Cray fish Growth and Water Quality. *Microbiol. Spectr.* **2022**, *10*, e02204-21. [CrossRef] [PubMed]
22. Patricia, B.; Pfisterer, A.B.; Nina, B.; Jing-Shen, H.; Tohru, N.; David, R.; Bernhard, S. Quantifying the evidence for biodiversity effects on ecosystem functioning and services. *Ecol. Lett.* **2010**, *9*, 1146–1156.
23. Richards, L.A. Diagnosis and Improvement of Saline and Alkali Soils. In *Agriculture Handbook*; United States Department of Agriculture: Washington, DC, USA, 1954; pp. 1–160.
24. National Soil Survey Center. Soil Survey Laboratory Methods Manual. In *Soil Survey Investigations Report No. 42, Version 3.0*; Natural Conservation Service: Washington, DC, USA, 1996; pp. 1–716.
25. Bray, R.A.; Kurtz, L.T. Determination of total organic and available form of phosphorus in soil. *Soil Sci.* **1945**, *59*, 39–45. [CrossRef]
26. Thomas, G.W. Method of Soil Analysis, Part 3: Chemical Methods. In *Soil pH and Soil Acidity*; Sparks, D.L., Page, A.L., Helmke, P.A., Loeppert, R.H., Soltanpour, P.N., Tabatabai, M.A., Johnston, C.T., Sumner, M.E., Eds.; ASA Inc.: Madison, WI, USA, 1996; pp. 475–490.
27. Walkley, A.; Black, J.A. An examination of the dichromate method for determining soil organic matter and a proposed modification of the chromic acid titration method. *Soil Sci.* **1934**, *37*, 29–38. [CrossRef]
28. Bolyen, E.; Rideout, J.R.; Dillon, M.R.; Bokulich, N.A.; Abnet, C.C.; Al-Ghalith, G.A.; Alexander, H.; Alm, E.J.; Arumugam, M.; Asnicar, F.; et al. Reproducible, Interactive, Scalable and Extensible Microbiome Data Science Using QIIME 2. *Nat. Biotechnol.* **2019**, *37*, 852–857. [CrossRef]
29. Martin, M. Cutadapt Removes Adapter Sequences from High-Throughput Sequencing Reads. *EMBnet J.* **2011**, *17*, 10–12. [CrossRef]
30. Callahan, B.J.; McMurdie, P.J.; Rosen, M.J.; Han, A.W.; Johnson, A.J.A.; Holmes, S.P. DADA2: High-Resolution Sample Inference from Illumina Amplicon Data. *Nat. Methods.* **2016**, *13*, 581–583. [CrossRef]
31. Rognes, T.; Flouri, T.; Nichols, B.; Quince, C.; Mahé, F. VSEARCH: A Versatile Open Source Tool for Metagenomics. *PeerJ* **2016**, *4*, e2584. [CrossRef]
32. Ivanova, A.A.; Oshkin, I.Y.; Danilova, O.V.; Philippov, D.A.; Ravin, N.V.; Dedysh, S.N. Rokubacteria in Northern Peatlands: Habitat Preferences and Diversity Patterns. *Microorganisms* **2021**, *10*, 11. [CrossRef]
33. Quast, C.; Pruesse, E.; Yilmaz, P.; Gerken, J.; Schweer, T.; Yarza, P.; Peplies, J.; Glöckner, F.O. The SILVA Ribosomal RNA Gene Database Project: Improved Data Processing and Web-Based Tools. *Nucleic Acids Res.* **2013**, *41*, D590–D596. [CrossRef] [PubMed]
34. Glöckner, F.O.; Yilmaz, P.; Quast, C.; Gerken, J.; Beccati, A.; Ciuprina, A.; Bruns, G.; Yarza, P.; Peplies, J.; Westram, R.; et al. 25 Years of Serving the Community with Ribosomal RNA Gene Reference Databases and Tools. *J. Biotechnol.* **2017**, *261*, 169–176. [CrossRef] [PubMed]
35. Louca, S.; Parfrey, L.W.; Doebeli, M. Decoupling Function and Taxonomy in the Global Ocean Microbiome. *Science* **2016**, *353*, 1272–1277. [CrossRef]
36. Sansupa, C.; Wahdan, S.F.M.; Hossen, S.; Disayathanoowat, T.; Wubet, T.; Purahong, W. Can We Use Functional Annotation of Prokaryotic Taxa (FAPROTAX) to Assign the Ecological Functions of Soil Bacteria? *Appl. Sci.* **2021**, *11*, 688. [CrossRef]
37. Douglas, G.M.; Maffei, V.J.; Zaneveld, J.R.; Yurgel, S.N.; Brown, J.R.; Taylor, C.M.; Huttenhower, C.; Langille, M.G.I. PICRUSt2 for Prediction of Metagenome Functions. *Nat. Biotechnol.* **2020**, *38*, 685–688. [CrossRef]
38. Hammer, O.; Harper, D.; Ryan, P. PAST: Paleontological Statistics Software Package for Education and Data Analysis. *Palaeontol. Electron.* **2001**, *4*, 9.
39. R Core Team. *R: A Language and Environment for Statistical Computing*; R Foundation for Statistical Computing: Vienna, Austria, 2018.
40. Segata, N.; Izard, J.; Waldron, L.; Gevers, D.; Miropolsky, L.; Garrett, W.S.; Huttenhower, C. Metagenomic biomarker discovery and explanation. *Genome Biol.* **2011**, *12*, R60. [CrossRef] [PubMed]
41. Mirhaj, M.; Razzak, M.A.; Wahab, M.A. Comparison of nitrogen balances and efficiencies in rice cum prawn vs. rice cum fish cultures in Mymensingh, North-Eastern Bangladesh. *Agric. Syst.* **2014**, *125*, 54–62. [CrossRef]
42. Bihari, P.; Nayak, A.K.; Gautam, P.; Lal, B.; Shahid, M.; Raja, R.; Tripathi, R.; Bhattacharyya, P.; Panda, B.B.; Mohanty, S.; et al. Long-term effect of rice-based farming systems on soil health. *Environ. Monit. Assess.* **2015**, *187*, 296. [CrossRef]

43. Nayak, P.; Panda, B.; Lal, B.; Gautam, P.; Poonam, A.; Shahid, M.; Tripathi, R.; Kumar, U.; Mohapatra, S.; Jambhulkar, N. Ecological mechanism and diversity in rice based integrated farming system. *Ecol. Indic.* **2018**, *91*, 359–375. [CrossRef]
44. Liang, G.; Liangliang, H.; Lufeng, Z.; Xiaoyu, S.; Zijun, J.; Lilian, D.; Weizheng, R.; Jian, Z.; Jianjun, T.; Xin, C. Coupling Rice with Fish for Sustainable Yields and Soil Fertility in China. *Rice Sci.* **2020**, *27*, 175–179. [CrossRef]
45. Li, P.; Wu, G.; Li, Y.; Hu, C.; Ge, L.; Zheng, X.; Zhang, J.; Chen, J.; Zhang, H.; Bai, N.; et al. Long-term rice-crayfish-turtle co-culture maintains high crop yields by improving soil health and increasing soil microbial community stability. *Geoderma* **2022**, *413*, 115745. [CrossRef]
46. Moore, P.A.; Patrick, W.H. Calcium and Magnesium Availability and Uptake by Rice in Acid Sulfate Soils. *Soil Sci. Soc. Am. J.* **1989**, *53*, 816–822. [CrossRef]
47. Si, G.H.; Peng, C.L.; Yuan, J.F.; Xu, X.Y.; Zhao, S.J.; Xu, D.B.; Wu, J.S. Changes in soil microbial community composition and organic carbon fractions in an integrated rice-crayfish farming system in subtropical China. *Sci. Rep.* **2017**, *7*, 28–56. [CrossRef] [PubMed]
48. Lin, K.; Wu, J. Effect of introducing frogs and fish on soil phosphorus availability dynamics and their relationship with rice yield in paddy fields. *Sci Rep.* **2020**, *10*, 21. [CrossRef]
49. Hassink, J. Preservation of plant residues in soils differing in unsaturated protectivecapacity. *Soil. Sci. Soc. Am. J.* **1996**, *60*, 487–491. [CrossRef]
50. Yi, X.; Yi, K.; Fang, K.; Gao, H.; Dai, W.; Cao, L. Microbial Community Structures and Important Associations Between Soil Nutrients and the Responses of Specific Taxa to Rice-Frog Cultivation. *Front. Microbiol.* **2019**, *10*, 1752. [CrossRef] [PubMed]
51. Chen, P.; Zhang, L.; Guo, X.; Dai, X.; Liu, L.; Xi, L.; Wang, J.; Song, L.; Wang, Y.; Zhu, Y.; et al. Diversity, biogeography, and biodegradation potential of Actinobacteria in the deep-sea sediments along the South-west Indian Ridge. *Front. Microbiol.* **2016**, *7*, 1340. [CrossRef] [PubMed]
52. Pankratov, T.A.; Dedysh, S.N.; Zavarzin, G.A. The leading role of Actinobacteria in aerobic cellulose degradation in Sphagnum peat bogs. *Dokl. Biol. Sci.* **2006**, *410*, 428–430. [CrossRef] [PubMed]
53. Bao, Y.; Dolfing, J.; Guo, Z.; Chen, R.; Wu, M.; Li, Z.; Lin, X.; Feng, Y. Important ecophysiological roles of non-dominant actinobacteria in plant residue decomposition, especially in less fertile soils. *Microbiome* **2021**, *9*, 84. [CrossRef]
54. Sorokin, D.Y.; Lücker, S.; Vejmelkova, D.; Kostrikina, N.A.; Kleerebezem, R.; Rijpstra, W.I.C.; Damsté, J.S.S.; Le Paslier, D.; Muyzer, G.; Wagner, M.; et al. Nitrification expanded: Discovery, physiology and genomics of a nitrite-oxidizing bacterium from the phylum Chloroflexi. *ISME J.* **2012**, *6*, 2245–2256. [CrossRef]
55. Trivedi, P.; Anderson, I.C.; Singh, B.K. Microbial modulators of soil carbon storage: Integrating genomic and metabolic knowledge for global prediction. *Trends Microbiol.* **2013**, *21*, 641–651. [CrossRef]
56. Liu, J.; Sui, Y.; Yu, Z.; Shi, Y.; Chu, H.; Jin, J.; Liu, X.; Wang, G. High throughput sequencing analysis of biogeographical distribution of bacterial communities in the black soils of northeast China. *Soil Biol. Biochem.* **2014**, *70*, 113–122. [CrossRef]
57. Lauber, C.L.; Hamady, M.; Knight, R.; Fierer, N. Pyrosequencing-based assessment of soil pH as a predictor of soil bacterial community structure at the continental scale. *Appl. Environ. Microb.* **2009**, *75*, 5111–5120. [CrossRef]
58. Dedysh, S.N.; Ivanova, A.A. Planctomycetes in boreal and subarctic wetlands: Diversity patterns and potential ecological functions. *FEMS Microbiol. Ecol.* **2019**, *95*, fiy227. [CrossRef]
59. García-Orenes, F.; Morugán-Coronado, A.; Zornoza, R.; Scow, K. Changes in soil microbial community structure influenced by agricultural management practices in a Mediterranean agro-ecosystem. *PLoS ONE* **2013**, *8*, e80522. [CrossRef]
60. Zhang, D.; Wang, C.; Li, X.L.; Yang, X.S.; Zhao, L.B.; Liu, L.; Zhu, C.; Li, R.H. Linking plant ecological stoichiometry with soil nutrient and bacterial communities in apple orchards. *Appl. Soil Ecol.* **2018**, *126*, 1–10. [CrossRef]
61. Fu, X.; Wang, J.; Sainju, U.M.; Zhao, F.; Liu, W. Soil microbial community and carbon and nitrogen fractions responses to mulching under winter wheat. *Appl. Soil Ecol.* **2019**, *139*, 64–68. [CrossRef]
62. Saxena, A.K.; Kumar, M.; Chakdar, H.; Anuroopa, N.; Bagyaraj, D.J. *Bacillus* species in soil as a natural resource for plant health and nutrition. *J. Appl. Microbiol.* **2019**, *128*, 1583–1594. [CrossRef] [PubMed]
63. Masuda, Y.; Yamanaka, H.; Xu, Z.-X.; Shiratori, Y.; Aono, T.; Amachi, S.; Senoo, K.; Itoh, H. Diazotrophic Anaeromyxobacter isolates from soils. *Appl. Environ. Microbiol.* **2020**, *86*, e00956-20. [CrossRef] [PubMed]
64. Xue, P.-P.; Carrillo, Y.; Pino, V.; Minasny, B.; McBratney, A.B. Soil Properties Drive Microbial Community Structure in a Large Scale Transect in South Eastern Australia. *Sci. Rep.* **2018**, *8*, 11725. [CrossRef]
65. Sui, X.; Zhang, R.; Frey, B.; Yang, L.; Liu, Y.; Ni, H.; Li, M.-H. Soil Physicochemical Properties Drive the Variation in Soil Microbial Communities along a Forest Successional Series in a Degraded Wetland in Northeastern China. *Ecol. Evol.* **2021**, *11*, 2194–2208. [CrossRef]
66. Viruel, E.; Fontana, C.A.; Puglisi, E.; Nasca, J.A.; Banegas, N.R.; Pier, S.; Cocconcelli, P.S. Land-use change affects the diversity and functionality of soil bacterial communities in semi-arid Chaco region, Argentina. *Appl. Soil Ecol.* **2022**, *172*, 104362. [CrossRef]
67. Hartmann, M.; Frey, B.; Mayer, J.; Mäder, P.; Widmer, F. Distinct soil microbial diversity under long-term organic and conventional farming. *ISME J.* **2015**, *9*, 1177–1194. [CrossRef]
68. Hou, J.; Cheng, X.; Li, J.; Dong, Y. Magnesium and nitrogen drive soil bacterial community structure under long-term apple orchard cultivation systems. *Appl. Soil Ecol.* **2021**, *167*, 104103. [CrossRef]

69. Zhang, L.; Zhao, Q.; Zhang, M.; Guo, J.; Zheng, J.; Chen, Z.; Jia, Y.; Zhang, J.; Li, Z.; Zhang, H. Mg^{2+} distribution in activated sludge and its effects on the nitrifying activity and the characteristics of extracellular polymeric substances and sludge flocs. *Process Biochem.* **2020**, *88*, 120–128. [CrossRef]
70. Chen, H.; Ma, K.; Huang, Y.; Yao, Z.; Chu, C. Stable Soil Microbial Functional Structure Responding to Biodiversity Loss Based on Metagenomic Evidences. *Front. Microbiol.* **2021**, *12*, 716764. [CrossRef] [PubMed]
71. Langille, M.G.I.; Zaneveld, J.; Caporaso, J.G.; McDonald, D.; Knights, D.; Reyes, J.A.; Clemente, J.C.; Burkepile, D.E.; Vega Thurber, R.L.; Knight, R.; et al. Predictive Functional Profiling of Microbial Communities Using 16S RRNA Marker Gene Sequences. *Nat. Biotechnol.* **2013**, *31*, 814–821. [CrossRef] [PubMed]

biology

MDPI

Article

Keystone Taxa and Predictive Functional Analysis of *Sphagnum palustre* Tank Microbiomes in Erxianyan Peatland, Central China

Baiying Man [1,*], Xing Xiang [1], Junzhong Zhang [2], Gang Cheng [1], Chao Zhang [1], Yang Luo [1] and Yangmin Qin [3]

1 College of Life Science, Shangrao Normal University, Shangrao 334001, China
2 Key Laboratory of Forest Disaster Warning and Control in Yunnan Higher Education Institutions, South West Forestry University, Kunming 650224, China
3 State Key Laboratory of Biogeology and Environmental Geology, China University of Geosciences, Wuhan 430074, China
* Correspondence: shmilying@126.com

Simple Summary: Deciphering the relationship between microbiome of keystone species *Sphagnum palustre* and potential function in the Erxianyan peatland ecosystems is important in the context of global peatland degradation. We evaluated the *S. palustre* tank microbiome and predicted the potential ecological functions. In total, 38 phyla, 55 classes, 122 orders and 490 genera were detected. Proteobacteria and Firmicutes are the dominant endophytes in *S. palustre*. Core microbiomes are mainly found in 7 phyla, 9 classes, 15 orders, 22 families and 42 genera. Functions predictive of microbial communities are involved in nitrogen fixation, carbon cycle, nitrate metabolism, sulfate respiration and chitinolysis, which may enable the *Sphagnum* to adapt to harsh environmental conditions. This study provides new insights into the relationship between *Sphagnum*-associated microbiomes and their potential ecological functions in subalpine peatlands.

Abstract: *Sphagnum* is a fundamental ecosystem of engineers, including more than 300 species around the world. These species host diverse microbes, either endosymbiotic or ectosymbiotic, and are key to carbon sequestration in peatland ecosystems. However, the linkages between different types of *Sphagnum* and the diversity and ecological functions of *Sphagnum*-associated microbiomes are poorly known, and so are their joint responses to ecological functions. Here, we systematically investigated endophytes in *Sphagnum palustre* via next-generation sequencing (NGS) techniques in the Erxianyan peatland, central China. The total bacterial microbiome was classified into 38 phyla and 55 classes, 122 orders and 490 genera. The top 8 phyla of Proteobacteria (33.69%), Firmicutes (11.94%), Bacteroidetes (9.42%), Actinobacteria (6.53%), Planctomycetes (6.37%), Gemmatimonadetes (3.05%), Acidobacteria (5.59%) and Cyanobacteria (1.71%) occupied 78.31% of total OTUs. The core microbiome of *S. palustre* was mainly distributed mainly in 7 phyla, 9 classes, 15 orders, 22 families and 43 known genera. There were many differences in core microbiomes compared to those in the common higher plants. We further demonstrate that the abundant functional groups have a substantial potential for nitrogen fixation, carbon cycle, nitrate metabolism, sulfate respiration and chitinolysis. These results indicate that potential ecological function of *Sphagnum palustre* in peatlands is partially rooted in its microbiomes, and that incorporating into functional groups of *Sphagnum*-associated microbiomes can promote mechanistic understanding of *Sphagnum* ecology in subalpine peatlands.

Keywords: Erxianyan peatland; *Sphagnum palustre*; core microbiome; function prediction

Citation: Man, B.; Xiang, X.; Zhang, J.; Cheng, G.; Zhang, C.; Luo, Y.; Qin, Y. Keystone Taxa and Predictive Functional Analysis of *Sphagnum palustre* Tank Microbiomes in Erxianyan Peatland, Central China. *Biology* **2022**, *11*, 1436. https://doi.org/10.3390/biology11101436

Academic Editor: Huizhong Chen

Received: 25 August 2022
Accepted: 28 September 2022
Published: 30 September 2022

1. Introduction

Peatlands play a crucial role in the global carbon cycle. *Sphagnum mosses* are considered to be engineers and contribute to the carbon sequestration in acidic, cold and water-saturated peatland ecosystems [1,2]. The genus *Sphagnum* includes 250–450 species around the world, covering about 4 × 10^6 km^2 of peatlands, making it the largest terrestrial carbon reservoir [3–5]. As a fundamental member of peatland ecosystems, *Sphagnum mosses* hosts diverse bacterial communities, either endosymbiotic or ectosymbiotic, and therefore, studying them should yield insights into *Sphagnum* biology and the function performed by microbials in peatland ecosystems [1]. Within those *Sphagnum*-associated bacterial communities, some functional groups have begun to be revealed in N_2 fixation, disease suppression, promote growth [6–8] of *Sphagnum* mosses and elemental cycling [9,10] in peatland ecosystems. In addition, the *Sphagnum* microbial biomass and microbial community are impacted by elevated temperatures and respond rapidly to temperature alterations [8,11,12]. Generally, *Sphagnum*-associated microbiomes differ from those of common plants due to the inhospitable acidic environments. Many microbes living in plants can endow hosts with the ability to adapt to extreme conditions, degrade organic pollutants [13], and improve plant growth [14], but little is known about the endosymbiotic nature of *Sphagnum* individuals [15–17]. Endophytes are microorganisms that inhabit the internal tissues of plants and have potentially biofunctional interactions with plants. Although *sphagnum* endophytes have been reported continuously in recent years [18,19], they are still relatively unknown compared with the more than 300 other species around the world.

With the increase in sequencing depth afforded by next-generation sequencing (NGS), studies began focusing on identifying the "core microbiome". The term "core microbiome" is used to describe the shared microbes common among the microbial communities [20–22]. Some core microbiomes are not only common to plant species or habitats [23], but also as important components to perform basic functions [24]. Although studies on core microbiomes are focusing on the key species within human, plant, lake, soil, and wastewater treatment systems [25–28], there is still a need to bridge the gap between *Sphagnum*-associated microbiomes and *Sphagnum* individuals in parallel with next-generation sequencing (NGS) [29].

Erxianyan, a typical subalpine peatland, is located in western China. So far, *Sphagnum-palustre*-associated microorganisms in Erxianyan have been demonstrated only for the diversity and ecology of testate amoebas [30]. In contrast, the endophytes of *S. palustre* in the Erxianyan peatland have yet to be characterized. This substantially limits knowledge acquisition of *Sphagnum*-associated microbiomes in situ as well as further exploration of the potential ecological functions. Thus, the *Sphagnum*-associated microbiomes and their function still remain to be elucidated. Therefore, the objectives of this work were to (1) investigate the diversity and ecology of *S.-palustre*-associated microbiomes in the Erxianyan peatland, central China, (2) determine the *S.-palustre*-associated core microbiomes, and (3) predict its potential ecological functions in subalpine peatlands.

2. Materials and Methods

2.1. Study Site Description and Sphagnum Collection

Erxianyan, a subalpine peatland, is located in the Middle Yangtze Reach, western Hubei Province, China (Figure 1a,b). This region is strongly influenced by the East Asian summer monsoon [30,31]. The Erxianyan peatland is considered to comprise ombrotrophic bogs, which receive most of their nutrients and water from precipitation. *Sphagnum palustre* is the dominant vegetation in the Erxianyan peatland based on field investigation combined with laboratory evaluation [32], while other plants include *Malus hupehensis* (Pamp.) Rehd, *Calamagrostis epigeios* (Linn.) Roth, *Carex* sp., *Echinochloa crusgalli* (L.) Beauv, *Hosta ventricosa* (Salisb) Stearm and *Reynoutria japonica* Houtt [30].

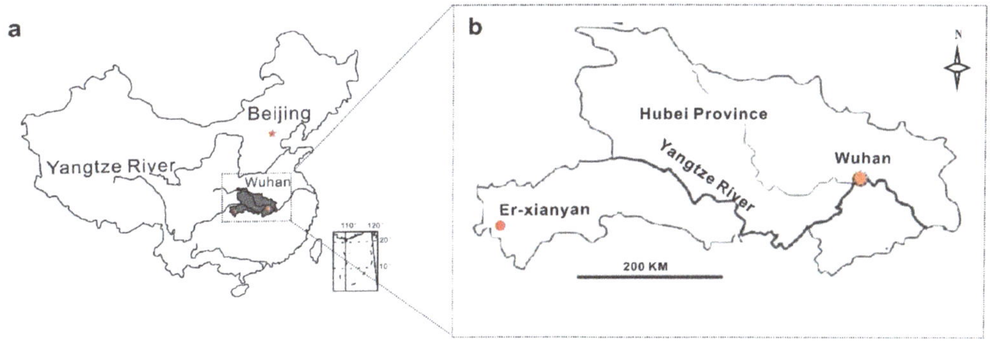

Figure 1. Location of study area (**a**) and sampling site (**b**) in the Erxianyan peatland, central China.

The *S. palustre* was collected aseptically at five sampling sites (200 m intervals). We collected 6 mixed samples (100 m apart) at each sampling site with sterile tubes (Corning) using the five-point sampling method. Altogether, 30 samples from five sites were collected and named as group A (E0W), group B (E2W), group C (E4W), group D (E10W) and group E (E18W). Temperature (22.1–28.4 °C), water pH (5.3–6.6), dissolved oxygen (0.22 to 0.45 mg/L), electric conductivity (13.2 to 44.1 μS/cm) and oxidation reduction potential (103.7 to 152.2 mV) were measured directly in the field using a multi-parameter water quality detector (HACH, Loveland, CO, USA). When the pore water was still, water table level (−11 to 0 cm) was measured. After being transported to the microbiology lab on ice within 24 h, the matrix was first cleaned by rinsing several times in sterile distilled water to remove the matrix and then washed 3 times for 5 min each time with sterile distilled water. Subsequently, samples were placed into 75% ethanol for 3 min, followed by washing 5 times with sterile distilled water. To ensure that the downstream experiments were all endophytes, the final sterile water was inoculated on the R2A to determine whether disinfection was complete. The residual water was finally absorbed by the sterilized filter paper, and the treated samples were stored at −80 °C in sterile plastic centrifuge tubes (Corning) until DNA extraction.

2.2. Genomic DNA Extraction and High-Throughput Sequencing

Total nucleic acids were extracted from 1 g of freeze-dried mixed samples using the PowerSoil DNA Kit (MoBio Laboratories, Inc., Carlsbad, CA, USA). Nucleic acids were eluted with 60 μL buffer and quantified with Nanodrop 2000 (Thermo Fisher Scientific, Waltham, MA, USA). Diluted DNA (1 ng/μL) was used as PCR templates and RNAase-free water as negative controls. To access the bacterial communities, three sets of index primers were used to amplify the V4 (515F–806R) region of the 16S rRNA gene of each sample [33]. PCR reactions contained a total volume of 50 μL, including 25 μL Premix Taq (Takara Biotechnology, Dalian Co., Ltd., Dalian, China), 1 μL of each primer (10 mM) and 3 μL DNA (20 ng/μL) template and 20 μL RNAase-free water. The PCR conditions were as follows: 95 °C for 3 min and 27 cycles of 95 °C for 30 s, 55 °C for 30 s and 72 °C for 45 s, with a final extension of 72 °C for 10 min. Products were purified and quantified with a Quit 2.0 fluorometer (Invitrogen, Carlsbad, CA, USA). Libraries of samples were generated using Ion Plus Fragment Library Kit 48 rxns (Thermo Fisher Scientific, Shanghai Co., Ltd., Shanghai, China) and assessed on the Qubit@ 2.0 Fluorometer (Invitrogen, Carlsbad, CA, USA) and Agilent Bioanalyzer 2100 system. Sequencing was then performed commercially on the IonS5TMXL sequencing platform at the Novogene Bioinformatics Technology (Beijing, China).

2.3. Bioinformatics Analysis

Briefly, the low-quality parts of reads were discarded by Cutadapt (V1.9.1, http://cutadapt.readthedocs.io/en/stable/, accessed on 5 March 2022) [34]. Raw reads were quality-filtered and de-multiplexed. Chimeric sequences were removed by VSEARCH [35]. Operational taxonomic units (OTUs) were assigned using UPARSE (version 7.1 http://drive5.com/uparse/, accessed on 5 March 2022) with a 97% cut-off [36]. The high-quality sequences were aligned against the SILVA database (http://www.arb-silva.de/, accessed on 13 September 2021) [36,37] for taxonomic classification. Diversity indices, such as Observed-species, Chao1, ACE, Shannon, Simpson and Good's coverage, were calculated with rarefied data using QIIME (Version 1.7.0). Species accumulation curves were created using R software. Differences in alpha diversity for all pairwise differences between means were compared via Tukey's and Wilcox tests. Dissimilarity comparisons in each two sampling sites were further examined under significant values ($\alpha = 0.05$) using PERMANOVA based on Bray-Curtis dissimilarities, and the *p*-value was adjusted via the Benjamini method. Differences in alpha diversity were tested with one-way analysis of variance (ANOVA) in SPSS 18. Community composition was performed via PCoA (principal co-ordinates analysis) in R software (Version 2.15.3). The distinctiveness of microbiomes in different sites was analyzed using the linear discriminant analysis effect size (LEfSe) method. The taxon with significant differences between groups and significance of the detected variations were analyzed via *t*-test (*p*-value).

The composition of the core microbiome (the OTUs observed in most ($\geq$90%) of the samples) was calculated to present the diversity of the bacterial community at a more refined taxonomic level. We submitted the OTU biome file to MetaCoMET (the Metagenomics Core Microbiome Exploration Tool) for discovery and visualization of the core microbiomes [38]. A significance test was performed using ANOVA, and a p value < 0.05 was considered statistically significant. In addition, FAPROTAX was constructed for functional annotation of microbiomes.

All the sequencing data were deposited in the National Omics Data Encyclopedia (NODE) under the project accession OEP001043 (https://www.biosino.org/node/project/detail/OEP001043, submitted on 3 July 2020).

3. Results

3.1. Community Composition and Biodiversity Assessment

A total of 30 *S. palustre* samples in the Erxianyan peatland were sequenced using the IonS5^{TM}XL platform, resulting in 984 Mb reads and 2,186,399 sequences. Of those, 2,110,195 (96%) high-quality chimera-free clean data passed the stringent quality control (Phred quality scores 20: 96.5% on average) and fell into 1,851,366 taxa tags (61,712 on average) and 3969 OTUs (Table 1). The mean OTUs ranged from 499 to 2152, with a 3% cutoff. A total of 3969 OTUs were annotated based on $\geq$97% nucleotide sequence identity between sequences, including 38 phyla, 55 classes, 122 orders and 490 genera. Only 0.05% of sequences belong to an unclassified group. Species accumulation curves were generated for each sample to assess whether the sample size provided sufficient OTU coverage of the *Sphagnum*-associated microbiomes (Figure 2). The species accumulation curves showed that the libraries could reflect the main bacterial information in each sample (Figure 2).

Table 1. The sequencing results of microbiomes in *S. palustre* in the Erxianyan peatland. Diversity indices and richness metrics are also presented.

Sample	Total Tags	Taxon Tags	OTUs	Observed Species	Shannon	Simpson	Chao1	ACE	Goods Coverage (%)
E0W01	70,028	64,973	1015	809	3.34	0.77	994.25	1000.77	99.60
E0W03	70,144	65,305	499	386	2.76	0.77	503.61	549.77	99.70
E0W1A	70,260	64,355	753	615	3.93	0.82	726.78	769.59	99.70
E0W3A	70,137	65,126	735	597	4.91	0.92	697.72	717.74	99.70
E0W1B	75,644	71,054	926	770	2.78	0.57	917.00	998.60	99.60
E0W3B	70,168	65,832	1466	1261	4.16	0.76	1415.29	1460.88	99.50
E2W01	70,086	58,693	2152	2028	9.19	1.00	2092.96	2108.92	99.70
E2W03	70,098	59,469	1109	914	4.25	0.82	1059.44	1106.76	99.60
E2W1A	70,197	66,422	615	497	2.40	0.57	613.04	653.66	99.70
E2W3A	70,736	66,471	779	624	2.95	0.69	768.01	782.86	99.70
E2W1B	62,775	55,600	1770	1610	5.59	0.91	1644.79	1703.36	99.70
E2W3B	68,365	63,327	748	601	4.42	0.91	722.26	753.98	99.70
E4W01	70,121	65,995	857	681	2.54	0.64	852.01	893.87	99.60
E4W03	70,039	64,692	911	762	5.00	0.91	907.76	955.94	99.60
E4W1A	70,135	30,225	1021	854	4.23	0.81	975.72	1015.33	99.60
E4W3A	75,810	38,562	775	644	4.11	0.86	791.02	831.05	99.70
E4W1B	70,114	32,545	927	773	4.25	0.81	920.86	952.33	99.60
E4W3B	70,141	65,377	714	576	3.22	0.77	728.80	754.16	99.70
E10W01	70,168	63,701	1995	1809	7.77	0.99	1996.37	2004.69	99.40
E10W03	70,113	66,461	1030	852	2.98	0.57	1007.37	1035.15	99.60
E10W1A	70,138	64,424	1057	901	6.40	0.98	1049.68	1081.34	99.60
E10W3A	70,131	64,665	1072	927	6.38	0.96	1046.72	1068.99	99.70
E10W1B	73,679	69,889	851	701	4.55	0.84	805.33	851.42	99.70
E10W3B	70,130	64,983	1136	1004	6.61	0.97	1157.46	1166.40	99.60
E18W01	70,162	65,309	1349	1169	5.74	0.91	1323.50	1367.93	99.50
E18W03	70,195	65,172	1833	1633	6.20	0.92	1804.13	1840.30	99.40
E18W1A	70,053	65,817	940	795	5.13	0.86	952.96	984.06	99.60
E18W3A	70,081	65,481	1114	921	5.31	0.92	1116.03	1167.87	99.50
E18W1B	70,213	64,520	1413	1227	7.00	0.97	1397.65	1435.94	99.50
E18W3B	70,131	66,921	704	602	5.09	0.89	714.13	735.06	99.70
Total	2,110,195	1,851,366	32,266	27,543					
Mean	70,340	61,712	1076	918	4.77	0.84	1056.75	1091.62	99.62

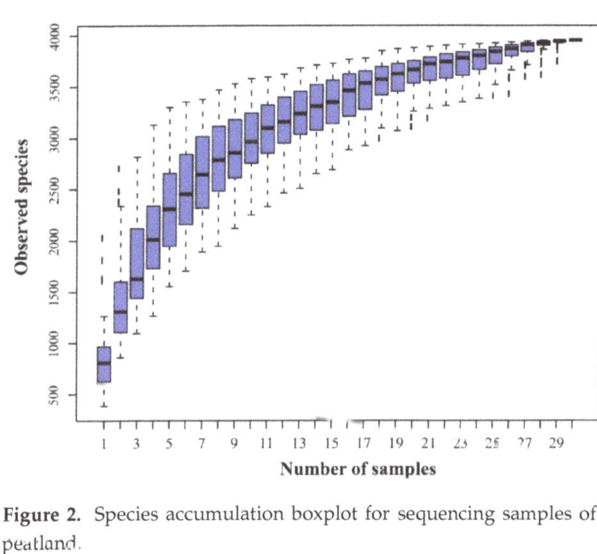

Figure 2. Species accumulation boxplot for sequencing samples of *S. palustre* in the Erxianyan peatland.

Of 38 detected phyla, Acidobacteria, Actinobacteria, Bacteroidetes, Cyanobacteria, Firmicutes, Proteobacteria and Gemmatimonadetes were distributed as common phyla in all samples. Further analysis indicated that Proteobacteria was the dominant phylum, accounting for 48.26 and 33.69% of total reads and OTUs. It was subdivided into three classes: gamma, alpha and delta. The top 8 phyla (the relative abundance > 1%) occupied 78.31% of the total OTUs, including Proteobacteria (33.69%, 1337 OTUs), Firmicutes (11.94%, 474 OTUs), Bacteroidetes (9.42%, 374 OTUs), Actinobacteria (6.53%, 259 OTUs), Planctomycetes (6.37%, 253 OTUs), Gemmatimonadetes (3.05%, 121 OTUs), Acidobacteria (5.59%, 222 OTUs) and Cyanobacteria (1.71%, 68 OTUs) (Figure 3). The remaining 30 bacterial taxa were rare and only accounted for 1.28% and 21.69% of the relative abundance and OTUs, respectively.

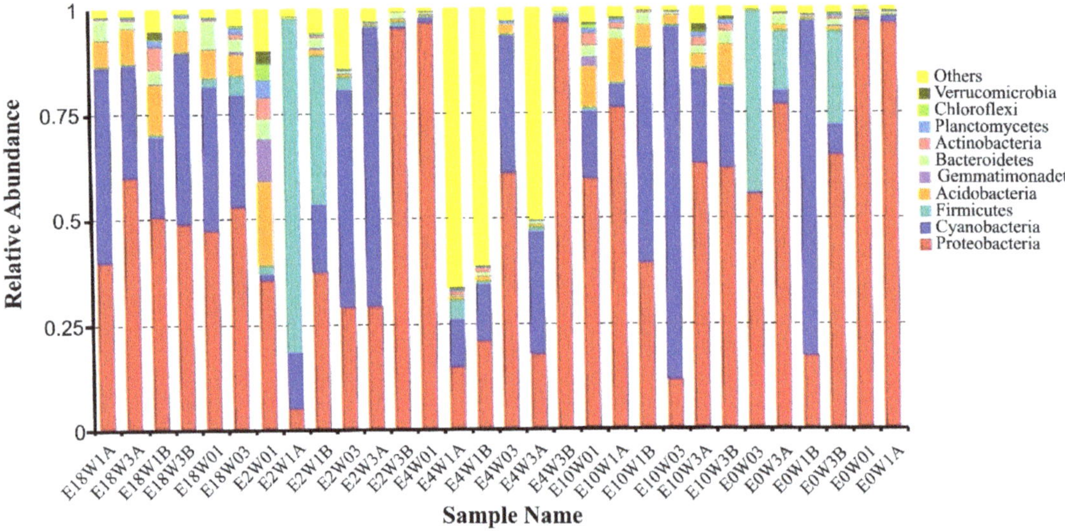

Figure 3. The relative abundance of top 10 phyla of endophytes detected from *S. palustre* in the Erxianyan peatland.

At the order level, a total of 122 orders were obtained from the Erxianyan peatland. The top 12 orders (the relative abundance > 1%) include 9 known orders and 3 unidentified groups. Of those, the relative abundance of unidentified_Cyanobacteria (27.16%) was the dominant group, followed by unidentified gamma Proteobacteria (18.19%), Xanthomonadales (12.71%), unidentified alpha Proteobacteria (12.17%), Lactobacillales (4.43%), Clostridiales (3.34%), Caulobacterales (3.00%), Acidobacteriales (2.85%), Rhizobiales (2.51%), Rickettsiales (2.24%), Sphingomonadales (1.66%) and Enterobacteriales (1.10%). The top 12 orders covered nearly 77.78% and 31.49% of total reads and OTUs, respectively. The remaining 106 bacterial taxa were rare (the relative abundance < 1%), and the relative abundance only accounted for 8.60%.

Microbiomes were highly diverse in *S. palustre*, as indicated by alpha diversity. The dominant phylotypes were fully captured by high-throughput sequencing, as evidenced by a high Good's coverage (from 99.40% to 99.70%, 99.62% on average, Table 1) (n = 30) and plateaued species accumulation curves (Figure 2). No significant difference was observed for the Chao1 (503.61 to 2092.96, Mean = 1056.75) and ACE (549.77 to 2108.92, Mean = 1091.62) indices. Shannon's index ranged from 2.40 to 9.19 (4.77 on average) among 30 samples (Table 1). Community diversity had a significant difference across sampling sites, especially in Shannon's index for the A and D (Wilcox test, $p = 0.0049$), A and E (Wilcox test, $p = 0.0027$) and C and E groups (Wilcox test, $p = 0.0043$). The Simpson index, however, had a significant difference only in the A and D (Wilcox test, $p = 0.0206$) and A and E groups (Wilcox test, $p = 0.0183$). Alpha diversity had a significant difference between the

B and E groups (ANOVA, $p = 0.004$). The microbiome between two sites had a significant difference, e.g., the A and E (PERMANOVA, $p = 0.03$, $F_{value} = 5.77$) groups and C and E (PERMANOVA, $p = 0.03$, $F_{value} = 13.49$) groups. PCoA based on the weighted-UniFrac distance explained 39.82% of the variation through Axis 1 and 25.21% through Axis 2 (Figure 4).

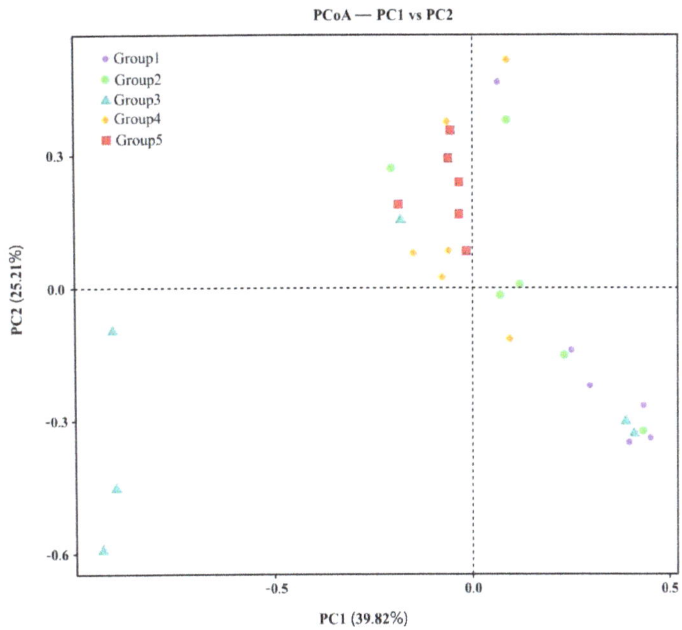

Figure 4. PCoA (weighted−UniFrac distance) of *S. palustre* endophytes in the Erxianyan peatland.

The application of LEfSe analysis can help to find indicator groups in various sites. In total, 30 indicator groups were distinguished from the 5 sampling sites and were mainly associated with four classes of gamma Protobacteria, alpha Protobacteria, Bacteroidia and Acidobacteriia. Group E identified 15 indicator groups, such as genera of *Acidisoma, Mucllaglnibacter, Bacteroidia, Granulicella, Rhodanobacter, Acidocella* and the family of Acetobacteraceae and Sphingobacteriaceae. Conversely, only seven indicator groups were detected in group B, including Bacilli, Lactobacillales, Enterobacteriales Enterobacteriaceae, Lactobacillaceae, Lactobacillus and Cupriavidus. Gamma Protobacteria (*t*-test, $p = 0.012$) and alpha Protobacteria (*t*-test, $p = 0.004$) had a relatively high abundance in group E, while Bacteroidia (*t*-test, $p = 0.003$) and Acidobacteriia (*t*-test, $p = 0.03$) had a relatively high abundance in group D.

3.2. The S.-palustre-Associated Core Microbiome

Microbiomes of *S. palustre* in the Erxianyan peatland are highly diverse, with 491 genera in total. We further selected core OTUs (OTU shared by most (≥90%) sample individuals) to present the bacterial community diversity at a more refined taxonomic level, such as known genus. In all, 183 core OTUs were selected and comprised up to 71.85% and 4.95% of total reads and OTUs, respectively. We found that the core microbiome was mainly distributed in 7 phyla, 9 classes, 15 orders, 22 families and 42 genera. Of them, 43 core known genera of 183 core OTUs were examined (Online Resource, Table S1). These core known genera, which in total accounted for 0.01–8.53% of total reads of core OTUs, showed clear dominant core known genera mainly in three phyla: Proteobacteria (24.98%, 50 OTUs), Firmicutes (4.54%, 11 OTUs) and Acidobacteria (1.59%, 4 OTUs). Proteobacteria was by far the most common and comprised up to 24.98% of the core bacterial microbiome in *S. palustre* at the phylum level (Figure 3). At the genus level (Online Resource, Table S1), the seven high-abundance genera (relative abundance >1%, bold in Online Resource, Table S1) of *Stenotrophomonas* (8.53%), *Dyella* (3.86%), *Lactobacillus* (3.39%), *Acidocella* (3.18%), *Acidisoma* (2.18%), *Granulicella* (1.42%) and *Rhodanobacter* (1.38%) and the six medium-abundance genera (relative abundance > 0.5%, bold in Online Resource, Table S1) of *Lactococcus, Sphingomonas, Acidisphaera, Roseiarcus, Serratia* and *Pseudomonas* encompassed nearly 27.52% and 19.13% of total reads and OTUs of 183 core OTUs, respectively, and relative abundances varied from 0.54% to 8.53%. The remaining genera were rare (relative abundance < 0.5%) and mainly distributed across seven phyla (Online Resource, Table S1).

In addition, we used the persistence method to identify the OTUs present in 30 samples and determine the core microbiome of *S. palustre* on the MetaCoMET platform. In all, 989 shared OTUs and unique OTUs for five sample groups were statistically shown in the Venn graph (Figure 5a). Unique or shared OTUs for each group are displayed in Figure 5b. Of them, 75 OTUs exhibited statistically significant differences between OTU abundance in different sample groups (ANOVA, $p < 0.05$); the hierarchical taxa are displayed in Figure 6. Of them, 13 known genera, i.e., *Acidipila, Acidisoma, Acidisphaera, Acidocella, Conexibacter, Granulicella, Methylocella, Mucilaginibacter, Novosphingobium, Phenylobacterium, Rhodanobacter, Roseiarcus* and *Singulisphaera*, showed statistically significant differences in different sample groups (ANOVA, $p < 0.05$), many of which were consistent with the known genus of 90% of the samples (Figure 6, Online Resource, Table S1). Notably, we found that 10 genera of *Acidicapsa, Alistipes, Aquisphaera, Arenimonas, Bryobacter, Gemmata, Inquilinus, Ralstonia, Rhodovastum* and *Terriglobus* still showed statistically significant differences for groups (ANOVA, $p < 0.05$), but they were not present in each individual sample.

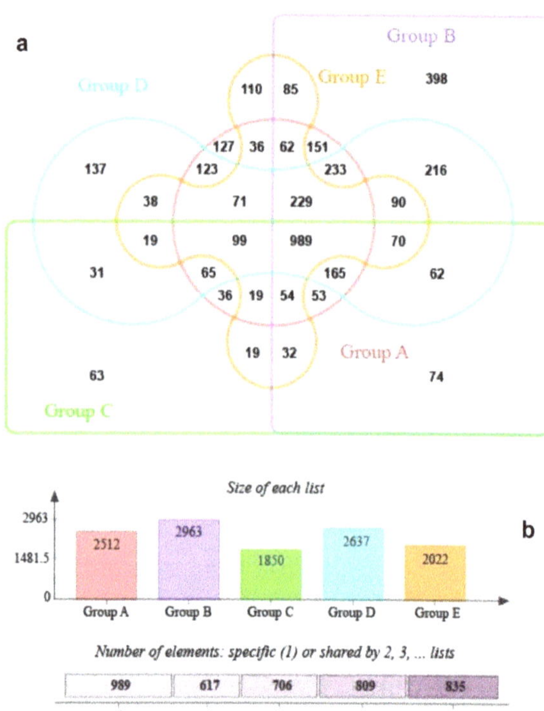

Figure 5. Venn diagram analysis of the total number of common and distinct OTUs (**a**) and sequences (**b**) for *S. palustre* endophytes in the Erxianyan peatland.

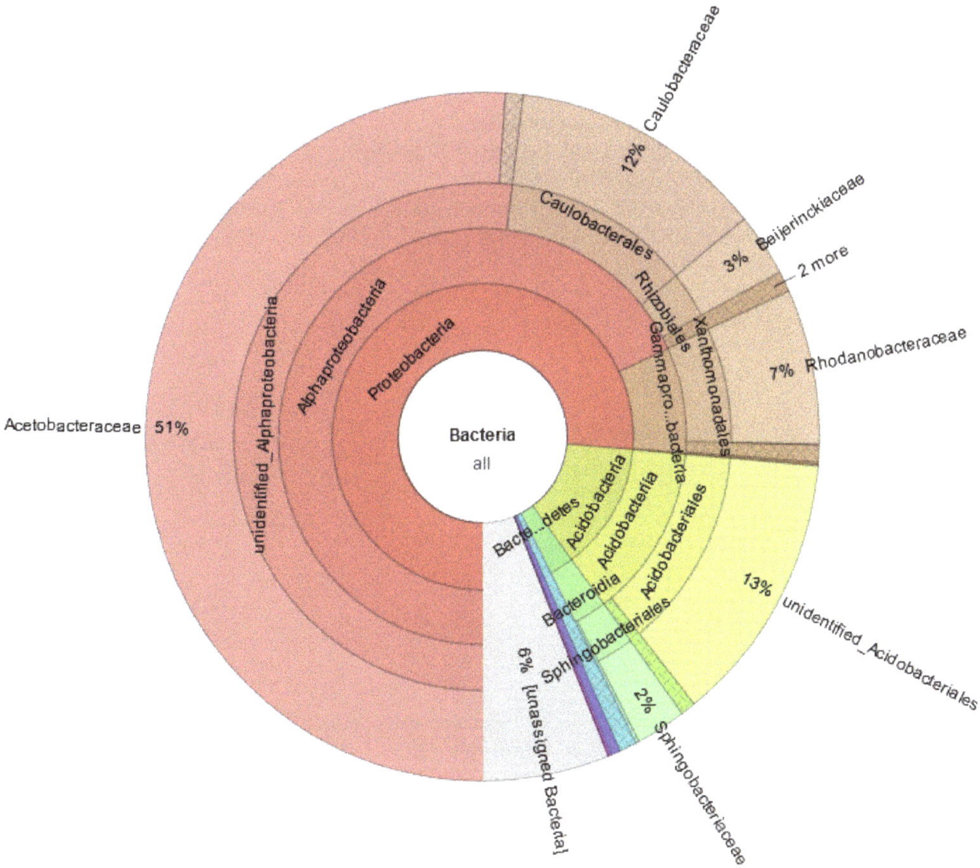

Figure 6. Taxonomic compositions of 75 OTUs at different phylogenetic levels using MetaCoMET. Circles go from inside to outside, indicating different taxa (kingdom to genus). The percentage represents the relative abundance of OTUs.

3.3. Functional Prediction of Microbial Communities

Functional annotation of *S.-palustre*-associated microbiomes revealed a rich repertoire of ecological function groups. A hierarchically clustered heatmap of the top 25 bacterial ecological functional groups was established (Figure 7a,b). The refined taxa of microbial groups and their potential functions are shown in an online resource, Table S2. Phototrophy, photoautotrophy, nitrogen fixation, nitrogen respiration, nitrification, nitrite respiration, nitrite ammonification, aerobic nitrite oxidation, nitrate reduction, nitrate respiration, fermentation, aerobic chemoheterotrophy, chemoheterotrophy, sulfate respiration, respiration of sulfur compounds and chitinolysis were generally abundant, indicating that both heterotrophs and autotrophs are important members of the *S. palustre* microbial communities (Figure 7a,b). We note that the predominant ecological function groups were aerobic chemoautotrophs and chemoheterotrophs, indicating that the oxidation of organic compounds is the main source of carbon and energy. In our study, we detected 13 ecological function groups related to the C cycles, including chemoheterotrophy, aerobic chemoheterotrophy, phototrophy, photoautotrophy, photoheterotrophy, cellulolysis, aromatic compound degradation, oxygenic photoautotrophy, methylotrophy, methanotrophy, methanol oxidation, hydrocarbon degradation and aromatic hydrocarbon degradation (Online Resource, Table S2).

Moreover, we note that the microbiomes of some samples were potentially involved in aerobic ammonia oxidation, denitrification and respiration of nitrogen compounds, including all nitrogen cycling steps, such as nitrification, denitrification, aerobic ammonia oxidation, nitrate reduction, nitrate respiration, nitrogen fixation, nitrogen respiration, aerobic nitrite oxidation, nitrite respiration, nitrous oxide denitrification, nitrate denitrification and nitrite denitrification (Figure 7a,b; Online Resource, Table S2). We also detected the functional group of nitrate denitrification and nitrite denitrification in 93% of the samples, which is the main step of the nitrogen loss pathway. Further, microbial communities in 93% of the samples also contained ureolysis, an ammonia-producing precursor to nitrification.

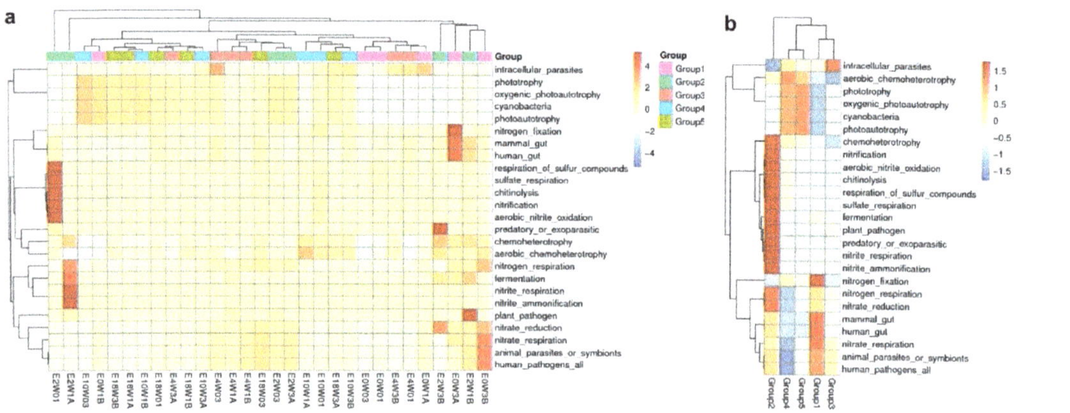

Figure 7. Clustering of functional prediction based on samples (**a**) and sample groups (**b**) for microbiomes of *S. palustre* based on the FAPROTAX database.

4. Discussion

4.1. Proteobacteria Dominant in S.-palustre-Associated Microbiomes

Bacterial communities play an important role in peatland ecosystems, and therefore, studying their dominant phyla should yield an in-depth understanding of their key roles. In our study, 491 genera, 122 orders and 55 classes were detected in 38 phyla. Of them, Proteobacteria dominated in *S.-palustre*-associated microbiomes, which corroborates prior works on the *Sphagnum* in other peatlands [18,39,40]. Recently, studies have implicated Proteobacteria populations as the dominant microbial taxa in peatlands [41] and found that they are inextricably linked to higher carbon availability in acidic environments [42]. Some of them have growth-promoting functions for *Sphagnum* and deeply affected the ecological function of peatland ecosystems [10]. The high abundance of Proteobacteria in *S.-palustre*-associated microorganisms could play a key role in the carbon cycle in the Erxianyan peatland.

In our survey, only three classes of Proteobacteria were detected in all samples, namely alpha, gamma and delta Proteobacteria, which accounted for 11.41%, 13.73% and 7.23% of total OTUs, respectively. *Sphagnum-palustre*-associated gamma Proteobacteria was dominant in Erxianyan. This result represents a significant difference from previous reports [19,43], which showed a predominance of alpha Proteobacteria in the Dajiuhu peatland. It has been demonstrated that gammaProteobacteria is ecologically diverse and richer in genera than all bacterial phyla except Firmicutes. This result is further echoed by our dataset of 205 known genera which belong to the class of gamma Proteobacteria, except for the unclassified groups. To note, the high relative abundance of alpha and gamma Proteobacteria in Erxianyan is consistent with that in tropical peatlands [44]. This result indicates that these populations may be specialized for subtropical and tropical peatlands.

4.2. The S.-palustre-Associated Core Microbiome

The core microbiome varies greatly in different niches but generally maintains the biodiversity and stability of ecosystems [22,27,45,46]. It has been demonstrated that the core microbiome does not only enhance resistance against environmental stress [47] and produce flavors for vinegar [38,48] but also produces a secondary metabolite [27,28]. In this study, the core microbiome was detected in more than 90% of *S. palustre* samples, including 43 known genera in 7 phyla. However, of the known genera in our dataset, Pseudomonadales mainly comprise *Roseiarcus*, *Acidocella*, *Serratia*, *Pseudomonas* and *Dyella*, and they seem to play a potential role in the degradation of aromatic compounds. As indicator groups for *S. palustre*, *Serratia* and *Pseudomona* were also detected in the Dajiuhu peatland [49]. Moreover, *Stenotrophomonas*, which had the highest relative abundance within the known genera, is prone to reduction of nitrate.

In addition to those listed above, 30 more known genera were shared by more than 90% of the samples, with a relative abundance that ranged from 0.01% to 0.45%. *Singulisphaera* is an aquatic bacterium that often inhabits fresh water and acidic *Sphagnum*-dominated wetlands [50], and it is still being examined in this study. *Lactobacillus*, *Lactococcus*, *Enterococcus*, *Bacillus*, *Sporosarcina* and *Butyrivibrio* were found in more than 90% of the samples; however, they are rarely detected in higher plants [51] (Alica et al., 2019). Therefore, there were many differences in the *Sphagnum*-associated core microbiome compared to those in the common higher plants.

Notably, the phylum Gemmatimonadetes was detected in all samples, although it was not listed in core microbiomes due to the lack of an identified genus at a more refined taxonomic level. Previous studies have also shown that this group is usually found in grassland soils, prairie soils and sediment environments [52–54] (Ma et al., 2018; Peng et al., 2017; Zinke et al., 2018). Its long-term evolution has made *S. palustre* an ideal host for Gemmatimonadetes [55]. Therefore, the possible functions of Gemmatimonadetes in *S. palustre* are still not clear and worth further study.

4.3. Methodological Limitations and Future Aspects

The *Sphagnum*-associated microbiome and its interactions provide an important reference for understanding the potential ecological functions of microbiomes in peatlands. Nevertheless, our results may be influenced by multiple factors, such as soil pH, organic matter, water table, climate and human practices. To uncover the relationship between microbes and *Sphagnum*, comprehensive environmental factors, more detailed monitoring of various parameters and the accumulation of annual data should be considered in the near future [2,40,49]. However, it should be noted that although the prediction tool FAPROTAX can be used for a fast-functional screening of 16S derived microbiome data from *S. palustre* (Figure 7), the actual ecological function in the Erxianyan peatland still needs to be verified after screening assays. Further isolation and identification of more potentially biofunctional endophytes in *S. palustre* will still be needed [18]. In addition, it is worthwhile to determine the ecological framework of microbiome, mycobiome and archaea in the future.

5. Conclusions

Microbial communities play a crucial role in peatland ecosystems. We detected the *S.-palustre*-associated core microbiomes in the Erxianyan peatland, which were mainly distributed across 7 phyla, 9 classes, 15 orders, 22 families and 42 genera. We analyzed the potential ecological functions that might enable the hosts to adapt to harsh environmental conditions. We found some common or unique taxa of *S.-palustre*-associated microbiomes in different groups. The potentially biofunctional endophytes in *S. palustre* are involved in nitrogen fixation, carbon cycling, nitrate metabolism, sulfate respiration and chitinolysis, which may enable *Sphagnum* to adapt to harsh environmental conditions. Although the predominant phylum is the same as others reported, there are still some subdominant groups that differ from those in other subalpine peatlands. Moreover, some groups were unique to *Sphagnum* individuals compared to those in the common higher plants. Certainly, further studies on the interaction mechanism between more *Sphagnum* genotype individuals and microbes are required, and the potential ecological functions in the hosts still need to be determined.

Supplementary Materials: The following supporting information can be downloaded at: https://www.mdpi.com/article/10.3390/biology11101436/s1, Table S1: The taxonomy information of 42 known genera in core microbiomes which shared with most (≥90%) of the samples in Erxianyan peatland, central China; Table S2: The predominant ecological function of the *Sphagnum palustre* associated microbiomes in Erxianyan peatland, central China.

Author Contributions: B.M., X.X., J.Z. and Y.Q. collected samples; B.M. designed the experiment, analyzed the data and wrote the paper; Y.Q. revised and edited the manuscript; X.X. and J.Z. performed the experiments; G.C., C.Z. and Y.L. analyzed parts of the data. All authors have read and agreed to the published version of the manuscript.

Funding: The project was jointly supported by the National Natural Science Foundation of China (31960026), Scientific Research Staring Foundation for Scholars of Shangrao Normal University (6000167), the Project of Science and Technology for Jiangxi Education Department, China (GJJ180889) and the General Project of Yunnan Science and Technology Department, China (2017FG001(-035)).

Institutional Review Board Statement: Not applicable.

Informed Consent Statement: Not applicable.

Data Availability Statement: The data are available from the National Omics Data Encyclopedia (accession OEP001043, https://www.biosino.org/node/project/detail/OEP001043).

Acknowledgments: We would like to thank two reviewers for their very helpful comments.

Conflicts of Interest: The authors declare no conflict of interest.

References

1. Vitt, D.H.; House, M. Bryophytes as key indicators of ecosystem function and structure of northern peatlands. *Bryophyt. Divers. Evol.* **2021**, *43*, 253–264. [CrossRef]
2. Ma, X.Y.; Xu, H.; Cao, Z.Y.; Shu, L.; Zhu, R.L. Will climate change cause the global peatland to expand or contract? Evidence from the habitat shift pattern of *Sphagnum* mosses. *Glob. Chang. Biol.* **2022**, 1–14. [CrossRef]
3. Dise, N.B. Peatland response to global change. *Science* **2009**, *326*, 810–811. [CrossRef] [PubMed]
4. Limpens, J.; Berendse, F.; Blodau, C.; Canadell, J.G.; Freeman, C.; Holden, J.; Roulet, N.; Rydin, H.; Schaepman-Strub, G. Peatlands and the carbon cycle: From local processes to global implications—A synthesis. *Biogeosciences* **2008**, *5*, 1475–1491. [CrossRef]
5. Shaw, A.J.; Schmutz, J.; Devos, N.; Shu, S.; Carrell, A.A.; Weston, D.J. The *sphagnum* genome project: A new model for ecological and evolutionary genomics. *Adv. Bot. Res.* **2016**, *78*, 1–18.
6. Shcherbakov, A.V.; Bragina, A.V.; Kuzmina, E.Y.; Berg, C.; Muntyan, A.N.; Makarova, N.M.; Malfanova, N.V.; Cardinale, M.; Berg, G.; Chebotar, V.K.; et al. Endophytic bacteria of *Sphagnum mosses* as promising objects of agricultural microbiology. *Microbiology* **2013**, *82*, 306–315. [CrossRef]
7. Shcherbakov, A.V.; Kuzmina, E.Y.; Lapshina, E.D.; Shcherbakova, E.N.; Chebotar, V.K. Taxonomic diversity of bacterial populations inhabiting gametophytes of *Sphagnum* mosses from different geographic regions of Russia. *Agron. Res.* **2015**, *13*, 192–201.
8. Carrell, A.A.; Kolton, M.; Glass, J.B.; Pelletier, D.A.; Warren, M.J.; Kostka, J.E.; Iversen, C.M.; Hanson, P.J.; Weston, D.J. Experimental warming alters the community composition, diversity, and N_2 fixation activity of peat moss (*Sphagnum fallax*) microbiomes. *Glob. Chang. Biol.* **2019**, *25*, 2993–3004. [CrossRef]
9. Weston, D.J.; Timm, C.M.; Walker, A.P.; Gu, L.; Muchero, W.; Schmutz, J.; Shaw, A.J.; Tuskan, G.A.; Warren, J.M.; Wullschleger, S.D. *Sphagnum* physiology in the context of changing climate: Emergent influences of genomics, modelling and host–microbiome interactions on understanding ecosystem function. *Plant Cell Environ.* **2015**, *38*, 1737–1751. [CrossRef]
10. Kostka, J.E.; Weston, D.J.; Glass, J.B.; Lilleskov, E.A.; Shaw, A.J.; Turetsky, M.R. The *Sphagnum* microbiome: New insights from an ancient plant lineage. *New Phytol.* **2016**, *211*, 57–64. [CrossRef]
11. Basińska, A.M.; Reczuga, M.K.; Gąbka, M.; Stróżecki, M.; Łuców, D.; Samson, M.; Urbaniak, M.; Leśny, J.; Chojnicki, B.H.; Gilbert, D.; et al. Experimental warming and precipitation reduction affect the biomass of microbial communities in a *Sphagnum* peatland. *Ecol. Indic.* **2020**, *112*, 106059. [CrossRef]
12. Carrell, A.A.; Lawrence, T.J.; Cabugao, K.G.M.; Carper, D.L.; Pelletier, D.A.; Lee, J.H.; Jawdy, S.S.; Grimwood, J.; Schmutz, J.; Hanson, P.J.; et al. Habitat-adapted microbial communities mediate *Sphagnum* peatmoss resilience to warming. *New Phytol.* **2022**, *234*, 2111–2125. [CrossRef] [PubMed]
13. Afzal, M.; Khan, Q.M.; Sessitsch, A. Endophytic bacteria: Prospects and applications for the phytoremediation of organic pollutants. *Chemosphere* **2014**, *117*, 232–242. [CrossRef]

14. Raymond, J.A. Dependence on epiphytic bacteria for freezing protection in an Antarctic moss, *Bryum argenteum*. *Environ. Microbiol. Rep.* **2016**, *8*, 14–19. [CrossRef]
15. Opelt, K.; Berg, C.; Schonmann, S.; Eberl, L.; Berg, G. High specificity but contrasting biodiversity of *Sphagnum*-associated bacterial and plant communities in bog ecosystems independent of the geographical region. *ISME J.* **2007**, *1*, 502–516. [CrossRef]
16. Opelt, K.; Chobot, V.; Hadacek, F.; Schonmann, S.; Eberl, L.; Berg, G. Investigations of the structure and function of bacterial communities associated with *Sphagnum* mosses. *Environ. Microbiol.* **2007**, *9*, 2795–2809. [CrossRef]
17. Opelt, K.; Berg, G. Diversity and antagonistic potential of bacteria associated with bryophytes from nutrient-poor habitats of the Baltic Sea Coast. *Appl. Environ. Microbiol.* **2004**, *70*, 6569–6579. [CrossRef]
18. Tang, J.Y.; Ma, J.; Li, X.D.; Li, Y.H. Illumina sequencing-based community analysis of bacteria associated with different bryophytes collected from Tibet, China. *BMC Microbiol.* **2016**, *16*, 276. [CrossRef]
19. Tian, W.; Xiang, X.; Ma, L.; Evers, S.; Wang, R.; Qiu, X.; Wang, H. Rare species shift the structure of bacterial communities across *sphagnum* compartments in a subalpine peatland. *Front. Microbiol.* **2020**, *10*, 3138. [CrossRef]
20. Stéphane, C.; Abdul, S.; Hanna, F.; Angela, S. A review on the plant microbiome: Ecology, functions, and emerging trends in microbial application. *J. Adv. Res.* **2019**, *19*, 29–37.
21. Risely, A. Applying the core microbiome to understand host–microbe systems. *J. Anim. Ecol.* **2020**, *89*, 1549–1558. [CrossRef]
22. Neu, A.T.; Allen, E.E.; Roy, K. Defining and quantifying the core microbiome: Challenges and prospects. *Proc. Natl. Acad. Sci. USA* **2021**, *118*, e2104429118. [CrossRef]
23. Vandenkoornhuyse, P.; Quaiser, A.; Duhamel, M.; Van, A.L.; Dufresne, A. The importance of the microbiome of the plant holobiont. *New Phytol.* **2015**, *206*, 1196–1206. [CrossRef]
24. Lemanceau, P.; Blouin, M.; Muller, D.; Moënne-Loccoz, Y. Let the core microbiota be functional. *Trends Plant Sci.* **2017**, *7*, 583–595. [CrossRef]
25. Pan, J.; Jeffrey, P.; Edwards, M.A.; Amy, P.; Mark, I.A. Impact of water chemistry, pipe material and stagnation on the building plumbing microbiome. *PLoS ONE* **2015**, *10*, e0141087.
26. Zarraonaindia, I.; Owens, S.M.; Weisenhorn, P.; West, K.; Hampton-Marcell, J.; Lax, S.; Bokulich, N.A.; Mills, D.A.; Martin, G.; Taghavi, S.; et al. The soil microbiome influences grapevine-associated microbiota. *mBio* **2015**, *6*, e02527-14. [CrossRef]
27. Chen, H.; Wu, H.; Yan, B.; Zhao, H.; Liu, F.; Zhang, H.; Sheng, Q.; Miao, F.; Liang, Z. Core microbiome of medicinal plant *salvia miltiorrhiza* seed: A rich reservoir of beneficial microbes for secondary metabolism? *Int. J. Mol. Sci.* **2018**, *19*, 672. [CrossRef]
28. Chen, S.Y.; Li, J.J.; Lin, J.; Bao, K.X.; Fan, J.Q.; Zhang, R.Q.; He, W. High-throughput sequencing fungal community structures in aging tobacco strips from different growing areas and stalk positions. *Tob. Sci. Technol.* **2018**, *51*, 12–19.
29. Kaul, S.; Sharma, T.; Dhar, M.K. "Omics" tools for better understanding the plant–endophyte interactions. *Front. Plant Sci.* **2016**, *7*, 955. [CrossRef]
30. Qin, Y.; Mitchell, E.A.; Lamentowicz, M.; Payne, R.J.; Lara, E.; Gu, Y.; Huang, X.; Wang, H. Ecology of testate amoebae in peatlands of central China and development of a transfer function for paleohydrological reconstruction. *J. Paleolimnol.* **2013**, *50*, 319–330. [CrossRef]
31. Qin, Y.; Payne, R.J.; Gu, Y.; Huang, X.; Wang, H. Ecology of testate amoebae in Dajiuhu peatland of Shennongjia Mountains, China, in relation to hydrology. *Front. Earth Sci.* **2012**, *6*, 57–65. [CrossRef]
32. Michaelis, D. *The Sphagnum Species of the World*; Schweizerbart, Science Publishers: Stuttgart, Germany, 2019.
33. Castrillo, G.; Teixeira, P.J.P.L.; Paredes, S.H.; Law, T.F.; De Lorenzo, L.; Feltcher, M.E.; Finkel, O.M.; Breakfield, N.W.; Mieczkowski, P.; Jones, C.D.; et al. Root microbiota drive direct integration of phosphate stress and immunity. *Nature* **2017**, *543*, 513–518. [CrossRef]
34. Louca, S.; Parfrey, L.W.; Doebeli, M. Decoupling function and taxonomy in the global ocean microbiome. *Science* **2016**, *353*, 1272–1277. [CrossRef]
35. Rognes, T.; Flouri, T.; Nichols, B.; Quince, C.; Mahe, F. VSEARCH: A versatile open source tool for metagenomics. *PeerJ* **2016**, *4*, e2584. [CrossRef]
36. Edgar, R.C. UPARSE: Highly accurate OTU sequences from microbial amplicon reads. *Nat. Methods* **2013**, *10*, 996–998. [CrossRef]
37. Wang, Q.; Garrity, G.M.; Tiedje, J.M.; Cole, J.R. Naïve Bayesian classifier for rapid assignment of rRNA sequences into the new bacterial taxonomy. *Appl. Environ. Microbiol.* **2007**, *73*, 5261–5267. [CrossRef] [PubMed]
38. Wang, Y.; Xu, L.; Gu, Y.; Coleman-Derr, D. MetaCoMET: A web platform for discovery and visualization of the core microbiome. *Bioinformatics* **2016**, *32*, 3469–3470. [CrossRef]
39. Kip, N.; Fritz, C.; Langelaan, E.S.; Pan, Y.; Bodrossy, L.; Pancotto, V.; Jetten, M.S.M.; Smolders, A.J.P.; Op den Camp, H.J.M. Methanotrophic activity and diversity in different *Sphagnum magellanicum* dominated habitats in the southernmost peat bogs of Patagonia. *Biogeosciences* **2011**, *9*, 47–55. [CrossRef]
40. Xiang, X.; Wang, H.; Tian, W.; Wang, R.; Gong, L.; Xu, Y. Composition and function of bacterial communities of bryophytes and their underlying sediments in the Dajiuhu Peatland, central China. *J. Earth Sci.* **2020**. Available online: https://kns.cnki.net/kcms/detail/42.1788.P.20201222.1817.004.html (accessed on 11 July 2021).
41. Danilova, O.V.; Belova, S.E.; Gagarinova, I.V.; Dedysh, S.N. Microbial community composition and methanotroph diversity of a subarctic wetland in Russia. *Microbiology* **2016**, *85*, 545–554. [CrossRef] [PubMed]

42. Leff, J.W.; Jones, S.E.; Prober, S.M.; Barberán, A.; Borer, E.T.; Firn, J.L.; Harpole, W.S.; Hobbie, S.E.; Hofmockel, K.S.; Knops, J.M.; et al. Consistent responses of soil microbial communities to elevated nutrient inputs in grasslands across the globe. *Proc. Natl. Acad. Sci. USA* **2015**, *112*, 10967–10972. [CrossRef] [PubMed]
43. Xiang, X.; Wang, H.; Gong, L.; Liu, Q. Vertical variations and associated ecological function of bacterial communities from *Sphagnum* to underlying sediments in Dajiuhu Peatland. *Sci. China-Earth Sci.* **2014**, *57*, 1013–1020. [CrossRef]
44. Mishra, S.; Lee, W.A.; Hooijer, A.; Reuben, S.; Sudiana, I.M.; Idris, A.; Swarup, S. Microbial and metabolic profiling reveal strong influence of water table and land-use patterns on classification of degraded tropical peatlands. *Biogeosciences* **2014**, *11*, 14009–14042. [CrossRef]
45. Toju, H.; Peay, K.G.; Yamamichi, M.; Narisawa, K.; Hiruma, K.; Naito, K.; Fukuda, S.; Ushio, M.; Nakaoka, S.; Onoda, Y.; et al. Core microbiomes for sustainable agroecosystems. *Nat. Plants* **2018**, *4*, 247–257. [CrossRef] [PubMed]
46. Zhou, J.; Yu, L.; Zhang, J.; Zhang, X.; Xue, Y.; Liu, J.; Xiao, Z. Characterization of the core microbiome in tobacco leaves during aging. *Microbiol. Open* **2020**, *9*, e984. [CrossRef] [PubMed]
47. Rui, J.; Li, J.; Zhang, S.; Xuefeng, Y.; Wang, Y.; Li, X. The core populations and co-occurrence patterns of prokaryotic communities in household biogas digesters. *Biotechnol. Biofuels* **2015**, *8*, 158. [CrossRef] [PubMed]
48. Wang, Z.-M.; Lu, Z.-M.; Shi, J.-S.; Xu, Z. Exploring flavour-producing core microbiota in multispecies solid-state fermentation of traditional Chinese vinegar. *Sci. Rep.* **2016**, *6*, 26818. [CrossRef]
49. Tian, W.; Wang, H.M.; Xiang, X.; Wang, R.C.; Xu, Y. Structural variations of bacterial community driven by *Sphagnum* microhabitat differentiation in a subalpine peatland. *Front. Microbiol.* **2019**, *10*, 1661. [CrossRef] [PubMed]
50. Kulichevskaya, I.S.; Ivanova, A.O.; Baulina, O.I.; Bodelier, P.L.; Damste, J.S.S.; Dedysh, S.N. *Singulisphaera acidiphila* gen. nov., sp. nov., a non-filamentous, Isosphaera-like planctomycete from acidic northern wetlands. *Int. J. Syst. Evol. Microbiol.* **2008**, *58*, 1186–1193. [CrossRef]
51. Alica, C.; Jií, B.; Eva, K.; Zuzana, U.; Tomá, P. Spatial heterogeneity of belowground microbial communities linked to peatland microhabitats with different plant dominants. *FEMS Microbiol. Ecol.* **2019**, *95*, fiz130.
52. Ma, B.; Cai, Y.; Bork, E.W.; Chang, S.X. Defoliation intensity and elevated precipitation effects on microbiome and interactome depend on site type in northern mixed-grass prairie. *Soil Biol. Biochem.* **2018**, *122*, 163–172. [CrossRef]
53. Peng, M.; Jia, H.; Wang, Q. The effect of land use on bacterial communities in saline–alkali soil. *Curr. Microbiol.* **2017**, *74*, 325–333. [CrossRef] [PubMed]
54. Zinke, L.A.; Kiel, R.B.; James, M.M.; Wheat, C.G.; Orcutt, B.N.; Amend, J.P. Sediment microbial communities influenced by cool hydrothermal fluid migration. *Front. Microbiol.* **2018**, *9*, 1249. [CrossRef]
55. Takaichi, S.; Maoka, T.; Takasaki, K.; Hanada, S. Carotenoids of Gemmatimonas aurantiaca (Gemmatimonadetes): Identification of a novel carotenoid, deoxyoscillol 2-rhamnoside, and proposed biosynthetic pathway of oscillol 2,2-dirhamnoside. *Microbiology* **2010**, *156*, 757–763. [CrossRef]

biology MDPI

Article

Assessing the Impact of Village Development on the Habitat Quality of Yunnan Snub-Nosed Monkeys Using the INVEST Model

Shuxian Zhu [1,2,3], Li Li [1,2,*], Gongsheng Wu [2], Jialan Liu [4], Timothy J. Slate [1], Hongyan Guo [2] and Dayong Li [1,2,*]

1 Key Laboratory of Southwest China Wildlife Resources Conservation (Ministry of Education), China West Normal University, Nanchong 637001, China
2 Wildlife Management and Ecosystem Health Center, Yunnan University of Finance and Economics, Kunming 650221, China
3 Natural Resources Bureau of Heping County, Heyuan 517200, China
4 Kunming Institute of Zoology, Chinese Academy of Sciences, Kunming 650201, China
* Correspondence: lilyzsu@126.com (L.L.); 980119lsc@163.com (D.L.)

Simple Summary: The Yunnan snub-nosed monkey is one of the most endangered species on the IUCN Red List. The study of its population and habitat quality is important in identifying opportunities for balancing socio-economic development against species conservation in the area's villages. Such balances are important to protecting and improving habitat diversity and biodiversity. Our habitat quality analysis indicates that increases in socio-economic developments in the villages around the habitat area have decreased both the habitat area and the habitat quality over time. This has resulted in a decline in biodiversity persistence, resilience, and breadth. It also has exacerbated the risk of declining species populations, potentially to extinction. Though focused on the Yunnan snub-nosed monkey, our approach toward the assessment of habitat quality based on species habitat suitability introduces a new perspective for assessing village development impacts on the habitat quality for the conservation of other species.

Abstract: The habitats of the already endangered Yunnan snub-nosed monkey (*Rhinopithecus bieti*) are degrading as village economies develop in and around these habitat areas, increasing the depopulation and biodiversity risk of the monkey. The paper aims to show the areas of these monkeys' high-quality habitats that are at highest risk of degradation by continued village development and hence be the focus of conservation efforts. Our analysis leveraged multiple tools, including primary component analysis, the InVEST Habitat-Quality model, and GIS spatial analysis. We enhanced our analysis by looking at habitat quality as it relates to the habitat suitability for the monkey specifically, instead of general habitat quality. We also focused on the impact of the smallest administrative scale in China—the village. These foci produced a clearer picture of the monkeys' and villages' situations, allowing for more targeted discussions on win–win solutions for both the monkeys and the village inhabitants. The results show that the northern habitat for the monkey is currently higher quality than the southern habitat, and correspondingly, the village development in the north is lower than in the south. Hence, we recommend conservation efforts be focused on the northern areas, though we also encourage the southern habitats to be protected from further degradation lest they degrade beyond the point of supporting any monkeys. We encourage developing a strategy that balances ecological protection and economic development in the northern region, a long-term plan for the southern region to reduce human disturbance, increase effective habitat restoration, and improve corridor design.

Keywords: snub-nosed monkey; village development; habitat quality; InVEST model; conservation policies; species conservation

Citation: Zhu, S.; Li, L.; Wu, G.; Liu, J.; Slate, T.J.; Guo, H.; Li, D. Assessing the Impact of Village Development on the Habitat Quality of Yunnan Snub-Nosed Monkeys Using the INVEST Model. *Biology* **2022**, *11*, 1487. https://doi.org/10.3390/biology11101487

Academic Editors: Daniel Puppe, Panayiotis Dimitrakopoulos and Baorong Lu

Received: 7 September 2022
Accepted: 8 October 2022
Published: 11 October 2022

1. Introduction

Northwestern Yunnan, with its rich natural resources, is among the most biodiverse areas in China. It is home to one of the most endangered species on Earth—the Yunnan snub-nosed monkey [1,2]. Yunnan snub-nosed monkeys are scattered in the Three Parallel Rivers region in China, spending most of their time in the dense fir-dominated coniferous forests and mixed coniferous forests [1,2]. They remain on high alert for approaching humans, moving quickly to avoid them [3–5]. This makes conducting accurate surveys difficult. In 2016, 15 isolated groups totaling approximately 3000 individuals were surveyed by tracking and photographing the monkeys and recording GPS information of the tracking route [6].

The continuing development of villages around the monkey habitat, as well as other factors, have led to overlap between the villages and the monkey habitats [7,8]. Human activities in these overlapping villages, such as deforestation and land cultivation for agriculture to develop their economies, have impacted the monkey habitat. Increase in the monkey population is limited by the loss, degradation, and fragmentation of its habitat, and risk of decline in the monkey population is increased [3–5].

The plight of the endangered Yunnan snub-nosed monkey highlights the impact of the imbalance of socio-economic development versus ecological development in villages, with the socio-economic development currently winning out [3,5]. Studying this monkey's habitat quality and population is an important foundation for balancing human economic development and species conservation in villages. This balance is an important way to protect an area's habitat diversity and biodiversity [9,10].

Village development has increased human deforestation and use of land areas, including cultivating the land for agriculture and grazing, expanding urban boundaries, and introducing non-native invasive species [11–13]. Such activities led to the decline of habitat quality, the increase of habitat fragmentation, and even the outright loss of wildlife habitats [14], which, in turn, leads to reduced or lost regional biodiversity and the threat of wildlife depopulation [12,14–16]. Therefore, by promoting the coordinated development of socio-economic and habitat quality in villages, habitat quality will be improved, and the durability and stability of ecosystems will be maintained, thus protecting habitat diversity and biodiversity [17,18].

Being able to model habitat quality in an area effectively is critical as habitat quality—the ability of an ecosystem to provide for the survival, reproduction, and development of organisms—reflects biodiversity [19,20]. Habitat quality depends on the proximity of the habitat area to human-developed land and on the intensity of human land use, among other factors. Increases in intensity of human land use have been shown to degrade the quality of nearby natural habitats [12,21]. There has been significant research on the use of the InVEST (Integrated Valuation of Ecosystem Servicer and Trades) Habitat Quality model in assessing biodiversity and in developing conservation measures [22,23].

The InVEST-Habitat Quality model is based on habitat suitability, combined with land cover and biodiversity threat factors, to evaluate habitat quality [20,24]. For example, the InVEST model has been used for habitat quality studies assessing the spatial vulnerability of natural habitats in Chaharmahal and Bakhtriari provinces [25], of bird communities in Central Italy [26], of bird species in Keoladeo National Park [27], and of the wildlife habitat quality in the Greater Serengeti Ecosystem [28]. This model can also describe different threat factors of biodiversity, such as climate change, population density, road density, land use intensity, urbanization (village development and village agglomeration), and changes in landscape patterns [29,30]. This model allows comparison of spatial patterns of biodiversity and ecosystem services, prioritizing species populations for conservation by evaluating multiple land-use change scenarios to find scenarios that best take advantage of the conservation of nature reserves while benefiting human economic development [31].

Using the socio-economic data of 2572 villages and 344 sample plots in northwestern Yunnan, we used principal component analysis of GIS spatial analysis and the InVEST

Habitat-Quality model to study the influence of village development on the habitat quality of Yunnan snub-nosed monkey distribution area in 2018, exploring these three challenges:

(1) Analyze and categorize the development status of villages;
(2) Analyze the spatial distribution of habitat quality in Yunnan snub-nosed monkey distribution area;
(3) With the results of those analyses, determine the impact of village development on the habitat quality of the Yunnan snub-nosed monkey population.

2. Materials and Methods

2.1. Study Area

The study area (Figure 1) is in the Three Parallel River region of northwest Yunnan Province (between 29.020 N, 98.030 E in the north and 25.053 N, 99.022 E in the south, with the elevation varying from 1200 m to 5500 m). It is one of the most ecologically significant areas of China in terms of biodiversity, covering approximately 17,000 km^2 across seven counties in Yunnan (Deqin, Weixi, Lanping, Shangri-La, Lijiang, Jianchuan, and Yunlong), with a total human population of about 1,182,500.

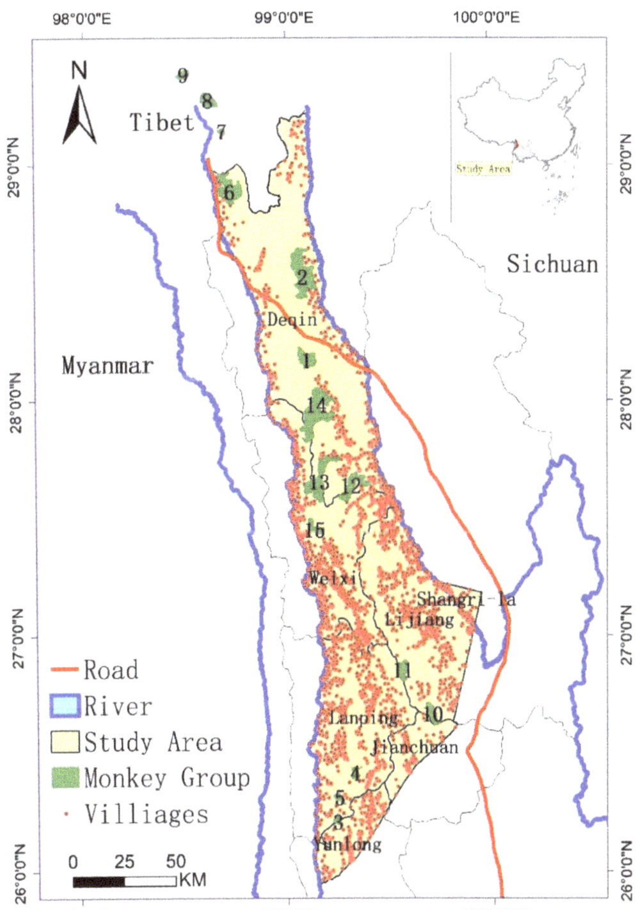

Figure 1. The study area and locations of monkey groups in Yunnan Province (China). The numbers labeling each green area represent the monkey group number (1–15).

The Yunnan snub-nosed monkey lives at a very high elevation (mainly above 3000 m) and is an endangered species on the IUCN Red List [32,33]; 15 groups of these monkeys have been identified previously. We obtained information on the monkey population size as of 2016, with the monkey population being approximately 3000 individuals at that time [6,32].

2.2. Land Use and Land Cover

Land Use and Land Cover (LULC) data in 2018 were obtained from a supervised classification on SPOT-5 images (Institute of Forest Inventory and Planning, Yunnan, 2012) with ground-truthing by the Conservation Information Centre of the Nature Conservancy's China program. All data were geo-corrected in ERDAS 9.2 with a root-mean-square (RMS) error < 1. LULC types (such as Armand pine and hemlock) were assigned one of five habitat categories, based on the Yunnan vegetation classification system and the monkey's habitat preferences. These five habitat categories, in declining quality, were optimal habitat, suboptimal habitat, suitable habitat, unfavorable habitat, and highly unfavorable habitat [5].

2.3. Villages and Rural Roads

Data on 2572 villages, related rural roads, and their socio-economic status were obtained from the National Geographic Information Resources Catalog Service system (https://www.webmap.cn/main/do?method=index, accessed on 13 December 2016) and Yunnan Digital Rural Network (http://www.ynszxc.gov.cn, accessed on 6 May 2018).

2.4. Principal Component Analysis

We selected 2572 villages located in the study area, obtaining 30 indicators for principal component analysis (Table 1). The 30 indicators were grouped into six categories: the natural resources, population, economy, infrastructure, energy, and educational factors. We used the principal component analysis [34] in R v.2.14.1 to calculate eigenvalues, using eigenvalues greater than 1 to identify the principal components of development factors of the village. This downscaled 30 variables into 8 principal components. We used weighted-sum method with these 8 principal components to calculate a comprehensive score for each village, which was used to evaluate the development level of village. Villages were then assigned into one of seven grades (I, II, III, IV, V, VI, and VII) using the Equal Interval Breaks Method (Table 2). Villages of grade I had the lowest village development, villages of grade VII had the highest.

Table 1. Socio-economic data index of villages.

Category	Variable	Code	Category	Variable	Code
Natural resources (X1)	Area of commonly used cultivated land (km^2)	X_{11}	Economics (X3)	Total economic income (million yuan)	X_{31}
	Paddy field area (km^2)	X_{12}		Income from farming (million yuan)	X_{32}
	Dry land area (km^2)	X_{13}		Income from graziery (million yuan)	X_{33}
	Area of cultivated land area per capita (km^2)	X_{14}		Forestry income (million yuan)	X_{34}
	Area of economic fruit woodland (km^2)	X_{15}		Income of secondary and tertiary industries (million yuan)	X_{35}
	Area of fruit woodland per capita (km^2)	X_{16}		Income per capita (yuan)	X_{36}

Table 1. *Cont.*

Category	Variable	Code	Category	Variable	Code
Population (X2)	Rural population (people)	X_{21}	Infrastructure (X4)	Distance to nearest station (km)	X_{41}
	Agricultural population (people)	X_{22}		Distance to nearest market (km)	X_{42}
	Labor force (people)	X_{23}		Cars (units)	X_{43}
	Number of people in the primary industry (people)	X_{24}		Agricultural transport vehicle (units)	X_{44}
	Number of people with tertiary education and above (people)	X_{25}		Tractors (units)	X_{45}
	Number of secondary schools (people)	X_{26}		Motorbike (units)	X_{46}
	Number of primary school students (people)	X_{27}	Energy resources (X5)	Biogas digester farmers (Households)	X_{51}
	Number of people not attending school (people)	X_{28}		Solar farmers (Households)	X_{52}
			Education (X6)	Primary school enrollment rate (%)	X_{61}
				Secondary school enrollment rate (%)	X_{62}

Table 2. Classification standard of villages development grades.

Comprehensive Score Range/Z	**(−∞,−0.5)**	**(−0.5,0)**	**(0,0.5)**	**(0.5,1)**	**(1.1.5)**	**(1.5,2)**	**(2,+∞)**
Scale grades of villages	I	II	III	IV	V	VI	VII

2.5. Plots

The 344 plots were obtained from the 2017 Forest Resources Survey Data of Yunnan Forestry Survey and Design Institute, which used the field survey method. The plots were laid out as squares, each with an area of 4 km^2.

By regularly updating the survey plot data every 5 years, dynamic changes in forest resource growth and decline were obtained. We used the ecological quality formula of plots $Y = \sum_{i=1}^{7} W_i X_i$ to calculate a comprehensive score for each plot, where:

i is the evaluation index (1, 2, and 3 for types I, II, and III, respectively);
X_i is the type score value of each evaluation index;
W_i is the weight of each evaluation index.

We then assigned the type score value of each evaluation index according to the classification criteria (Table 3). The weights (as a percentage) were determined by via the expert scoring method for each evaluation index against the whole.

According to the comprehensive score of the plots, we used Equal Interval Natural Breaks Method to divide the range to four ecological quality grades (Table 4).

Table 3. Plot evaluation indicators and weights.

Evaluation Index	Classification Criteria			Weight (%)
	I	II	III	
Forest naturalness	1, 2	3, 4	5	0.19
Forest community structure	1	2	3	0.18
Tree species structure	6, 7	3, 4, 5	1, 2	0.17
Total vegetation coverage	[70%, 100%]	[50%, 70%)	[0%, 50%)	0.14
Crown density	[0.7, 1.0]	[0.4, 0.7)	[0.2, 0.4)	0.13
Average tree height	[15.0, +∞)	[5.0, 15.0)	[0.9, 5.0)	0.13
Litter depth grade	1	2	3	0.06

Table 4. Classification criteria for the four ecological grades of plots.

Ecological Grade	Comprehensive Ecological Quality Score of Plots	Code
Excellent	<1.4	1
Good	1.4–1.8	2
Medium	1.8–2.2	3
Poor	>2.2	4

2.6. *Habitat Quality Evaluation and Spatial Analysis*

InVEST-Habitat Quality model [35] was used to evaluate snub monkey's habitat quality based on LULC map from 2018 and biodiversity threat factors including villages, village roads, other non-forestry land, economic forest, cropland, and artificial construction [28,36]. We analyzed the correlation between villages, rural roads, and plots to get maximum distance thresholds of villages and rural roads in ArcGIS 10.6 and R [3,8]. Finally, we scored the parameters related to threat factor weights, suitability, and sensitivity of land use types and obtained threat factor attributes (Table 5) and sensitivity of land use types to threats (Table 6). We also applied Spatial Autocorrelation (Moran's I) and Hot Spot Analysis (Getis-Ord Gi*) to explore the spatial variation characteristics of the influence of villages on Yunnan snub-nosed monkey habitat quality [37,38].

Table 5. The threat factors and related coefficients, including maximum effective distance of threats (km), weight, decay type.

Threat Factors	Maximum Effective Distance of Threats (km)	Weight	Decay Type
Village I	2	0.4	Exponential
Village II	2	0.5	Exponential
Village III	2	0.6	Exponential
Village IV	2	0.7	Exponential
Village V	2	0.8	Exponential
Village VI	2	0.9	Exponential
Village VII	2	0.95	Exponential
Village road	4	0.7	Linear
Other non-forestry land	1	0.6	Exponential
Economic forest	1	0.7	Exponential
Cropland	1	0.5	Exponential
Artificial construction	3	0.8	Exponential

Table 6. Sensitivity of land cover types to each threat factor. (Vil1 = Village I, Vil2 = Village II, Vil3 = Village III, Vil4 = Village IV, Vil5 = Village V, Vil6 = Village VI, Vil7 = Village VII, Vr = Village road, Onfl = Other non-forestry land, Ef = Economic forest, Cr = Cropland, Ac = Artificial construction.).

Land Cover Type Code	Land Cover Types	Habitat Suitability	Vil1	Vil 2	Vil 3	Vil 4	Vil 5	Vil 6	Vil 7	Vr	Onfl	Ef	Cr	Ac
1	other non-forest land	0	0	0	0	0	0	0	0	0	0	0	0	0
2	cold coniferous forest (Alpine coniferous forests)	0.6	0.7	0.75	0.8	0.9	0.9	0.9	0.9	0.95	0.4	0.3	0.4	0.8
3	shrublands	0.8	0.3	0.3	0.3	0.3	0.3	0.3	0.3	0.3	0.5	0.3	0.4	0.8
4	Armand pine and hemlock	1	0.9	0.9	0.9	0.9	0.9	0.9	0.9	0.9	0.5	0.4	0.5	0.9
5	barren land	0.2	0.5	0.5	0.5	0.5	0.5	0.5	0.5	0.5	0.2	0.1	0.1	0.2
6	broad-leaved forests	0.6	0.6	0.6	0.6	0.6	0.7	0.7	0.8	0.9	0.4	0.3	0.4	0.8
7	cropland	0	0	0	0	0	0	0	0	0	0	0	0	0
8	planted economic forests	0	0	0	0	0	0	0	0	0	0	0	0	0
9	water body	0.2	0.2	0.2	0.2	0.2	0.2	0.2	0.3	0.4	0.2	0.1	0.1	0.2
10	sclerophyllous evergreen broad-leaved forest	0.8	0.7	0.8	0.8	0.8	0.8	0.8	0.8	0.8	0.5	0.3	0.4	0.8
11	fir-spruce forest	1	0.8	0.8	0.8	0.8	0.8	0.8	0.8	0.8	0.5	0.4	0.5	0.9
12	warm coniferous forest (Yunnan pine forest)	0.2	0.3	0.3	0.3	0.3	0.4	0.4	0.4	0.4	0.2	0.3	0.1	0.2
13	coniferous broad-leaved mixed forest	1	0.8	0.8	0.8	0.8	0.8	0.8	0.85	0.85	0.5	0.4	0.5	0.9
14	Artificial construction	0	0	0	0	0	0	0	0	0	0	0	0	0

2.7. Impact of Village Development on Habitat Quality

We used the kernel density in ArcGIS to analyze the spatial pattern of the socio-economic development of the villages [39,40]. Then the impact of villages on the habitat quality of Yunnan snub-nosed monkeys was evaluated based on GIS spatial analysis.

3. Results

3.1. Analysis the Development of Villages

Our village development analysis revealed that more than half the villages (59.84%) were of Grade II (the second-lowest economic development grade). Grade III (just slightly higher economically developed than Grade II) represented 24.49% of the villages. The remaining Grades represented less than 7% of the villages each (Table 7).

Table 7. The quantities and weight of grades of villages.

Village Grade	Quantities	Weight (%)
Grade I	138	4.20
Grade II	1874	59.84
Grade III	810	24.49
Grade IV	220	6.26
Grade V	74	2.45
Grade VI	57	2.18
Grade VII	21	0.58

Figure 2 shows higher density development of villages in the southern part of the study area than in the north, with the development trend showing a multi-center radiation pattern. This corresponds directly with a lower quality monkey habitat in the south than in the north as shown in Figure 3.

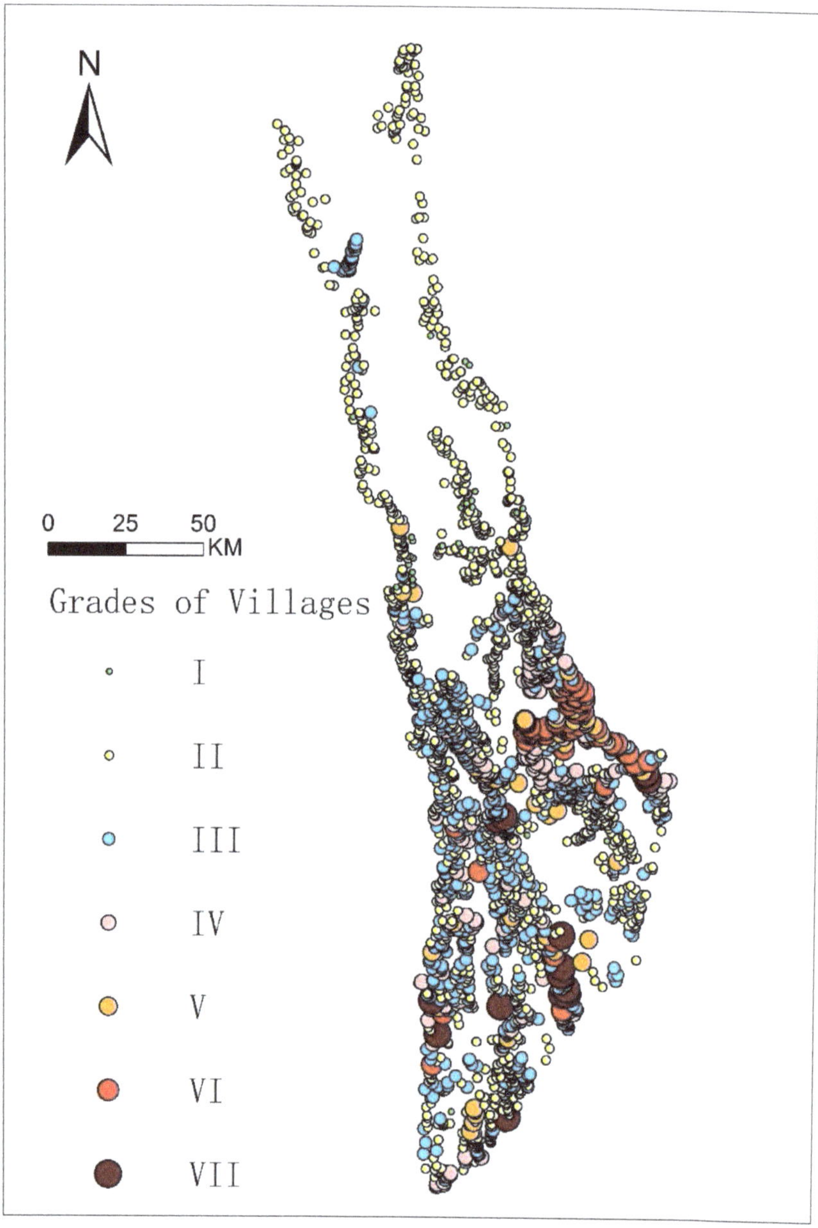

Figure 2. Distribution of villages and their grades in Northwest Yunnan.

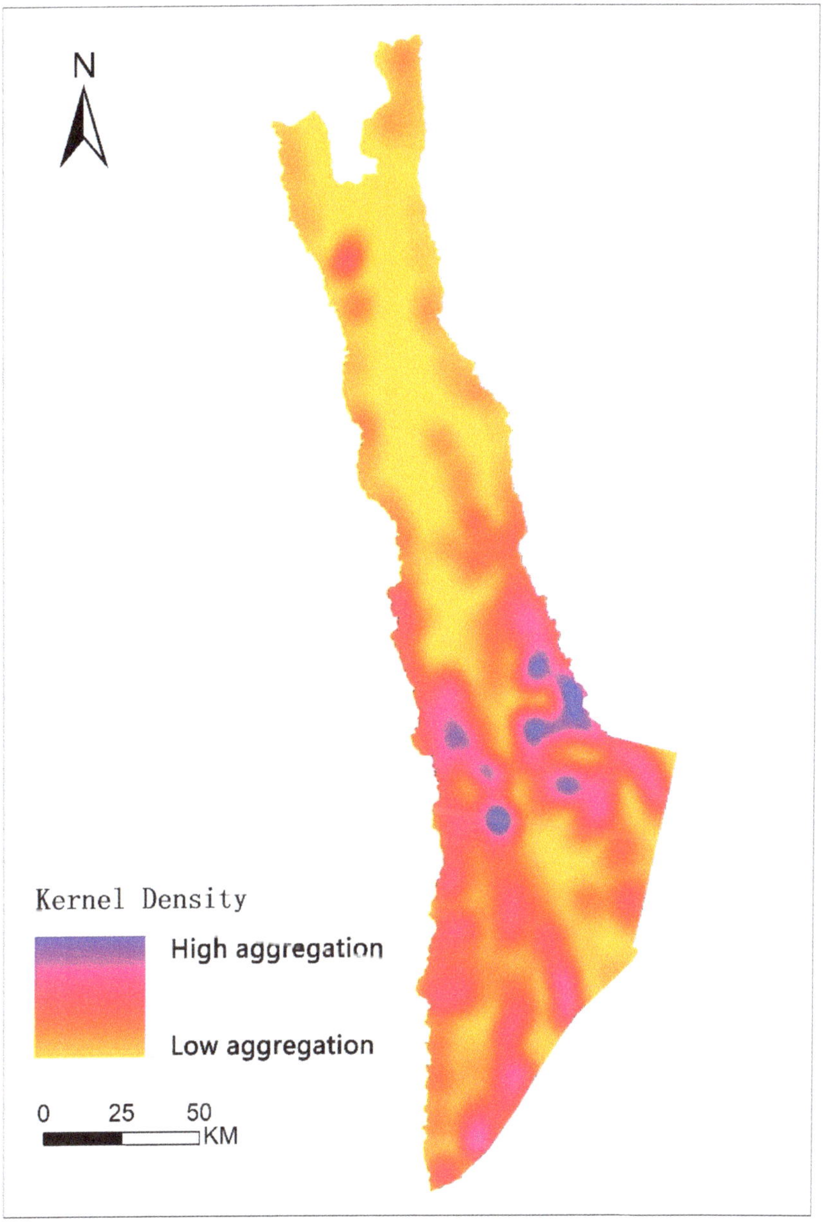

Figure 3. Distribution of kernel density in Northwest Yunnan.

3.2. Analysis of the Impact of Village Development on Habitat Quality

Figure 4 shows that habitat quality is higher in the northern part of the study area than in the southern part, that the development of villages in the north is less impactful on the habitat quality of snub-nosed monkeys than in the south, that ecological stability is poor, and that ecological protection needs to be prioritized.

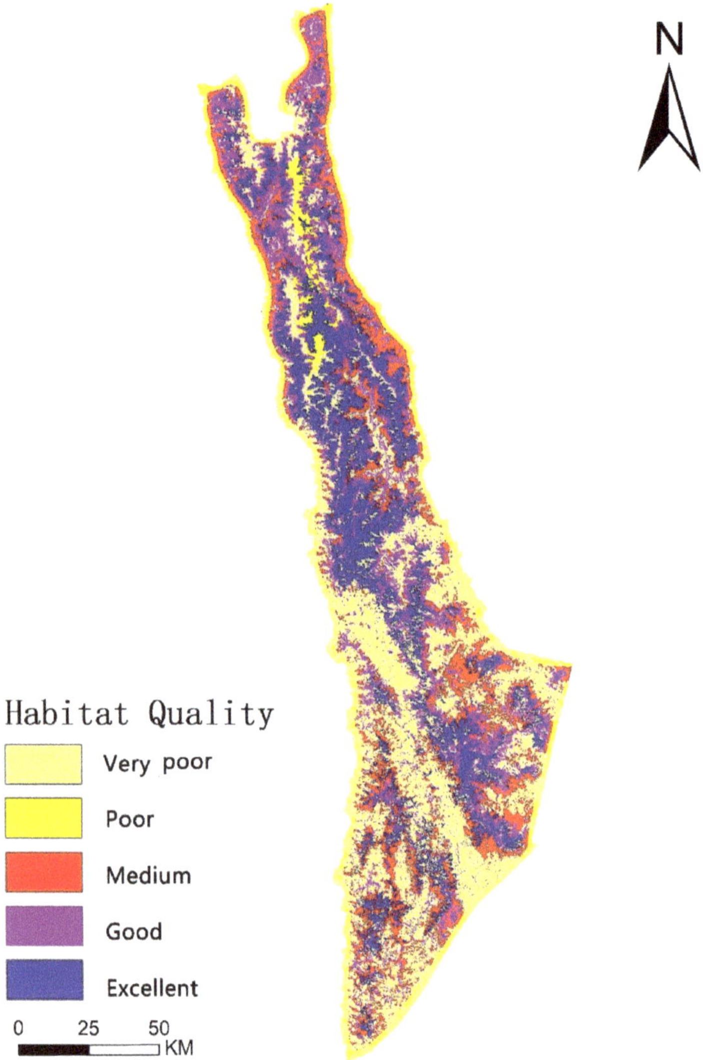

Figure 4. Spatial distribution pattern of habitat quality in 2018.

The mean value of the habitat quality index for Yunnan snub-nosed monkeys in 2018 was 0.4679, representing a medium habitat quality. Areas of very poor-quality habitat represented over half the total area (7759.39 km^2, or 51.94%). These very poor-quality habitats were mainly located in the outer and southern parts of the region and had villages densely distributed with socio-economic activities significantly degrading the habitat. The areas of good and excellent habitat quality were less than 30% of the total area and were mainly located in the northern part of the region. Those areas had more suitable habitat areas, less village distribution, and less human interference, resulting in slower habitat degradation (Table 8). Differing socio-economic development in the villages has caused differing degrees of habitat destruction, but overall socio-economic development has resulted in more low-quality habitat area and less high-quality habitat area.

Table 8. Percentage of area and mean value of habitat quality at grades in 2018.

Habitat Quality Grade	Value Interval	Area (km^2)	Area Weight (%)
Very poor	[0, 0.2)	7759.39	51.94
Poor	[0.2, 0.4)	506.31	3.08
Medium	[0.4, 0.6)	2534.52	15.13
Good	[0.6, 0.8)	2585.69	13.24
Excellent	[0.8, 1)	3038.35	16.60

3.3. *Analysis of the Habitat Quality of the Yunnan Snub-Nosed Monkey*

The analysis in Table 9 shows 12 monkey groups with a mean habitat quality value of 0.7408, which is relatively high for the overall habitat. The highest habitat quality was found in monkey group 11 (C11), with a value of 0.9047, while groups C3, C6, and C14 had the lowest habitat quality (Table 9).

Table 9. The value of habitat quality in each monkey reserve.

Code of Monkey	Habitat Quality
C1	0.8107
C2	0.7113
C3	0.5943
C4	0.7139
C5	0.7723
C6	0.4982
C10	0.8074
C11	0.9047
C12	0.7653
C13	0.8238
C14	0.6943
C15	0.7931

4. Discussion

4.1. *Impacts of Village Development on Habitat Quality*

There were significant spatial differences in habitat quality distribution patterns. The spatial autocorrelation of habitat quality revealed characteristics of weak agglomeration distribution in the space. The statistical parameters of spatial analysis indicated that the global Moran's I index of habitat quality was 0.0907, and the possibility of agglomeration distribution was less than 1%.

Figure 5 reveals that the hot spots and sub-hot-spots of habitat quality were in the northern and southern region as opposed to the central region, and the habitat quality changed greatly, as formerly forested land was converted into non-forested land such as crop land and villages or converted into ecological restoration projects. Conversely, the cold spots and sub-cold-spots were mainly distributed in the central region, which is a relatively suitable habitat for Yunnan snub-nosed monkey with great conversation and little change in habitat quality. Comparing Figures 4 and 5, we found that there was significant overlap between the Very Poor habitat quality area from Figure 4 and the Not Significant area from Figure 5.

If habitat or land use changes are representative of genetic, species, or ecosystem changes, then a low habitat quality will mean a decline in the biodiversity in the habitat and will mean habitat change unfavorable for species survival [16]. As the population of a village increases dramatically, people expand its boundaries, increase the use of cropland, and develop unused land to carry the increased population and provide food [7,41]. In addition, to develop the economy, people plant large amounts of economic forests, leading to habitat fragmentation [42]. Therefore, we need to carry out ecological restoration projects in these areas through cropland reforestation. We need to promote the restoration of the

Huashan pine hemlock arrow bamboo forests, spruce fir forests, and mixed coniferous and broad-leaved forests to improve habitat quality [4]. Bamboo and lichen are important food sources for Yunnan snub-nosed monkeys [10]. Forest rangers should restore bamboo plants and lichen, which will provide adequate food conditions for the Yunnan snub-nosed monkeys.

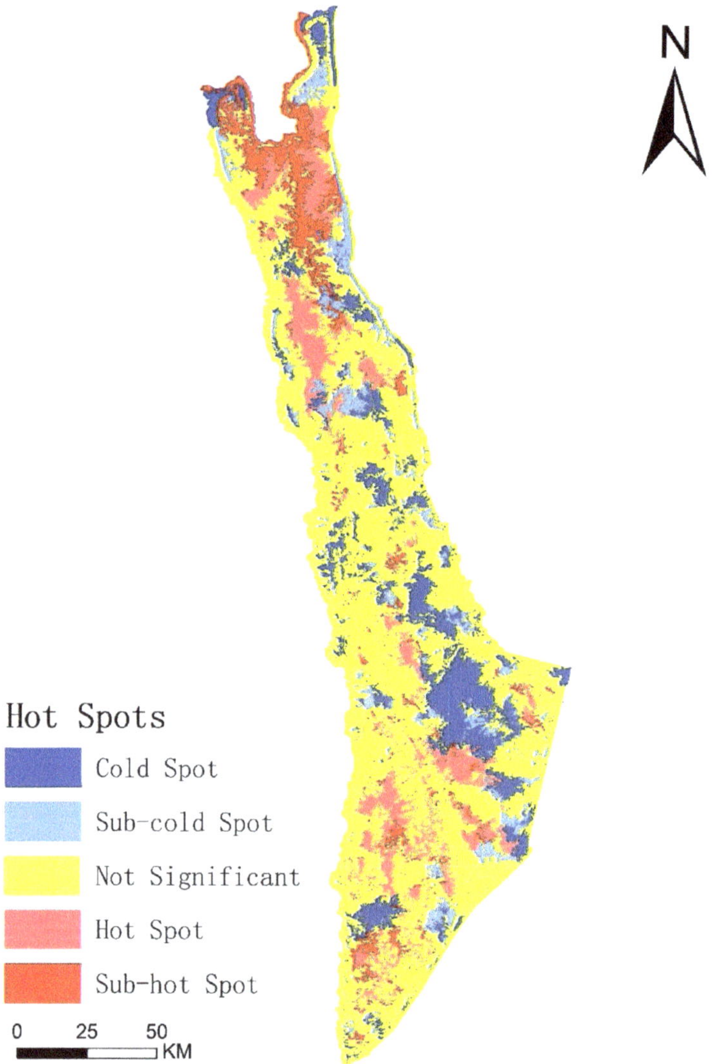

Figure 5. Hot spot analysis map of habitat quality in 2018.

4.2. Impacts of Village Development on Yunnan Snub-Nosed Monkeys

Coordinating the relationship between the socio-economic development of the village and the protection of the ecological environment can promote a harmonious symbiosis between humans and Yunnan snub-nosed monkeys. In places where the socio-economic development of villages is lagging or in decline, local villagers currently rely on massive timber cutting and occupying forest land to develop agriculture and grazing to sustain their livelihoods, threatening the survival of monkeys. There were 10 villages (Zanzhuaro,

Amu Shiguang, Muguang Aji, Sebu Feiha, Zaopo, Guilong, Cuding, Zengda, Yidoushe, and Lilinong) directly competing with the C3, C6, and C14 monkey groups (Figure 6). The government should implement poverty alleviation policies for these villages, removing the need of deforestation to drive economic development, instead developing monkey habitat–friendly eco-tourism and minority cultural industries.

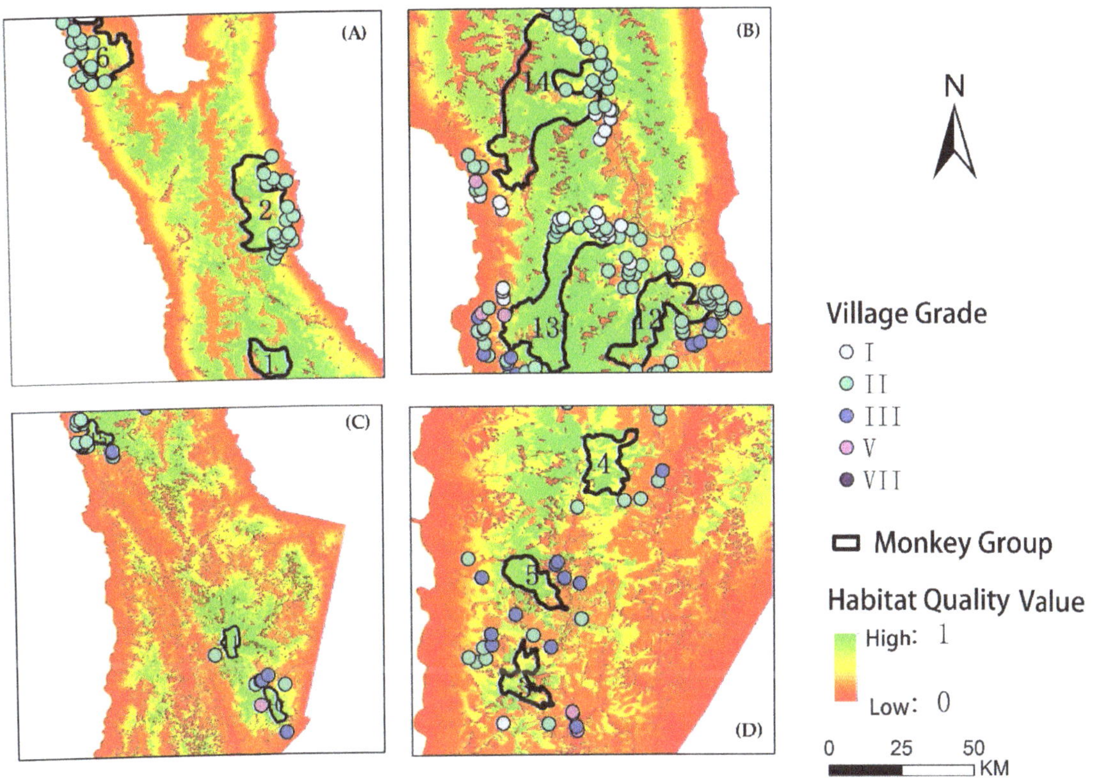

Figure 6. Zoom-in of four sections of the research area, showing the distribution of villages and habitat quality in relation to each monkey group. (**A**) Northern part of the area showing groups C1, C2, and C6. (**B**) North-central part of the area showing groups C12, C13, and C14. (**C**) South-central part of the area showing groups C10, C11, and C15. (**D**) Southern part of the area showing groups C3, C4, and C5.

Habitat quality, habitat patch size, and habitat connectivity mutually trade off against each other [43]. The habitat quality of the C3, C6, and C14 monkey groups is relatively low due to the development on villages. This indicates that village development had a strong influence on habitat quality, with the higher development level of a village translating to a lower habitat quality of the monkey group. We should enhance habitat connectivity and build up ecological corridors to promote gene exchange and conserve genetic diversity in these areas where connectivity with other monkey groups is impeded by agriculture and grazing land, roads, and villages. This is especially true for the isolated monkey groups (C3, C6, and C14) [4–6,44].

4.3. Implications for Conservation

In addition to protecting Yunnan snub-nosed monkeys through traditional measures such as ecological restoration, enhancing habitat connectivity, building corridors, relocation, and poverty alleviation, we need to take additional, creative measures. Illegal hunting is still an important factor threatening species conservation [45]. It is necessary to combat the illegal wildlife trade, curbing wildlife consumption through interventions and/or shifting consumption to more sustainable alternatives [46,47].

The government must provide ecological compensation to villages with high levels of ecological protection and restoration and establish ecological compensation mechanisms. These are a form of compensation for the damage cause to ecological functions and quality during development. The purpose of these forms of compensation is to improve the environmental quality of damaged areas or to be used to create new areas with similar ecological functions and environmental quality that can further protect, and sustainably use, ecosystem services [48,49]. We also need to strengthen science and education on biodiversity and wildlife conservation so that villagers can participate in species conservation and reap the benefits from it. In general, through a variety of measures, such as education, social influence, legal regulation, and behavioral facilitation, emphasis must be on multidisciplinary applications, multi-subject cooperation, and multi-scale considerations for species conservation [46,47].

The InVEST models is relatively mature and superior to transitional methods in spatial expression and dynamic research. We preliminarily discussed the improvement of adding quantitative village factors to the conventional model and have conducted analysis on the development on villages to include the habitat quality for Yunnan snub-nosed monkeys.

Due to the protection of some data and limited access to other data, the study area only has the distribution area of the Yunnan snub-nosed monkey in northwestern Yunnan. For example, data for some areas in Tibet province is lacking. The research is also lack dynamicity, integrity, and comprehensiveness. Data for some areas of Tibet will be supplemented in future work, and the influence of villages on the habitat quality of monkeys in the whole territory will be studied quantitatively. We also plan to improve the InVEST model parameters and strengthen parameter verification to obtain more effective variable factors, such as climate change factors. We also plan to further improve the variables that reflect the socio-economic development of villages, such as roads at all levels, population spatial distribution km grid data, nighttime lighting data, and GDP spatial distribution km grid data sets.

5. Conclusions

By using the smallest administrative scale in China—the village—we were able to accurately study the impact of village development on habitat quality, with focus on the Yunnan snub-nosed monkey. The modeling of habitat quality using the InVEST-Habitat Quality Model provided information about the relationship between the spatio-temporal distribution of habitat quality and potential biodiversity. The results of the habitat quality analysis indicate that decreases in habitat area and quality over time meant a decline in biodiversity persistence, resilience, and breadth and exacerbated the risk of declining species populations. In addition, the assessment of habitat quality based on species habitat suitability provides a new perspective to assess the impact of village developments on the habitat quality for the conservation of specific species. While we used the Yunnan snub-nosed monkey in our paper, these approaches can be applied for the conservation of other species.

Author Contributions: Data curation, S.Z.; formal analysis, T.J.S.; investigation, S.Z., L.L., G.W., J.L., H.G. and D.L.; methodology, L.L., G.W. and D.L.; supervision, L.L.; validation, T.J.S. and D.L.; visualization, H.G.; writing—original draft, S.Z.; writing—review and editing, L.L., G.W., T.J.S. and D.L. All authors have read and agreed to the published version of the manuscript.

Funding: Financial support was provided by the project of the National Natural Science Foundation of China (No. 32070454), the Second Tibetan Plateau Scientific Expedition and Research Program (No. 2019QZKK0501), and the Biodiversity Survey and Assessment Project of the Ministry of Ecology and Environment, China (No. 2019HJ2096001006).

Institutional Review Board Statement: Not applicable.

Informed Consent Statement: Not applicable.

Data Availability Statement: Data are available on request.

Acknowledgments: We thank Xiqing Deng from the Forestry Department of Yunnan Province for his guidance on field work. We also thank the Administration of Baima Snow Mountain Nature Conservation Bureau for its support for field work and their valuable suggestions on data integration and processing. This research has been conducted within the context of the GDRI (international Research Network) Ecosystem Health and Environmental Disease Ecology.

Conflicts of Interest: The authors declare no conflict of interest.

References

1. Baoping, R.; Ming, L.; Yongcheng, L.; Fuwen, W. Influence of day length, ambient temperature, and seasonality on daily travel distance in the Yunnan snub-nosed monkey at Jinsichang, Yunnan, China. *Am. J. Primatol.* **2008**, *71*, 233–241. [CrossRef] [PubMed]
2. Liu, Z.; Ren, B.; Wu, R.; Zhao, L.; Hao, Y.; Wang, B.; Wei, F.; Long, Y.; Li, M. The effect of landscape features on population genetic structure in Yunnan snub-nosed monkeys (*Rhinopithecus bieti*) implies an anthropogenic genetic discontinuity. *Mol. Ecol.* **2009**, *18*, 3831–3846. [CrossRef] [PubMed]
3. Clauzel, C.; Xiqing, D.; Gongsheng, W.; Giraudoux, P.; Li, L. Assessing the impact of road developments on connectivity across multiple scales: Application to Yunnan snub-nosed monkey conservation. *Biol. Conserv.* **2015**, *192*, 207–217. [CrossRef]
4. Li, W.; Clauzel, C.; Dai, Y.; Wu, G.; Giraudoux, P.; Li, L. Improving landscape connectivity for the Yunnan snub-nosed monkey through cropland reforestation using graph theory. *J. Nat. Conserv.* **2018**, *38*, 46–55. [CrossRef]
5. Li, L.; Xue, Y.; Wu, G.; Li, D.; Giraudoux, P. Potential habitat corridors and restoration areas for the black-and-white snub-nosed monkey *Rhinopithecus bieti* in Yunnan, China. *Oryx* **2014**, *49*, 719–726. [CrossRef]
6. Zhao, X.; Ren, B.; Li, D.; Xiang, Z.; Garber, P.A.; Li, M. Effects of habitat fragmentation and human disturbance on the population dynamics of the Yunnan snub-nosed monkey from 1994 to 2016. *PeerJ* **2019**, *7*, e6633. [CrossRef]
7. Quan, R.-C.; Huang, Y.; Warren, M.W.; Zhao, Q.-K.; Ren, G.-P.; Huo, S.; Long, Y.; Zhu, J. How Human Household Size Affects the Habitat of Black-and-White Snub-Nosed Monkeys (*Rhinopithecus bieti*) in Hongla Snow Mountain Nature Reserve in Tibet, China. *Int. J. Primatol.* **2011**, *32*, 1190–1202. [CrossRef]
8. Zhao, X.; Ren, B.; Garber, P.A.; Li, X.; Li, M. Impacts of human activity and climate change on the distribution of snub-nosed monkeys in China during the past 2000 years. *Divers. Distrib.* **2017**, *24*, 92–102. [CrossRef]
9. Nüchel, J.; Bøcher, P.K.; Xiao, W.; Zhu, A.-X.; Svenning, J.-C. Snub-nosed monkeys (*Rhinopithecus*): Potential distribution and its implication for conservation. *Biodivers. Conserv.* **2018**, *27*, 1517–1538. [CrossRef]
10. Xiang, Z.-F.; Huo, S.; Wang, L.; Cui, L.-W.; Xiao, W.; Quan, R.-C.; Tai, Z. Distribution, status and conservation of the black-and-white snub-nosed monkey *Rhinopithecus bieti* in Tibet. *Oryx* **2007**, *41*, 525–531. [CrossRef]
11. Liu, M.; Tian, H. China's land cover and land use change from 1700 to 2005: Estimations from high-resolution satellite data and historical archives. *Glob. Biogeochem. Cycles* **2010**, *24*, GB3003. [CrossRef]
12. Mckinney, M.L. Urbanization, Biodiversity, and Conservation the impacts of urbanization on native species are poorly studied, but educating a highly urbanized human population about these impacts can greatly improve species conservation in all ecosystems. *BioScience* **2002**, *52*, 883–890. [CrossRef]
13. Jeschke, J.M.; Genovesi, P. Do biodiversity and human impact influence the introduction or establishment of alien mammals? *Oikos* **2010**, *120*, 57–64. [CrossRef]
14. Mashizi, A.K.; Escobedo, F.J. Socio-ecological assessment of threats to semi-arid rangeland habitat in Iran using spatial models and actor group opinions. *J. Arid Environ.* **2020**, *177*, 104136. [CrossRef]
15. Haddad, N.M.; Brudvig, L.A.; Clobert, J.; Davies, K.F.; Gonzalez, A.; Holt, R.D.; Lovejoy, T.E.; Sexton, J.O.; Austin, M.P.; Collins, C.D.; et al. Habitat fragmentation and its lasting impact on Earth's ecosystems. *Sci. Adv.* **2015**, *1*, e1500052. [CrossRef] [PubMed]
16. Newbold, T.; Hudson, L.N.; Hill, S.L.L.; Contu, S.; Lysenko, I.; Senior, R.A.; Börger, L.; Bennett, D.J.; Choimes, A.; Collen, B.; et al. Global effects of land use on local terrestrial biodiversity. *Nature* **2015**, *520*, 45. [CrossRef]
17. Vögeli, M.; Serrano, D.; Pacios, F.; Tella, J.L. The relative importance of patch habitat quality and landscape attributes on a declining steppe-bird metapopulation. *Biol. Conserv.* **2010**, *143*, 1057–1067. [CrossRef]
18. Wiegand, T.; Naves, J.; Garbulsky, M.F.; Fernández, N. Animal habitat quality and ecosystem functioning: Exploring seasonal patterns using ndvi. *Ecol. Monogr.* **2008**, *78*, 87–103. [CrossRef]
19. He, J.; Huang, J.; Li, C. The evaluation for the impact of land use change on habitat quality: A joint contribution of cellular automata scenario simulation and habitat quality assessment model. *Ecol. Model.* **2017**, *366*, 58–67. [CrossRef]
20. Nelson, E.; Mendoza, G.; Regetz, J.; Polasky, S.; Tallis, H.; Cameron, D.R.; Chan, K.M.A.; Daily, G.C.; Goldstein, J.; Kareiva, P.M.; et al. Modeling multiple ecosystem services, biodiversity conservation, commodity production, and tradeoffs at landscape scales. *Front. Ecol. Environ.* **2009**, *7*, 4–11. [CrossRef]
21. Vistnes, I.; Nellemann, C.; Jordhøy, P.; Strand, O. Effects of infrastructure on migration and range use of wild reindeer. *J. Wildl. Manag.* **2004**, *68*, 101–108. [CrossRef]

22. Terrado, M.; Sabater, S.; Chaplin-Kramer, B.; Mandle, L.; Ziv, G.; Acuña, V. Model development for the assessment of terrestrial and aquatic habitat quality in conservation planning. *Sci. Total Environ.* **2016**, *540*, 63–70. [CrossRef] [PubMed]
23. Thomas, J.A.; Bourn, N.; Clarke, R.T.; Stewart, K.E.; Simcox, D.J.; Pearman, G.S.; Curtis, R.; Goodger, B. The quality and isolation of habitat patches both determine where butterflies persist in fragmented landscapes. *Proc. R. Soc. B Boil. Sci.* **2001**, *268*, 1791–1796. [CrossRef] [PubMed]
24. Ding, Q.; Chen, Y.; Bu, L.; Ye, Y. Multi-Scenario Analysis of Habitat Quality in the Yellow River Delta by Coupling FLUS with InVEST Model. *Int. J. Environ. Res. Public Health* **2021**, *18*, 2389. [CrossRef]
25. Nematollahi, S.; Fakheran, S.; Kienast, F.; Jafari, A. Application of InVEST habitat quality module in spatially vulnerability assessment of natural habitats (case study: Chaharmahal and Bakhtiari province, Iran). *Environ. Monit. Assess.* **2020**, *192*, 487. [CrossRef]
26. Di Febbraro, M.; Sallustio, L.; Vizzarri, M.; De Rosa, D.; De Lisio, L.; Loy, A.; Eichelberger, B.; Marchetti, M. Expert-based and correlative models to map habitat quality: Which gives better support to conservation planning? *Glob. Ecol. Conserv.* **2018**, *16*, e00513. [CrossRef]
27. Choudhary, A.; Deval, K.; Joshi, P.K. Study of habitat quality assessment using geospatial techniques in Keoladeo National Park, India. *Environ. Sci. Pollut. Res.* **2020**, *28*, 14105–14114. [CrossRef]
28. Kija, H.K.; Ogutu, J.O.; Mangewa, L.J.; Bukombe, J.; Verones, F.; Graae, B.J.; Kideghesho, J.R.; Said, M.Y.; Nzunda, E.F. Spatio-Temporal Changes in Wildlife Habitat Quality in the Greater Serengeti Ecosystem. *Sustainability* **2020**, *12*, 2440. [CrossRef]
29. Krauss, J.; Bommarco, R.; Guardiola, M.; Heikkinen, R.K.; Helm, A.; Kuussaari, M.; Lindborg, R.; Öckinger, E.; Pärtel, M.; Pino, J.; et al. Habitat fragmentation causes immediate and time-delayed biodiversity loss at different trophic levels. *Ecol. Lett.* **2010**, *13*, 597–605. [CrossRef]
30. Ng, C.N.; Xie, Y.J.; Yu, X.J. Measuring the spatio-temporal variation of habitat isolation due to rapid urbanization: A case study of the Shenzhen River cross-boundary catchment, China. *Landsc. Urban Plan.* **2011**, *103*, 44–54. [CrossRef]
31. Nelson, E.; Polasky, S.; Lewis, D.J.; Plantinga, A.J.; Lonsdorf, E.; White, D.; Bael, D.; Lawler, J.J. Efficiency of incentives to jointly increase carbon sequestration and species conservation on a landscape. *Proc. Natl. Acad. Sci. USA* **2008**, *105*, 9471–9476. [CrossRef] [PubMed]
32. Long, Y.; Kirkpatrick, C.R.; Zhong, T.; Xiao, L. Report on the distribution, population, and ecology of the yunnan snub-nosed monkey (*Rhinopithecus bieti*). *Primates* **1994**, *35*, 241–250. [CrossRef]
33. Wong, M.H.G.; Li, R.; Xu, M.; Long, Y. An integrative approach to assessing the potential impacts of climate change on the Yunnan snub-nosed monkey. *Biol. Conserv.* **2012**, *158*, 401–409. [CrossRef]
34. Mokarram, M.; Pham, T.M. CA-Markov model application to predict crop yield using remote sensing indices. *Ecol. Indic.* **2022**, *139*, 108952. [CrossRef]
35. Zhu, C.; Zhang, X.; Zhou, M.; He, S.; Gan, M.; Yang, L.; Wang, K. Impacts of urbanization and landscape pattern on habitat quality using OLS and GWR models in Hangzhou, China. *Ecol. Indic.* **2020**, *117*, 106654. [CrossRef]
36. Wu, L.; Sun, C.; Fan, F. Estimating the Characteristic Spatiotemporal Variation in Habitat Quality Using the InVEST Model-A Case Study from Guangdong-Hong Kong-Macao Greater Bay Area. *Remote Sens.* **2021**, *13*, 1008. [CrossRef]
37. Ulak, M.; Kocatepe, A.; Ozguven, E.E.; Horner, M.W.; Spainhour, L. Geographic Information System–Based Spatial and Statistical Analysis of Severe Crash Hotspot Accessibility to Hospitals. *Transp. Res. Rec. J. Transp. Res. Board* **2017**, *2635*, 90–97. [CrossRef]
38. Premo, L. Local spatial autocorrelation statistics quantify multi-scale patterns in distributional data: An example from the Maya Lowlands. *J. Archaeol. Sci.* **2004**, *31*, 855–866. [CrossRef]
39. Cai, X.; Wu, Z.; Cheng, J. Using kernel density estimation to assess the spatial pattern of road density and its impact on landscape fragmentation. *Int. J. Geogr. Inf. Sci.* **2013**, *27*, 222–230. [CrossRef]
40. Kang, Y.; Cho, N.; Son, S. Spatiotemporal characteristics of elderly population's traffic accidents in Seoul using space-time cube and space-time kernel density estimation. *PLoS ONE* **2018**, *13*, e0196845. [CrossRef]
41. Liu, J.; Li, D.; Matsuzawa, T.; Hirata, S. The Lisu people's traditional natural philosophy and its potential impact on conservation planning in the Laojun Mountain region, Yunnan Province, China. *Primates* **2020**, *62*, 153–164. [CrossRef] [PubMed]
42. Wilson, M.C.; Chen, X.-Y.; Corlett, R.; Didham, R.K.; Ding, P.; Holt, R.D.; Holyoak, M.; Hu, G.; Hughes, A.C.; Jiang, L.; et al. Habitat fragmentation and biodiversity conservation: Key findings and future challenges. *Landsc. Ecol.* **2015**, *31*, 219–227. [CrossRef]
43. Donaldson, L.; Wilson, R.; Maclean, I. Old concepts, new challenges: Adapting landscape-scale conservation to the twenty-first century. *Biodivers. Conserv.* **2016**, *26*, 527–552. [CrossRef] [PubMed]
44. Zhang, Y.; Clauzel, C.; Li, J.; Xue, Y.; Zhang, Y.; Wu, G.; Giraudoux, P.; Li, L.; Li, D. Identifying refugia and corridors under climate change conditions for the Sichuan snub-nosed monkey (*Rhinopithecus roxellana*) in Hubei Province, China. *Ecol. Evol.* **2019**, *9*, 1680–1690. [CrossRef]
45. Zhao, X.; Ren, B.; Li, D.; Garber, P.A.; Zhu, P.; Xiang, Z.; Grueter, C.C.; Liu, Z.; Li, M. Climate change, grazing, and collecting accelerate habitat contraction in an endangered primate. *Biol. Conserv.* **2019**, *231*, 88–97. [CrossRef]
46. Jiao, Y.; Yeophantong, P.; Lee, T.M. Strengthening International Legal Cooperation to Combat the Illegal Wildlife Trade Between Southeast Asia and China. *Front. Ecol. Evol.* **2021**, *9*, 645427. [CrossRef]
47. Thomas-Walters, L.; Hinsley, A.; Bergin, D.; Burgess, G.; Doughty, H.; Eppel, S.; MacFarlane, D.; Meijer, W.; Lee, T.M.; Phelps, J.; et al. Motivations for the use and consumption of wildlife products. *Conserv. Biol.* **2020**, *35*, 483–491. [CrossRef]

48. Cuperus, R.; Kalsbeek, M.; De Haes, H.A.U.; Canters, K.J. Preparation and Implementation of Seven Ecological Compensation Plans for Dutch Highways. *Environ. Manag.* **2002**, *29*, 736–749. [CrossRef]
49. Vaissière, A.-C.; Meinard, Y. A policy framework to accommodate both the analytical and normative aspects of biodiversity in ecological compensation. *Biol. Conserv.* **2020**, *253*, 108897. [CrossRef]

biology

MDPI

Article

Geology Can Drive the Diversity–Ecosystem Functioning Relationship in River Benthic Diatoms by Selecting for Species Functional Traits

Evangelia Smeti [1,*], George Tsirtsis [2] and Nikolaos Theodor Skoulikidis [1]

1 Institute of Marine Biological Resources and Inland Waters, Hellenic Centre for Marine Research, 46.7 km Athinon-Souniou Ave., 19013 Anavyssos, Greece
2 Department of Marine Sciences, University of the Aegean, University Hill, 81100 Mytilene, Greece
* Correspondence: evasmeti@hcmr.gr; Tel.: +30-229-107-6439

Simple Summary: The way that diversity affects ecosystem functioning is of great importance, as it helps us understand the health state of an ecosystem. Primary producers contribute to ecosystem functioning through biomass production, which is considered to be a proxy of ecosystem functioning. In rivers, the primary producers of the biofilm are diatoms, unicellular algae with cell walls of silica. In this study, we tested the way diatom species affect biomass production across nine rivers in Greece. Nutrient concentrations that drive primary production are linked to river geology. We found that the geological substrate of a river could be responsible for the diversity–biomass relationship: in rivers with a siliceous substrate, more diatom species increased biomass, whereas in rivers with a calcareous substrate, a change in diatom species number did not change biomass. By using model simulations, we found that this difference could be attributed to the different stages of the biofilm in time. Our results show the importance of different factors that affect diatom species, their functional traits and biomass production and what we should consider when testing for ecosystem functioning.

Abstract: The biodiversity–ecosystem functioning (BEF) relationship has been studied extensively for the past 30 years, mainly in terrestrial plant ecosystems using experimental approaches. Field studies in aquatic systems are scarce, and considering primary producers, they mainly focus on phytoplankton assemblages, whereas benthic diatoms in rivers are considerably understudied in this regard. We performed a field study across nine rivers in Greece, and we coupled the observed field results with model simulations. We tested the hypothesis that the diversity–biomass (as a surrogate of ecosystem functioning) relationship in benthic diatoms would be affected by abiotic factors and would be time-dependent due to the highly dynamic nature of rivers. Indeed, geology played an important role in the form of the BEF relationship that was positive in siliceous and absent in calcareous substrates. Geology was responsible for nutrient concentrations, which, in turn, were responsible for the dominance of specific functional traits. Furthermore, model simulations showed the time dependence of the BEF form, as less mature assemblages tend to present a positive BEF. This was the first large-scale field study on the BEF relationship of benthic diatom assemblages, offering useful insights into the function and diversity of these overlooked ecosystems and assemblages.

Keywords: biofilm; diatoms; rivers; Greece; model simulations

Citation: Smeti, E.; Tsirtsis, G.; Skoulikidis, N.T. Geology Can Drive the Diversity–Ecosystem Functioning Relationship in River Benthic Diatoms by Selecting for Species Functional Traits. *Biology* **2023**, *12*, 81. https://doi.org/10.3390/biology12010081

Academic Editor: José Carlos Hernández

Received: 13 December 2022
Revised: 29 December 2022
Accepted: 30 December 2022
Published: 4 January 2023

1. Introduction

Ecosystem functioning comprises multiple processes that account for ecosystem health and sustain ecosystem services. For the past 30 years, research has been focusing on proving the pivotal role of diversity in driving ecosystem functioning [1]. In particular, the study of the form of the biodiversity–ecosystem functioning (BEF) relationship is considered important in light of global change and species extinctions [2]. It could be further used as a proxy of ecosystem health, resilience and species interactions, providing great insight into an ecosystem's need for conservation [3,4]. BEF studies, especially early ones, have focused on experimental work, mainly on terrestrial plants, whereas aquatic environments and especially freshwater, remain understudied [5]. Field studies are rare, and regarding microalgae, they have focused on phytoplankton in Scandinavian and USA lakes [6,7] and the Baltic Sea [6,8]. A few studies on biofilm are limited to estuaries [9,10], but studies on the productivity of the river biofilm, especially on benthic diatoms, are almost missing (see [11]).

Diatoms, a major component of phytobenthos in rivers and the most diverse group of protists, are unicellular algae with silica cell walls, responsible for 20% of O_2 production, and are important indicators of water quality [12]. As primary producers, their growth depends on nutrient concentrations and light, contributing immensely to primary biofilm productivity, an important ecosystem function. The importance of benthic diatoms on biofilm biomass production, along with their high diversity, renders them an ideal group of organisms for studying the BEF relationship in river biofilms. Furthermore, in recent years, functional traits related to cell size, to adherence to substrates and life forms are increasingly used in describing benthic diatom assemblages [13,14]. Despite the growing evidence that functional richness could be more important in driving ecosystem functions than taxonomic richness [15], functional diversity metrics are not widely used in BEF studies.

Although initial research focused on positive BEF relationships (i.e., an increase in ecosystem functions with increasing diversity), recent research and meta-analysis suggest different relationships (e.g., negative or hump-shaped relationships), depending on the type and duration of the study (i.e., observational field or experiment), the ecosystem type and the taxa studied [5]. Furthermore, the form of the BEF relationship can be affected by multiple diversity levels in space and time, such as the regional and local species pools or the initial and realized diversity [16]. Although they have the advantage of the large scale and the natural world, observational field studies can give ambiguous results. Natural systems are very complex and dynamic, and patterns could be masked by different diversity scales that are part of different assembly phases. On the other hand, modeling studies could provide mechanistic interpretations of the form of the BEF relationship [17,18]. For example, modeling studies on phytoplankton have suggested that when species utilize all the available resource space (e.g., when the system is at a mature steady-state), there is no BEF relationship, whereas when a part of the resource space stays unutilized (e.g., after a species extinction), then a strong positive BEF is apparent [17]. Therefore, coupling field observations with numerical modeling could give better insights into the drivers of the BEF relationship.

The aim of this study was to test the form of the BEF relationship in benthic diatoms in rivers and the drivers of this relationship. Toward this aim, we collected samples along nine Greek rivers, varying in their geographical location, geology, drainage area and nutrient concentrations. As rivers are highly dynamic ecosystems and Greece is a country with diverse landscapes and geology, we hypothesized that a general BEF relationship would be hard to observe and that it would be driven by additional, possibly abiotic, factors. We further investigated species traits (size, attachment to the substrate) that can be responsible for the observed relationship. To further understand species coexistence and the consecutive biomass production during a succession of the biofilm in time, we ran model simulations and checked the BEF relationship at different maturity levels of the biofilm. We hypothesized that a less mature biofilm (at the beginning of time succession) would present a positive BEF, as resources are still unutilized, whereas a mature biofilm, where all available resources are used, would not present a BEF. To the best of our knowledge, this is the first large-scale observational study of the BEF relationship in benthic river diatoms, and although incomplete, it can give important insights into the function and diversity of these overlooked systems.

2. Materials and Methods

This study combined field observations with numerical modeling. Field sampling was conducted in river biofilm, comprising samples for both microscopic observation and biomass. Water physico-chemical parameters and nutrient concentrations were measured at each site. Diversity (i.e., species richness and evenness) of diatoms in the biofilm was defined using microscopic counts of species abundances, and functional traits were assigned to species to account for functional diversity. Biomass was measured in the lab as chlorophyll a concentration and was used as a proxy of ecosystem functioning. The shape of the BEF relationship was tested at different spatial scales, and nutrient concentrations and functional traits were investigated as possible drivers of the observed shape. Model simulations, using a well-known numerical model on species competition for available resources, were run, and the BEF relationship was observed at different time points to test for the dependence of the relationship to the maturity of the assemblage.

2.1. Field Sampling

Nine Greek rivers (Nestos, Lissos, Fonias, Spercheios, Mornos, Alfeios, Arkadikos, Neda and Evrotas) were sampled in the summer of 2020, at a low flow period, when no major disturbances would cause shifts in the assemblages and their biomass (Figure A1, Table A1). These rivers were selected based on accessibility and appropriate sampling substrate (stones) as well as due to their differences in terms of size, geology and environmental conditions. In each river, five sampling sites were sampled from upstream to downstream, apart from Arkadikos and Lissos, where only four samples were taken. In order to ensure replication, in each site, three spots were sampled, each comprising three stones. From each stone, two surfaces of the defined area were scraped, the first used for chlorophyll analysis (immediately put in a dark bag and frozen) and the other for species identification and counting (preserved with 70% ethanol). This ensured the direct comparison between species diversity and biomass production. At each site, physico-chemical parameters (Temperature, DO, pH, Conductivity, Turbidity) were also measured in situ using a Portable multiparameter Aquaprobe, and water samples were collected for the determination of nutrients (NO_3, NO_2, NH_4, TN, PO_4, TP and SiO_2).

2.2. Analysis of Samples

In the laboratory, after filtration through 0.45 μm pore size membrane filters, nutrients were determined by a Skalar San++ Continuous Flow Analyzer [19]. For the determination of chlorophyll, the trichromatic equations were applied [20], where all three main chlorophylls were measured (Chl-a, Chl-b, Chl-c), and their concentrations were in mg/cm^2. Chl-a is a measure of the whole phytobenthos biomass production, whereas Chl-c is more indicative of the biomass produced by benthic diatoms.

Diatom species samples were treated with hot hydrogen peroxide to remove organic matter and obtain clean frustules, to be used for diatom species identification [21]. Clean frustules were mounted with Naphrax®, identified to species level with a light microscope (Nikon Eclipse Ci-L, Nikon Microscope Solutions, Europe) at 1000× magnification and counted until no more new species were detected in each sample. As the surface scraped out of the stone was defined, and the volume at each step of the procedure was also measured, the counting reflected the absolute abundance of cells per cm^2. For the taxonomy, the work of [22] was mainly used.

2.3. Data Analysis

In order to ensure that the sampling effort was adequate for all rivers examined, species accumulation curves (SACs) were constructed, showing that, indeed, most species were observed under the specific sampling and analysis procedures (Figure A2). Furthermore, to check that species richness counts were not biased due to macroecological patterns and large differences in the drainage areas, species–area relationships (SAR) for the selected rivers were performed, demonstrating the absence of a relationship between the area and the observed species richness (Figure A3). Another potentially confounding factor in natural systems is pollution. In the present study, pollution levels slightly differed, even between sites of the same river, based on a biological quality diatom index, but quality classes did not play an important role in the BEF relationship (interaction term p-value = 0.07).

Taxonomic diversity was calculated using both species richness (S) and evenness (J), to account for the abundance distribution of individuals among species (i.e., assemblage structure). For calculating functional diversity, the functional richness index was used, defined as the total branch length of a functional dendrogram based on species' functional traits [23]. Functional traits used were cell size (L/W ratio, biovolume), substrate adherence (high profile, low profile, motile and planktonic guilds), life forms (colonial, singular) and nitrogen fixation [14]. For the calculation of biovolume, equations of geometric shapes were used [24], and dimensions that could not be measured in our samples (e.g., cell height) were defined based on the literature [14]. The total biovolume of each sample was divided by the total abundance of the sample for the calculation of the average cell size of each assemblage, aiming to compare cell size between different groups of rivers [25].

Biomass metrics tested were Chl-a, expressing biomass production of the entire biofilm, Chl-c and Total biovolume, linked to benthic diatoms. Chl-a and Chl-c were highly correlated (Spearman r = 0.86, p-value < 0.001) and presented the same trends. Therefore, only Chl-a was used as a surrogate of biomass production. In order to show more clearly linear trends, Chl-a concentrations were ln-transformed.

The form of the relationship between the different diversity metrics and Chl-a was determined for the whole dataset, searching for a general pattern in the examined Greek rivers, as well as for each river separately to test for possible differences in the BEF relationship between rivers. Rivers were further grouped in two previously defined hydrochemical zones in Greece [26] with distinct silicate and phosphorus concentration ranges, affected by geology; in zone A, siliceous substrates are more prominent, and silicate and phosphorus concentrations in water are higher, whereas, in zone B, calcareous substrates dominate and silicate and phosphorus concentrations in water are lower, the latter due to adsorption on carbonate-rich particles and sediments [27,28]. Substrate geology (i.e., siliceous vs. calcareous) is known to select for species diatom species [22]. Therefore, as phosphorus and silica are important nutrients for diatom growth, their different concentrations in these two zones could affect assemblage characteristics and, thus, the corresponding BEF relationships. For the rest of the manuscript, when we refer to substrates (siliceous or calcareous), we refer to the geologic substrate of a river basin.

For testing the significance of the BEF relationship when used in different groups in the dataset, generalized linear mixed-effects models were used, with the river as a random factor. Data analyses and illustrations were performed in R (v. 4.0.3) [29], using packages vegan v. 2.5-7 [30], BAT v. 2.7.1 [31], lme4 v. 1.1-28 [32], ggplot2 v. 3.3.5 [33] and plotly 4.10.1 [34].

2.4. Model Simulations

Model simulations were performed in an effort to understand the importance of temporal succession and the maturity of the biofilm on the BEF relationship. Applied models were based on well-known models for phytoplankton competition for resources, assuming a continuous inflow of nutrients [35]. This model describes the population dynamics of 400 diatom species (N_i) competing for two nutrients (R_j), namely nitrogen and phosphorus. The initial species number (n = 400) is based on the total diatom species observed in all field samples.

$$\frac{dN_i}{dt} = N_i\left(min\left(\frac{\mu_{max_i} \times R_1}{K_{1i} + R_1}, \frac{\mu_{max_i} \times R_2}{K_{2i} + R_2}\right) - m_i\right), \quad i = 1 - n \tag{1}$$

$$\frac{dR_i}{dt} = D(S_j - R_j) - \sum_{i=1}^{n} c_{ji} \times min\left(\frac{\mu_{max_i} \times R_1}{K_{1i} + R_1}, \frac{\mu_{max_i} \times R_2}{K_{2i} + R_2}\right)N_i, \quad j = 1,2 \tag{2}$$

N_i is biomass of species i, and R_j is the concentration of nutrient j; μ_{max_i} is the specific maximum growth rate of species i, and K_{ji} is the half-saturation constant of resource j for species i, based on the Monod model of growth limitation; m_i is the mortality induced by flushing, and it was calculated as the flushing rate (D) divided by the maximum growth rate of each species in the model; D is the nutrient flushing rate; S_j is the input nutrient j concentration; and c_{ji} is the intracellular content of nutrient j in species i. In diatoms, the maximum growth rate is linked to species size, with smaller species presenting a higher growth rate [36]. As larger species in a biofilm tend to be more affected by flushing than smaller species that tend to adhere to the substrate stronger, we assumed that they are more affected by flushing, which increases larger species mortality.

The two nutrients used in the model are phosphorus and nitrogen, as they are both essential nutrients for growth. Based on field observations, phosphorus was mainly the limiting nutrient, whereas nitrogen limitation was also observed in some cases. The two nutrients in the model are added synchronously and in a continuous manner during the simulations at concentrations following the Redfield ratio. This synchronous and continuous flow simulates an ideal river environment, from upstream (nutrients entering the system) to downstream (nutrients flushing). Following the N:P:Si ratio in field observations, Si was never found to be limiting; therefore, even though an important nutrient for diatom growth, it was not considered in model simulations.

Life history traits were assigned to species based on the literature values and on species functional traits that we observed in the field samples. The three main life history traits we focused on were the specific maximum growth rate (μ_{max}), the competitive ability for Phosphorus (K_P) and the competitive ability for Nitrogen (K_N). Based on field data, smaller species tended to be at low nutrient concentrations; therefore, we assigned three groups of species, with each group being superior for two life history traits: one group consisted of fast-growing species with the increased competitive ability for phosphorus but not for nitrogen (high μ_{max} and low K_P but high K_N), one group consisted of fast-growing species with the increased competitive ability for nitrogen (high μ_{max} and low K_N but high K_P) and one group consisted of slow-growing species with the increased competitive ability both for phosphorus and nitrogen (low μ_{max} and low K_P and K_N). Keeping a trade-off was important as the presence of a "superspecies", superior for all traits, would exclude all other species, and thus, species richness in an assemblage would be extremely low. When assigning traits to virtual species, we made sure that there was a trade-off between R^*_P and R^*_N, with a level of complementarity equal to 0.49 [37]. R* is the minimum concentration of a resource at which a species could keep its population stable, and it is a summary value of both growth rate and K_j. Life history traits were assigned to species using R (v. 4.0.3).

The mathematical equations were solved numerically using a specially developed Fortran code following [37] and adapted to meet the characteristics of the studied systems. The BEF relationship was tested at each time step using the species richness and evenness against the log-transformed abundance of the 100 replicates. For each replicate, the initial biomass of each species and the total initial abundance varied randomly. The model parameters values, ranges and initial conditions are detailed in Table A2.

3. Results

3.1. Field Observations

The general BEF relationship (when all samples were pooled together) when using species richness (S) as the diversity predictor of biomass was positive, albeit rather weak (p-value < 0.01, Figure 1a). A seemingly similar but not significant trend was apparent when functional richness was used as a biomass predictor (p-value = 0.235, Figure 1c). The lack of a significant relationship was also present when evenness (J) was used as a diversity predictor of biomass (p-value = 0.676, Figure 1b).

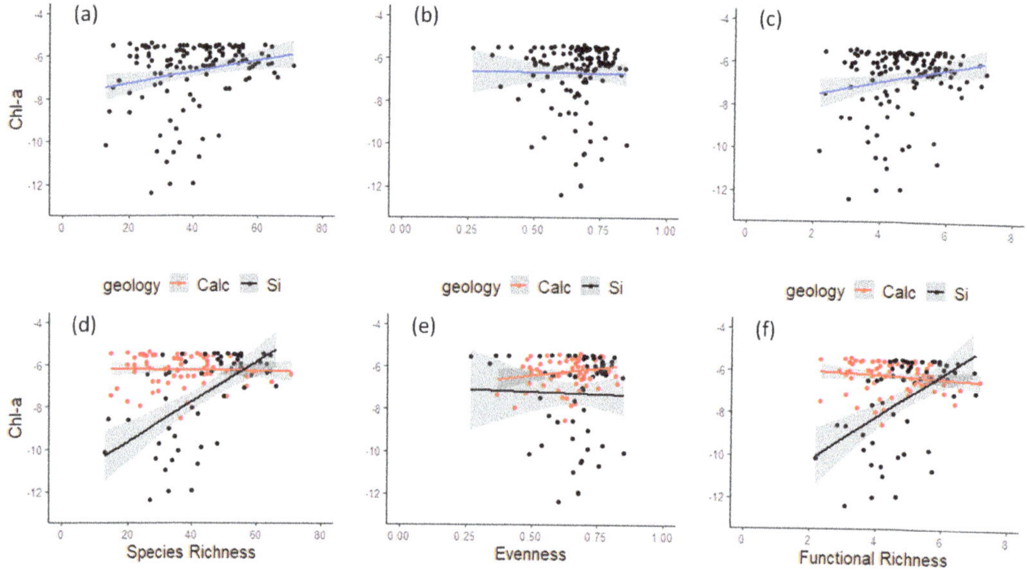

Figure 1. Diversity (expressed as (**a**,**d**) species richness S, (**b**,**e**) evenness J and (**c**,**f**) functional richness) and ecosystem functioning (presented as ln(Chl-a) on y-axis) relationships in each river on the whole dataset (**a**–**c**) and at different geological substrate levels (**d**–**f**).

When each river was considered separately, the relationship between species richness and biomass production was variable between the different rivers sampled (Figure A4a). Indeed, Chl-a was best explained when the interaction between species richness and the river was also considered (adjusted $R^2 = 0.55$, p-value < 0.001). This variation is also apparent when considering other diversity metrics (evenness J and functional richness, Figure A4b,c). Variability among rivers was also evident in environmental conditions, as depicted in the physico-chemical parameters and nutrient concentrations measured (Figure A5).

A strong interaction effect is apparent when testing for the substrate geology (interaction term p-value < 0.001). In siliceous substrate, an increase in species richness resulted in an increase in biomass production (positive BEF-slope = 0.097, p-value < 0.001), whereas, in the calcareous substrate, an increase in species richness did not have any effect on biomass production (no BEF relationship-slope = −0.0014, p-value = 0.9—Figure 1d). Functional richness presented the same trend (interaction term p-value < 0.05, Figure 1f), but evenness (J) had no effect on predicting biomass ((interaction term p-value = 0.204, Figure 1e). There was no significant difference between species richness or biomass for the two groups of rivers.

In rivers with a siliceous substrate, all tested nutrients (TinN (i.e., sum of NO_2, NO_3, NH_4), PO_4, SiO_2) presented higher concentrations than in rivers with a calcareous substrate (p-value < 0.05–Figure 2a–c). Regarding species traits, rivers in siliceous substrates have diatom assemblages comprised of bigger and motile species, whereas rivers in calcareous substrates have diatom assemblages comprised of smaller, low-profile species ($p < 0.05$–Figure 2d–f). Overall, motile species tended to increase with increased phosphorus concentrations, whereas low-profile species tended to decrease with increased phosphorus concentrations (Figure 3a,b). The other guilds (high-profile and planktonic species) did not present any consistent relationship between geology or nutrient concentrations. Furthermore, a higher relative abundance of low-profile species resulted in high dominance assemblages (Figure 3d), whereas higher evenness was observed when more motile species were present (Figure 3c).

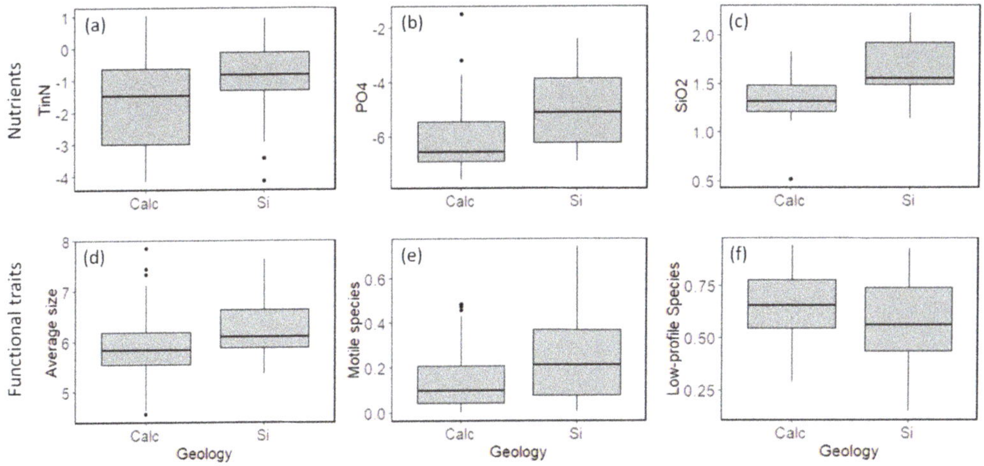

Figure 2. Difference between nutrient concentrations: (**a**) total inorganic nitrogen; (**b**) orthophosphates; (**c**) silicates and species functional traits; (**d**) average size in the assemblage; (**e**) relative abundance of motile species; (**f**) relative abundance of low-profile species between calcareous and siliceous substrates. Y-axis in (**a–d**) is ln-transformed.

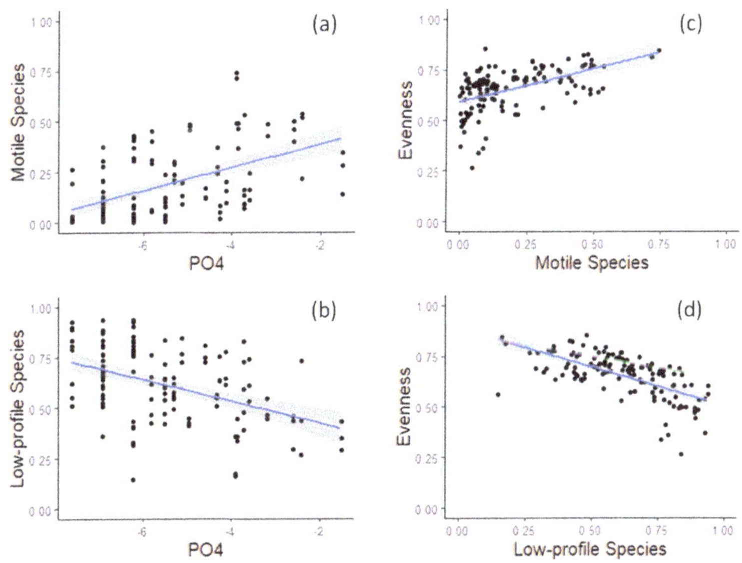

Figure 3. Relationship between PO4 concentration and diatom guilds: (**a**) relative abundance of motile species; (**b**) relative abundance of low profile species; and (**c**,**d**) the same diatom guilds and evenness J.

3.2. Model Results

The above field results on nutrient concentrations and species guild and size indicate that small, fast-growing cells are also good competitors for phosphorus. This was the assumption we used in the model parameterization regarding life history traits of the initial species pool (explained in the methods above).

Model results varied with time during the simulation. At the very beginning of the simulation period, species richness started to increase, along with total biomass, and the

BEF relationship was positive for these initial time steps (Figure 4a, Table 1). During succession, once all the species presented detectable biomass, no significant relationship was apparent between species richness and total biomass production (Figure 4a, Table 1). Even later in succession, when species started to go extinct and the total biomass started to reach the maximum carrying capacity of the system, there was still no significant relationship, or a negative one, between species richness and total biomass production (Figure 4a, Table 1). However, species richness and total biomass did not vary a lot between replicates at later stages of succession. On the other hand, as species started to go extinct and the system reached its maximum biomass, evenness presented a higher variability between replicates and a negative relationship with total biomass, whereas assemblages with higher dominance also presented higher biomass (Figure 4b, Table 1).

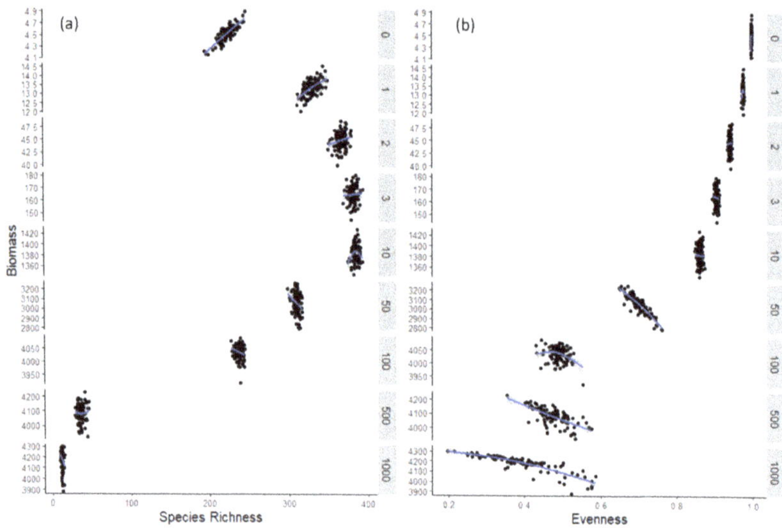

Figure 4. Model results between (**a**) species richness S and (**b**) evenness J (on x-axis) and biomass (y-axis) for different days (0–1000) during simulations. For each day, the 100 replicates are plotted.

Table 1. Slopes and their statistical significance for the equations. ln(biomass) = aS + b and ln(biomass) = aJ + b.

	Species Richness (S)		Evenness (J)	
Day	**Slope (a)**	***p*-Value**	**Slope (a)**	***p*-Value**
0	0.00286	***	31.01131	**
1	0.002282	***	2.686801	0.278
2	0.001267	*	0.447689	0.742
3	0.000544	0.569	−0.87027	0.445
10	0.00018	0.585	−0.05598	0.823
50	−0.00278	**	−1.21366	***
100	0.000241	0.170	−0.08981	**
500	0.000103	0.787	−0.24929	***
1000	−0.00434	**	−0.19818	***

Note: *** < 0.001, ** < 0.01, * < 0.05.

Nutrient concentrations started to decline and reached their minimum fast, with the system being phosphorus-limited early in succession, whereas nitrogen concentrations took longer to decline (Figure 5). It was during this period of nitrogen depletion that total biomass increased further and reached its maximum when both nutrients reached their minimum values (Figure 5). During succession, the species that first went extinct were the slow-growing species (Figure 6), whereas, at the end of the simulation period, the species that survived and contributed the most to the total biomass were the ones with high growth rate and high competitive ability for phosphorus (i.e., low K_P) (Figure 6).

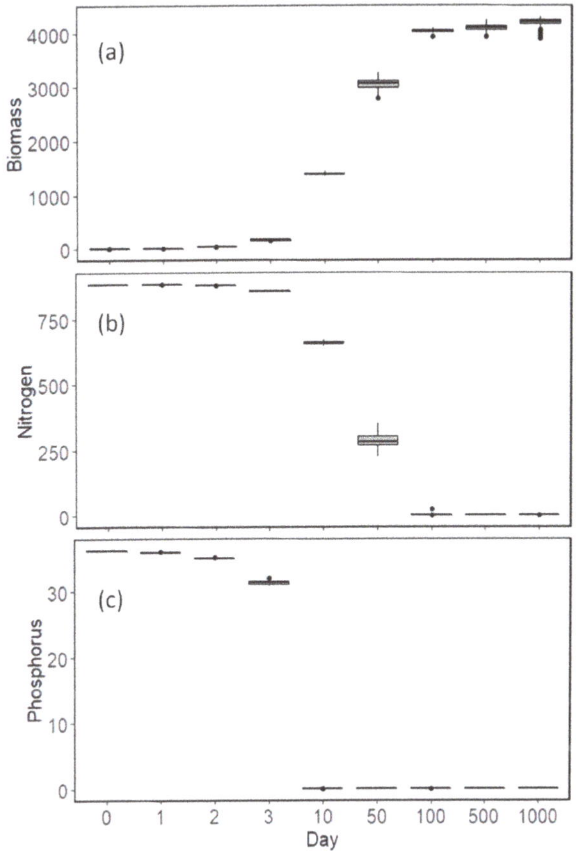

Figure 5. (**a**) Total biomass and nutrient concentrations; (**b**) nitrogen and (**c**) phosphorus during time succession in model simulations.

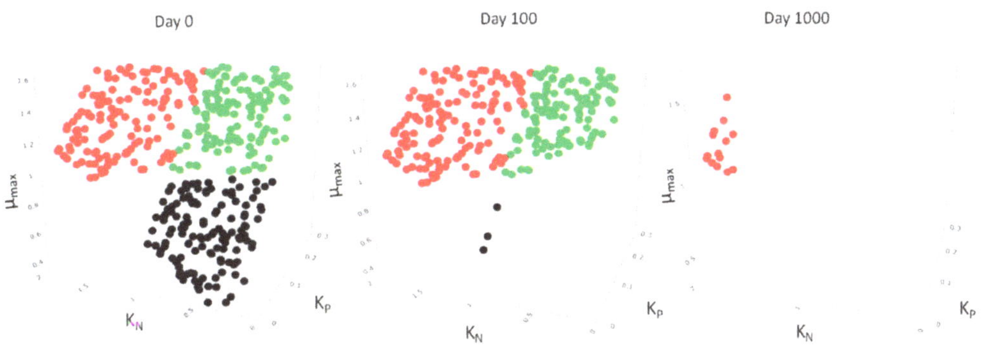

Figure 6. Three-dimensional diagram with the species life-history traits (μ_{max}—maximum specific growth rate, K_N-half saturation constant for nitrogen, K_P-half saturation constant for phosphorus) in the model at day 0 (initial species pool), day 100 and day 1000 (end of simulation). Different colors are the three groups of species based on their competitive abilities (see methods for details).

4. Discussion

Overall, our study suggests that the BEF relationship in river benthic diatoms, although it could be regarded as positive, it seems to be a function of different factors. The main driver of the BEF seems to be geology, directly linked to nutrient availability, which, in turn, selects for specific functional traits. Furthermore, the maturity of the assemblage (i.e., time point during the succession of the biofilm) seems to be an important factor in the observed relationship, as suggested by the model simulations. This is the first attempt to generalize the BEF relationship in river benthic diatoms, using large-scale field observations and numerical modeling, and although the conclusion should be driven with caution, it offers valuable insight into these ecologically important assemblages.

The high variability among rivers regarding their environmental conditions (physico-chemical parameters, nutrient concentrations, drainage area) led to variable diatom assemblages and biomass production. Therefore, it was not surprising that the BEF relationship would also vary among rivers, greatly masking the effect of diversity on biomass in the whole dataset. However, when rivers were split into two groups based on the geology of the substrate, the two patterns were very clear: positive BEF in a siliceous substrate and no BEF in a calcareous substrate. The two substrate groups differed in nutrient concentrations, which were higher in sites with a siliceous substrate. This was expected for silica, as it originates from silicate rock weathering [38]. Regarding phosphorus, this pattern has already been shown in calcareous substrates, as phosphorus is being removed from the water column due to adsorption mechanisms on carbonate material [27,39,40]. Nitrogen was also lower in calcareous substrates, although this trend was not so pronounced. Although most of the sites were phosphorus-limited, there were some sites that were nitrogen limited, belonging, though, to both geological groups. This is consistent with previous studies in Greek rivers, suggesting that the limiting nutrient was site-dependent [26].

Nutrient concentrations largely affected diatom guilds. More specifically, motile species were more abundant in increased phosphate concentrations (and thus siliceous substrates), whereas low-profile species were more abundant in low phosphate concentrations (and thus calcareous substrates). This is in agreement with previous studies, where low-profile species showed a preference for low nutrient concentrations, whereas motile species abundance started to increase with increased nutrient concentration [13]. The fact that low-profile species (i.e., species that adhere strongly to the substrate) were more abundant in calcareous substrates could indicate that species in this functional group take advantage of the precipitated phosphorus. Furthermore, most of the low-profile species found in the study (especially *Achnanthidium* spp.) have a small size, are fast growers and tend to present high populations, increasing dominance [11].

The difference in the BEF relationship between the two substrates could be explained by the combination of nutrient concentrations and traits predominance and by the maturity of the biofilm. According to field data, at higher nutrient concentrations (siliceous substrate), the addition of species could increase biomass, suggesting that species do not occupy all available niches and new arriving species make use of available space, increasing biomass [41]. On the other hand, a stable BEF relationship suggests that the species present occupy all the available niches, consuming all the available resources, and the system has reached a saturated state, even from a few species [17,18]. Model results suggest that assemblages with a positive BEF could be at an early assembly process, whereas a stable relationship could be an indication of a later in succession, more mature assemblage. This is in agreement with previous modeling studies [17] for phytoplankton, using similar models but with different parameters regarding species' life history traits. This could be an indication of a general trend in microalgae assemblages.

Model outcomes suggest that the predominant species traits are related to fast growth and strong phosphorus competition. This is related to our field results, where phosphorus limitation was more predominant, and it would select species with low phosphorus requirements [35]. The selection for fast-growing species was also highly enforced by the penalty induced in slow-growing species, a rather simplistic function that selects for specific traits. The model applied in the present study followed many assumptions and generalizations and could not capture the complexity of a natural system. For example, nutrient inputs follow similar ratios as observed field nutrient concentrations but could be different in many cases, such as, for example, in highly polluted systems or when point-source pollution increases the concentration of a particular nutrient [42]. However, when comparing observational and model results, it is important to remember that it is not the absolute values that are being compared but rather the trends that could give indications on mechanisms underlying observed patterns. Therefore, we believe that our model results reinforce our field findings and assumptions on the time-dependent BEF observations.

The lack of studies on the BEF relationship in benthic diatoms can be explained by a number of challenges and restrictions that it entails, some limitations of which were also apparent in the present study. Specifically, in rivers that are highly dynamic environments, biofilm assemblages can be highly affected by incidents such as heavy rains and floods and point source pollution that could make results evaluation harder. This was one of the reasons that sampling took place during summer, at low flow conditions, when there was a lower probability of heavy rain events, and we expected to collect a more mature biofilm. However, other stressors, such as pollution and desiccation, could be affecting our results [11]. Another limitation of benthic studies is the quantification of benthic concentrations and abundances and the overall sampling effort. In our study, we tried to eliminate this by scraping the biofilm of a defined surface and by using the same stone for both biomass measurement and diversity quantification. The use of chlorophyll a as a surrogate of ecosystem function is widespread in the literature, and it focuses on the biomass of primary producers of the biofilm and the general ecosystem state [43]. On the other hand, photosynthetic biofilm (i.e., phytobenthos) is a complex formation comprising many different groups of photosynthetic organisms apart from diatoms, including cyanobacteria. Therefore, different groups of species and pigments should be carefully considered in order to cover the full spectrum of the BEF relationship of the biofilm. Moreover, as water samples for the quantification of nutrients were from the water above the biofilm, and this differed from nutrient concentrations on the biofilm [44], the use of other ecosystem function metrics, such as the resource use efficiency (accounting for both biomass production and nutrient assimilation in cells, [6]), could not be directly related to our study.

5. Conclusions

This was the first large-scale field study searching for a BEF relationship in benthic diatoms in rivers. Despite the limitations recognized in a field study on benthic microorganisms, it offers important insights into species' contribution to biomass production. It highlights the importance of geology and nutrient concentrations on the form of BEF relationship and indicates species functional traits that could be responsible. The coupled modeling approach demonstrates the time-dependence of the BEF relationship during the succession of the biofilm formation and agrees with field observations on species functional traits. Further experimental work and application of different model scenarios could expand our knowledge and understanding of the ecosystem function of this ecologically important group of organisms.

Author Contributions: Conceptualization, E.S.; methodology, E.S.; software, G.T.; formal analysis, E.S.; investigation, E.S.; resources, E.S.; data curation, E.S.; writing—original draft preparation, E.S.; writing—review and editing, G.T. and N.T.S.; visualization, E.S.; supervision, N.T.S.; funding acquisition, E.S. All authors have read and agreed to the published version of the manuscript.

Funding: This research is co-financed by Greece and the European Union (European Social Fund-ESF) through the Operational Programme «Human Resources Development, Education and Lifelong Learning» in the context of the project "Reinforcement of Postdoctoral Researchers—2nd Cycle" (MIS-5033021), implemented by the State Scholarships Foundation (IKY).

Institutional Review Board Statement: Not applicable.

Informed Consent Statement: Not applicable.

Data Availability Statement: Data are available on request.

Acknowledgments: We thank K. Gritzalis and I. Karaouzas for assisting in river and site selectio A. Lampou, A. Masouras and Yiorgos Maskalidis for assisting in field sampling, S. Laschou, I Kapetanaki and G. Filippi for performing nutrient analyses, A. Abonyi and R. Ptacnik for fruitf discussions and suggestions.

Conflicts of Interest: The authors declare no conflict of interest. The funders had no role in the desig of the study; in the collection, analyses, or interpretation of data; in the writing of the manuscript; c in the decision to publish the results.

Appendix A

In Appendix A are figures and tables related to the methodology and detailed rive results. All the available Figures and Tables are mentioned in the text.

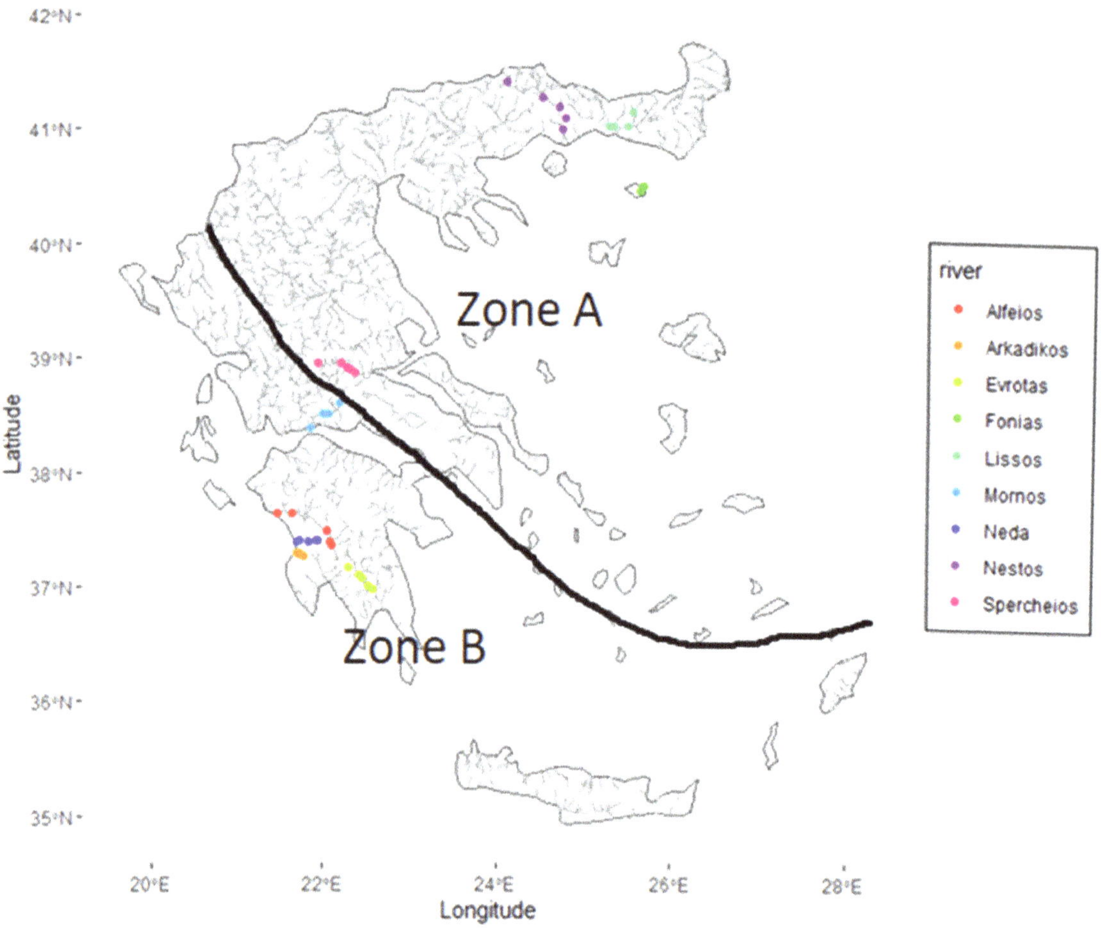

Figure A1. Map with the hydrographic network of Greece and the sampling points for the nine rivers. Hydrographic network is from https://geodata.gov.gr/en/dataset/udrographiko-diktuo (accessed on 6 December 2022). Map was created in R (v. 4.0.3), using packages rnaturalearth (v. 0.1.0), sf (v. 1.0-6) and ggplot2 (v. 3.5.3).

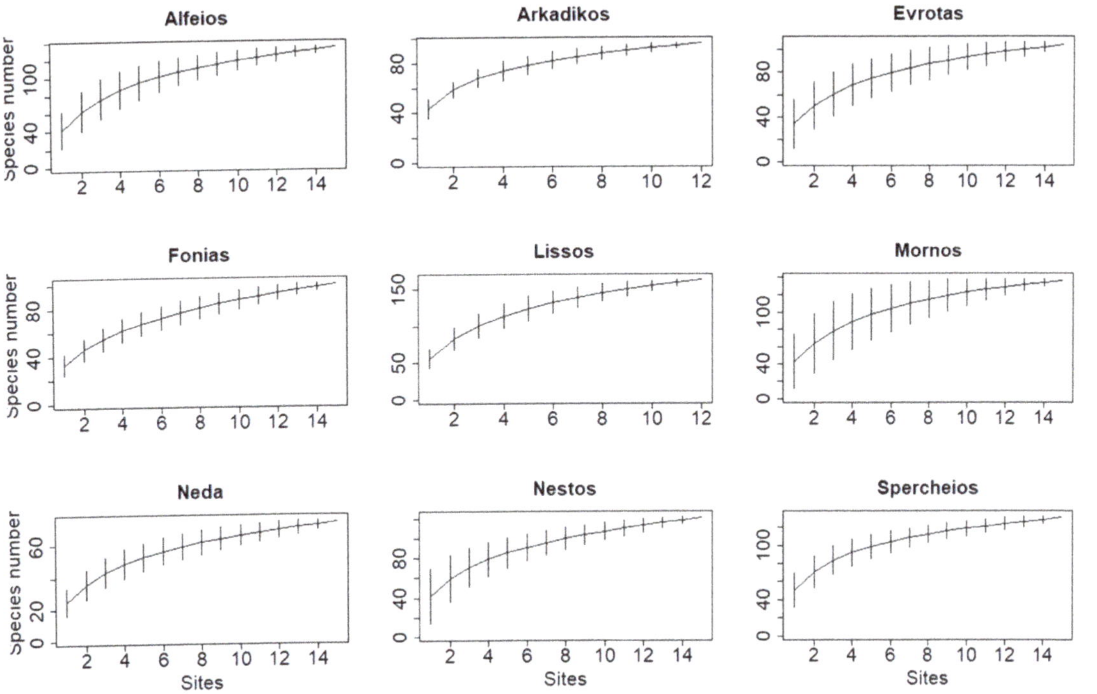

Figure A2. Species–accumulation plots for each river. When the curve reaches a plateau, we assume that sampling effort was enough to detect most of the species present. Analysis was performed in R (v. 4.0.3), using package vegan (v. 2.5-7).

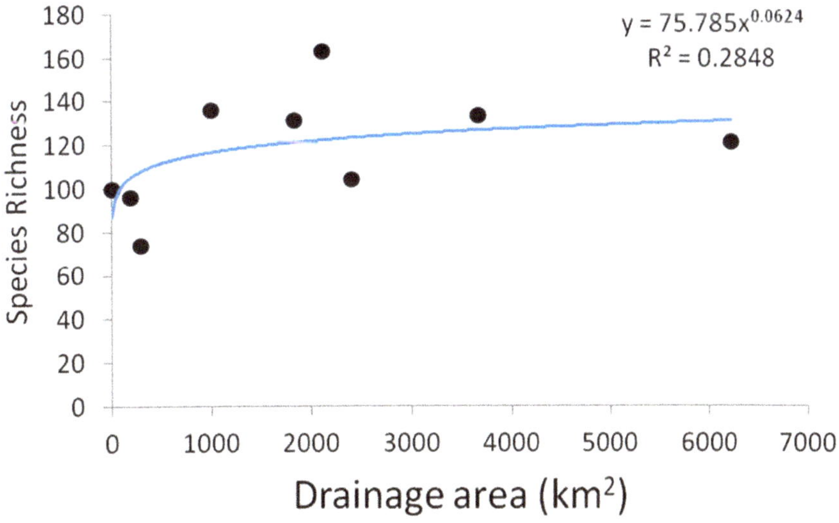

Figure A3. Species–area relationship for the sampled rivers. There is no clear evidence that species richness increases with increased drainage area.

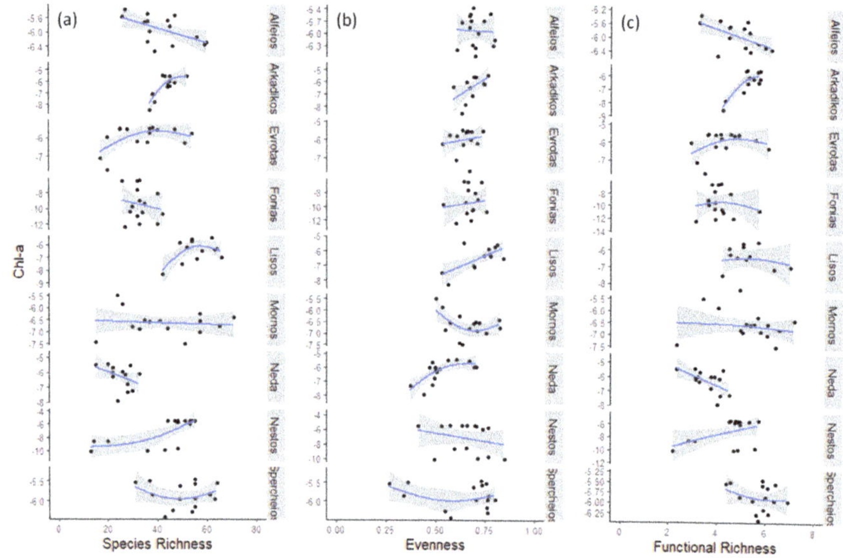

Figure A4. Diversity (expressed as (**a**) species richness-S, (**b**) evenness-J, (**c**) functional richness-FRichness in x-axis) and biomass (ln Chl-a in y-axis) relationship for each river.

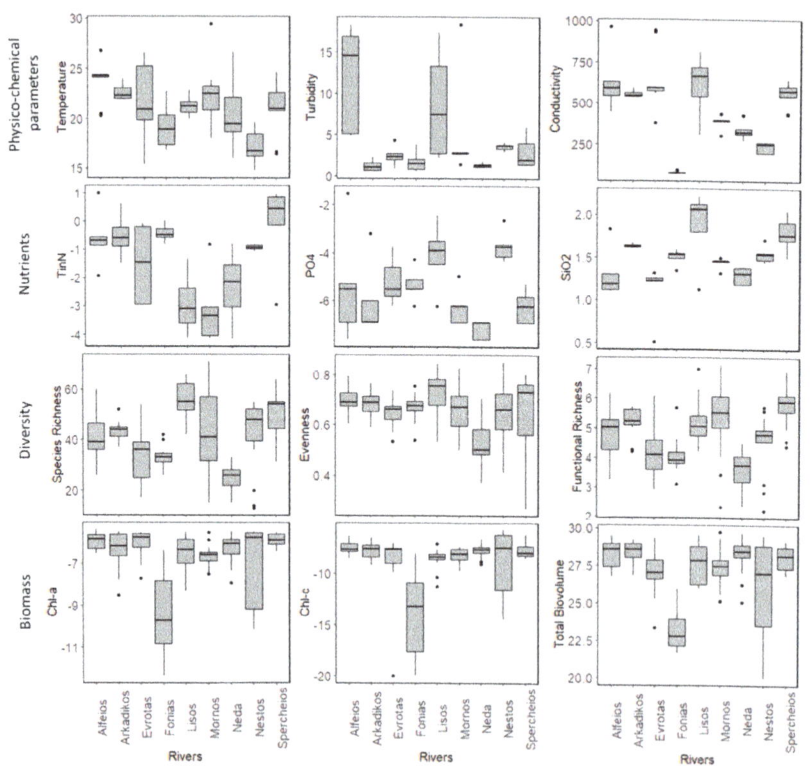

Figure A5. Physico-chemical (temperature, turbidity, conductivity), nutrient concentration (Si, TN, TP), diversity (species richness, functional richness, evenness) and biomass (Chl-a, Chla-c, Total biovolume) variation between the nine rivers of the study.

Table A1. Coordinates and sampling dates of each site. Site numbers 1–5 correspond to sites from upstream to downstream.

River	Site	Latitude	Longitude	Sampling Date
Alfeios	1	37.36144	22.0947	2 July 2020
Alfeios	2	37.39018	22.08635	2 July 2020
Alfeios	3	37.47971	22.04998	2 July 2020
Alfeios	4	37.63421	21.64196	3 July 2020
Alfeios	5	37.64135	21.47642	3 July 2020
Arkadikos	1	37.26862	21.78473	21 July 2020
Arkadikos	2	37.2796	21.7412	21 July 2020
Arkadikos	3	37.28775	21.72578	21 July 2020
Arkadikos	4	37.29342	21.697	21 July 2020
Evrotas	1	37.17217	22.30336	1 July 2020
Evrotas	2	37.09295	22.42634	1 July 2020
Evrotas	3	37.06522	22.45116	1 July 2020
Evrotas	4	36.99387	22.51856	1 July 2020
Evrotas	5	36.97334	22.58183	1 July 2020
Fonias	1	40.45111	25.6258	22 August 2020
Fonias	2	40.4561	25.62369	22 August 2020
Fonias	3	40.45862	25.62405	22 August 2020
Fonias	4	40.48059	25.64669	22 August 2020
Fonias	5	40.49182	25.65536	22 August 2020
Lissos	1	41.13642	25.53514	7 September 2020
Lissos	2	41.02474	25.3223	7 September 2020
Lissos	3	41.0249	25.48959	7 September 2020
Lissos	4	41.0148	25.26305	7 September 2020
Mornos	1	38.59818	22.18833	10 July 2020
Mornos	2	38.51151	22.07488	10 July 2020
Mornos	3	38.50764	21.99866	10 July 2020
Mornos	4	38.50438	22.02188	10 July 2020
Mornos	5	38.38779	21.86056	10 July 2020
Neda	1	37.40079	21.9485	21 July 2020
Neda	2	37.4053	21.92258	21 July 2020
Neda	3	37.39259	21.84667	21 July 220
Neda	4	37.39526	21.72911	21 July 2020
Neda	5	37.38446	21.68998	21 July 2020
Nestos	1	41.41019	24.10549	4 September 2020
Nestos	2	41.26262	24.50997	4 September 2020
Nestos	3	41.17856	24.70111	4 September 2020
Nestos	4	41.08417	24.77134	5 September 2020
Nestos	5	40.99428	24.7438	5 September 2020
Spercheios	1	38.94828	21.94711	27 August 2020
Spercheios	2	38.94361	22.21083	27 August 2020
Spercheios	3	38.90667	22.28583	27 August 2020
Spercheios	4	38.89611	22.3225	27 August 2020
Spercheios	5	38.86722	22.36333	27 August 2020

Table A2. Model parameters.

Parameter	Explanation	Value/Range	Unit
i	species number	400	
j	number of resources	2	
μ_{max}	maximum growth rate	0.3–1.7	d^{-1}
K_P	half-saturation constant for phosphorus	0.02–0.2	μM
K_N	half-saturation constant for nitrogen	0.2–2	μM
c_P	intracellular content for P	0.00397–0.055	μM
c_N	intracellular content for N	0.055–0.244	μM
S_N	input nitrogen concentration	882	μM
S_P	input phosphorus concentration	36.2	μM
D	nutrient flushing rate	0.1	d^{-1}
m_i	species-specific flushing-induced mortality (D/μ_{max})	0.3–0.06	d^{-1}
	total initial biomass	4×10^6–5×10^6	cells/cm^2
	threshold abundance for a species' survival	0.01×10^6	cells/cm^2
	range of initial abundance of each species	0.000009×10^6–0.0225×10^6	cells/cm^2

References

1. van der Plas, F. Biodiversity and Ecosystem Functioning in Naturally Assembled Communities. *Biol. Rev.* **2019**, *94*, 1220–1245. [CrossRef] [PubMed]
2. Dudgeon, D. Prospects for Sustaining Freshwater Biodiversity in the 21st Century: Linking Ecosystem Structure and Function. *Curr. Opin. Environ. Sustain.* **2010**, *2*, 422–430. [CrossRef]
3. Srivastava, D.S.; Vellend, M. Biodiversity-Ecosystem Function Research: Is It Relevant to Conservation? *Annu. Rev. Ecol. Evol. Syst.* **2005**, *36*, 267–294. [CrossRef]
4. Truchy, A.; Angeler, D.G.; Sponseller, R.A.; Johnson, R.K.; McKie, B.G. Linking Biodiversity, Ecosystem Functioning and Services, and Ecological Resilience: Towards an Integrative Framework for Improved Management. *Adv. Ecol. Res.* **2015**, *53*, 55–96. [CrossRef]
5. Daam, M.A.; Teixeira, H.; Lillebø, A.I.; Nogueira, A.J.A. Establishing Causal Links between Aquatic Biodiversity and Ecosystem Functioning: Status and Research Needs. *Sci. Total Environ.* **2019**, *656*, 1145–1156. [CrossRef]
6. Ptacnik, R.; Solimini, A.G.; Andersen, T.; Tamminen, T.; Brettum, P.; Lepistö, L.; Willén, E.; Rekolainen, S. Diversity Predicts Stability and Resource Use Efficiency in Natural Phytoplankton Communities. *Proc. Natl. Acad. Sci. USA* **2008**, *105*, 5134–5138. [CrossRef]
7. Filstrup, C.T.; King, K.B.S.; McCullough, I.M. Evenness Effects Mask Richness Effects on Ecosystem Functioning at Macro-Scales in Lakes. *Ecol. Lett.* **2019**, *22*, 2120–2129. [CrossRef]
8. Olli, K.; Ptacnik, R.; Klais, R.; Tamminen, T. Phytoplankton Species Richness along Coastal and Estuarine Salinity Continua. *Am. Nat.* **2019**, *194*, E41–E51. [CrossRef]
9. Ribeiro, L.; Benyoucef, I.; Poulin, M.; Jesus, B.; Rosa, P.; Méléder, V.; Du, G.Y.; Barillé, L. Spatio-Temporal Variation of Microphytobenthos Biomass, Diversity and Assemblage Structure in the Loire Estuary, France. *Aquat. Microb. Ecol.* **2021**, *87*, 61–77. [CrossRef]
10. Virta, L.; Gammal, J.; Järnström, M.; Bernard, G.; Soininen, J.; Norkko, J.; Norkko, A. The Diversity of Benthic Diatoms Affects Ecosystem Productivity in Heterogeneous Coastal Environments. *Ecology* **2019**, *100*, e02765. [CrossRef]
11. Smeti, E.; von Schiller, D.; Karaouzas, I.; Laschou, S.; Vardakas, L.; Sabater, S.; Tornés, E.; Monllor-Alcaraz, L.S.; Guillem-Argiles, N.; Martinez, E.; et al. Multiple Stressor Effects on Biodiversity and Ecosystem Functioning in a Mediterranean Temporary River. *Sci. Total Environ.* **2019**, *647*, 1179–1187. [CrossRef] [PubMed]
12. Spaulding, S.A.; Potapova, M.G.; Bishop, I.W.; Lee, S.S.; Gasperak, T.S.; Jovanoska, E.; Furey, P.C.; Edlund, M.B. Diatoms.Org: Supporting Taxonomists, Connecting Communities. *Diatom Res.* **2021**, *36*, 291–304. [CrossRef] [PubMed]
13. Passy, S.I. Diatom Ecological Guilds Display Distinct and Predictable Behavior along Nutrient and Disturbance Gradients in Running Waters. *Aquat. Bot.* **2007**, *86*, 171–178. [CrossRef]
14. Rimet, F.; Bouchez, A. Life-Forms, Cell-Sizes and Ecological Guilds of Diatoms in European Rivers. *Knowl. Manag. Aquat. Ecosyst.* **2012**, *406*, 01. [CrossRef]
15. Abonyi, A.; Horváth, Z.; Ptacnik, R. Functional Richness Outperforms Taxonomic Richness in Predicting Ecosystem Functioning in Natural Phytoplankton Communities. *Freshw. Biol.* **2018**, *63*, 178–186. [CrossRef]
16. Crawford, M.S.; Barry, K.E.; Clark, A.T.; Farrior, C.E.; Hines, J.; Ladouceur, E.; Lichstein, J.W.; Maréchaux, I.; May, F.; Mori, A.S.; et al. The Function-Dominance Correlation Drives the Direction and Strength of Biodiversity–Ecosystem Functioning Relationships. *Ecol. Lett.* **2021**, *24*, 1762–1775. [CrossRef]
17. Smeti, E.; Roelke, D.L.; Tsirtsis, G.; Spatharis, S. Species Extinctions Strengthen the Relationship between Biodiversity and Resource Use Efficiency. *Ecol. Modell.* **2018**, *384*, 75–86. [CrossRef]
18. Vallina, S.M.; Cermeno, P.; Dutkiewicz, S.; Loreau, M.; Montoya, J.M. Phytoplankton Functional Diversity Increases Ecosystem Productivity and Stability. *Ecol. Modell.* **2017**, *361*, 184–196. [CrossRef]
19. APHA. *Standard Methods for the Examination of Water and Wastewater*, 15th ed.; American Public Health Association: Washington, DC, USA, 1980.
20. Jeffrey, S.W.; Humphrey, G.F. New Spectrophotometric Equations for Determining Chlorophylls a, b, C1 and C2 in Higher Plants, Algae and Natural Phytoplankton. *Biochem. Physiol. Pflanz.* **1975**, *167*, 191–194. [CrossRef]
21. Battarbee, R.W. Diatom Analysis. In *Handbook of Holocene Palaeoecology and Palaeohydrology*; Berglund, B.E., Ed.; Wiley-Interscience; John Wiley & Sons Ltd: Chichester, UK, 1986; pp. 527–570.
22. Cantonati, M.; Kelly, M.G.; Lange-Bertalot, H. *Freshwater Benthic Diatoms of Central Europe: Over 800 Common Species Used in Ecological Assessment*; Koeltz Botanical Books: Hessen, Germany, 2017.
23. Petchey, O.L.; Gaston, K.J. Functional Diversity (FD), Species Richness and Community Composition. *Ecol. Lett.* **2002**, *5*, 402–411. [CrossRef]
24. Hillebrand, H.; Dürselen, C.D.; Kirschtel, D.; Pollingher, U.; Zohary, T. Biovolume Calculation for Pelagic and Benthic Microalgae. *J. Phycol.* **1999**, *35*, 403–424. [CrossRef]
25. Sommer, U.; Charalampous, E.; Genitsaris, S.; Moustaka-Gouni, M. Benefits, Costs and Taxonomic Distribution of Marine Phytoplankton Body Size. *J. Plankton Res.* **2017**, *39*, 494–508. [CrossRef]
26. Skoulikidis, N.T.; Amaxidis, Y.; Bertahas, I.; Laschou, S.; Gritzalis, K. Analysis of Factors Driving Stream Water Composition and Synthesis of Management Tools—A Case Study on Small/Medium Greek Catchments. *Sci. Total Environ.* **2006**, *362*, 205–241. [CrossRef] [PubMed]

27. Skoulikidis, N.; Amaxidis, Y. Origin and Dynamics of Dissolved and Particulate Nutrients in a Minimally Disturbed Mediterranean River with Intermittent Flow. *J. Hydrol.* **2009**, *373*, 218–229. [CrossRef]
28. Karaouzas, I.; Theodoropoulos, C.; Vardakas, L.; Zogaris, S.; Skoulikidis, N. The Evrotas River Basin: 10 Years of Ecological Monitoring. In *The Rivers of Greece*; Springer: Berlin/Heidelberg, Germany, 2018; pp. 279–326.
29. R Core Team. *R: A Language and Environment for Statistical Computing*; R Foundation for Statistical Computing; R Core Team: Vienna, Austria, 2020.
30. Oksanen, J.; Blanchet, F.G.; Friendly, M.; Kindt, R.; Legendre, P.; Minchin, P.R.; O'hara, R.B.; Simpson, G.L.; Solymos, P.; Stevens, M.H.H.; et al. *Vegan: Community Ecology Package*; Wiley Online Library: Hoboken, NJ, USA, 2020.
31. Cardoso, P.; Mammola, S.; Rigal, F.; Carvalho, J. *BAT: Biodiversity Assessment Tools*; University of Helsinki: Helsinki, Finland, 2021.
32. Bates, D.; Mächler, M.; Bolker, B.M.; Walker, S.C. Fitting Linear Mixed-Effects Models Using Lme4. *J. Stat. Softw.* **2015**, *67*, 1–48. [CrossRef]
33. Wickham, H. *Ggplot2: Elegant Graphics for Data Analysis*; Springer: New York, NY, USA, 2016.
34. Sievert, C. *Interactive Web-Based Data Visualization with R, Plotly, and Shiny*; Chapman and Hall/CRC: London, UK, 2020.
35. Tilman, D. *Resource Competition and Community Structure*; Princeton University Press: Princeton, NJ, USA, 1982.
36. Sarthou, G.; Timmermans, K.R.; Blain, S.; Tréguer, P. Growth Physiology and Fate of Diatoms in the Ocean: A Review. *J. Sea Res.* **2005**, *53*, 25–42. [CrossRef]
37. Roelke, D.L.; Spatharis, S. Phytoplankton Succession in Recurrently Fluctuating Environments. *PLoS ONE* **2015**, *10*, e0121392. [CrossRef]
38. Dürr, H.H.; Meybeck, M.; Hartmann, J.; Laruelle, G.G.; Roubeix, V. Global Spatial Distribution of Natural Riverine Silica Inputs to the Coastal Zone. *Biogeosciences* **2011**, *8*, 597–620. [CrossRef]
39. von Wandruszka, R. Phosphorus Retention in Calcareous Soils and the Effect of Organic Matter on Its Mobility. *Geochem. Trans.* **2006**, *7*, 1–8. [CrossRef]
40. Vitousek, P.M.; Porder, S.; Houlton, B.Z.; Chadwick, O.A. Terrestrial Phosphorus Limitation: Mechanisms, Implications, and Nitrogen-Phosphorus Interactions. *Ecol. Appl.* **2010**, *20*, 5–15. [CrossRef]
41. Loreau, M. *From Populations to Ecosystems: Theoretical Foundations for a New Ecological Synthesis (MPB-46)*; Princeton University Press: Princeton, NJ, USA, 2010; ISBN 9781400834167.
42. Li, T.; Zhou, P.; Ding, Y.; Tang, Q.; Zhou, S.; Liu, Y. Distribution Characteristics and Source Analysis of Nitrogen and Phosphorus in Different Rivers in Two Water Period: A Case Study of Pi River and Shiting River in the Upper Reaches of Tuo River in China. *Int. J. Environ. Res. Public Health* **2022**, *19*, 12433. [CrossRef] [PubMed]
43. von Schiller, D.; Acuña, V.; Aristi, I.; Arroita, M.; Basaguren, A.; Bellin, A.; Boyero, L.; Butturini, A.; Ginebreda, A.; Kalogianni, E.; et al. River Ecosystem Processes: A Synthesis of Approaches, Criteria of Use and Sensitivity to Environmental Stressors. *Sci. Total Environ.* **2017**, *596–597*, 465–480. [CrossRef] [PubMed]
44. Nimick, D.A.; Gammons, C.H.; Parker, S.R. Diel Biogeochemical Processes and Their Effect on the Aqueous Chemistry of Streams: A Review. *Chem. Geol.* **2011**, *283*, 3–17. [CrossRef]

biology MDPI

Article

Giant Panda Microhabitat Study in the Daxiangling Niba Mountain Corridor

Wei Jia [1,2], Shasha Yan [3], Qingqing He [3], Ping Li [4], Mingxia Fu [4] and Jiang Zhou [1,3,*]

1 School of Life Sciences, Guizhou Normal University, Guiyang 550001, China
2 Key Laboratory of Animal Ecology and Conservation Biology, Institute of Zoology, Chinese Academy of Sciences, Beijing 100101, China
3 School of Karst Sciences, Guizhou Normal University, Guiyang 550001, China
4 Administration of Daxiangling Nature Reserve, Yaan 625000, China
* Correspondence: zhoujiang@ioz.ac.cn

Simple Summary: The giant panda is an endemic species in China and the flagship species of global wildlife conservation. Habitat studies of the giant panda corridor can reveal their survival status, habitat environment, and the threats they face, which is crucial for giant panda population recovery and habitat conservation. The results of this study show that due to the opening of the National Road 5 (G5) Niba Mountain tunnel and the completion of the Niba Mountain giant panda corridor plan, the recovery of vegetation within the Niba Mountain giant panda corridor has led to the emergence of giant panda activity in the area, which may have spread to the central part of the reserve through the corridor. The habitat selection characteristics of the giant pandas in the corridor were clarified by investigating the microhabitats of the giant panda corridor in Niba Mountain. These findings can provide a reference for scientists to formulate practical habitat conservation and management measures for giant pandas in the study area.

Abstract: Habitat reduction and increased fragmentation are urgent issues for the survival and recovery of the giant panda (*Ailuropoda melanoleuca*). However, changes in the distribution and microhabitat selection of giant panda habitats in different seasons in the same region have rarely been assessed. To further understand giant panda habitat requirements, this study analyzed the giant panda habitat selection characteristics and differences using the sample data of the giant panda occurrence sites collected during 2020–2022. The results showed that the giant panda in both seasons selected medium altitudes (2000–2400 m), southeastern slopes, slopes less than 15°, taller tree layers (8–15 m) with a larger diameter at breast height (17–25 cm) and medium density (25–55%), shorter shrub layers (<4 m) with sparse density (<30%), and taller bamboo (>2 m) with high density (>35%). The giant panda microhabitat survey in the Niba Mountain corridor clarified the characteristics of suitable habitat selection for the giant panda in the corridor. The findings of the study can provide scientific references for the development of practical habitat conservation and management measures for giant pandas in the study area.

Keywords: *Ailuropoda melanoleuca*; Niba Mountain corridor; suitable habitat; habitat selection; principal component analysis

Citation: Jia, W.; Yan, S.; He, Q.; Li, P.; Fu, M.; Zhou, J. Giant Panda Microhabitat Study in the Daxiangling Niba Mountain Corridor. *Biology* **2023**, *12*, 165. https://doi.org/10.3390/biology12020165

Academic Editors: Daniel Puppe, Panayiotis Dimitrakopoulos and Baorong Lu

Received: 11 December 2022
Revised: 17 January 2023
Accepted: 18 January 2023
Published: 20 January 2023

1. Introduction

The giant panda is an endemic species in China and the flagship species of global wildlife conservation. Currently, the species is distributed in six mountains: Qinling [1], Minshan [2], Qionglai [3], Daxiangling [4], Xiaoxiangling [5], and Liangshan [6,7]. In recent decades, the giant panda population has decreased, with suitable habitat areas becoming increasingly shrunken and fragmented due to an imbalance between economic development and ecological conservation [8,9]. Habitat reduction and increased fragmentation are urgent challenges for the survival and recovery of the giant panda today [10].

To effectively protect the giant panda's habitats, the establishment of habitat corridors for the giant panda has been widely considered [11]. In 2007, with the support of the World Wildlife Fund, the earliest exploration of giant panda habitat corridor construction began in the Tudeling area, connecting the A and B populations of the giant panda in Minshan [12]. According to the Fourth National Giant Panda Survey (2011–2014), four giant panda corridor zones were identified in the distribution area of the giant panda. The Niba Mountain giant panda corridor zone, located in the Daxiangling Mountains, was identified as a priority ecological corridor for construction [13]. There has been no historical distribution of the giant panda in Niba Mountain, and no traces of giant panda activity were found in the area during the Fourth National Giant Panda Survey. Niba Mountain was classified as a key corridor area connecting the Daxiangling giant panda population and the Qionglai population. The construction of giant panda corridors is important for alleviating the declining genetic diversity of giant panda populations among different mountains [14]. In recent years, Gong et al. found that the habitat pattern of the giant panda is an important basis for corridor site selection. Additionally, they suggested that the study of all habitat-related microhabitat factors should be focused on [15]. Habitat studies of the giant panda corridor can more fully reveal the giant panda's survival status, habitat environment, and threats they face. This is crucial for the population recovery and habitat conservation of wild giant pandas [16].

Various constraints affect the habitat selection of the giant panda, including topography (elevation, slope, and aspect) and community structure (tree size and bamboo cover) [17]. These constraints may affect giant panda mobility, shelter availability, and the palatability of edible bamboo for the panda [18]. Previous studies showed that between 1999–2000 and 2011–2014, the giant panda in several mountains in Sichuan experienced a shift in habitat use. The giant panda increasingly utilized secondary forests, which had recovered due to protective measures. The giant panda migrated to higher elevations, despite the availability of bamboo food sources at lower elevations [10]. The implementation of natural forest conservation programs, infrastructure construction, livestock encroachment, and a range of emerging threats may have affected giant panda habitat selection [10].

Previous studies have also investigated the giant panda's habitat selection in the Daxiangling Mountains and reported changes in the habitat selection of this species between 2001 and 2020 [4,19,20]. However, these studies only described the overall habitat selection of the giant panda throughout the year. The habitat selection and utilization of the giant panda under the influence of different environmental conditions in different seasons were not further investigated. The habitat selection of the giant panda varies over time. Furthermore, they found that highly edible bamboo and good shelter sites had an important influence on the habitat selection of the giant panda [18]. The giant pandas feed mainly on *Chimonobambusa szechuanensis* and *Qiongzhuea multigemmia* at lower elevations (1500–2200 m) and *Arundinaria faberi* at higher elevations (2000–2600 m) in the Daxiangling Mountains [21,22]. However, changes in the distribution and microhabitat selection of giant panda habitats in different seasons in the same region have rarely been assessed. Determining a suitable habitat distribution pattern and selection characteristics within the protected area is essential for formulating a scientific and reasonable giant panda conservation and management plan [8]. Therefore, based on the sample data of giant panda occurrence sites collected during 2020–2022, this study aimed to apply MaxEnt and other methods to evaluate the habitat selection characteristics, including the spatial distribution pattern of suitable habitats for the giant panda in the Daxiangling Niba Mountain corridor [23]. The research conclusions can provide a scientific reference for giant panda habitat conservation and management in the area.

2. Materials and Methods

2.1. Study Area

The Daxiangling Mountains are located in the transition zone between the Sichuan Basin and the Qinghai–Tibet Plateau in the eastern part of the Hengduan Mountains, covering an area of about 6440 km^2. The vegetation in the region has obvious vertical distribution zones, and the distribution order is evergreen broad-leaved forest (below 1400 m), deciduous broadleaf forest (1400–1800 m), mixed coniferous forest (1800–2600 m), coniferous forest (2600–3100 m), and alpine scrub meadow (above 3100 m) [24].

According to the results of the Fourth National Giant Panda Survey, there are 14 wild giant pandas distributed in the Daxiangling Reserve and Longcang Valley [18]. Daxiangling is the main conservation area of the Yingjing Area of Giant Panda National Park and one of the key areas of Giant Panda National Park [25,26]. The Daxiangling Niba Mountain giant panda ecological corridor (102°29′–102°52′ E, 29°28′–29°43′ N) is located in the southwestern part of the reserve in Yingjing County and Hanyuan County, Sichuan Province, covering an area of about 38.5 km^2. The National Road 108 (G108) Ya'an section and the G5 Yaxi Expressway Niba Mountain tunnel were completed and opened to traffic in 2000 and 2012, respectively. Both cross the corridor from north to south. For the construction of these highways, many bamboo trees were cut down, which are major food sources for pandas. Furthermore, frequent human activities have caused the separation of the giant panda habitat in Daxiangling [24]. The Niba Mountain corridor has also become a key area for the exchange of giant panda populations in the Daxiangling Mountains [13].

2.2. Giant Panda Habitat Selection Analysis

From October 2020 to April 2022, the data of giant panda traces were recorded using the sample line method and infrared camera monitoring method, and a sample survey was conducted in March–April and October–November (from 2020 to 2022) in the areas where giant panda traces were recorded to determine the preferred environmental factors of the giant panda habitat. A total of 34 sample lines $\geq$3 km were set at intervals of $\geq$500 m. The sample lines covered as many vegetation types and as many potential giant panda distribution areas as possible. Combining data from 158 infrared cameras placed in the study area, the entire Daxiangling Reserve was divided into 145 square grids of 2 km^2 each, with each camera spaced at least 500 m apart to ensure uniform camera coverage (Figure 1). Ten microhabitat variables were recorded in a 10 $\times$ 10 m sample square centered on the site of giant panda traces. The classification criteria for different environmental variables are shown in Table 1. A control sample was randomly set up along the sample line for every 500 m of walking or 100 m of elevation climb without traces of giant panda activity to reflect the environmental background information, and the setting and habitat variables of the control sample were recorded in the same way as the utilization sample [27]. A total of 348 samples were set up [23].

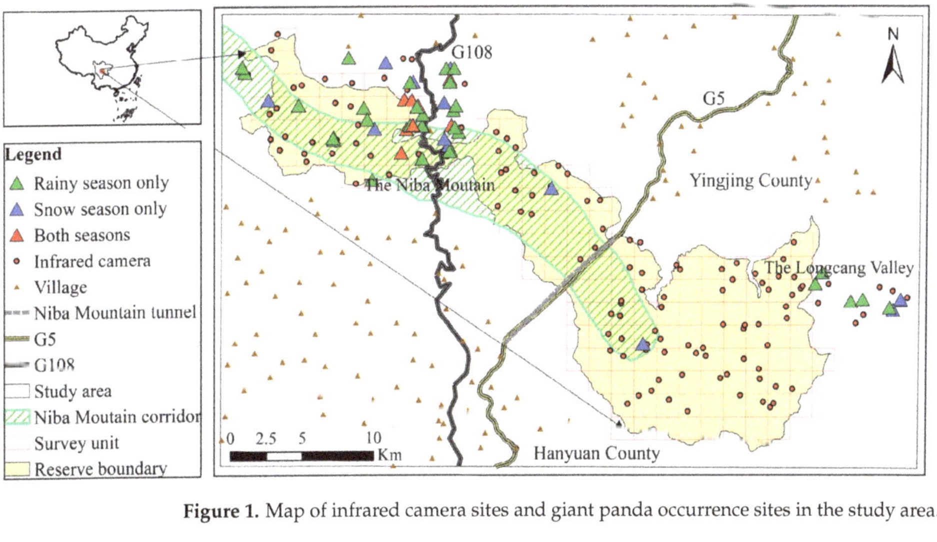

Figure 1. Map of infrared camera sites and giant panda occurrence sites in the study area.

Table 1. Classification criteria for microhabitat variables.

Microhabitat Variable	Classification Criteria
Altitude	<1400 m (evergreen broadleaf forests); 1400–1800 m (deciduous broadleaf forests); 1800–2600 m (mixed coniferous forests); 2600–3100 m (coniferous forests); >3100 m (alpine scrub meadows)
Aspect	<22.5°, >337.5°(N); 22.5°–67.5°(NE); 67.5°–112.5°(E); 112.5°–157.5°(SE); 157.5°–202.5°(S); 202.5°–247.5°(SW); 247.5°–292.5°(W); 292.5°–337.5°(NW)
Slope	<15°; 15°–30°; >30°
Average height of trees	<8 m; 8–12 m; >12 m
Average diameter at breast height of trees	<17 cm; 17–25 cm; >25 cm
Tree coverage	<25%; 25–55%; >55%
Shrub height	<3 m; 3–4 m; >4 m
Shrub coverage	<30%; 30–50%; >50%
Average height of bamboo	<1 m; 1–2 m; >2 m
Bamboo coverage	<20%; 20–35%; >35%

These habitat selection and ecological niche data were input into Excel for the relevant conversions. Following the conversion, the data were entered into SPSS13.0 for normality testing via the one-sample K-S test. Data that conformed to a normal distribution were tested through one-way analysis of variance (ANOVA), and data that did not conform to a normal distribution were tested using the Mann–Whitney *U* test.

2.3. Construction of a Suitable Habitat Model for the Giant Panda in the Niba Mountain Corridor

The estimation of suitable habitats for giant pandas in the study area was performed using the MaxEnt model. A total of 62 giant panda occurrence sites were obtained in the field, with 44 and 18 occurrence sites in the rainy and snow seasons, respectively. To reduce autocorrelation, an 1125 m radius buffer was generated in ArcGIS 10.2 with giant panda occurrence sites in the rainy and snow seasons. When the occurrence site buffers overlapped with each other, one of them was randomly retained, and the rest were eliminated, resulting in 20 and 10 occurrence sites retained in the rainy and snow seasons, respectively [27]. The giant panda has rigorous requirements for habitat, usually choosing primary forests with low human interference [28,29]. Climate and land-use types are also important factors that influence the spatial distribution of the giant panda [9,30]. Therefore, climate, topography, vegetation, and human disturbance are important factors affecting the spatial distribution of the giant panda. In the construction of the model, the above variables were selected to evaluate the habitat. Access to each variable is shown in Table 2 [23]. Because the prediction accuracy of the model was affected by the correlation between environmental factors, the "caret" function in R 4.2.1 was used to remove the highly correlated variables, and the factors with Pearson correlation coefficients greater than 0.8 were removed. Finally, nine factors were retained (Table 2).

Table 2. List of environmental factors for habitat suitability evaluation of the giant panda in the study area.

Types	Factor Codes	Description of Factors	Unit	Data Sources
Climate	Bio2	Mean Diurnal Range	°C	Worldclim (http://www.worldclim.org/, accessed on 7 November 2021)
	Bio7	Temperature Annual Range	°C	
	Bio11	Mean Temperature of Coldest Quarter	°C	

Table 2. *Cont.*

Types	Factor Codes	Description of Factors	Unit	Data Sources
Topography	Aspect	Aspect	°	Geospatial data cloud (http://www.gscloud.cn/, accessed on 6 November 2021)
	Slope	Slope	°	
Interfere	Road	Distance to roads	m	The 4th Survey Report on Giant Panda in Sichuan Province
	Village	Distance to villages	m	
Resources	River	Distance to rivers	m	Resource and environment science and data center (http://www.resdc.cn/, accessed on 6 November 2021)
	Vegetation	Vegetation types categorical variable, divided into six categories: evergreen broadleaf forest, deciduous broadleaf forest, mixed coniferous forest, coniferous forest, alpine scrub meadows, and other lands	/	

In total, 75% of the occurrence sites of the giant panda were selected for modeling, and 25% of the occurrence sites were retained for validation. The importance of each environmental factor was assessed using the Jackknife method, and the output was in the logistic format. The model prediction results were tested using the receiver operating characteristic (ROC) curve [31]. The evaluation criteria were as follows: the area enclosed by the ROC curve and the area under the curve (AUC value) was 0.5–0.6 for failure, 0.6–0.7 for poor, 0.7–0.8 for fair, 0.8–0.9 for good, and 0.9–1.0 for excellent [32]. The means of 10 calculation results were averaged to gain the habitat suitability index (HSI). The suitable habitat range in the study area was divided using Youden's index as the threshold.

2.4. PCA of Microhabitat Factors for Giant Panda Habitat Selection in the Niba Mountain Corridor

PCA can project the high-dimensional original data onto the low-dimensional mutually orthogonal principal components. This process can maximize the information content of the original data while reducing the dimensionality of the data and can effectively overcome the correlation or multicollinearity problem between variables through mutually orthogonal principal components [27]. The raw data of giant panda microhabitat variables were analyzed using PCA; the mean and covariance matrices of the sample data matrix were calculated, and the eigenvalue of the correlation matrix was found. The principal components with eigenvalues >1 were extracted, and each principal component and its contribution rate were derived from the eigenvalue to determine the factors that play a major role in giant panda habitat selection [33]. PCA was performed in IBM SPSS Statistics 26.0. In the statistical analysis, $p < 0.05$ was considered statistically significant.

3. Results

3.1. Results of Giant Panda Habitat Selection Analysis

The giant panda occurrence sites included 18 sites recorded using the sample line method and 26 sites photographed by infrared cameras in the rainy season. All occurrence sites were photographed by infrared cameras during the snow season. The total number of control samples was 155 in the rainy season and 131 in the snow season (Table 3).

Table 3. Comparison of microhabitat characteristics of the giant panda during the rainy and snow seasons (mean ± standard deviation).

Microhabitat Variable	Rainy Season			Snow Season		
	Microhabitat	Control	Advantageous Plants	Microhabitat	Control	Advantageous Plants
Altitude (m)	2147.84 ± 232.44	2153.98 ± 386.02		2209.17 ± 227.09	2074.75 ± 294.52	
Aspect (°)	158.98 ± 96.05	184.65 ± 106.14		141.39 ± 115.69	177.97 ± 104.63	
Slope (°)	13.41 ± 9.32	14.41 ± 10.18		8.89 ± 5.87	12.93 ± 10.29	
Average height of trees (m)	10.98 ± 4.60	12.61 ± 5.36	*Acer oliverianum* Pax (9.41 ± 4.17)	12.28 ± 3.88	11.83 ± 4.22	*Abies fabri* (13.27 ± 3.80)
Average diameter at breast height of trees (cm)	24.27 ± 14.32	19.23 ± 8.80	27.41 ± 19.46	17.72 ± 6.06	19.88 ± 8.78	19.27 ± 6.23
Tree coverage (%)	38.30 ± 16.21	40.14 ± 18.16	37.94 ± 18.63	40.56 ± 16.35	38.00 ± 14.64	42.73 ± 11.26
Average height of shrub (m)	2.47 ± 1.60	3.21 ± 1.19		3.58 ± 0.58	3.34 ± 0.87	
Shrub coverage (%)	18.75 ± 16.54	26.84 ± 16.36		27.22 ±7.71	24.42 ± 10.85	
Average height of bamboo (m)	2.12 ± 1.11	1.77 ± 1.45	*Qiongzhuea multigemmia* (1.91 ± 0.47)	2.39 ± 0.54	1.78 ± 1.37	*Arundinaria faberi* (2.36 ± 0.46)
Bamboo coverage (%)	59.32 ± 21.20	41.87 ± 31.71	57.78 ± 21.85	63.06 ± 24.32	35.00 ± 29.51	64.50 ± 23.51

The one-sample K–S test was performed for 10 ecological factors in the habitat selection and control plots. The results showed that six variables—namely, the altitude, average height of trees, diameter at breast height of trees, tree coverage, the average height of shrub, and bamboo height—were normally distributed, while the other four variables, including aspect, were not. The plot type was used as the control variable for the ANOVA. In normally distributed variables, there were significant differences between altitude, the average height of trees, and diameter at the breast height of trees ($p < 0.05$), but no significant differences between tree coverage, average height of trees, and the average height of bamboo ($p > 0.05$). In the control and habitat selection plots, the Mann–Whitney *U* test revealed significant differences between two variables, namely shrub cover and bamboo cover ($p < 0.05$). However, there was no significant difference between the aspect and slope ($p > 0.05$; Table 4).

Table 4. Comparison of variables in habitat selection plots and control plots in the study area by analysis of variance (ANOVA) and Mann–Whitney *U* tests.

Microhabitat Variable	Rainy Season		Snow Season	
	ANOVA	Mann–Whitney *U* Test	ANOVA	Mann–Whitney *U* Test
	F (*p*)	*U* (*p*)	*F* (*p*)	*U* (*p*)
Altitude (m)	40.04 (0.00 ***)		36.07 (0.00 ***)	
Aspect (°)		171,173 (0.17)		165,673 (0.08)
Slope (°)		167,424 (0.10)		197,344 (0.11)
Average height of trees (m)	35.70 (0.00 ***)		45.50 (0.00 ***)	
Average diameter at breast height of trees (cm)	9.95 (0.01 **)		10.03 (0.01 **)	
Tree coverage (%)	2.76 (0.97)		2.67(0.93)	
Average height of shrub (m)	0.15 (0.70)		0.21 (0.83)	
Shrub coverage (%)		163,905 (0.04 *)		164,335 (0.04 *)
Average height of bamboo (m)	6.20 (0.13)		6.17 (0.11)	
Bamboo coverage (%)		146,994 (0.00 ***)		133,885 (0.00 ***)

* $p < 0.05$, ** $p < 0.01$, *** $p < 0.001$.

In both seasons, giant pandas preferred medium altitudes (2000–2400 m), southeastern slopes, and areas with slopes of 15°. The average elevation preferred by the giant panda in the snow season (2209 m) was higher than that in the rainy season (2147 m), i.e., a difference of 62 m. The giant panda was captured at an average aspect of 158.98° during the rainy season and 141.39° during the snow season. Furthermore, the giant panda was captured at an average slope of 13.41° during the rainy season and 8.89° during the snow season.

The preferred community structure of the giant panda habitat was characterized by a preference for tall (8–15 m), large (17–25 cm) diameter at breast height, and moderately dense (25–55%) trees; short (<4 m) and sparse (<30%) shrub; and tall (>2 m) and dense (>35%) bamboo trees in both seasons. The mean height of trees during the rainy season (10.98 m) and the snow season (12.28 m) were both greater than 8 m. The mean diameter at the breast height of trees during the rainy season (24.27 cm) and the snow season (17.72 cm) were both greater than 17 cm. Furthermore, the depression of trees during the rainy season (38.30%), and the snow season (40.56%), were both greater than 25%. The mean height of the shrub during the rainy season (2.47 m) and the snow season (3.58 m) were both less than 4 m. Additionally, the shrub cover during the rainy season (18.75%) and the snow season (27.22%) were both less than 30%. The mean height of the bamboo during the rainy season (2.12 m) and the snow season (2.39 m) were both greater than 2 m. Furthermore, the bamboo cover during the rainy season (59.32%) and the snow season (63.06%) were both greater than 30%.

3.2. Assessment of Suitable Giant Panda Habitat in the Niba Mountain Corridor

According to the ROC test results of the Maxent model, the AUC values were 0.964 and 0.967 for the rainy and the snow seasons, respectively, the prediction accuracy reached the level of "excellent," and the maximum Youden's index values were 0.316 and 0.527, respectively. The prediction results were transformed in ArcGIS 10.2. The results showed that the suitable habitat areas for the giant panda in the rainy and snow seasons were about 95.61 km^2 and 41.56 km^2, respectively, accounting for 7.30% and 3.17% of the total area of the study area, respectively. In addition, most of the suitable habitat areas for the giant panda in the rainy and snow seasons were located between 1800–2600 m (mixed coniferous and broad forest). The area of suitable habitat for the giant panda in the snow season overlapped with that in the rainy season, while the overlapping area of suitable habitat in the rainy and snow seasons was about 26.85 km^2, with an area overlap rate of 24.34%. Suitable habitats in the rainy season were located in the western part of the Daxiangling Reserve, in Niba Mountain, and the Longcang Valley (Figure 2), while suitable habitats in the snow season were mainly located in Niba Mountain (Figure 3). The area of the west side of G108 in the rainy and snow seasons was about 55.88 km^2 and 19.86 km^2, respectively, while the area of the east side of G108 was approximately 22.23 km^2 and 14.12 km^2, respectively. The suitable habitat during the rainy season was located in the eastern periphery of the Daxiangling Reserve in the Longcang Valley, with an area of about 17.14 km^2; the suitable habitat during the snow season was also scattered in the eastern part of the Reserve, and the periphery of the Longcang Valley, with an area of about 7.07 km^2. The distribution of suitable habitats in both the rainy and snow seasons showed obvious fragmented characteristics.

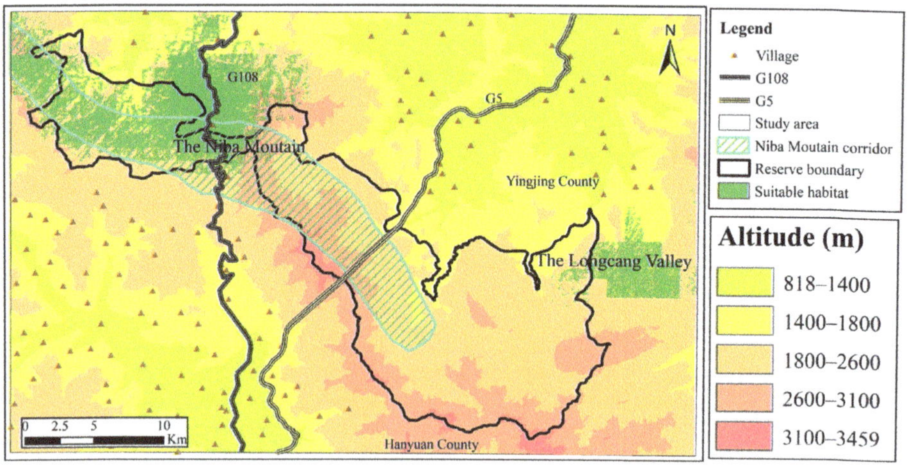

Figure 2. Suitable giant panda habitat during the rainy season.

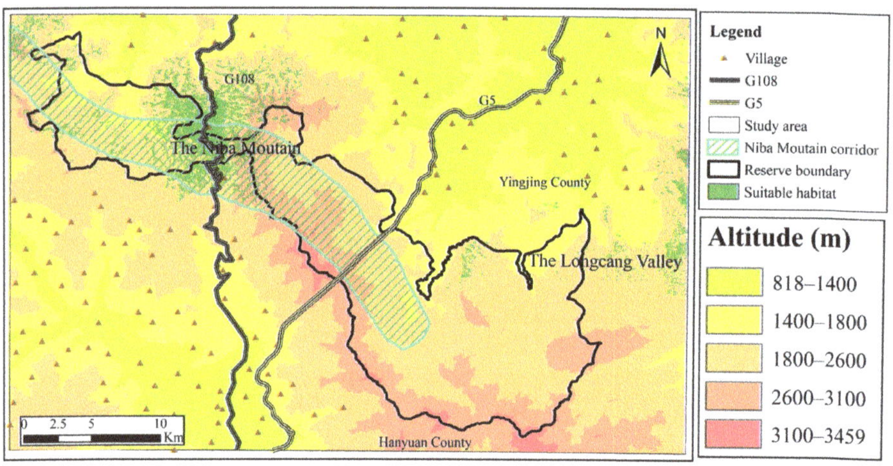

Figure 3. Suitable giant panda habitat during the snow season.

3.3. Main Factors Affecting Microhabitat Selection in the Giant Panda

Among the 10 principal components of the rainy season and snow season, the eigenvalues of the first four principal components were >1, and the cumulative contributions reached 62.71% and 58.64% during the rainy and snow seasons, respectively. The first four principal components were extracted, and their corresponding eigenvectors were calculated. The variables with higher loadings in PC1 during the rainy and snow seasons were the mean height of the tree layer (10.98 m), the mean height of bamboo (2.12 m), the mean height of the tree layer (12.28 m), and the mean diameter at breast height of the tree layer (17.72%). The variables with higher loadings in PC2 were the mean height of the shrub layer (2.47 m), the shrub layer cover (18.75%), the mean height of bamboo (2.39 m), and the bamboo cover (63.06%). The variables with higher loadings in PC3 were the mean height of bamboo (59.32%) and the shrub layer cover (27.22%). Furthermore, the variables with higher loadings in PC4 for both the rainy and snow seasons were altitudes of 2147.84 m and 2209.17 m, respectively (Tables 5 and 6).

Table 5. Principal component analysis loadings of habitat factors at the microhabitat scale of the giant panda during the rainy season.

Microhabitat Variable	Rainy Season			
	PC1	PC2	PC3	PC4
Altitude (m)	−0.023	0.000	−0.400	0.516
Aspect (°)	0.074	0.170	0.031	−0.781
Slope (°)	−0.244	0.176	−0.302	0.148
Average height of trees (m)	0.711	0.413	−0.254	−0.007
Average diameter at breast height of trees (cm)	0.679	0.129	−0.307	−0.176
Tree coverage (%)	0.567	0.277	−0.367	0.059
Average height of shrub (m)	0.065	0.790	0.387	0.203
Shrub coverage (%)	−0.267	0.776	0.347	0.055
Average height of bamboo (m)	0.714	−0.229	0.372	0.165
Bamboo coverage (%)	0.552	−0.343	0.569	0.145
Eigenvalue	2.245	1.721	1.279	1.026
Contribution (%)	22.450	17.213	12.790	10.258
Cumulative contribution (%)	22.450	39.663	52.453	62.711

Table 6. Principal component analysis loadings of habitat factors at the microhabitat scale of the giant panda during the snowy season.

Microhabitat Variable	Snow Season			
	PC1	PC2	PC3	PC4
Altitude (m)	−0.154	0.257	−0.242	0.616
Aspect (°)	−0.235	−0.018	0.204	−0.672
Slope (°)	−0.038	−0.231	−0.324	0.360
Average height of trees (m)	0.810	−0.051	0.206	0.062
Average diameter at breast height of trees (cm)	0.785	−0.109	0.000	−0.055
Tree coverage (%)	0.642	−0.170	0.098	0.084
Average of shrub height (m)	−0.083	0.365	0.604	0.274
Shrub coverage (%)	−0.136	0.227	0.706	0.216
Average height of bamboo (m)	0.252	0.802	−0.160	−0.149
Bamboo coverage (%)	0.153	0.792	−0.295	−0.133
Eigenvalue	1.877	1.617	1.233	1.136
Contribution (%)	18.772	16.175	12.330	11.359
Cumulative contribution (%)	18.772	34.947	47.277	58.636

4. Discussion

4.1. Main Activity Area of the Giant panda in the Niba Mountain Corridor

According to the data from previous giant panda surveys, there were no traces of giant panda activity in Niba Mountain before 2015 [18]. A trail of giant panda activity was discovered in Niba Mountain in 2016. Infrared cameras began to be placed in 2017, and giant panda activity was photographed in 2018. The traces of giant panda activity were recorded during different seasons in the Niba Mountain corridor (Figure 1). Furthermore, the giant pandas were spread out in the middle of the corridor. Therefore, the giant panda may have been able to use the corridor to migrate, and the planning of the giant panda corridor in Niba Mountain is reasonable.

In 2012, the G5 Niba Mountain tunnel was completed and put into use, and the traffic flow of the G108 dropped sharply, which greatly reduced the transportation disruption in Niba Mountain. However, the vegetation in the area is still in a slow recovery stage, and no giant panda activity was found during the Fourth Survey [34]. In 2015, the planning of the Niba Mountain giant panda corridor was completed, and vegetation restoration work began in Niba Mountain [13]. Traces of giant panda activity were recorded in 2016, which indicates that it will take at least 4 years for the vegetation within an 1125 m radius of giant panda activity on the left and right sides of G108 to recover to a stage that can be utilized by the giant panda. We found that some human activities, such as bamboo shoot collection, grazing, and medicinal plant collection, occurred within an 1125 m radius of giant panda activity. However, human activities are controlled on a limited scale. Therefore, no major disturbance impacting the activities of the giant panda in the area has occurred [35].

4.2. Giant Panda Microhabitat Characteristics in the Niba Mountain Corridor

Our study showed that there were distinctive features of giant panda habitat selection in the study area. The giant panda chose to move in southeastern, gently sloping areas at medium altitudes with tall trees in both seasons, which is consistent with the results reported by Fu et al. and Bai et al. [17,18]. The giant panda always chose habitats with lower energy consumption requirements as well as higher nutritional value and net energy gain [36]. A survey of the microhabitats of the giant panda in Niba Mountain clarified that the suitable habitat characteristics of the giant panda in this area were similar to those in other areas of Sichuan Province [17,23,37].

The giant panda preferred to move at moderate altitudes in both seasons, but the average distribution elevation in the snow season was about 62 m higher than that in the rainy season. We believe that this may be related to their preference to feed on bamboo species in different seasons. The giant panda prefers to feed on *Qiongzhuea multigemmia* and *Chimonobambusa szechuanensis* at relatively low altitudes (1500–2200 m) in the rainy season and *Arundinaria faberi* at relatively high altitudes (2000–2600 m) in the snow season. This indicates that the giant panda constantly migrates according to the seasons to meet their energy needs [38], similar to the findings of Chen et al. and Liu et al. on the existence of "seasonal vertical movement" in the giant panda [39,40]. Human activities in the Daxiangling Mountains are mainly concentrated in valley areas at low altitudes (<1400 m). There are few human activities in the high-altitude areas (>1800 m) where the giant panda is active, and the giant panda can avoid anthropogenic disturbance by choosing middle- and high-altitude areas as their habitat [10,11]. The results of PCA showed that altitude and bamboo growth status were significantly associated with the habitat preference of the giant panda. This is because different altitudes can provide different cumulative temperatures for the bamboo. For example, *Qiongzhuea multigemmia* only reaches the conditions for shoot development at an altitude of 1500–2200 m above sea level and when its accumulated temperature reaches a certain standard, coupled with the high ambient precipitation in March [41]. Our results indicated that altitude affects the bamboo shoot collection, the staple food of the giant panda, and thus influences the habitat selection of the giant panda in different seasons.

The Maxent model was used to predict the exact distribution and area of suitable giant panda habitats in the region during different seasons. The distribution of suitable giant panda habitats was more fragmented in snow than in rainy seasons. With the construction of the corridor, the suitable habitat area for the giant panda in the Daxiangling Reserve and Longcang Valley reached about 95.61 km^2 in the rainy season and 41.56 km^2 in the snow season, which can promote further rejuvenation of the giant panda population. The suitable habitat area for giant pandas in the snow season overlapped with that during the rainy season. The overlapping area was about 26.85 km^2, which accounts for 24.34% of the total suitable habitat area. The overlapping area, which was primarily distributed in Niba Mountain, was also the core area for the giant panda's year-round activities [40]. This is because the giant panda prefers to select different bamboo species distributed at different elevations and in different seasons. Furthermore, this verifies the research result that the average distribution elevation of the giant panda in the snow season is about 62 m higher than in the rainy season. The next step will be to introduce a finer level (four levels: most suitable, suitable, less suitable, and unsuitable) of assessment criteria for giant panda habitat assessment in the study area to explore the potential habitat and dispersal pathways of giant pandas in the study area.

The results showed significant seasonal differences in the selection of microhabitats by the giant panda, reflecting that the species has different resource requirements in different seasons [42].

4.3. Newly Recorded Site for Giant Panda Activity in the Middle of the Niba Mountain Corridor

A giant panda activity trail was discovered in the central part of the Niba Mountain corridor in 2019. Two infrared cameras captured giant panda activity in February 2022. Both sites are located within the Niba Mountain giant panda corridor, and it is possible that the giant panda in Niba Mountain is spreading westward along the corridor or the giant panda in the Longcang Valley is spreading eastward. The Niba Mountain corridor is a dispersal channel for the giant panda populations on the left and right sides, creating conditions for the exchange of giant panda populations between the two sides. However, at present, the Niba Mountain corridor is only planned to be about 5 km east of the G5 in the central part of the reserve and does not reach the Longcang Valley. Therefore, it is suggested that another ecological corridor for the giant panda can be planned from the central part of the reserve to the Longcang Valley. Alternatively, the eastern part of the Niba Mountain corridor can be extended to the Longcang Valley to strengthen the exchange of giant panda populations [18]. In 2016, a new point of giant panda activity was discovered in Niba Mountain, and in 2019, it was discovered in the middle of the Niba Mountain corridor. This indicates that it takes about 3 years for the giant panda to move from the west to the east of the reserve and from the north to the south [35].

We are aiming to better verify the actual role of the Niba Mountain giant panda corridor in the dispersal of the giant panda. In the next step, biological samples such as fresh feces will be collected from the giant panda in different areas. Then, we will explore the genetic diversity of the giant panda population by combining microsatellite and mitochondrial control regions, using feces as the main experimental material. Furthermore, we will explore the origin of newly recorded giant panda occurrence sites in the central part of the country to determine the dispersal path of the giant panda population in the Daxiangling Mountains [43].

As one of the important methods used to connect local populations of the giant panda and restore their habitat, giant panda corridors have been established in key areas under the requirements of China's giant panda habitat management plan. The establishment of these corridors has restored vegetation in the relevant areas, connected the more severely fragmented giant panda populations, and promoted exchanges between various giant panda populations [44]. The construction of the G5 Niba Mountain tunnel also provides a reference for the conservation of giant panda habitat in other mountain systems in Sichuan Province, demonstrating that it is possible to minimize the impact on the connectivity of the original giant panda habitat using elevated roadways or tunnels. This strategy not only meets the transportation development needs of human society but also maintains the normal distribution and dispersal of giant panda populations to a certain extent. We recommend monitoring the movement of the giant panda near roads and their use of corridors (e.g., road tunnels) to evaluate the impact of corridors on the giant panda. Further research should assess the effects of bamboo cover on the foraging and movement of the giant panda. In addition, it is suggested that field patrols be strengthened to decrease human activities in the giant panda habitat, including grazing, herb collection, bamboo shoot collection, and hunting. The future production and living modes of residents can be driven by ecological tourism development and the exploration of agricultural and sideline products with regional characteristics, which may change the negative attitudes of residents toward ecological conservation. This method can promote the industrial transformation to realize the balanced development of economic construction and ecological conservation [18]. Additionally, the Daxiangling Mountains should focus on the continuous conservation of major habitats for the giant panda, decrease habitat fragmentation caused by anthropogenic interference, and strengthen habitat restoration for the giant panda [23].

5. Conclusions

The present study shows that the giant panda prefers to move to medium-altitude areas; southeastern slopes; and areas with slopes less than 15°, with an overlap of suitable habitats in both seasons. Due to the opening of the G5 Niba Mountain tunnel and the completion of the Niba Mountain giant panda corridor plan, the recovery of vegetation within the Niba Mountain giant panda corridor has led to the emergence of giant panda activity in the area, which may have spread to the central part of the reserve through the corridor. The findings can provide a reference for the future planning and construction of giant panda corridors.

Author Contributions: Data curation, W.J. and M.F.; formal analysis, W.J.; investigation, W.J. and P.L.; methodology, W.J.; supervision, J.Z.; validation, W.J., S.Y. and Q.H.; visualization, W.J.; writing—original draft preparation, W.J.; writing—review and editing, S.Y., Q.H. and J.Z.; All authors have read and agreed to the published version of the manuscript.

Funding: This work was funded by the Strategic Priority Research Program of the Chinese Academy of Sciences (XDA23080000).

Institutional Review Board Statement: Not applicable.

Informed Consent Statement: Not applicable.

Data Availability Statement: Data are available on request.

Acknowledgments: We thank Hong Yang, Daxiangling Nature Reserve Administration, for his guidance on the fieldwork and the members of the Daxiangling Patrol and Monitoring Team for their support of the fieldwork and their valuable suggestions on data integration and processing.

Conflicts of Interest: The authors declare no conflict of interest.

References

1. Fan, J.; Li, J.; Xia, R.; Hu, L.; Wu, X.; Li, G. Assessing the Impact of Climate Change on the Habitat Distribution of the Giant Panda in the Qinling Mountains of China. *Ecol. Model.* **2014**, *274*, 12–20. [CrossRef]
2. Wang, X.; Xu, W.; Ouyang, Z. Integrating Population Size Analysis into Habitat Suitability Assessment: Implications for Giant Panda Conservation in the Minshan Mountains, China. *Ecol. Res.* **2009**, *24*, 1101–1109. [CrossRef]
3. Huang, J.; Zhou, S.; Tan, Y.; Zhou, X.; Wang, P.; Zhang, H. Study on the Species Diversity of Plant Community in the Giant Panda Habitat of Wolong Natural Reserve:Species Richness, Species Diversity and Evenness. *Sci. Silvae Sin.* **2007**, *43*, 73–78. [CrossRef]
4. Ran, J.; Zeng, Z.; Wang, H.; Liu, S. A Survey of the Giant Panda Population and Habitats in the Daxiangling Mountains. *J. Sichuan Univ. (Nat. Ural Sci. Ed.)* **2006**, *43*, 889–893. [CrossRef]
5. Ran, J.; Zeng, Z.; Wang, H.; Liu, S.; Fu, J.; Liu, S. A Survey of the Giant Panda Population and Habitats in the Xiaoxiangling Mountains. *Acta Theriol. Sin.* **2005**, *4*, 35–40. [CrossRef]
6. Forestry Department of Sichuan Province. *Giant Pandas in Sichuan Province: The Fourth Giant Pandas Survey Report of Sichuan Province*; Sichuan Science and Technology Press: Chengdu, China, 2015.
7. Wang, X.; Huang, J.; Connor, T.A.; Bai, W.; Zhang, J.; Wei, W.; Zhang, Z.; Liu, D.; Zhou, C. Impact of Livestock Grazing on Biodiversity and Giant Panda Habitat. *J. Wild. Mgmt.* **2019**, *83*, 1592–1597. [CrossRef]
8. Xu, W.; Viña, A.; Kong, L.; Pimm, S.L.; Zhang, J.; Yang, W.; Xiao, Y.; Zhang, L.; Chen, X.; Liu, J.; et al. Reassessing the Conservation Status of the Giant Panda Using Remote Sensing. *Nat. Ecol. Evol.* **2017**, *1*, 1635–1638. [CrossRef]
9. Tang, J.; Swaisgood, R.R.; Owen, M.A.; Zhao, X.; Wei, W.; Pilfold, N.W.; Wei, F.; Yang, X.; Gu, X.; Yang, Z.; et al. Climate Change and Landscape-Use Patterns Influence Recent Past Distribution of Giant Pandas. *Proc. R. Soc. B.* **2020**, *287*, 20200358. [CrossRef]
10. Wei, W.; Swaisgood, R.R.; Dai, Q.; Yang, Z.; Yuan, S.; Owen, M.A.; Pilfold, N.W.; Yang, X.; Gu, X.; Zhou, H.; et al. Giant Panda Distributional and Habitat-use Shifts in a Changing Landscape. *Conserv. Lett.* **2018**, *11*, e12575. [CrossRef]
11. Wei, W.; Swaisgood, R.R.; Pilfold, N.W.; Owen, M.A.; Dai, Q.; Wei, F.; Han, H.; Yang, Z.; Yang, X.; Gu, X.; et al. Assessing the Effectiveness of China's Panda Protection System. *Curr. Biol.* **2020**, *30*, 1280–1286.e2. [CrossRef]
12. Cheng, J.; Gao, T.; Zhang, W. Analysis of the Characteristics and Effective Factors of the Giant Panda Habitat Corridor Establishment in Tudeling Area. *J. Sichuan For. Sci. Technol.* **2011**, *32*, 52–56. [CrossRef]
13. Hou, N.; Dai, Q.; Ran, J.; Jiao, Y.; Chen, Y.; Zhao, C. A Corridor Design for the Giant Panda in the Niba Mountain of China. *Chin. J. Appl. Environ. Biol.* **2014**, *20*, 1039–1045. [CrossRef]
14. Huang, F.; He, L.; He, K.; Dai, Q.; Zhang, K.; Tang, B.; Gu, X.; Yang, Z. Spatial and Temporal Distribution of Human Disturbance in Tuowushan Giant Panda Corridor: Survey by Camera Trap Array. *Chin. J. Zool.* **2017**, *52*, 403–410. [CrossRef]
15. Gong, M.H.; Ouyang, Z.Y.; Xu, W.H.; Song, Y.L.; Dai, B. The Location of Wildlife Corridors under the Impact of Road Disturbance: Case Study of a Giant Panda Conservation Corridor. *Shengtai Xuebao* **2015**, *35*, 3447–3453. [CrossRef]

16. Shi, X.; Zhang, J.; Ouyang, Z. Research progress on population investigation methods for wild giant panda. *Acta Ecol. Sin.* **2016**, *36*, 7528–7537. [CrossRef]
17. Bai, W.; Huang, Q.; Zhang, J.; Stabach, J.; Huang, J.; Yang, H.; Songer, M.; Connor, T.; Liu, J.; Zhou, S.; et al. Microhabitat Selection by Giant Pandas. *Biol. Conserv.* **2020**, *247*, 108615. [CrossRef]
18. Fu, M.; Pan, H.; Song, X.; Dai, Q.; Qi, D.; Ran, J.; Hou, R.; Yang, X.; Gu, X.; Yang, B.; et al. Back-and-Forth Shifts in Habitat Selection by Giant Pandas over the Past Two Decades in the Daxiangling Mountains, Southwestern China. *J. Nat. Conserv.* **2022**, *66*, 126129. [CrossRef]
19. Xu, W. Assessment of giant panda habitat in the Daxiangling Mountain Range, Sichuan, China. *Biodivers. Sci.* **2006**, *14*, 223. [CrossRef]
20. He, K.; Dai, Q.; Gu, X.; Zhang, Z.; Zhou, J.; Qi, D.; Gu, X.; Yang, X.; Zhang, W.; Yang, B.; et al. Effects of Roads on Giant Panda Distribution: A Mountain Range Scale Evaluation. *Sci. Rep.* **2019**, *9*, 1110. [CrossRef]
21. Yu, J.; Fu, M.; Song, X.; Gao, F.; Yang, B.; Li, S. MaxEnt Modeling-Based Habitat Suitability Assessment of *Macaca thibetana* in Daxiangling Nature Reserve, Sichuan Province. *J. Sichuan For. Sci. Technol.* **2020**, *41*, 45–50. [CrossRef]
22. Zhang, L.; Gan, X.; Hou, Z.; Yang, Z.; Zhang, Z. Grazing by Wild Giant Pandas Does Not Affect the Regeneration of *Arundinaria spanostachya*. *J. For. Res.* **2019**, *30*, 1513–1520. [CrossRef]
23. Ruan, T.; Han, H.; Wei, W.; Qiu, L.; Hong, M.; Tang, J.; Zhou, H.; Zhang, Z. Habitat Suitability Evaluation for Giant Panda in Liziping National Nature Reserve, Sichuan Province. *Glob. Ecol. Conserv.* **2021**, *30*, e01780. [CrossRef]
24. Wang, Y.; He, X.; Zhang, S.; Zhang, Y.; He, Q.; Wang, B.; Wang, B.; Song, X.; Fu, M.; Zhu, M.; et al. Diversity and Community Structure of Birds in Breeding Season in Daxiangling of Yingjing, Sichuan. *Sichuan J. Zool.* **2021**, *40*, 344–360. [CrossRef]
25. Li, B.V.; Pimm, S.L. China's Endemic Vertebrates Sheltering under the Protective Umbrella of the Giant Panda: China's Protected Areas and Biodiversity. *Conserv. Biol.* **2016**, *30*, 329–339. [CrossRef] [PubMed]
26. Yang, B.; Qin, S.; Xu, W.; Busch, J.; Yang, X.; Gu, X.; Yang, Z.; Wang, B.; Dai, Q.; Xu, Y. Gap Analysis of Giant Panda Conservation as an Example for Planning China's National Park System. *Curr. Biol.* **2020**, *30*, 1287–1291.e2. [CrossRef] [PubMed]
27. Wang, B.; Xu, Y.; Zhang, B.; Wu, Y.; He, X.; Ran, J.; Zeng, T. Overlap and Selection of Dust-Bathing Sites among Three Sympatric Montane Galliform Species. *Auk* **2018**, *135*, 1076–1086. [CrossRef]
28. Zhang, Z.; Swaisgood, R.R.; Zhang, S.; Nordstrom, L.A.; Wang, H.; Gu, X.; Hu, J.; Wei, F. Old-Growth Forest Is What Giant Pandas Really Need. *Biol. Lett.* **2011**, *7*, 403–406. [CrossRef]
29. He, K.; Dai, Q.; Foss-Grant, A.; Gurarie, E.; Fagan, W.F.; Lewis, M.A.; Qing, J.; Huang, F.; Yang, X.; Gu, X.; et al. Movement and Activity of Reintroduced Giant Pandas. *Ursus* **2019**, *29*, 163. [CrossRef]
30. Songer, M.; Delion, M.; Biggs, A.; Huang, Q. Modeling Impacts of Climate Change on Giant Panda Habitat. *Int. J. Ecol.* **2012**, *2012*, 1–12. [CrossRef]
31. Soberón, J. Grinnellian and Eltonian Niches and Geographic Distributions of Species. *Ecol. Lett.* **2007**, *10*, 1115–1123. [CrossRef]
32. Chen, J.; Yao, Z.; Shi, R.; Gao, H.; Liu, Z. Habitat suitability assessment of rodents on the west slope of the Helan Mountain based on MAXENT model. *Acta Ecol. Sin.* **2022**, *42*, 4209–4216. [CrossRef]
33. Zhang, Z.; Wang, Q.; Cao, L.; Zhang, W. Principal Component Analysis of Different Origins of *Platycodon grandiflorum* (Jacq.)A.DC.Based on Element Profile. *J. Gansu Agric. Univ.* **2018**, *53*, 191–196. [CrossRef]
34. Tang, J.; Zhang, J.; Zhao, X.; Wei, W.; Hong, M.; Zhou, H.; Zhang, J.; Zhang, Z. The Fate of Giant Panda and Its Sympatric Mammals under Future Climate Change. *Biol. Conserv.* **2022**, *274*, 109715. [CrossRef]
35. Zhao, C.; Yue, B.; Ran, J.; Moermond, T.; Hou, N.; Yang, X.; Gu, X. Relationship between Human Disturbance and Endangered Giant Panda *Ailuropoda melanoleuca* Habitat Use in the Daxiangling Mountains. *Oryx* **2017**, *51*, 146–152. [CrossRef]
36. Nie, Y.; Speakman, J.R.; Wu, Q.; Zhang, C.; Hu, Y.; Xia, M.; Yan, L.; Hambly, C.; Wang, L.; Wei, W.; et al. Exceptionally Low Daily Energy Expenditure in the Bamboo-Eating Giant Panda. *Science* **2015**, *349*, 171–174. [CrossRef]
37. Bai, W.; Zhang, J.; Dong, X.; Li, C.; Gu, X.; Zhou, C. Dynamics of the space use patterns of giant pandas in Wolong Nature Reserve. *Acta Theriol. Sin.* **2017**, *37*, 327–335. [CrossRef]
38. Zhang, W.; Tang, Z.; Qi, D.; Hu, Y.; Hu, J. Habitat Assessment for Giant Pandas (*Ailuropoda melanoleuca*) on the Northern Slope of the Daxiangling Mountains. *Acta Theriol. Sin.* **2007**, *27*, 146–152. [CrossRef]
39. Liu, X.; Wang, T.; Wang, P.; Yang, J. Application research of radio-tracking data on giant panda movements in Wolong Nature Reserve, China. *Acta Theriol. Sin.* **2008**, *28*, 180–186. [CrossRef]
40. Chen, Y.; Wang, X.; Zheng, X.; Gong, Y.; Chen, M.; Qiu, L.; Zhou, H.; Wei, W.; Han, H. Space Use and Microhabitat Selection of Wild Giant Pandas in Meigu Dafengding National Nature Reserve, China. *Front. Ecol. Evol.* **2022**, *10*, 1000841. [CrossRef]
41. Zhang, J.; Pan, S.; Che, Q.; Wei, W.; Zhao, X.; Tang, J. Impacts of Climate Change on the Distributions and Diversity of the Giant Panda with Its Sympatric Mammalian Species. *Ecol. Indic.* **2022**, *144*, 109452. [CrossRef]
42. Qiu, L.; Han, H.; Zhou, H.; Hong, M.; Zhang, Z.; Yang, X.; Gu, X.; Zhang, W.; Wei, W.; Dai, Q. Disturbance Control Can Effectively Restore the Habitat of the Giant Panda (*Ailuropoda melanoleuca*). *Biol. Conserv.* **2019**, *238*, 108233. [CrossRef]

43. Qin, Q.; Huang, Y.; Liu, J.; Chen, D.; Zhang, L.; Qiu, J.; Tan, H.; Wen, Y. The Landscape Patterns of the Giant Panda Protection Area in Sichuan Province and Their Impact on Giant Pandas. *Sustainability* **2019**, *11*, 5993. [CrossRef]
44. Chen, X.; Wang, X.; Li, J.; Kang, D. Integrating Livestock Grazing and Sympatric Takin to Evaluate the Habitat Suitability of Giant Panda in the Wanglang Nature Reserve. *Animals* **2021**, *11*, 2469. [CrossRef] [PubMed]

biology

MDPI

Article

Effect of Rice Straw and Stubble Burning on Soil Physicochemical Properties and Bacterial Communities in Central Thailand

Noppol Arunrat [1,*], Sukanya Sereenonchai [1], Chakriya Sansupa [2], Praeploy Kongsurakan [3] and Ryusuke Hatano [4]

1 Faculty of Environment and Resource Studies, Mahidol University, Nakhon Pathom 73170, Thailand
2 Department of Biology, Faculty of Science, Chiang Mai University, Chiang Mai 50200, Thailand
3 Graduate School of Fisheries and Environmental Sciences, Nagasaki University, Nagasaki 852-8521, Japan
4 Laboratory of Soil Science, Research Faculty of Agriculture, Hokkaido University, Sapporo 060-8589, Japan
* Correspondence: noppol.aru@mahidol.ac.th

Simple Summary: Fire is traditionally used by farmers for clearing fields when the fallow period is short. Soil chemical properties changed significantly after burning and returned to prefire levels after 1 year. *Bacillus*, HSB OF53-F07, *Conexibacter*, and *Acidothermus* abundances increased immediately after burning and then significantly declined, with lower levels 1 year after burning. *Anaeromyxobacter* and Candidatus *Udaeobacter* dominated at 1 year after burning. Burning under high soil moisture conditions and within a very short time caused no effect to the bacterial soil communities.

Abstract: Rice straw and stubble burning is widely practiced to clear fields for new crops. However, questions remain about the effects of fire on soil bacterial communities and soil properties in paddy fields. Here, five adjacent farmed fields were investigated in central Thailand to assess changes in soil bacterial communities and soil properties after burning. Samples of soil prior to burning, immediately after burning, and 1 year after burning were obtained from depths of 0 to 5 cm. The results showed that the pH, electrical conductivity, NH_4-N, total nitrogen, and soil nutrients (available P, K, Ca, and Mg) significantly increased immediately after burning due to an increased ash content in the soil, whereas NO_3-N decreased significantly. However, these values returned to the initial values. Chloroflexi were the dominant bacteria, followed by Actinobacteria and Proteobacteria. At 1 year after burning, Chloroflexi abundance decreased remarkably, whereas Actinobacteria, Proteobacteria, Verrucomicrobia, and Gemmatimonadetes abundances significantly increased. *Bacillus*, HSB OF53-F07, *Conexibacter*, and *Acidothermus* abundances increased immediately after burning, but were lower 1 year after burning. These bacteria may be highly resistant to heat, but grow slowly. *Anaeromyxobacter* and Candidatus *Udaeobacter* dominated 1 year after burning, most likely because of their rapid growth and the fact that they occupy areas with increased soil nutrient levels after fires. Amidase, cellulase, and chitinase levels increased with increased organic matter levels, whereas β-glucosidase, chitinase, and urease levels positively correlated with the soil total nitrogen level. Although clay and soil moisture strongly correlated with the soil bacterial community's composition, negative correlations were found for β-glucosidase, chitinase, and urease. In this study, rice straw and standing stubble were burnt under high soil moisture and within a very short time, suggesting that the fire was not severe enough to raise the soil temperature and change the soil microbial community immediately after burning. However, changes in soil properties due to ash significantly increased the diversity indices, which was noticeable 1 year after burning.

Keywords: paddy field; soil organic carbon; soil total nitrogen; microbial diversity; fire

Citation: Arunrat, N.; Sereenonchai, S.; Sansupa, C.; Kongsurakan, P.; Hatano, R. Effect of Rice Straw and Stubble Burning on Soil Physicochemical Properties and Bacterial Communities in Central Thailand. *Biology* **2023**, *12*, 501. https://doi.org/10.3390/biology12040501

Academic Editors: Daniel Puppe, Panayiotis Dimitrakopoulos and Baorong Lu

Received: 27 February 2023
Revised: 22 March 2023
Accepted: 22 March 2023
Published: 26 March 2023

1. Introduction

Rice straw comprises a major agricultural residue in Thailand. However, large amounts of rice straw are left in fields after harvesting, which are then often burned to prepare the land for the following crops [1,2]. The production of rice straw in Thailand has been estimated at over 20 million tons per year [3]. Among the rice cultivation areas in Thailand, the central region has the highest potential, with two harvests a year. According to the agricultural statistics of Thailand, the total area of the first rice harvest in central Thailand was 1.31 million ha in the crop year of 2020/2021 (3.74 tons ha^{-1} in average yield), whereas the second rice harvest area covered 0.54 million ha (4.33 tons ha^{-1} in average yield) [4]. This could be explained by its location in the floodplains of the Chao Phraya river basin, which facilitates water availability, thus, encouraging farmers to prepare their lands for the following crops as soon as possible. In this sense, burning is the method of choice for rapidly eliminating rice straw and stubble. If left in the field, rice straw and stubble can represent obstacles for land preparation, and using fire is also the most efficient method for controlling weeds and pests [5,6].

Direct burning in fields significantly changes the soil temperature, moisture, and organic matter content, especially in the topsoil layer [7,8], resulting in an abrupt decline in microbial biomass and diversity [9]. Mickovski [10] reported that burning rice straw and stubble resulted in an increase in soil temperatures by up to 50–70 °C in the uppermost 0 to 3 cm of soil, causing a 77% decrease in heterotrophic microorganisms. Biederbeck et al. [11] also found that bacterial populations in the topsoil layer (0–2.5 cm) were reduced by >50% in rice straw burning areas compared to areas where the rice straw remained in the fields. Furthermore, Kumar et al. [12] reported that paddy straw burning reduced the populations of bacteria, fungi, actinomycetes, phosphate-solubilizing microorganisms, potassium-solubilizing microorganisms, cellulose, and microbial enzymes.

Changes to the physical and chemical properties of soil after burning also affect the structure of soil bacterial communities due to the properties of rice straw ash [13]. Duan et al. [14] stated that rice straw ash is alkaline and mainly contains potassium oxide and silicon dioxide; silicon plays a crucial role as a biological stimulant for plant growth [15]. The ash causes changes in the soil pH, affecting the abundance of soil microorganisms. Guo et al. [16] and Zhao et al. [17] reported that changing the soil pH resulted in changes in the abundance of Acidobacteria in rice–wheat cropping systems, whereas the abundances of Proteobacteria, Gemmatimonadetes, and Nitrospirae in soil were modified due to changes in nitrogen, phosphorus, and potassium levels. Avoiding the burning of rice straw by incorporating it into the soil can improve soil nutrient contents by increasing carbon and nitrogen concentrations, thus, affecting the structure of soil bacterial communities. In a study by Zhao et al. [17], the population of Proteobacteria increased following the addition of rice straw due to an increase in soil nitrogen levels. Currently, the incorporation of rice straw into soil is not always cost-effective for farmers in central Thailand, and, thus, burning is the most effortless and cheapest practice for clearing fields when the fallow period is short. However, the effects of burning rice straw and stubble on soil properties and soil bacterial communities remain poorly understood. We hypothesized that soil nutrients would increase, but that soil bacterial communities would decrease after a fire. The objectives of the present study were as follows: (1) to determine the effects of fire on soil organic carbon (SOC), soil total nitrogen (STN), and soil nutrients in paddy fields; (2) to examine changes in the composition and diversity of soil bacterial communities as a result of rice straw and stubble burning; and (3) to analyze the relationship between soil nutrients and soil bacterial communities. The results of this study provide the scientific knowledge for the minimization of postfire risks in paddy fields, which could contribute to implementing suitable management practices for maintaining biodiversity and ecosystem functioning.

2. Materials and Methods

2.1. Study Site

The study was conducted in the Taluk subdistrict, Sapphaya district, Chainat province, central Thailand. The area is located in the floodplain of the Chao Phraya river basin and has a tropical savanna (Köppen 'Aw') climate, with an average annual temperature of 27.8 °C. April and May are the hottest months, with maximum temperatures in the range of 38–40 °C, whereas the lowest temperatures occur during December–February (18–23 °C). The average annual precipitation ranges from 1000 to 1200 mm. The highest precipitation is usually recorded in September, with a total monthly precipitation of 200–300 mm and 10–18 rainy days. The soil belongs to the Chainat series (Cn), which consists of fine, mixed, active, nonacidic, and isohyperthermic Aeric (Vertic) Endoaquepts, and is predominantly dark grayish brown and dark gray. The pH ranges from 5.5 to 6.5, and the soil texture is characterized by silty clay or clay loam. The slopes range from 0 to 2%.

2.2. Field Management Practices and Fire Measurements

To avoid the effects of variation in environmental conditions, five adjacent farmed fields were investigated in the Taluk subdistrict. The "RD 43" (95 days), "RD 57" (110 days), and "RD 41" (105 days) rice varieties were planted twice a year. The pregerminated rice seeds were sown using the broadcasting method, and chemical fertilizers were applied using 46-0-0 (62.5 kg ha^{-1}) and 16-20-0 (156.3 kg ha^{-1}). Glyphosate (48% *w*/*v* SL) and alachlor (48% *w*/*v* EC) were applied to control weeds, whereas acephate (75% S) and chlorpyrifos (40% EC) were used for disease and insect control. A harvesting machine was used for rice harvesting. Rice straw and stubble were burnt in the field once a year after 20–25 days of sun-drying in August. After 1–3 days of burning, tillage was started for the second rice cultivation, which was harvested in November. As there was not sufficient water for a third rice cultivation, the field was left to fallow for approximately 5 months (December–April) to allow sufficient time for the natural decomposition of rice straw and stubble, and the first rice cultivation was started in May, with harvesting in September.

Five replicated plots were investigated in each field to assess the effects of rice straw and stubble burning on soil properties and bacterial communities, with an area of 5 × 5 m for each plot (Figure 1). Burning was conducted at 10.00–12.00 am and took approximately 98–130 s for each plot (the average burning speed was 6.5 s m^{-2}). Air, fire, and soil temperatures as well as soil moisture were measured in each plot at three periods: before burning (19 August 2021), immediately after burning (19 August 2021), and 1 year after burning (27 August 2022). The fire temperature during burning was measured using an infrared thermometer. The soil temperature and soil moisture were measured at a depth of 5 cm with a Thermocouple Type K and a soil moisture meter, respectively. Each plot was burnt only once, meaning that there was no repeat burning even if rice straw and stubble remained, reflecting the current burning practice.

(a) (b) (c)

Figure 1. Rice straw and stubble burning practices: (**a**) during burning, (**b**) postburning, and (**c**) one year after burning. Photos were taken by Noppol Arunrat.

2.3. Sample Collection and Analysis

Soil samples before burning (preburning), immediately after burning (postburning), and 1 year after burning were obtained from five adjacent fields. In each field, five replicated plots at depths of 0–5 cm were investigated. In each plot, the soil samples were collected from five positions (four positions at the four corners of the plot and one in the center). Roots, grasses, stones, and residues were removed manually from the samples, which were then mixed to obtain one composite sample for each field. Approximately 1 kg of soil was placed into a plastic bag for the analysis of the soil physical and chemical properties. Additionally, 100 g samples of soil were placed into zip-lock plastic bags, cooled, and transported to the laboratory for the analysis of bacterial communities. To determine the soil bulk density, a 5.0 × 5.5 cm soil core was taken from each layer, and bulk density was measured after drying at 105 °C for 24 h.

Soil texture was determined using a hydrometer. The soil pH was measured with a pH meter using 1:1 suspensions of solids in water [18]. Electrical conductivity (EC_e) was measured in saturation paste extracts using an EC meter [19]. The available calcium (Ca), magnesium (Mg), and potassium (K) levels were measured with atomic absorption spectrometry (NH_4OAc pH 7.0 extraction) [20]. The available phosphorus (P) concentration was determined using the molybdate blue method (Bray II extraction) [21]. The ammonium nitrogen (NH_4-N) and nitrate-nitrogen (NO_3-N) levels were measured with the KCl extraction method, and total nitrogen (TN) was measured with the micro-Kjeldahl method. The cation exchange capacity (CEC) was analyzed using the NH_4OAc pH 7.0 method. The organic carbon (OC) contents were determined following the method described by Walkley and Black [22]. The SOC stock was estimated using the following equation:

$$SOC\ stock = OC \times BD \times L \times 10{,}000, \tag{1}$$

where SOC is the soil organic carbon stock (Mg C ha^{-1}), OC is the organic carbon (%), BD is the soil bulk density (Mg m^{-3}), and L is the soil thickness (m).

The STN stock was calculated using the following equation:

$$STN\ stock = TN \times BD \times L \times 10{,}000, \tag{2}$$

where STN is the amount of soil total nitrogen (Mg N ha^{-1}), TN is the total nitrogen (%), BD is the soil bulk density (Mg m^{-3}), and L is the soil thickness (m).

2.4. DNA Extraction, Bacterial 16s Amplification, and Sequencing

The DNA was extracted from approximately 0.25 g of soil using a DNeasy PowerSoil Pro DNA Kit (Qiagen, USA). The hypervariable V3–V4 region was amplified with the 16s rRNA gene using primers 341F (5′-CCTAYGG-GDBGCWSCAG) and 805R (5′-GGACTAC-NVGGGTHTCTAAT-3′) [23]. Subsequently, the PCR products were sequenced using the Paired-end Illumina Miseq platform (2 × 250 bp) at the Omics Sciences and Bioinformatics Center of Chulalongkorn University (Bangkok, Thailand). All sequencing data associated with this study can be found in the National Center for Biotechnology Information (NCBI) under the BioProject accession number PRJNA881635.

2.5. Bacterial Taxonomic and Functional Identification

The bioinformatic analysis of the bacterial 16s rRNA gene was conducted on QIIME2 v. 2022.2 [24]. Raw sequence data were quality-filtered and merged, and chimera were removed using the DADA2-plugin [25]. Amplicon sequence variants (ASVs) with less than two sequence reads (singletons) were eliminated. Bacterial taxonomy was assigned using the Silva v. 138 database [26,27], and ASVs that were assigned to mitochondria or chloroplasts were removed. The remaining ASVs were then resampled and normalized to a minimum number of sequences from each sample using the rarefy plugin. This rarefied dataset was functionally assigned using PICRUSt2 [28] to predict the bacterial functions based on marker genes. The gene families for the bacterial sequences were annotated corresponding to the enzyme classification numbers (E.C. numbers). In this study, we highlighted 15 soil enzymes that potentially indicated soil health [29]. The E.C. numbers and names of these enzymes are presented in Supplementary Materials, Table S1.

2.6. Statistical Analysis

The soil properties before burning, immediately after burning, and after harvesting (1 year after burning) were compared with a one-way ANOVA and post hoc Tukey's HSD tests. The visual graphics were generated with the 'ggplot2' package in the R environment (v. 4.0.2) [30]. The alpha diversity indices, which included the observed richness, Chao-1, Simpson, and Shannon indices, were computed and statistically compared among the sites using the ANOVA (for normal distribution data) or Kruskal–Wallis tests (for non-normal distribution data). Bacterial community and functional compositions were analyzed and visualized through a principal coordinate analysis (PCoA) based on the Bray–Curtis distance. The differences in compositions were tested using permutational multivariate analyses of variance (PERMANOVAs). A redundancy analysis (RDA) was employed to determine the influence of soil properties on soil bacterial community compositions, and the significance of the correlation between them was confirmed using the Mantel test.

3. Results

3.1. Soil Physical and Chemical Properties

No significant differences in soil moisture and soil temperature were found amon the three periods (preburning, postburning, and 1 year after burning) (Table 1). At a dept of 5 cm, the soil moisture ranged from 45.1% to 48.4% in the preburning sites, and a sligh decrease to 44.5–46.0% was detected in the postburning samples. A rise in soil temperatur was measured at 25.9–26.8 °C after burning compared with preburning (25.7–26.5 °C During burning, the fire temperature in the litter layer ranged from 415.5 to 469.5 °C.

Table 1. Soil moisture and soil, air, and fire temperatures at the study sites (minimum–maximum).

Variables	Preburning	Postburning	One Year after Burning
Soil moisture (%)	45.1–48.4 a	44.5–46.0 a	49.0–55.3 a
Soil temperature (°C)	25.7–26.5 a	25.9–26.8 a	25.2–26.1 a
Fire temperature in the litter layer (°C)		415.5–469.5	

a letters denote significant statistical differences ($p \leq 0.05$).

There were no significant differences of soil physicochemical properties were detectec between preburning and 1 year after burning (Table 2). At a depth of 0–5 cm in paddy soils the bulk density, organic matter (OM), CEC, and soil texture showed no significant changes after burning. The soils had a significantly lower acidity as well as lower NO_3-N levels after burning. Conversely, the burned soils showed higher levels of TN, NH_4-N, available P, K, Ca, and Mg, and had higher ECe values (Table 2). At 1 year after burning, the soil pH, ECe, CEC, NH_4-N, NO_3-N, and available P, K, Ca, and Mg were higher than the initial values (preburning), whereas OM and TN were reduced and lower than the preburning values, but the significant differences were not found.

Table 2. Physicochemical properties of paddy soil samples before burning, after burning, and 1 year after burning.

Soil Properties	Preburning	Postburning	One Year after Burning
pH (1:1)	5.08 ± 0.03 a	6.89 ± 0.16 b	5.15 ± 0.01 a
Bulk density (g cm^{-1})	1.42 ± 0.01 a	1.40 ± 0.03 a	1.41 ± 0.01 a
Organic matter (%)	4.15 ± 0.05 a	4.22 ± 0.10 a	4.13 ± 0.03 a
Organic carbon (%)	2.39 ± 0.03 a	2.44 ± 0.07 a	2.38 ± 0.04 a
CEC (meq 100 g^{-1})	24.10 ± 0.68 a	27.16 ± 0.96 a	25.01 ± 0.51 a
ECe (dS m^{-1})	2.45 ± 0.05 a	5.24 ± 0.64 b	3.11 ± 0.04 a
NH_4N (mg kg^{-1})	48.48 ± 1.59 a	58.06 ± 2.69 b	51.34 ± 1.88 a
NO_3N (mg kg^{-1})	55.22 ± 2.69 a	19.90 ± 2.84 b	57.59 ± 3.66 a
Total nitrogen (%)	0.29 ± 0.01 a	0.32 ± 0.03 b	0.26 ± 0.03 a
Available P (mg kg^{-1})	50.52 ± 1.24 a	56.73 ± 3.28 b	53.36 ± 2.11 a
Available K (mg kg^{-1})	297.37 ± 19.35 a	505.91 ± 13.10 b	319.37 ± 21.33 a
Available Ca (mg kg^{-1})	1766.81 ± 96.58 a	2191.13 ± 65.86 b	1898.81± 77.11 a
Available Mg (mg kg^{-1})	288.67 ± 23.16 a	365.73 ± 21.06 b	312.67± 16.66 a
Sand (%)	36.20 ± 2.01 a	37.21 ± 1.98 a	35.68 ± 2.33 a
Silt (%)	36.19 ± 2.54 a	35.29 ± 2.66 a	34.91 ± 2.21 a
Clay (%)	27.61 ± 1.88 a	27.50 ± 2.69 a	29.41 ± 1.73 a
Texture	Clay Loam	Clay Loam	Clay Loam

a–b letters denote significant statistical differences ($p \leq 0.05$).

The SOC stock was 17.10 ± 0.3 Mg C ha^{-1} in the preburning samples, which increased insignificantly to 17.19 ± 0.1 Mg C ha^{-1} after burning. A slightly declined SOC stock was observed at 1 year after burning, with 16.91 ± 0.23 Mg C ha^{-1}, which was slightly lower than the preburning value (Figure 2a). A significantly higher STN stock, with an increase from 2.03 ± 0.09 Mg N ha^{-1} to 2.27 ± 0.081 Mg N ha^{-1}, was also identified in the postburning soils. A significant reduction in the STN stock was detected at 1 year after burning (1.85 ± 0.07 Mg N ha^{-1}) (Figure 2b).

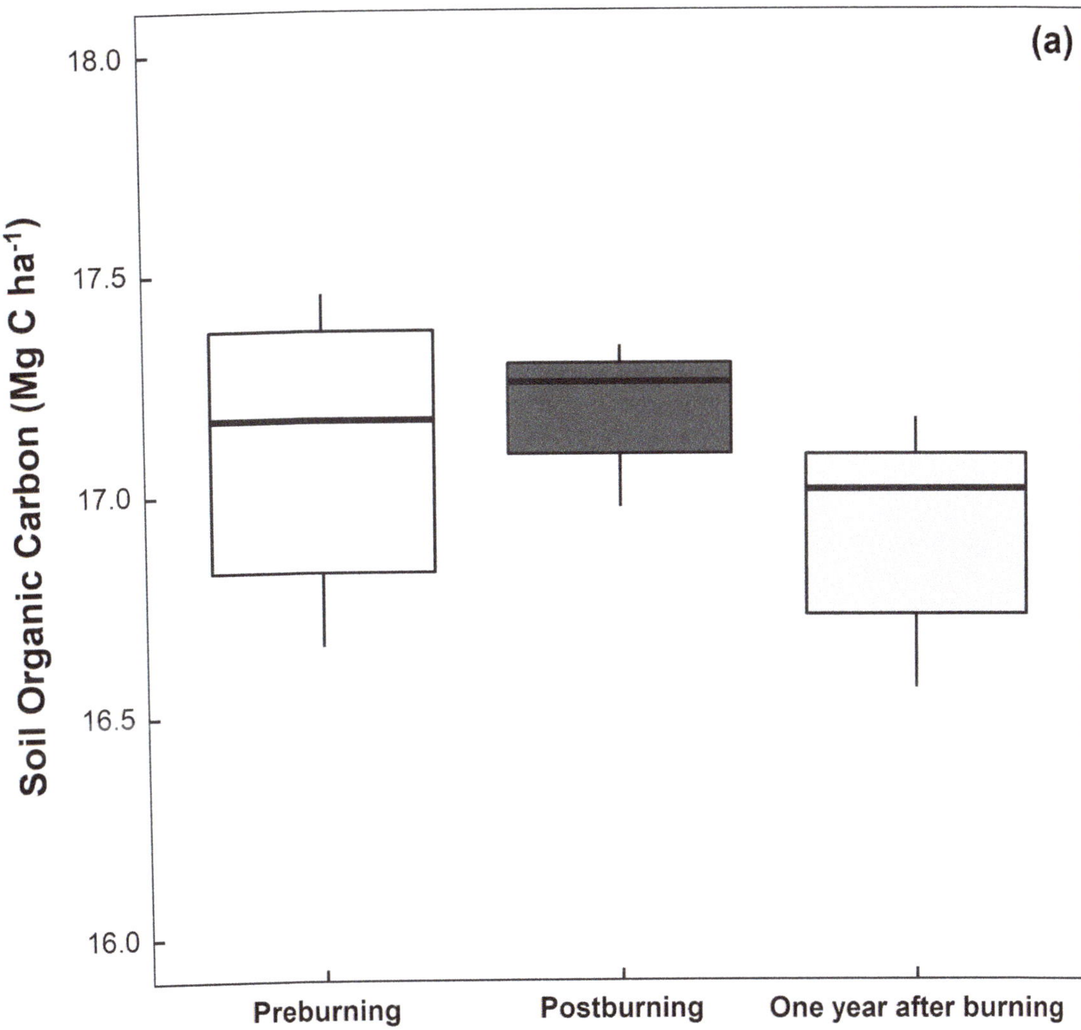

Figure 2. *Cont.*

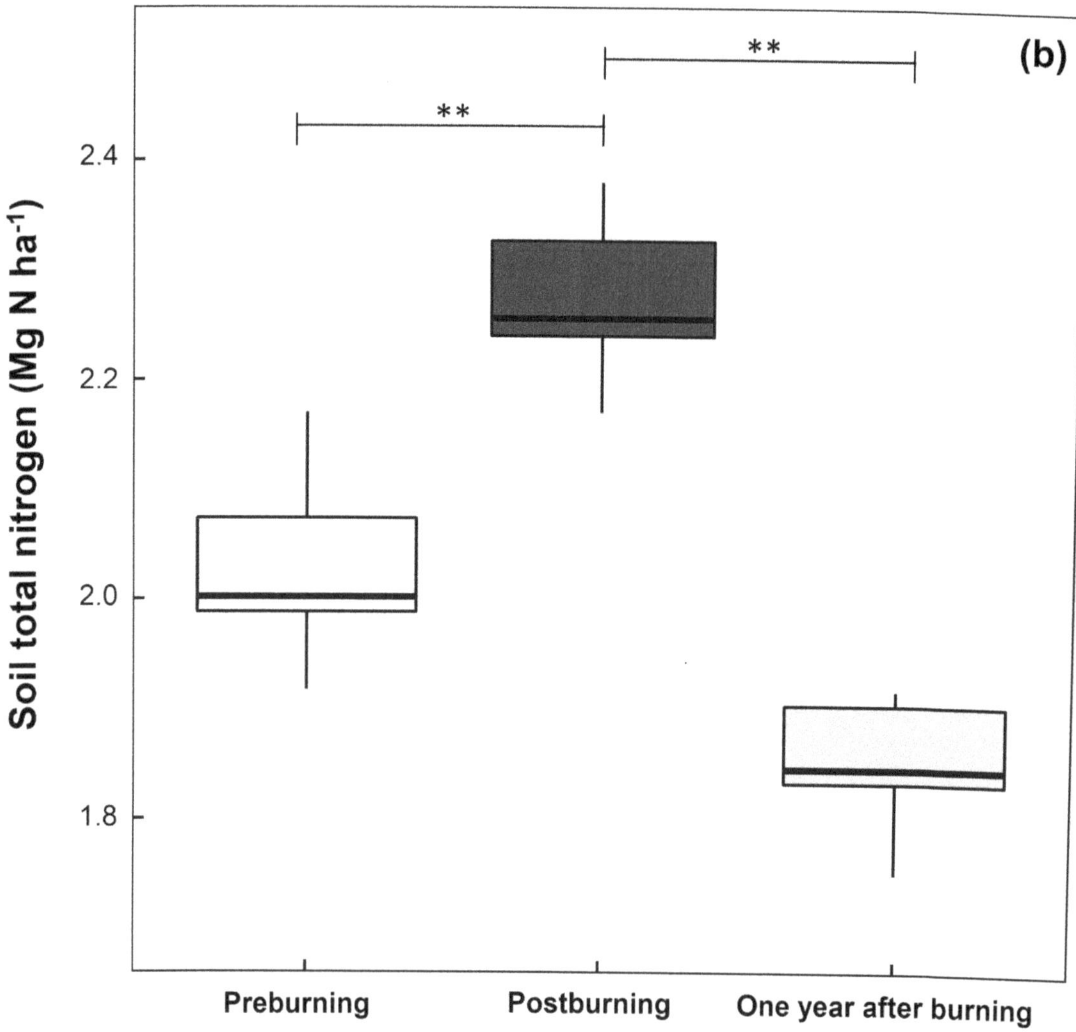

Figure 2. Soil organic carbon (SOC; (**a**) and soil total nitrogen (STN; (**b**) levels in paddy soils. ** denotes significant statistical differences ($p \leq 0.05$).

3.2. Overview of the Sequencing Analysis

A total of 616,815 (41,121 sequences/sample) clean sequences were obtained in this study. As shown in Figure 3, the rarefaction curves of all samples gradually flattened, indicating that the number of sequences obtained in this study could reflect the bacterial community in the study sites. Here, the sequences were grouped into 18,715 ASVs, which were then classified into 52 phyla, 119 classes, 245 orders, 312 families, and 543 genera.

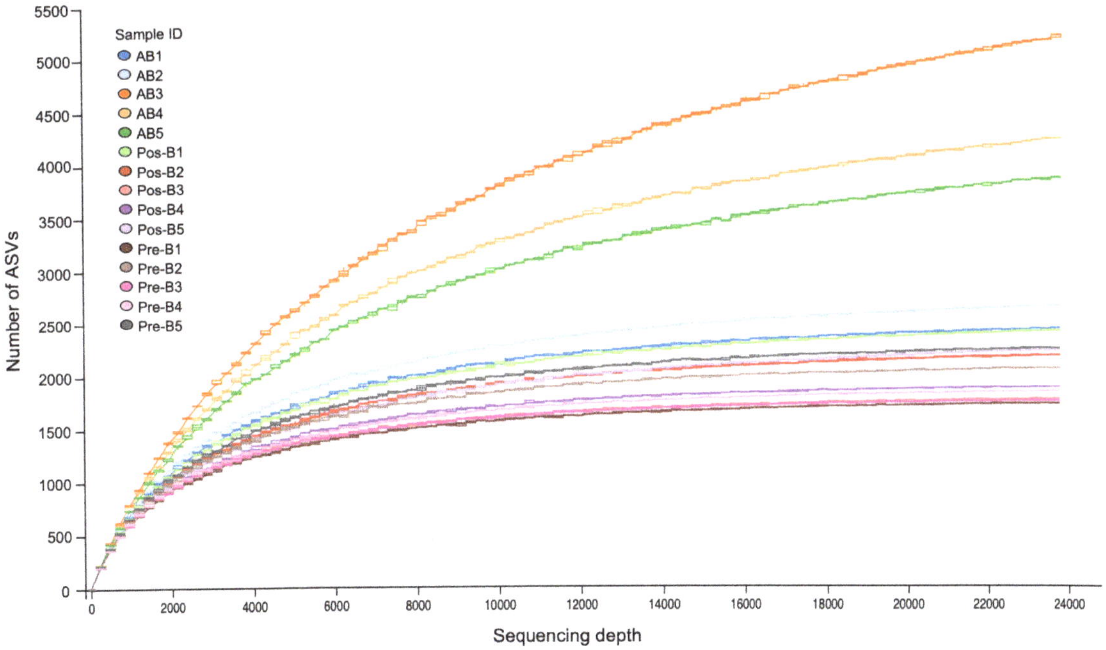

Figure 3. Rarefaction curves of all samples in rice fields at preburning (pre-B), postburning (pos-B), and 1 year after burning (AB).

3.3. Taxonomic Distribution

As shown in Figure 4, whilst the most abundant taxa in the preburning and postburning samples were Chloroflexi (33%), followed by Actinobacteria (Pre-B = 15%; Pos-B = 18%), and Proteobacteria (Pre-B = 12%; Pos-B = 9%), those in the site 1 year after burning were Actinobacteria (22%), followed closely by Proteobacteria (20%) and Chloroflexi (16%) (Figure 4a). Overall, 11 phyla, 15 order, and 45 genera were indicated as differentially abundant taxa ($p < 0.05$; LDA score > 3). The ANOVA results showed that the relative abundances of 7 out of 10 abundant phyla (average relative abundance > 1%) were notably different among the study samples (Figure 4a). Chloroflexi abundance dramatically decreased by approximately 15–17% at 1 year after burning, whereas the abundance of Planctomycetes increased immediately after burning and then decreased, reaching a level similar to that before burning. The abundances of Actinobacteria, Proteobacteria, Verrucomicrobia, and Gemmatimonadetes significantly increased by 7%, 8%, 3%, and 1%, respectively, at 1 year after burning compared to the pre- and postburning soils. At the order level, significant changes were found to have occurred for several taxa (Figure 4b). The abundances of Ktedonobacterales, Bacillales, and Betaproteobacteria increased slightly immediately after burning, but significantly decreased by 13–14%, 3–5%, and 2–3%, respectively, at 1 year after burning. Myxococcales and Gaiellales abundances significantly increased by 3–4% at 1 year after burning, compared to the two other timepoints.

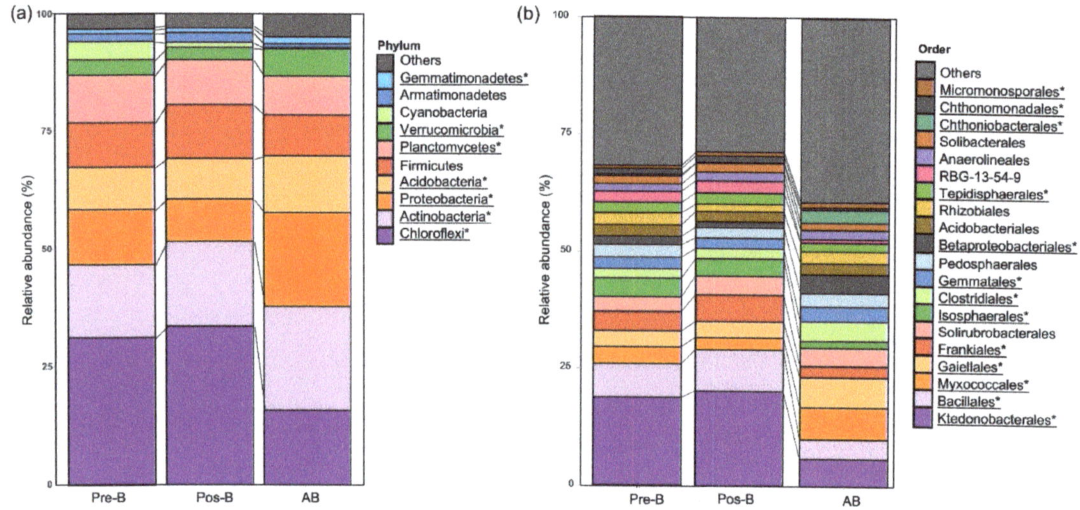

Figure 4. Stacked bar plot showing the relative abundances of the bacterial phyla (**a**) and orders (**b**) in rice fields at preburning (pre-B), postburning (pos-B), and 1 year after burning (AB). Asterisks beside the phylum name indicate statistical significance ($p < 0.05$).

Ten abundant genera (average relative abundance > 1%) were detected in this study. *Bacillus* and HSB OF53-F07 dominated at the pre- and postburning soils, accounting for 5% and 4% in the preburning sites and 6% and 4% in postburning soils. As shown in Figure 5, the abundances of these two taxa increased immediately after burning, but decreased significantly 1 year after burning. This trend was also found for *Conexibacter* and Acidothermus. On the other hand, the abundances of *Anaeromyxobacter* and *Candidatus* Udaeobacter increased significantly 1 year after burning, accounting for 3.1%, 2.7%, and 2.5%, respectively, of the taxa (Figure 5).

3.4. Bacterial Diversity, Community Compositions, and Correlations to Soil Properties

As shown in Table 3, all alpha diversity indices, including observed richness, Chao-1, Simpson, and Shannon indices, presented similar trends. The diversity indices did not change immediately after burning, but there were significant increases after 1 year. The beta diversity, presented by the PCoA ordination based on the Bray–Curtis distance, overlapped between the pre- and postburning samples, but these two groups were separate from those 1 year after burning (Figure 6a). According to the PERMANOVA results, the bacterial community compositions in the rice fields were significantly differed at 1 year after burning (Figure 6a).

Table 3. Alpha diversity indices.

Site	Observed Richness	Chao-1	Shannon	Simpson
Preburning	1936.80 ± 203.24 b	1954.39 ± 243.83 b	7.03 ± 0.10 b	0.9984 ± 0.0001 b
Postburning	2114.60 ± 273.81 b	2169.11 ± 315.25 b	7.09 ± 0.12 b	0.9984 ± 0.0002 b
One year after burning	3761.40 ± 1203.85 a	4176.56 ± 1547.81 a	7.66 ± 0.30 a	0.9991 ± 0.0002 a

a–b letters denote significant statistical differences ($p \leq 0.05$).

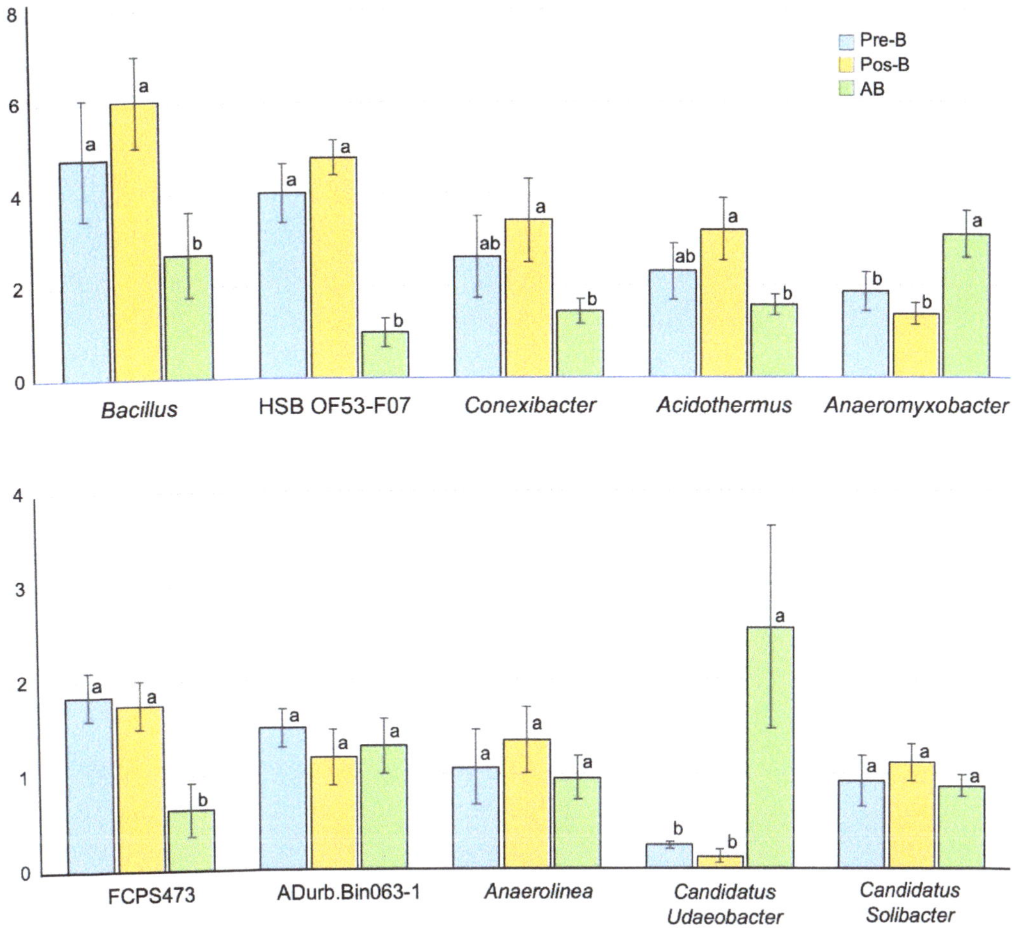

Figure 5. Bar plots for the most abundant genera in this study. Plots with different letters were statistically different. Pre B—preburning; Pos B—postburning; AB—1 year after burning.

As shown in Figure 6b, the redundancy analysis (RDA) revealed that the soil properties could explain 67.4% of the total variations in the bacterial community compositions. According to the Mantel test, the TN, STN, sand, clay, soil moisture, and soil temperature were significant soil parameters, shaping the bacterial communities (Table 4). Among these parameters, only clay and soil moisture presented a strong correlation (correlation coefficient > 0.7) (Table 4).

Table 4. Correlations and significant values of the bacterial communities and soil properties determined with the Mantel test.

Soil Properties	Correlation Coefficient	*p*-Value
pH	0.0274	0.259
Bulk density	−0.1338	0.975
Organic matter	0.2153	0.056
Organic carbon	0.2357	0.052

Table 4. *Cont.*

Soil Properties	Correlation Coefficient	*p*-Value
Soil organic carbon	0.1802	0.071
CEC	−0.0739	0.712
Ece	−0.0194	0.452
NH_4N	−0.1115	0.813
NO_3N	0.0291	0.283
Total nitrogen	0.3780	0.010 *
Soil total nitrogen	0.4268	0.006 *
Available P	−0.1008	0.812
Available K	0.0292	0.260
Available Ca	−0.0704	0.758
Available Mg	−0.0252	0.510
Sand	0.3377	0.011 *
Silt	0.1602	0.094
Clay	0.7666	0.001 *
Soil moisture	0.7223	0.001 *
Air temperature	0.1176	0.142
Soil temperature	0.2259	0.028 *

* denotes significant statistical differences ($p \leq 0.05$).

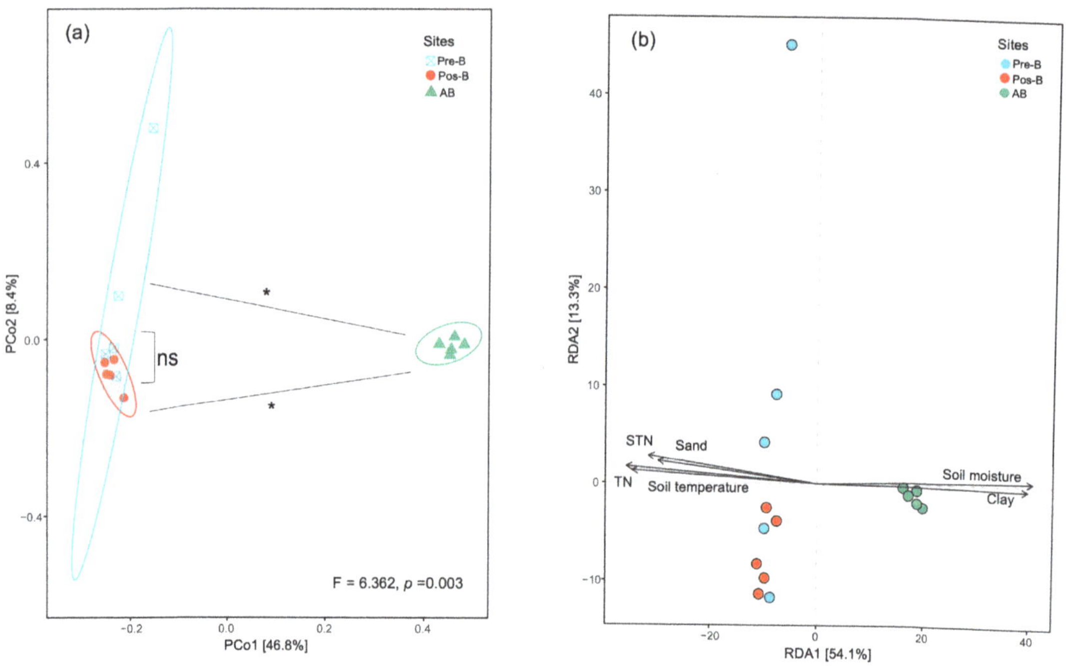

Figure 6. Bacterial community composition and correlations to soil properties. (**a**) Principal coordinate analysis (PCoA) ordination based on the Bray–Curtis distance, showing the community composition of bacteria detected in the study sites. (**b**) RDA ordination presents soil properties that significantly correlated with community composition. Significant parameters indicated with the Mantel test. Pre-B—preburning; Pos-B—postburning; AB—1 year after burning. * indicates statistically difference.

3.5. Predictive Functions

We used PICRUSt2 to predict the functions of the bacterial community based on enzymatic genes. In total, 2372 predictive enzymes were detected across all samples. As shown in Figure 7a, PCoA, which explained 90.1% of the total functional composition, showed that the functional compositions of the bacterial communities found in preburning and postburning sites were largely similar, although they differed significantly 1 year after burning (PERMANOVA test, $p > 0.05$). In addition, 15 enzymes involved in the carbon, nitrogen, and phosphorus cycles were selected to highlight the potential enzyme activities in soils. The ANOVA results revealed that 10 out of 15 selected enzymes differed significantly among the study sites (Figure 7b). Whilst the abundances of β-glucosidase, chitinase, and urease were higher in pre- and postburning soils compared to the sites 1 year after burning, alpha-N-acetylglucosaminidase and endo-1,4-beta-xylanase presented inconsistent trends. Cellulase and nitrogenase levels increased 1 year after burning compared to the postburning soils. Pectin lyase and nitrate reductase showed similar levels at preburning and 1 year after burning.

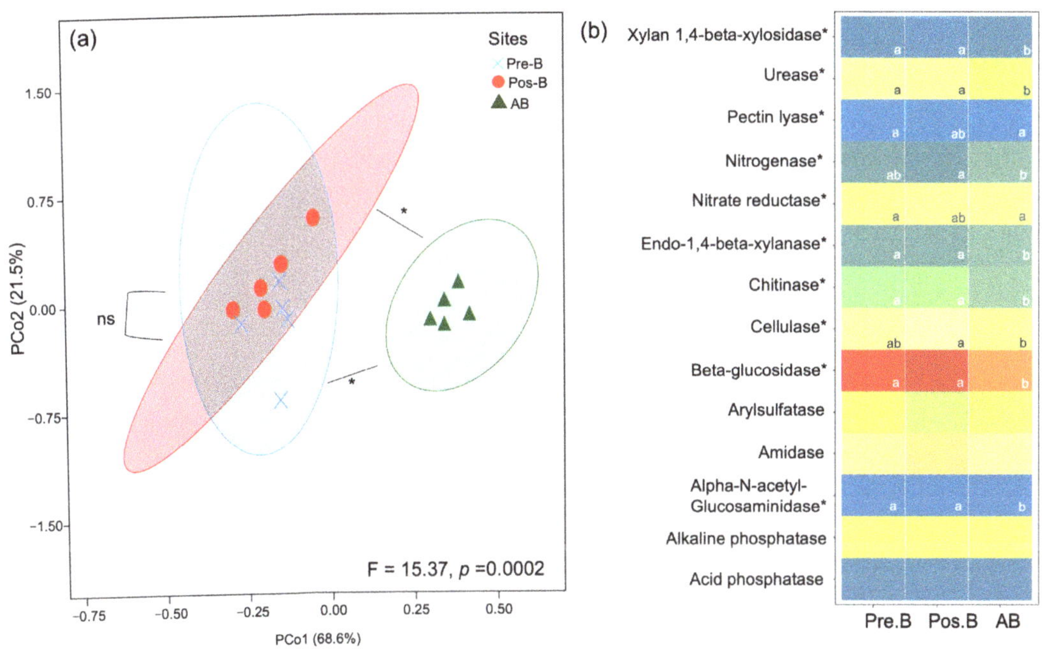

Figure 7. Bacterial functions predicted using PICRUSt2. (**a**) Principal coordinate analysis (PCoA) ordination, based on the Bray–Curtis distance, shows the functional composition of bacteria. (**b**) Heatmap shows the mean abundances of soil enzymes potentially produced by bacteria detected in the study sites. Pre-B—preburning; Pos-B—postburning; AB—1 year after burning. * indicates statistically difference.

As shown in Figure 8, Spearman's rank correlation analysis indicated correlations between soil properties and the selected soil enzymes. The parameters OM, OC, TN, STN, soil texture, moisture, and temperature were significantly correlated with several soil enzymes. Positive correlations were found, for example, for OM and amidase, cellulase, and chitinase, for SOC and amidase, cellulase, and TN, and for STN and β-glucosidase, chitinase, and urease. In contrast, negative correlations were found for these enzymes and the clay and soil moisture levels.

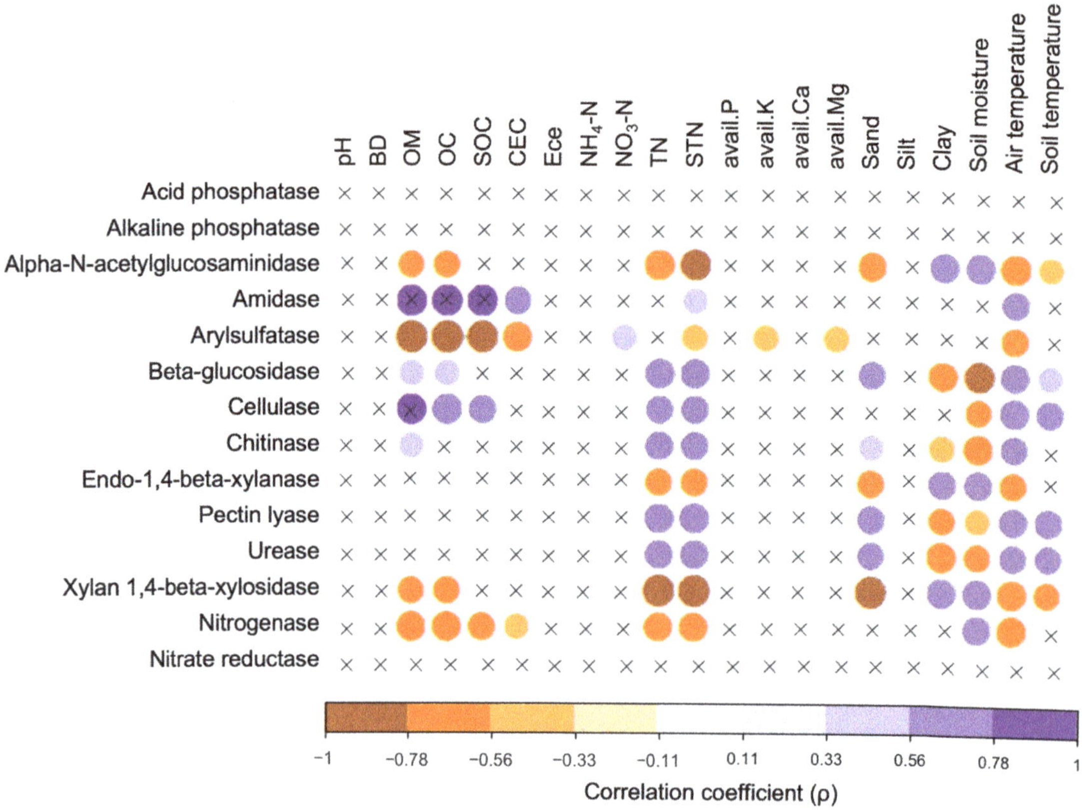

Figure 8. Spearman's rank correlation between soil properties and soil enzymes predicted with PICRUSt2. X marks indicate insignificant correlation, whereas circles indicate a significant correlation ($p < 0.05$). Circle color corresponds to the correlation coefficient.

4. Discussion

4.1. Effects of Burning Rice Straw and Stubble on Soil Physicochemical Properties

Fire can alter the physical and chemical properties of soils through ash deposit. In our study, significant increases in pH, ECe, TN, NH_4-N, and soil nutrients (available P, K, Ca, and Mg) were found (Table 2) after fire was used. An elevated soil pH following a fire was also reported in the meta-analysis of Ribeiro Filho et al. [31]. Granged et al. [32] also detected an increase in soil pH from 6.2 to 7.5 immediately after a fire. This increase could be explained by the loss of OH-groups from clay minerals, the formation of oxides, and the incorporation of ash into the soil, leading to an increase in the base cations in soils [33,34]. Kumar et al. [12] and Nigussie and Kissi [35] found increased ECe values after crop residue burning due to increased ash contents in the soil, along with higher available P levels after burning, possibly derived from the available P in ash. Kumar et al. [12] reported that the available P increased from 68.39 to 76.31 kg ha^{-1} after the use of fire. Nigussie and Kissi [35] detected an increase in the available P by 73.41% after burning.

Similarly, burning rice straw and stubble provided ash containing high K levels, resulting in an increase in the available K after burning, which was consistent with the results of the study by Gangwar et al. [36]. Burning facilitated the release of nitrogen from rice straw and stubble, increasing the TN content. Pellegrini et al. [37] and Parro et al. [38] also reported an increase in nitrogen after the use of fire. Ammonium (NH_4-N) and nitrate (NO_3-N) are the plant-available forms of nitrogen, and are generated via the decomposition of organic N compounds [39]. Previous studies [40–42] have reported high NH_4-N concentrations after high-severity burning, whereas NO_3-N did not form directly through heating [43], but was produced after burning through the nitrification of NH_4-N [44]. This might explain the significant decrease in NO_3-N of the paddy soils after burning (Table 2). According to Wan et al. (2001), NH_4-N increases immediately after burning and returns to a preburning level after 1 year, whereas NO_3-N can recover to preburning levels after 5 years. Covington et al. [45] also found that changes in NO_3-N could not be detected immediately after a fire, but the levels increased 1 year after burning, exceeding the preburning levels.

Interestingly, the bulk density, OM, OC, CEC, and soil texture remained unaffected by the burning of rice straw and stubbles (Table 2), most likely because the fire was not severe enough to alter these soil properties. In contrast, Baldock and Smernik [46], Oguntunde et al. [47], and Ayodele et al. [48] detected a decreased bulk density due to the conversion of residues to a char form and the residue remaining from incomplete combustion. Although previous studies reported a decline in OM after burning [7,39,49], this was not the case in the present study. As shown in Figure 1, burning was practiced while the rice stubbles mostly stood in the fields, resulting in a lower impact of the fire on the soil surface. Due to the insignificant changes in OM and clay content after burning, the CEC was largely unaffected, as it is closely related to the OM and clay content. Similarly, Fonseca et al. [50] detected unchanged CEC levels after shrub burning in the northeast of Portugal. As shown in Figure 2, burnt paddy soils contained higher SOC stocks than unburnt paddy soils. The slight increase in the SOC stock after burning may have occurred as a result of the charred material and ash. However, a significant difference was not detected, indicating that burning rice straw and stubble did not alter the SOC stock. We, therefore, hypothesized that the fire was not severe enough to consume the soil carbon. This was in agreement with the findings of Neill et al. [51], who detected no significant changes in soil carbon levels in a prescribed burning in a Cape Cod oak–pine forest. In contrast, the STN stock increased significantly in the postburning soils (Figure 2), which was also observed in previous studies [52,53].

Due to remaining ash in the paddy fields, the soil pH, ECe, CEC, NH_4-N, NO_3-N, and available P, K, Ca, and Mg levels were higher 1 year after burning than before burning. On the other hand, the OM levels were lower after burning, most likely because of the slow decomposition of residues. A similar trend was observed for TN, which may have occurred due to plant uptake (Table 2). Burning rice straw and stubble resulted in temporary changes in SOC and STN stocks, as well as soil nutrient levels. Several studies have also reported that changes in soil nutrient levels were temporary and that, generally, the levels returned to those measured before burning [54–56].

4.2. Soil Bacterial Community Composition and Diversity Responses Immediately after Burning

Bacteria represent the most abundant and diversified population of microorganisms worldwide [57], and play an important role in the decomposition of OM and the cycling of nutrients in agricultural ecosystems [58,59]. Fire directly compromises the survival of soil microbial communities through soil heating [9]; in our study, straw burning caused a decline in soil microorganisms at the soil surface due to an increase in soil surface temperatures. Mickovski [10] reported that burning straw and stubble heated the soil temperature in the uppermost 0–3 cm to approximately 50–70 °C immediately after a fire. The heat resulted in the mortality of 77% and 9% of the heterotrophic microorganisms in the topsoil (0–5 cm) and the 5–10 cm layer, respectively. Biederbeck et al. [11] investigated the effects of burning cereal straw and reported that repeated burning in the field caused a reduction in the bacterial population by more than 50%. Kumar et al. [12] also reported that the soil temperature increased to 55 °C immediately after rice straw burning, resulting in a significant decrease in bacteria, fungi, actinomycetes, phosphate-solubilizing microorganisms, potassium-solubilizing microorganisms, and cellulose degraders after burning. The results of the abovementioned studies were, however, in contrast to our findings, namely, concerning the relative abundances of bacterial phyla, which did not significantly differ ($p > 0.05$) between preburning and postburning sites (Figure 4a). Our sites had a high initial soil moisture; therefore, the burning finished after only a short period of time (6.5 s m^{-2} on average), resulting in the fire not being severe enough to kill the soil microbial communities. Although the fire temperatures in the residue layer reached between 298.2 and 603.0 °C, the soil temperature after burning only increased by 1.6–3.8 °C compared to the preburning samples (Table 1). Moreover, it was hypothesized that whilst the fire could affect the soil microbial communities in the uppermost layer (0–1 cm), no effect was observed in the deeper layer (2–5 cm) because the high soil moisture contents may have had limited the heat transfer to the deeper soil layers. According to Busse et al. [60], heat transfer can decrease when the soil water content exceeds 20%. This was also supported by Whelan et al. [61], who reported that the prevalence of dormant soil microorganisms decreased with low soil moisture contents and high temperatures.

Previous studies reported that fire has a significant effect on the thin topsoil layer, such as the 0–1 cm layer [62], the 0–2.5 cm layer [11], and the 0–3 cm layer [63]. This could explain the insignificant difference between the soil microbial communities of the preburning and postburning sites observed here at a depth of 0–5 cm (Figure 4a). Similarly, Li et al. [64] investigated the responses of soil Acidobacteria to a wildfire disturbance in the topsoil (0–10 cm) and subsoil (10–20 cm), and found no significant differences in Acidobacterial α-diversity between these two soil layers across different fire severities. However, studies on the impacts of rice straw burning on soil microbial dynamics are still limited, since most studies have focused on wildfires or prescribed fires. Kumar et al. [12] investigated the responses of soil microbial communities to paddy straw burning in sandy loam soil. The authors discovered that soil microorganisms and microbial enzymes temporarily decreased after burning, before recovering 30–60 days after burning.

Chloroflexi were the dominant bacteria in both preburning and postburning fields, followed by Actinobacteria, Firmicutes, Planctomycetes, Proteobacteria, and Acidobacteria, although no significant difference was detected among the sites (Figure 4a). Previous studies also reported the dominance of these bacterial phyla in paddy soils [65–68]. Chloroflexi act as primary degraders of polysaccharides under the anaerobic conditions of rice fields, whereas Actinobacteria can also degrade OM under such conditions [69] and produce enzymes involved in carbon cycling for plant residue decomposition and carbon sequestration in soils [70]. Firmicutes comprise a range of thermophilic, antibiotic, and endospore-producing members, many of which are extremely resistant to desiccation, heat, and radiation, and can survive in extreme conditions [71]. Planctomycetes and Proteobacteria are important phyla involved in nitrification and denitrification processes in the soil [72,73]. Trivedi et al. [74] reported that Proteobacteria are classified as "copiotrophs" (R-strategists), which are more abundant in nutrient-rich conditions. Proteobacteria also play a key role in OM decomposition and produce several types of glycosyl hydrolases, thus, promoting plant growth [75]. Acidobacteria use diverse metabolic pathways in ecological processes, including biogeochemical cycles, biopolymer decomposition, and exopolysaccharide secretion [76]. Stinca et al. [77] investigated the soil in a beech forest 2 years after a wildfire, and found that the most abundant phyla (Proteobacteria, Acidobacteria, Bacteroidetes, Planctomycetes, Firmicutes, Gemmatimonadetes, and Chloroflexi) had remained relatively unaffected by the fire, except for Actinobacteria.

At the genus level, *Bacillus* and HSB OF53-F07 were the dominant taxa in both the preburning and postburning sites, but significant differences were not detected. Moreover, *Anaeromyxobacter* was more abundant in the preburning than postburning sites ($p > 0.05$), whereas *Acidothermus* was more abundant in the postburning sites than the preburning ones ($p > 0.05$) (Figure 5); *Acidothermus* is more resistant to higher temperatures than other bacteria. This was consistent with the findings of Mohagheghi et al. [78], who reported that the *Acidothermus* bacteria could remain active at high temperatures, with an optimum growth temperature of 55 °C. Furthermore, the *Acidothermus* genome contains genes that encode for thermostable enzymatic cellulose degradation.

Most soil microorganisms die at temperatures exceeding 50 °C due to changes in their cells and enzymes [79]. In the present study, β-glucosidase, chitinase, and urease were the dominate soil enzymes produced by the soil bacterial communities. However, the relative abundances of those enzymes did not differ significantly between preburning and postburning soils (Figure 7b). Stott et al. [80] reported that β-glucosidase is an important enzyme for the decomposition of crop residues and provides the energy source for heterotrophic bacteria.

4.3. Changes in Soil Physicochemical Properties and Soil Bacterial Community Composition and Diversity

Fires can promote the growth of soil bacteria with heat-resistance capacities and enhance those with potential fast-growth strategies [81]. Increased abundances of Actinobacteria, Proteobacteria, Verrucomicrobia, and Gemmatimonadetes were observed 1 year after burning, whereas the abundance of Chloroflexi decreased, and could not reach the initial value (Figure 4a). Although the Planctomycetes abundance increased immediately after burning, the relative abundance decreased and returned to that of the preburning soils. This was consistent with previous studies reporting that changes in soil physical and chemical properties after a fire could indirectly affect soil microbial communities [82–84]. Zhao et al. [17] reported that the relative abundances of Proteobacteria, Betaproteobacteria, and Actinobacteria significantly increased with increased levels of STN, available N, and available P. Our study also pointed out that STN, TN, sand, clay, soil moisture, and soil temperature significantly shaped the bacterial community composition (Table 2). The increased pH after burning (Table 2) was a crucial factor in determining microbial activities, as it led to changes in community composition and diversity.

At the genus level, the abundances of *Bacillus*, HSB OF53-F07, *Conexibacter*, and *Acidothermus* increased immediately after burning and then significantly decreased, whereas *Anaeromyxobacter* and Candidatus *Udaeobacter* dominated 1 year after burning (Figure 5). We hypothesized that *Bacillus*, HSB OF53-F07, *Conexibacter*, and *Acidothermus* were dominant genera in paddy soils, and may be highly resistant to heat, but with low growth rates. In contrast, *Anaeromyxobacter* and Candidatus *Udaeobacter* may have high potential growth rates and mainly occupy areas with increased soil nutrient levels after a fire. However, for a better understanding of the behavior of these bacteria, long-term studies are necessary.

Bacillus can promote plant growth through nitrogen fixation, the solubilization and mineralization of phosphorus, zinc, and iron [85]. *Conexibacter* can transform nitrate into nitrite via denitrification under limited oxygen conditions [86]. Seki et al. [87] reported that *Conexibacter* is a slow-growing microorganism and can survive in arid environments. *Acidothermus* is involved in cellulose degradation and enhances plant growth [88], whereas *Anaeromyxobacter* can fix and assimilate N_2 to NH_4N with the use of nitrogenase, and is involved in the reduction of N_2O to N_2 [89,90]. This could explain the increase in the abundance of *Anaeromyxobacter* (Figure 5) and in the levels of nitrogenase (Figure 7) 1 year after burning. Moreover, *Anaeromyxobacter* is an iron-reducing bacterium, using OM as the electron donors [91], and can denitrify NO_2^- to NO through Fe^{2+} oxidation [92]. Candidatus *Udaeobacter* can secrete antibiotics in the soil and has the potential to remove trace gases, especially H_2 [93]. Brewer et al. [94] revealed that Candidatus *Udaeobacter* is an aerobic heterotroph with numerous amino acid and vitamin transporters, minimizing the cellular architecture and sacrificing metabolic versatility to become dominant in the soil.

The cellulase and nitrogenase levels were increased 1 year after burning (Figure 7), most likely because the increased soil nutrient levels accelerated the microbial enzyme production. Cellulase is generally produced for cellulose decomposition mechanisms [95], whereas nitrogenase can reduce N_2 to NH_3 [96]. Soil enzymes play the most active part in all biochemical processes in soil. In the present study, positive correlations were found between OM and amidase, cellulase, and chitinase, and between SOC and amidase. Moreover, cellulase increased with increasing TN levels, and an increase in STN influenced β-glucosidase, chitinase, and urease (Figure 8). This led us to infer that soil enzymes were more related to OM levels and directly involved in OM mineralization, thus, affecting carbon and nitrogen cycles. To further reveal the effects of repeated fires on soil bacterial activities and soil properties, long-term studies are needed. Furthermore, it should be noted that the enzymatic results from this study based on the prediction tool, the actual measurements of the enzymatic activities, should be further investigated in the future.

5. Conclusions

Soil chemical properties (pH, ECe, NH_4N, NO_3N, TN, and available P, K, Ca, and Mg) of paddy soils significantly increased after burning. However, these values largely returned to prefire levels after 1 year. The most abundant taxa in preburning and post-burning soils were Chloroflexi, Actinobacteria, and Proteobacteria. At 1 year after burning, Chloroflexi abundance decreased dramatically, whereas the abundances of Actinobacteria, Proteobacteria, Verrucomicrobia, and Gemmatimonadetes significantly increased. At the genus level, *Bacillus*, HSB OF53-F07, *Conexibacter*, and *Acidothermus* abundances increased immediately after burning, and then significantly declined, with lower levels 1 year after burning. *Anaeromyxobacter* and Candidatus *Udaeobacter* dominated at 1 year after burning. Fires tend to directly and indirectly affect soil microbial communities and functions. Direct effects are exerted through soil heating, killing some bacterial species, whereas indirect effects are due to changes in soil physicochemical properties. In this study, rice straw and standing stubble were burnt under high soil moisture conditions, and burning finished within a very short period of time, indicating that the fire was not severe enough to sufficiently heat the soil and kill the soil microorganisms.

Supplementary Materials: The following supporting information can be downloaded at: https://www.mdpi.com/article/10.3390/biology12040501/s1, Table S1: the enzyme classification numbers and their descriptions.

Author Contributions: Conceptualization, N.A. and S.S.; methodology, N.A., C.S. and P.K.; investigation, N.A. and S.S.; writing—original draft preparation, N.A., C.S., P.K., S.S. and R.H.; writing—review and editing, N.A., C.S., P.K., S.S. and R.H.; supervision, R.H. All authors have read and agreed to the published version of the manuscript.

Funding: This research project was supported by the Mahidol University (Basic Research Fund: fiscal year 2021). We would also like to thank Thailand Science Research and Innovation (TSRI) for their funding of this project, under contract number BRF1-A8/2564.

Institutional Review Board Statement: The study was conducted according to the guidelines of the Declaration of Helsinki and approved by the institutional review board of the Institute for Population and Social Research, Mahidol University (IPSR-IRB) (COA. no. 2021/01-003; date of approval: 28 January 2021).

Informed Consent Statement: Not applicable.

Data Availability Statement: Raw sequence data generated for this study are available in the sequence read archives (SRAs) of the National Center for Biotechnology Information (NCBI) under BioProject accession number PRJNA819169.

Acknowledgments: We would like to thank Mahidol University (Basic Research Fund: fiscal year 2021) and Thailand Science Research and Innovation (TSRI) for their funding. Moreover, we would like to express special thanks to all key informants for their information, including the reviewers for their comments and suggestions to improve this paper.

Conflicts of Interest: The authors declare no conflict of interest.

References

1. Tipayarom, D.; Oanh, N.K. Effects from open rice straw burning emission on air quality in the Bangkok metropolitan region. *Sci. Asia* **2007**, *33*, 339–345. [CrossRef]
2. Oanh, N.T.K.; Ly, B.T.; Tipayarom, D.; Manandhar, B.R.; Prapat, P.; Simpson, C.D.; Liu, L.J.S. Characterization of particulate matter emission from open burning of rice straw. *Atmos. Environ.* **2011**, *45*, 493–502. [CrossRef] [PubMed]

3. Gadde, B.; Menke, C.; Wassmann, R. Rice straw as a renewable energy source in India, Thailand, and the Philippines: Overall potential and limitations for energy contribution and greenhouse gas mitigation. *Biomass Bioenergy* **2009**, *33*, 1532–1546. [CrossRef]
4. Office of Agricultural Economics (OAE). Agricultural Statistics of Thailand 2021. Centre for Agricultural Information, Office of Agricultural Economics (in Thai), 2021. Available online: https://www.oae.go.th/assets/portals/1/files/jounal/2565/yearbook2564.pdf (accessed on 21 August 2022).
5. Arunrat, N.; Pumijumnong, N. Practices for reducing greenhouse gas emissions from rice production in Northeast Thailand. *Agriculture* **2017**, *7*, 4. [CrossRef]
6. Arunrat, N.; Sereenonchai, S.; Hatano, R. Impact of burning on soil organic carbon of maize-upland rice system in Mae Chaem Basin of Northern Thailand. *Geoderma* **2021**, *392*, 115002. [CrossRef]
7. Certini, G. Effects of fire on properties of forest soils: A review. *Oecologia* **2005**, *143*, 1–10. [CrossRef]
8. Arunrat, N.; Sereenonchai, S.; Hatano, R. Effects of fire on soil organic carbon, soil total nitrogen, and soil properties under rotational shifting cultivation in northern Thailand. *J. Environ. Manag.* **2022**, *302*, 113978. [CrossRef]
9. Dooley, S.R.; Treseder, K.K. The effect of fire on microbial biomass: A meta-analysis of field studies. *Biogeochemistry* **2012**, *109*, 49–61. [CrossRef]
10. Mickovski, M. Effect of burned straw on the microflora of the soil. *Annu. Fac. Agric. Univ. Skopje* **1967**, *20*, 55–68.
11. Biederbeck, V.O.; Campbell, C.A.; Bowren, K.E.; Schnitzer, M.; McIver, R.N. Effect of Burning Cereal Straw on Soil Properties and Grain Yields in Saskatchewan. *Soil Sci. Soc. Am. J.* **1980**, *44*, 103–111. [CrossRef]
12. Kumar, A.; Kushwaha, K.K.; Singh, S.; Shivay, Y.S.; Meena, M.C.; Nain, L. Effect of paddy straw burning on soil microbial dynamics in sandy loam soil of Indo-Gangetic plains. *Environ. Technol. Innov.* **2019**, *16*, 100469. [CrossRef]
13. Luo, X.S.; Han, S.; Lai, S.S.; Huang, Q.Y.; Chen, W.L. Long-term straw returning affects *Nitrospira*-like nitrite oxidizing bacterial community in a rapeseed-rice rotation soil. *J. Basic Microbiol.* **2017**, *57*, 309–315. [CrossRef] [PubMed]
14. Duan, G.; Zhang, H.; Shen, Y.; Li, G.; Wang, H.; Cheng, W. Mitigation of heavy metal accumulation in rice grain with silicon in animal manure fertilized field. *Environ. Eng. Manag. J.* **2016**, *15*, 2223–2229.
15. Van Oosten, M.J.; Pepe, O.; De Pascale, S.; Silletti, S.; Maggio, A. The role of biostimulants and bioeffectors as alleviators of abiotic stress in crop plants. *Chem. Biol. Technol. Agric.* **2017**, *4*, 5. [CrossRef]
16. Guo, L.J.; Zheng, S.X.; Cao, C.G.; Li, C.F. Tillage practices and straw-returning methods affect topsoil bacterial community and organic C under a rice-wheat cropping system in central China. *Sci. Rep.* **2016**, *6*, 33155. [CrossRef] [PubMed]
17. Zhao, J.; Ni, T.; Xun, W.; Huang, X.; Huang, Q.; Ran, W.; Shen, B.; Zhang, R.; Shen, Q. Influence of straw incorporation with and without straw decomposer on soil bacterial community structure and function in a rice-wheat cropping system. *Appl. Microbiol. Biotechnol.* **2017**, *101*, 4761–4773. [CrossRef] [PubMed]
18. National Soil Survey Center. Soil Survey Center. Soil Survey Laboratory Methods Manual. In *Soil Survey Investigations Report No. 42*; Version 3.0; Natural Conservation Service: Washington, DC, USA, 1996.
19. United States Department of Agriculture (USDA). *Diagnosis and Improvement of Saline and Alkali Soils, Agriculture. Handbook No. 60*; U.S. Salinity Laboratory; Government Printing Office: Washington, DC, USA, 1954.
20. Thomas, G.W. *Soil pH and Soil Acidity. Method of Soil Analysis, Part 3: Chemical Methods*; Sparks, D.L., Page, A.L., Helmke, P.A., Loeppert, R.H., Soltanpour, P.N., Tabatabai, M.A., Johnston, C.T., Sumner, M.E., Eds.; SSSA Inc.: Madison, WI, USA; ASA Inc.: Madison, WI, USA, 1996; pp. 475–490.
21. Bray, R.A.; Kurtz, L.T. Determination of total organic and available form of phosphorus in soil. *Soil Sci.* **1945**, *59*, 39–45. [CrossRef]
22. Walkley, A.; Black, I.A. An examination of the Degtjareff method for determining soil organic matter, and a proposed modification of the chromic acid titration method. *Soil Sci.* **1934**, *37*, 29–38. [CrossRef]
23. Klindworth, A.; Pruesse, E.; Schweer, T.; Peplies, J.; Quast, C.; Horn, M.; Glöckner, F.O. Evaluation of General 16S Ribosomal RNA Gene PCR Primers for Classical and Next-Generation Sequencing-Based Diversity Studies. *Nucleic Acids Res.* **2013**, *41*, e1. [CrossRef]
24. Bolyen, E.; Rideout, J.R.; Dillon, M.R.; Bokulich, N.A.; Abnet, C.C.; Al-Ghalith, G.A.; Alexander, H.; Alm, E.J.; Arumugam, M.; Asnicar, F.; et al. Reproducible, Interactive, Scalable and Extensible Microbiome Data Science Using QIIME 2. *Nat. Biotechnol.* **2019**, *37*, 852–857. [CrossRef]
25. Callahan, B.J.; McMurdie, P.J.; Rosen, M.J.; Han, A.W.; Johnson, A.J.A.; Holmes, S.P. DADA2: High-Resolution Sample Inference from Illumina Amplicon Data. *Nat. Methods* **2016**, *13*, 581–583. [CrossRef]
26. Quast, C.; Pruesse, E.; Yilmaz, P.; Gerken, J.; Schweer, T.; Yarza, P.; Peplies, J.; Glöckner, F.O. The SILVA Ribosomal RNA Gene Database Project: Improved Data Processing and Web-Based Tools. *Nucleic Acids Res.* **2013**, *41*, D590–D596. [CrossRef] [PubMed]
27. Glöckner, F.O.; Yilmaz, P.; Quast, C.; Gerken, J.; Beccati, A.; Ciuprina, A.; Bruns, G.; Yarza, P.; Peplies, J.; Westram, R.; et al. 25 Years of Serving the Community with Ribosomal RNA Gene Reference Databases and Tools. *J. Biotechnol.* **2017**, *261*, 169–176. [CrossRef] [PubMed]
28. Douglas, G.M.; Maffei, V.J.; Zaneveld, J.R.; Yurgel, S.N.; Brown, J.R.; Taylor, C.M.; Huttenhower, C.; Langille, M.G.I. PICRUSt2 for Prediction of Metagenome Functions. *Nat. Biotechnol.* **2020**, *38*, 685–688. [CrossRef]
29. Das, S.K.; Varma, A. Role of Enzymes in Maintaining Soil Health. In *Soil Enzymology*; Shukla, G., Varma, A., Eds.; Soil Biology; Springer: Berlin/Heidelberg, Germany, 2011; pp. 25–42.
30. Wickham, H. *ggplot2*; Springer: New York, NY, USA, 2009.

31. Ribeiro Filho, A.A.; Adams, C.; Manfrendini, S.; Aguilar, R.; Neves, W. Dynamics of soil chemical properties in shifting cultivation systems in the tropics: A Meta analysis. *Soil Use Manag.* **2015**, *31*, 474–482. [CrossRef]
32. Granged, A.J.P.; Zavala, L.M.; Jordan, A.; Moreno, G.B. Post- fire evolution of soil properties and vegetation cover in a Mediterranean heathland after experimental burning: A 3-year study. *Geoderma* **2011**, *164*, 85–94. [CrossRef]
33. Raison, R.J. Modification of the soil environment by vegetation fires, with particular reference to nitrogen transformations: A review. *Plant Soil.* **1979**, *51*, 73–108. [CrossRef]
34. Zavala, L.; De Celis, R.; Jordán, A. How wildfires affect soil properties. *Brief Rev. Cuad. Investig. Geográf.* **2014**, *40*, 311–332.
35. Nigussie, A.; Kissi, E. Impact of biomass burning on physicochemical properties of Nitisol in the southwestern Ethiopia. *Asian J. Agric. Res.* **2011**, *5*, 223–233. [CrossRef]
36. Gangwar, K.S.; Singh, K.K.; Sharma, S.K.; Tomar, O.K. Alternative tillage and crop residue management in wheat after rice in sandy loam soils of Indo-Gangetic plains. *Soil Tillage Res.* **2006**, *88*, 242–252. [CrossRef]
37. Pellegrini, A.F.; Ahlström, A.; Hobbie, S.E.; Reich, P.B.; Nieradzik, L.P.; Staver, A.C.; Scharenbroch, B.C.; Jumpponen, A.; Anderegg, W.R.; Randerson, J.T.; et al. Fire frequency drives decadal changes in soil carbon and nitrogen and ecosystem productivity. *Nature* **2018**, *553*, 194. [CrossRef] [PubMed]
38. Parro, K.; Köster, K.; Jõgiste, K.; Seglinš, K.; Sims, A.; Stanturf, J.A.; Metslaid, M. Impact of post-fire management on soil respiration, carbon and nitrogen content in a managed hemiboreal forest. *J. Environ. Manag.* **2019**, *233*, 371–377. [CrossRef] [PubMed]
39. Knicker, H.; Müller, P.; Hilscher, A. How useful is chemical oxidation with dichromate for the determination of "Black Carbon" in fire affected soils? *Geoderma* **2007**, *142*, 178–196. [CrossRef]
40. Riggan, P.J.; Lockwood, R.N.; Jacks, P.M.; Colver, C.G.; Weirich, F.; DeBano, L.F.; Brass, J.A. Effects of Fire Severity on Nitrate Mobilization in Watersheds Subject to Chronic Atmospheric Deposition. *Environ. Sci. Technol.* **1994**, *28*, 369–375. [CrossRef] [PubMed]
41. Wan, S.; Hui, D.; Luo, Y. Fire effects on nitrogen pools and dynamics in terrestrial ecosystems: A meta-analysis. *Ecological* **2001**, *11*, 711–730. [CrossRef]
42. Neary, D.G.; Ryan, K.C.; DeBano, L.F. *Wildland Fire in Ecosystems: Effects of Fire on Soils and Water*; General Technical Report RMRS-GTR-42; United States Department of Agriculture, Forest Service, Rocky Mountain Research Station: Fort Collins, CO, USA, 2005; Volume 4, pp. 1–262.
43. Delač, D.; Pereira, P.; Bogunović, I.; Kisić, I. Short-Term Effects of Pile Burn on N Dynamic and N Loss in Mediterranean Croatia. *Agronomy* **2020**, *10*, 1340. [CrossRef]
44. Caon, L.; Vallejo, V.R.; Coen, R.J.; Geissen, V. Effects of wildfire on soil nutrients in Mediterranean ecosystems. *Earth Sci. Rev.* **2014**, *139*, 47–58. [CrossRef]
45. Covington, W.W.; DeBano, L.F.; Huntsberger, T.G. Soil nitrogen changes associated with slash pile burning in pinyon-juniper woodlands. *For. Sci.* **1991**, *37*, 347–355.
46. Baldock, J.A.; Smernik, R.J. Chemical composition and bioavailability of thermally altered *Pinus resinosa* (red pine) wood. *Org. Geochem.* **2002**, *33*, 1093–1109. [CrossRef]
47. Oguntunde, P.G.; Abiodun, B.J.; Ajayi, A.E.; van de Giesen, N. Effects of charcoal production on soil physical properties in Ghana. *J. Plant Nutr. Soil Sci.* **2008**, *171*, 591–596. [CrossRef]
48. Ayodele, A.; Oguntunde, P.; Joseph, A.; De Souza Dias Junios, M. Numerical analysis of the impact of charcoal production on soil hydrological behavior, runoff response and erosion susceptibility. *Rev. Bras. Cienc. Solo* **2009**, *33*, 137–145. [CrossRef]
49. Giovannini, G.; Lucchesi, S.; Giachetti, M. Effect of heating on some physical and chemical parameters related to soil aggregation and erodibility. *Soil Sci.* **1988**, *146*, 255–261. [CrossRef]
50. Fonseca, F.; de Figueiredo, T.; Nogueira, C.; Queirós, A. Effect of prescribed fire on soil properties and soil erosion in a Mediterranean mountain area. *Geoderma* **2017**, *307*, 172–180. [CrossRef]
51. Neill, C.; Patterson, W.A.; Crary, D.W. Responses of soil carbon, nitrogen and cations to the frequency and seasonality of prescribed burning in a Cape Cod oak-pine forest. *For. Ecol. Manag.* **2007**, *250*, 234–243. [CrossRef]
52. Stock, W.D.; Lewis, O.A.M. Soil Nitrogen and the Role of Fire as a Mineralizing Agent in a South African Coastal Fynbos Ecosystem. *J. Ecol.* **1986**, *74*, 317–328. [CrossRef]
53. Badía, D.; Martí, C. Plant ash and heat intensity eff ects on chemical and physical properties of two contrasting soils. *Arid Land Res. Manag.* **2003**, *17*, 23–41. [CrossRef]
54. Johnson, B.G.; Johnson, D.W.; Chambers, J.C.; Blank, R.R. Fire effects on the mobilization and uptake of nitrogen by cheatgrass (*Bromus tectorum* L.). *Plant Soil* **2011**, *341*, 437–445. [CrossRef]
55. Castelli, L.; Lazzari, M. Impact of fire on soil nutrients in central semiarid Argentina. *Arid Land Res. Manag.* **2002**, *16*, 349–364. [CrossRef]
56. Duguy, B.; Rovira, P.; Vallejo, R. Land-use history and fire effects on soil fertility in eastern Spain. *Eur. J. Soil Sci.* **2007**, *58*, 83–91. [CrossRef]
57. Acosta-Martínez, V.; Dowd, S.; Sun, Y.; Allen, V. Tag-encoded pyrosequencing analysis of bacterial diversity in a single soil type as affected by management and land use. *Soil Biol. Biochem.* **2008**, *40*, 2762–2770. [CrossRef]

58. Edwards, J.; Johnson, C.; Santos-Medellín, C.; Lurie, E.; Podishetty, N.K.; Bhatnagar, S.; Eisen, J.A.; Sundaresan, V. Structure, variation, and assembly of the root-associated microbiomes of rice. *Proc. Natl. Acad. Sci. USA* **2015**, *112*, E911–E920. [CrossRef] [PubMed]
59. Ding, L.J.; Su, J.Q.; Sun, G.X.; Wu, J.X.; Wei, W.X. Increased microbial functional diversity under long-term organic and integrated fertilization in a paddy soil. *Appl. Microbiol. Biotechnol.* **2018**, *102*, 1969–1982. [CrossRef] [PubMed]
60. Busse, M.D.; Shestak, C.J.; Hubbert, K.R.; Knapp, E.E. Soil physical properties regulate lethal heating during burning of woody residues. *Soil Sci. Soc. Am. J.* **2010**, *74*, 947–955. [CrossRef]
61. Whelan, R.J.; Landedyk, W.; Pashby, A.S. The effects of Wildfire on arthropod Populations in Jerrah Banksia Woodland. *West. Aust. Nat.* **1980**, *14*, 214–220.
62. Badía, D.; Martí, C.; Aguirre, A.J.; Aznar, J.M.; González-Pérez, J.A.; De la Rosa, J.M.; León, J.; Ibarra, P.; Teresa Echeverría, T. Wildfire effects on nutrients and organic carbon of a Rendzic Phaeozem in NE Spain: Changes at cm-scale topsoil. *CATENA* **2014**, *113*, 267–275. [CrossRef]
63. Armas-Herrera, C.M.; Martí, C.; Badía, D.; Ortiz-Perpiñá, O.; Girona-García, A.; Mora, J.L. Short-term and midterm evolution of topsoil organic matter and biological properties after prescribed burning for pasture recovery (Tella, Central Pyrenees, Spain). *Land Degrad. Dev.* **2018**, *29*, 1545–1554. [CrossRef]
64. Li, W.; Liu, X.; Niu, S. Differential responses of the acidobacterial community in the topsoil and subsoil to fire disturbance in *Pinus tabulaeformis* stands. *PeerJ* **2019**, *7*, e8047. [CrossRef]
65. Li, W.T.; Chen, X.F.; Liu, M.; Kuzyakov, Y.; Jiang, C.Y.; Wu, M.; Li, Z.P. Shifts in microbial communities with increasing soil fertility across a chronosequence of paddy cultivation in subtropical China. *Appl. Soil Ecol.* **2017**, *120*, 153–159. [CrossRef]
66. Wu, Z.; Liu, Q.; Li, Z.; Cheng, W.; Sun, J.; Guo, Z.; Li, Y.; Zhou, J.; Meng, D.; Li, H.; et al. Environmental factors shaping the diversity of bacterial communities that promote rice production. *BMC Microbiol.* **2018**, *18*, 51. [CrossRef]
67. Herlambang, A.; Murwantoko, M.; Istiqomah, I. Dynamic change in bacterial communities in the integrated rice– fish farming system in Sleman, Yogyakarta, Indonesia. *Aquac. Res.* **2021**, *52*, 5566–5578. [CrossRef]
68. Chen, L.; Xu, J.; Wan, W.; Xu, Z.; Hu, R.; Zhang, Y.; Zheng, J.; Gu, Z. The Microbiome Structure of a Rice-Cray fish Integrated Breeding Model and Its Association with Cray fish Growth and Water Quality. *Microbiol. Spectr.* **2022**, *10*, e0220421. [CrossRef] [PubMed]
69. Ahn, J.H.; Song, J.; Kim, B.Y.; Kim, M.S.; Joa, J.H.; Weon, H.Y. Characterization of the bacterial and archaeal communities in rice field soils subjected to long-term fertilization practices. *J. Microbiol.* **2012**, *50*, 754–765. [CrossRef] [PubMed]
70. Bao, Y.; Dolfing, J.; Guo, Z.; Chen, R.; Wu, M.; Li, Z.; Lin, X.; Feng, Y. Important ecophysiological roles of non-dominant *Actinobacteria* in plant residue decomposition, especially in less fertile soils. *Microbiome* **2021**, *9*, 84. [CrossRef] [PubMed]
71. Bukar, M.; Sodipo, O.; Dawkins, K.; Ramirez, R.; Kaldapa, J.T.; Tarfa, M.; Esiobu, N. Microbiomes of Top and Sub-Layers of Semi-Arid Soils in North-Eastern Nigeria Are Rich in Firmicutes and Proteobacteria with Surprisingly High Diversity of Rare Species. *Adv. Microbiol.* **2019**, *9*, 102–118. [CrossRef]
72. Hayatsu, M.; Tago, K.; Saito, M. Various players in the nitrogen cycle: Diversity and functions of the microorganisms involved in nitrification and denitrification. *Soil Sci. Plant Nutr.* **2008**, *54*, 33–45. [CrossRef]
73. Shridhar, B.S. Review: Nitrogen fixing microorganisms. *Int. J. Microbiol. Res.* **2012**, 3, 46–52.
74. Trivedi, P.; Anderson, I.C.; Singh, B.K. Microbial modulators of soil carbon storage: Integrating genomic and metabolic knowledge for global prediction. *Trends Microbiol.* **2013**, *21*, 641–651. [CrossRef] [PubMed]
75. Aislabic, J.; Deslippe, J.; Dymond, J. Soil microbes and their contribution to soil services. In *Ecosystem Services in New Zealand: Conditions and Trends*; Manaaki Whenua Press: Lincoln, New Zealand, 2013; pp. 143–161.
76. Kalam, S.; Basu, A.; Ahmad, I.; Sayyed, R.Z.; El-Enshasy, H.A.; Dailin, D.J.; Suriani, N.L. Recent Understanding of Soil Acidobacteria and Their Ecological Significance: A Critical Review. *Front. Microbiol.* **2020**, *11*, 580024. [CrossRef] [PubMed]
77. Stinca, A.; Ravo, M.; Marzaioli, R.; Marchese, G.; Cordella, A.; Rutigliano, F.A.; Esposito, A. Changes in Multi-Level Biodiversity and Soil Features in a Burned Beech Forest in the Southern Italian Coastal Mountain. *Forests* **2020**, *11*, 983. [CrossRef]
78. Mohagheghi, A.; Grohmann, K.; Himmel, M.; Leighton, L.; Updegraff, D.M. Isolation and characterization of *Acidothermus cellulolyticus* gen. nov., sp. nov., a new genus of thermophilic, acidophilic, cellulolytic bacteria. *Int. J. Syst. Evol. Microbiol.* **1986**, *36*, 435–443. [CrossRef]
79. Bárcenas-Moreno, G.; Gomez-Brandon, M.; Rousk, J.; Bååth, E. Adaptation of soil microbial communities to temperature: Comparison of fungi and bacteria in a laboratory experiment. *Glob. Chang. Biol.* **2009**, *15*, 2950–2957. [CrossRef]
80. Stott, D.E.; Andrews, S.S.; Liebig, M.A.; Wienhold, B.J.; Karlen, D.L. Evaluation of β-Glucosidase Activity as a Soil Quality Indicator for the Soil Management Assessment Framework. Soil Sci. *Am. J.* **2010**, *74*, 107–119.
81. Bárcenas-Moreno, G.; García-Orenes, F.; Mataix-Solera, J.; Mataix-Beneyto, J.; Bååth, E. Soil microbial recolonisation after a fire in a Mediterranean forest. *Biol. Fertil. Soils* **2011**, *47*, 261–272. [CrossRef]
82. Thirukkumaran, C.M.; Parkinson, D. Microbial respiration, biomass, metabolic quotient and litter decomposition in a lodgepole pine forest floor amended with nitrogen and phosphorous fertilizers. *Soil Biol. Biochem.* **2000**, *32*, 59–66. [CrossRef]
83. van der Heijden, M.G.A.; Bardgett, R.D.; Van Straalen, N.M. The unseen majority: Soil microbes as drivers of plant diversity and productivity in terrestrial ecosystems. *Ecol. Lett.* **2008**, *11*, 296–310. [CrossRef] [PubMed]
84. Ginzburg, O.; Steinberger, Y. Effects of forest wildfire on soil microbial community activity and chemical components on a Temporal-seasonal scale. *Plant Soil.* **2012**, *360*, 243–257. [CrossRef]

85. Goswami, D.; Thakker, J.N.; Dhandhukia, P.C. Portraying mechanics of plant growth promoting rhizobacteria (PGPR): A review. *Cogent Food Agric.* **2016**, *2*, 1127500. [CrossRef]
86. Monciardini, P.; Cavaletti, L.; Schumann, P.; Rohde, M.; Donadio, S. *Conexibacter woesei* gen. nov., sp. nov., a novel representative of a deep evolutionary line of descent within the class *Actinobacteria*. *Int. J. Syst. Evol. Microbiol.* **2003**, *53*, 569–576. [CrossRef]
87. Seki, T.; Matsumoto, A.; Shimada, R.; Inahashi, Y.; Omura, S.; Takahashi, Y. *Conexibacter arvalis* sp. nov., isolated from a cultivated field soil sample. *Int. J. Syst. Evol. Microbiol.* **2012**, *62*, 2400–2404. [CrossRef]
88. Talia, P.; Sede, S.M.; Campos, E.; Rorig, M.; Principi, D.; Tosto, D.; Hopp, H.E.; Grasso, D.; Cataldi, A. Biodiversity characterization of cellulolytic bacteria present on native Chaco soil by comparison of ribosomal RNA genes. *Res. Microbiol.* **2012**, *163*, 221–232. [CrossRef]
89. Sanford, R.A.; Wagner, D.D.; Wu, Q.; Chee-Sanford, J.C.; Thomas, S.H.; Cruz-García, C.; Rodríguez, G.; Massol-Deyá, A.; Krishnani, K.K.; Ritalahti, K.M.; et al. Unexpected nondenitrifier nitrous oxide reductase gene diversity and abundance in soils. *Proc. Natl. Acad. Sci. USA* **2012**, *109*, 19709–19714. [CrossRef] [PubMed]
90. Masuda, Y.; Itoh, H.; Shiratori, Y.; Isobe, K.; Otsuka, S.; Senoo, K. Predominant but previously-overlooked prokaryotic drivers of reductive nitrogen transformation in paddy soils, revealed by metatranscriptomics. *Microbes Environ.* **2017**, *32*, 180–183. [CrossRef] [PubMed]
91. Tang, S.; Liao, Y.; Xu, Y.; Dang, Z.; Zhu, X.; Ji, G. Microbial coupling mechanisms of nitrogen removal in constructed wetlands: A review. *Bioresour. Technol.* **2020**, *314*, 123759. [CrossRef] [PubMed]
92. Onley, J.R.; Ahsan, S.; Sanford, R.A.; Löffler, F.E. Denitrification by Anaeromyxobacter dehalogenans, a common soil bacterium lacking the nitrite reductase genes nirS and nirK. *Appl. Environ. Microbiol.* **2017**, *84*, e01985-17. [CrossRef] [PubMed]
93. Willms, I.M.; Rudolph, A.Y.; Göschel, I.; Bolz, S.H.; Schneider, D.; Penone, C.; Poehlein, A.; Schöning, I.; Nacke, H. Globally abundant 'Candidatus Udaeobacter' benefits from release of antibiotics in soil and potentially performs trace gas scavenging. *Msphere* **2020**, *5*, e00186-20. [CrossRef] [PubMed]
94. Brewer, T.E.; Handley, K.M.; Carini, P.; Gilbert, J.A.; Fierer, N. Genome reduction in an abundant and ubiquitous soil bacterium 'Candidatus Udaeobacter copiosus'. *Nat. Microbiol.* **2016**, *2*, 16198. [CrossRef] [PubMed]
95. Ahmed, A.; Bibi, A. Fungal cellulase; production and applications: Minireview. *Int. J. Health Life-Sci.* **2018**, *4*, 1–19. [CrossRef]
96. Masepohl, B.; Forchhammer, K. Chapter 9—Regulatory cascades to express nitrogenases. In *Biology of the Nitrogen Cycle*; Elsevier: Amsterdam, The Netherlands, 2007; pp. 131–145.

biology

MDPI

Article

Phenotypic Plasticity Strategy of *Aeluropus lagopoides* Grass in Response to Heterogenous Saline Habitats

Abdulaziz M. Assaeed , Basharat A. Dar , Abdullah A. Al-Doss , Saud L. Al-Rowaily, Jahangir A. Malik and Ahmed M. Abd-ElGawad *

Plant Production Department, College of Food and Agriculture Sciences, King Saud University, Riyadh 11451, Saudi Arabia
* Correspondence: aibrahim2@ksu.edu.sa; Tel.: +966-5626-80864

Simple Summary: Plants adapt themselves to harsh environmental conditions by changing morphological parameters through phenotypic plasticity. Plants modify their functional traits and allocate biomass to either tolerate or resist the stress caused by their variable habitats. In this study, we observed that *Aeluropus lagopoides,* being among the few halophytic palatable species of salt marshes, adapt to the harsh salt marshes of different eco-regions by significantly modifying its morphological and physiological traits. Due to this structural modification, this plant showed great potential to rehabilitate different inland and coastal saline flat areas (sabkha), taking saline agriculture and soil remediation into consideration.

Abstract: Understanding the response variation of morphological parameters and biomass allocation of plants in heterogeneous saline environments is helpful in evaluating the internal correlation between plant phenotypic plasticity mechanism and biomass allocation. The plasticity of plants alters the interaction among individuals and their environment and consequently affects the population dynamics and aspects of community and ecosystem functioning. The current study aimed to assess the plasticity of *Aeluropus lagopoides* traits with variation in saline habitats. Understanding the habitat stress tolerance strategy of *A. lagopoides* is of great significance since it is one of the highly palatable forage grass in the summer period. Five different saline flat regions (coastal and inland) within Saudi Arabia were targeted, and the soil, as well as the morphological and physiological traits of *A. lagopoides,* were assessed. Comprehensive correlation analyses were performed to correlate the traits with soil, region, or among each other. The soil analysis revealed significant variation among the five studied regions for all measured parameters, as well as among the soil layers showing the highest values in the upper layer and decreased with the depth. Significant differences were determined for all tested parameters of the morphological and reproductive traits as well as for the biomass allocation of *A. lagopoides*, except for the leaf thickness. In the highly saline region, Qaseem, *A. lagopoides* showed stunted aerial growth, high root/shoot ratio, improved root development, and high biomass allocation. In contrast, the populations growing in the low saline region (Jizan) showed the opposite trend. Under the more stressful condition, like in Qaseem and Salwa, *A. lagopoides* produce low spikes in biomass and seeds per plant, compared to the lowest saline habitats, such as Jouf. There was no significant difference in physiological parameters except stomatal conductance (g_s), which is highest in the Jizan region. In conclusion, the population of *A. lagopoides* is tolerant of harsh environments through phenotypic plasticity. This could be a candidate species to rehabilitate the saline habitats, considering saline agriculture and saline soil remediation.

Keywords: functional traits; saline flat regions; halophytes; biomass allocation; desalination

Citation: Assaeed, A.M.; Dar, B.A.; Al-Doss, A.A.; Al-Rowaily, S.L.; Malik, J.A.; Abd-ElGawad, A.M. Phenotypic Plasticity Strategy of *Aeluropus lagopoides* Grass in Response to Heterogenous Saline Habitats. *Biology* **2023**, *12*, 553. https://doi.org/10.3390/biology12040553

Academic Editors: Daniel Puppe, Panayiotis Dimitrakopoulos, Baorong Lu and Caifu Jiang

Received: 23 February 2023
Revised: 28 March 2023
Accepted: 3 April 2023
Published: 5 April 2023

1. Introduction

Different eco-regions with different climatic conditions can alter the available resources and conditions crucial to plant performance. The response of the plants to these environmental changes is through induced phenotypic changes [1]. Plant species with wide distribution among different eco-regions show large intraspecific variations in most functional and phenotypic traits [2,3]. The spatial variation in functional traits and their phenotypic plasticity can help plants persist under global climate change [4,5]. Variations in biotic and abiotic factors in different geographical regions can lead to morphologically and functionally different ecotypes [6]. Plants have to adjust to environmental heterogeneity through the plasticity of adaptive traits and respond to changes in light availability [7–9], water availability [9], nutrients [9,10], salinity [11,12], and temperature to survive and sustain in the soil-plant atmospheric continuum environment [13].

One major dependency for plant species to maintain their populations under variable environmental conditions is phenotypic plasticity [14,15]. Plants persist through this potential mechanism of phenotypic plasticity when faced with faster environmental changes and lead it toward homeostasis levels, thus allowing their proper functioning [16]. In the face of global warming, phenotypic plasticity has become a benchmark for understanding its potential for population persistence and adaptation [17]. Alterations in environmental conditions like light regimes, soil properties, humidity, and rainfall may shift several phenotypic traits [18–20]. Different plant populations exhibit adaptive plasticity in response to spatial variability of environmental conditions, such as climate and edaphic factors [21–26]. In heterogeneous environments, both abiotic and biotic factors can influence plant life-history traits (seed germination, growth, flowering, and reproduction) and adaptation (plasticity or differentiation) [15,27]. Measuring these phenotypic trait variations (both reproductive success and vegetative growth) and investigating the environmental variability of sites in which the population occurs is necessary to assess the adaptability and conservation status of the target species [28,29]. The plasticity of plants has a crucial impact on the community structure and dynamics, where plasticity alters many interactions between organisms and their abiotic and biotic factors of the environments [30]. The plant species characterized by phenotypic plasticity can colonize a wide range of habitats and modify the community structure as they tolerate different environmental factors.

Plants respond to variable environments by adjusting their multiple aspects of allocation and architecture, morphology, and physiology [1,31] to mitigate stress levels and increase the uptake of the limiting resources [32]. Biomass allocation among plant parts is driven by environmental conditions, which define many plant growth processes [33,34] and is related to the phenotypic characteristics of plants. Therefore, plant phenotypic plasticity can be used as a potential covariate for understanding biomass allocation [35,36].

In arid and semi-arid regions, water loss due to evapotranspiration increases the salt concentration in soil components [37], leading to more severe salinization issues. Natural saline habitats vary in salinity levels both spatially and temporally due to topography, soil properties, and micro-climate differences [38]. One such saline habitat is sabkha i.e.., a flat area of clay, silt, or sand with an overlying crust of soil [39]. The salt stress, moisture content, physio-chemical soil characteristics, and other environmental factors in saline areas tend to show relative stability with time [40], which has an extensive influence on community structure, plant morphological structure, and biomass allocation [41]. The biomass allocation of plants represents their growth and metabolism and affects the plant's functional attributes [42].

Most salt marsh plant species are halophytes with a high degree of phenotypic variations, occupying a broad range of environmental conditions and possessing various traits to adapt to saline conditions [43–45]. Halophytic species have developed different mechanisms for regulating growth and development to ensure their survival in highly-saline inland or coastal areas, salt marshes, dunes, and desert habitats [46,47]. The distribution of some halophytic species is best correlated along a gradient of soil variables, such as salinity, moisture content, soil texture, organic matter, and calcium carbonate [48]. Halophytic grasses can tolerate salinity at a species-specific level and vary with the ecotype, region's habitat, and specific environmental factors [14,49]. These halophytic species show adaptive phenotypic plasticity, enabling them to cope with different saline environments [50], as most traits exhibited considerable plasticity in response to different salinity stresses [51].

Aeluropus lagopoides (L.) Thwaites (Poaceae) is a stoloniferous halophytic perennial C_4 photosynthetic grass ranging in distribution from Northern Africa (Morocco to Somalia), Italy, and Cyprus, through the deserts of the Middle East to Central Asia, Pakistan, and India [52]. In Saudi Arabia, it is found in different regions of saline coastal environments and inland areas. *A. lagopoides* was recorded from the inland wadi (valley) of Qareenah, Riyadh, saltmarsh areas of Qaseem and Jouf, and coastal zones of Salwa and Jizan region [53–56]. *A. lagopoides* is of economic importance as it is a palatable summer forage in arid areas as well as a sand stabilizer [14] and can be used for landscaping the urban areas of desert regions [57]. The plant withstands high salinity stresses up to 25 $dS{\cdot}m^{-1}$ and can adapt to heterogeneous environments due to structural adaptations and phenotypic trait modifications [58]. There is a considerable variation in environmental conditions of *A. lagopoides* habitats between different coastal and inland regions of Saudi Arabia, with variable effects on water relations, salinity, light, ambient temperature, and edaphic factors [55]. Consequently, the only dependency to maintain its populations under stressful environmental conditions is adaptive plasticity [14,15]. The ability of *A. lagopoides* to grow in different regions provides an excellent opportunity to study its phenotypic trait variations with respect to the regions in which it grows. However, the relationship between biomass allocation and root/shoot morphological strategies of *A. lagopoides* growing in different saline regions is unclear. Therefore, in this study, we aimed to explore the linkage of the variation in the functional traits of vegetative and root parts of *A. lagopoides* (i.e., phenotypic plasticity) with the differences in the habitats (edaphic factors). We aimed to clarify the following questions (1) how do morphological parameters of *A. lagopoides* synergistically change in response to habitat conditions? (2) what biomass allocation strategies did *A. lagopoides* have under different saline regions?

2. Materials and Methods

2.1. Surveyed Regions and Soil Analysis

The populations of *A. lagopoides* were studied along Saudi Arabia during the years 2020–2021 and were found in five saline regions (Figure S1) were identified as follows: (1) Salwa (lowland coastal saline flat area), (2) Jizan (southern coastal saline flat area), (3) Qareenah (inland saline flat areas of wadi Hargan, Riyadh region), (4) Qaseem (an inland saline flat area of Al-Aushazia) and (5) Jouf (an inland saline flat area in Domat Aljandal). The regions' details are presented in Table S1. Each region was geographically different from the others as the shortest point-to-point distance between them was more than 300 km (Table S1 and Figure S2). Within each region, five distinct *A. lagopoides* patches were randomly selected for soil sampling and plant morphological traits measurements. From September to March (when *A. lagopoides* become fully flourished), three random plots (5 × 5 m) were selected within each patch (Figure 1). To assess the relationship between morphological and biomass allocation of the plant and resource allocation of the rhizosphere soil properties, three core soil sampling was selected. At three random points, three core soil samples were collected from three soil layers (0–15 cm, 15–30 cm, and 30–45 cm) within each plot. Each corresponding soil layer of these three soil samples was merged as one composite sample. A total of three composite soil samples represented each plot, and subsequently, a total of 9 samples from each patch. Hence, a total of 225 composite soil samples from all the studied region (5 regions × 5 patches × 3 plots × 3 layers) were collected. For soil moisture content, part of each sample was collected in duly labeled moisture tin, and the moisture content was immediately determined by the weight-loss method for all the samples.

Figure 1. Different studied flat saline regions, (**A**) Qareenah, (**B**) Qaseem, (**C**) Salwa, (**D**) Jouf, and (**E**) Jizan. The left is an overview, and the right is a close view of *A. lagopoides*.

For further analysis, all the soil samples were collected in plastic bags, duly labeled, and transferred to Range Science Lab., College of Food and Agriculture Sciences, King Saud

University, Riyadh, Saudi Arabia. All the soil samples were spread over separate plastic sheets, air-dried at room temperature, and sieved through a 2 mm sieve to remove any debris, and the soil samples were analyzed similarly to the previously reported approach of Dar et al. [55]. In brief, the soil texture was determined using the hydrometer method [59]. Soil organic matter was determined by wet combustion with dichromate at 450 °C [60]. Soil water extracts (1:5) were prepared for the estimation of soil electrical conductivity (EC) and pH [60]. Soluble inions (Cl^- and SO_4^{2-}) were determined by the titration method, while soluble cations (Ca^{2+}, Mg^{2+}, Na^+, and K^+) were determined using a flame photometer according to Rhoades [61].

2.2. Morphological Traits Measurements

Within each plot, five fully matured individuals were randomly selected for morphological parameters, including both on-field and off-field functional trait measurements were recorded. A total of 375 individuals (5 regions × 5 patches × 3 plots × 5 individuals) were targeted for the measurements. In the field, shoot length, number of tiller/plant, number of leaves/tiller, number of spikes/plant, spike length, top internode length of the main tiller, number of stolon/plant, and top internode stolon length were measured.

On the other hand, after taking the field measurement, the same individuals were excavated and collected in labeled plastic bags. The bags were brought in an ice-cool box to the Range Science Laboratory, College of Food and Agriculture Sciences, King Saud University, Riyadh, Saudi Arabia, for other measurements like leaf area, biomass, and root morphological parameters. Plant samples were separated into root and shoot systems. The leaf area of five leaves of each individual was measured using the WinDIAS system (Delta-T Devices Ltd., Cambridge, UK). Also, the average area and biomass of five spikes of each individual were measured. The root system of all the individuals was thoroughly washed, and their main root length, root hair length, total root area, and root dry matter were measured. Based on these measurements, specific leaf area (SLA) was determined as the ratio of leaf area to leaf dry mass. Leaf dry matter content (LDMC) was calculated as the ratio of leaf dry mass to saturated fresh mass [62]. Leaf thickness was calculated as the ratio ($SLA \times LDMC^{-1}$). For resource allocation, root/shoot ratio, root mass fraction, and shoot mass fraction were calculated. These functional traits were selected to assess the response and plasticity of *A. lagopoides* to the environmental factors within different regions, according to Perez-Harguindeguy et al. [63].

2.3. Determination of Ecophysiological Traits

Before the targeted plant individuals were excavated, chlorophyll fluorescence, chlorophyll content, leaf temperature, and stomatal conductance were measured. Chlorophyll fluorescence was measured with an Opti-Sciences OS30p+ chlorophyll fluorometer (Opti-Sciences, Hudson, NY, USA). Fluorescence was measured at midday (11.00–13.00 h, solar time). Chlorophyll fluorescence, initial fluorescence (F0), maximum fluorescence (Fm), and variable fluorescence (Fv) were determined, and the ratios of Fv/F0 and Fv/Fm were calculated using MINI-PAM fluorometer (Heinz Walz GmbH, Effeltrich, Germany). The minimum and maximum dark-adapted fluorescence (F0, Fm) and Fv/Fm (where Fv = Fm—F0) were obtained after the leaves of the plants were dark-adapted for at least 20–25 min [64].

In-situ stomatal conductance (gs) was measured using a steady-state diffusion porometer (model SC-1, Decagon Devices, Inc., Pullman). Each day before measurements, the porometer was calibrated, and gs was measured on the adaxial surface of a fully developed penultimate leaf in the afternoon (13:00–15:00 h). The chlorophyll content was measured according to the method of Lichtenthaler and Wellburn [65] with some modifications. About 0.5 g of the fresh plant sample was extracted by acetone, and the content of chlorophyll a (Chl. a), chlorophyll b (Chl. b), and total chlorophyll were determined by measurements of the absorbance at 663 and 645 nm with the UV-VIS spectrophotometer (SHIMADZU, Kyoto, Japan, UV1800).

2.4. Statistical Analysis

To compare the various traits of *A. lagopoides* and determine the significant variations among regions, the data for the plant functional traits and ecophysiological parameters were analyzed by one-way ANOVA with the region as a factor. However, the soil properties were analyzed for three-way ANOVA with the regions, soil layers, and samples as factors. The ANOVA was performed using Statistix 8.1 software. The soil samples were also used as a factor in the soil analysis to check the homogeneity of soil samples within the studied region. Mean values were compared by the Duncan Multiple Range (DMR) test using SAS 9.1.3. The standard error (SE) was calculated for each mean value.

To correlate the various plant traits (vegetative and reproductive) with each region, a data matrix of the shoot, root, and reproductive traits from the five studied regions was subjected to principal component analysis (PCA). The plant morphological traits were plotted as loading vectors in a bi-plot, while the region was plotted as observations.

On the other side, in order to assess the relationship between the soil parameters of each region and morphological traits, two datasets were prepared; one of the various morphological traits and the second of the soil variables of the studied regions at the three layers (0–15 cm, 15–30 cm, and 30–45 cm). These two datasets were subjected to ordination analysis using canonical correspondence analysis (CCA). Also, the agglomerative hierarchical clustering (AHC) and heatmap were performed based on the data of the top layer soil parameters and the morphological traits of *A. lagopoides* populations within the five studied regions. The AHC was performed based on the Pearson correlation coefficient and weighted pair-group average agglomeration method. PCA and AHC were performed using the XLSTAT software program (version 2018, Addinsoft, NY, USA), while CCA was produced using the MVSP software program, ver. 3.1 [66].

3. Results

3.1. Soil Layer Variations among the Regions of A. lagopoides

Soil analysis revealed significant variation ($p < 0.05$) among the five studied regions supporting the growth of *A. lagopoides* for all measured parameters, except for HCO_3 as well as among the soil layers (Table 1). The Qaseem, Salwa, and Qareenah regions had the highest and comparable pH values for the top layer soil (0–15 cm), while the Jizan and Jouf regions attained the lowest values of the pH (Table 1). In general, the pH values decreased significantly ($p = 0.0219$) with the soil depth in all studied regions.

The soil salinity is highly significantly varied among soil layers for all regions ($p = <0.001$). The soil of the Qaseem region generally had the highest values of EC (25.95 dS/m for the top layer, 10.28 dS/m for the middle layer, and 6.32 dS/m for the lower layer), cation (Ca^{2+}, Mg^{2+}, and Na^{+}) and anions (Cl^{-}) of all the regions as well as for all the soil layers. However, the values of K^{+} and SO_4^{2-} varied in significance from layer to layer among the locations, the highest value of K^{+} for the top layer (20.22 meq/L) being in the Qaseem region and the below two layers (9.95 meq/L for the top layer and 3.34 meq/L for the lower layer in Jouf region (Table 1). The highest value of SO_4^{2-} for the top layer (71.8 meq/L) was recorded in the Jizan region. However, Qaseem attained the highest amount of Cl^{-} (237.60, 73.30, and 51.00 meq/L for the top middle and lower layers, respectively) for all three layers compared to other locations. The cations and anions of the Jouf region soil showed a trend of lower concentration with soil depth. The soil of the Qareenah region attained the highest content of $CaCO_3$ among all regions, and the content increased with the increase in soil depth. On the other hand, the Qaseem region attained the highest organic matter content for the top layer (1.78%), followed by the Qareenah region for all the soil layers (1.63%, 0.98%, and 0.91% for the top, middle, and lower layers, respectively).

Table 1. Physical and chemical properties of different soil layers supporting *Aeluropus lagopoides* in different regions.

Parameters	Layer	Region					*p*-Value
		Qareenah	Qaseem	Salwa	Jouf	Jizan	
pH	0–15 cm	8.38 ± 0.18 A,a	8.39 ± 0.141 A,a	8.39 ± 0.140 A,a	8.19 ± 0.108 B,C,a	8.09 ± 0.220 C,a	0.0011 **
	15–30 cm	8.47 ± 0.04 A,a	8.16 ± 0.027 B,b	8.22 ± 0.022 AB,b	8.02 ± 0.051 C,b	8.10 ± 0.192 C,a	
	30–45 cm	8.19 ± 0.07 B,b	8.15 ± 0.034 B,b	8.27 ± 0.115 A,B,a,b	8.02 ± 0.097 C,b	7.88 ± 0.207 C,b	
p-value	0.0219 *						
EC (dS·m^{-1})	0–15 cm	15.39 ± 0.92 B,a	25.95 ± 3.87 A,a	7.29 ± 0.17 C,a	2.74 ± 0.13 C,b	1.032 ± 0.07 D,a	<0.0001 ***
	15–30 cm	5.17 ± 0.78 B,b	10.28 ± 1.83 A,b	5.25 ± 0.26 B,b	4.52 ± 1.03 C,a	1.10 ± 0.16 D,a	
	30–45 cm	3.87 ± 0.46 B,c	6.32 ± 1.14 A,c	3.53 ± 0.58 C,c	3.59 ± 1.12 C,a	0.94 ± 0.07 D,b	
p-value	<0.0001 ***						
Ca^{2+} (meq/L)	0–15 cm	19.90 ± 2.69 C,a	39.86 ± 3.74 A,a	31.87 ± 0.43 B,a	16.45 ± 1.38 D,a	2.76 ± 0.36 E,a	<0.0001 ***
	15–30 cm	14.08 ± 1.45 C,b	23.66 ± 3.51 A,b	24.55 ± 0.84 B,b	11.12 ± 3.16 D,b	3.14 ± 0.65 E,a	
	30–45 cm	15.30 ± 1.54 B,b	20.17 ± 3.46 A,b	19.30 ± 2.74 A,c	9.02 ± 3.06 C,c	2.15 ± 0.45 E,b	
p-value	<0.0001 ***						
Mg^{2+} (meq/L)	0–15 cm	35.40 ± 2.73 B,a	60.63 ± 8.30 A,a	8.25 ± 0.43 C,a	5.77 ± 0.39 C,b	1.67 ± 0.25 D,b	<0.0001 ***
	15–30 cm	16.10 ± 0.89 B,b	33.74 ± 3.28 A,b	8.56 ± 0.36 D,a	13.51 ± 0.95 C,a	2.80 ± 0.24 E,a,b	
	30–45 cm	6.96 ± 0.53 C,c	11.49 ± 0.55 A,b	5.83 ± 0.60 D,b	10.34 ± 1.03 B,a	2.55 ± 0.32 E,a	
p-value	<0.0001 ***						
Na^{+} (meq/L)	0–15 cm	45.05 ± 6.52 B,a	134.81 ± 33.36 A,a	30.57 ± 0.71 B,C,a	5.12 ± 0.06 C,b	5.38 ± 0.61 C,a	<0.0001 ***
	15–30 cm	25.18 ± 4.89 B,b	44.34 ± 13.19 A,b	18.04 ± 1.09 B,C,b	15.53 ± 4.70 C,a	3.76 ± 0.60 D,b	
	30–45 cm	15.43 ± 2.41 B,b	27.34 ± 6.36 A,b	9.73 ± 2.83 C,c	13.49 ± 5.22 B,C,a	4.54 ± 0.18 c	
p-value	<0.0001 ***						
K^{+} (meq/L)	0–15 cm	14.63 ± 4.24 B,a	20.22 ± 4.97 A,a	2.16 ± 0.04 C,a	0.61 ± 0.11 D,c	0.51 ± 0.09 D,b	0.0001 ***
	15–30 cm	2.78 ± 0.47 B,b	1.42 ± 0.34B C,b	1.75 ± 0.08B C,b	9.95 ± 2.98 A,a	2.57 ± 0.08 B,a	
	30–45 cm	1.88 ± 0.49 B,c	1.52 ± 0.36 B,C,b,	1.25 ± 0.40 B,C,b	3.34 ± 2.36 A,b	0.52 ± 0.05 C,b	
p-value	<0.0001 ***						
Cl^{-} (meq/L)	0–15 cm	96.90 ± 13.08 B,a	237.60 ± 41.19 A,a	66.36 ± 1.27 B,C,a	13.04 ± 1.24 C,D,c	8.65 ± 0.78 D,a	<0.0001 ***
	15–30 cm	48.02 ± 4.62 B,b	73.30 ± 18.32 A,b	48.00 ± 1.76 B,C,b	41.60 ± 10.16 C,a	7.25 ± 1.42 D,b	
	30–45 cm	30.94 ± 3.40 B,c	51.00 ± 11.90 A,c	25.67 ± 8.35 B,C,c	32.58 ± 9.77 B,a,b	7.07 ± 0.78 C,b	
p-value	<0.0001 ***						
HCO_3^{-} (meq/L)	0–15 cm	3.00 ± 0.137 A,a	3.58 ± 0.54 A,a	2.43 ± 0.21 A,a	1.13 ± 0.03 B,b	1.22 ± 0.03 B,b	0.2300 ns
	15–30 cm	1.51 ± 0.130Bb	2.36 ± 0.18Aa	0.9 ± 0.12Cb	1.38 + 0.15Bb	2.09 ± 0.27Aa	
	30–45 cm	1.14 ± 0.091 B,b	2.13 ± 0.19 A,a	1.09 ± 0.05 B,a	4.14 ± 2.55 A,a	1.47 ± 0.19 B,b	
p-value	0.3849						
SO_4^{2-} (meq/L)	0–15 cm	15.04 ± 4.32 B,a	14.19 ± 3.03 B,C,b	3.92 ± 0.61 D,b	12.84 ± 0.16 C,a	71.80 ± 0.05 A,a	<0.0001 ***
	15–30 cm	6.36 ± 1.39 B,b	27.13 ± 1.14 A,a	3.05 ± 0.59 C,b	1.86 ± 0.43 D,b	0.96 ± 0.19 E,b	
	30–45 cm	10.97 ± 1.77 A,a,b	9.42 ± 2.52 A,c	8.14 ± 2.60 B,a	1.13 ± 0.56 C,b	0.60 ± 0.07 D,b	
p-value	0.0058 **						
OM (%)	0–15 cm	1.63 ± 0.19 A,a	1.78 ± 0.37 A,a	0.55 ± 0.02 B,b	0.81 ± 0.12 B,a	0.33 ± 0.04 B,b	<0.0001 ***
	15–30 cm	0.98 ± 0.17 A,b	0.87 ± 0.20 A,b	0.35 ± 0.02 C,a	0.45 ± 0.14 B,b	0.44 ± 0.12 B,a	
	30–45 cm	0.91 ± 0.33 A,b	0.54 ± 0.06 B,c	0.98 ± 0.29 A,a	0.39 ± 0.13 C,c	0.26 ± 0.03 D,c	
p-value	0.0003 ***						
$CaCO_3$ (%)	0–15 cm	34.81 ± 2.10 A,b	18.02 + 2.01 D,a	20.77 ± 0.77 C,b	32.58 ± 0.14 B,a	1.23 ± 0.09 E	<0.0001 ***
	15–30 cm	36.62 ± 0.94 A,b	16.76 ± 1.04 C,b	18.31 ± 0.55 B,c	4.08 ± 0.56 D,b	0.00 ± 0.00 E	
	30–45 cm	45.35 ± 1.68 A,a	17.01 ± 0.88 C,c	29.36 ± 2.17 B,a	4.03 ± 0.78 D,b	0.00 ± 0.00 E	
p-value	<0.0001 ***						
Clay (%)	0–15 cm	12.48 ± 065 C,a	16.80 ± 1.26 A,a	15.14 ± 1.64 B,a	12.67 ± 0.08 C,a	14.33 ± 1.03B,a	0.0023 **
	15–30 cm	10.84 ± 0.64 C,b	12.72 ± 0.41 B,b	7.79 ± 0.13 D,c	11.20 ± 0.53 B,b	13.58 ± 1.76 A,b	
	30–45 cm	10.60 ± 0.54 C,b	10.84 ± 0.83 C,c	11.20 ± 0.56 B,b	10.80 ± 0.83 C,c	12.32 ± 0.86 A,b	
p-value	<0.0001 ***						
Silt (%)	0–15 cm	17.28 ± 1.86 C,b	46.40 ± 2.15 A,b	6.88 ± 1.17 E,b	15.25 ± 1.04 D,c	36.70 ± 1.88 B,b	<0.0001 ***
	15–30 cm	22.48 ± 1.49 D,a	54.77 ± 1.46 A,a	2.04 ± 0.27 E,c	32.28 ± 1.37 C,a	41.30 ± 2.58 B,a	
	30–45 cm	22.68 ± 1.51 D,a	32.54 ± 1.75 A,b	12.10 ± 1.41 E,a	26.08 ± 2.56 C,b	29.61 ± 2.26 B,c	
p-value	<0.0001 ***						
Sand (%)	0–15 cm	70.24 ± 2.50 C,a	36.80 ± 1.634 E,b	78.58 ± 1.46 A,b	74.41 ± 2.52 B,a	48.68 ± 2.11 D,b	<0.0001 ***
	15–30 cm	66.68 ± 1.23 B,b	34.51 ± 0.99 E,b	90.17 ± 0.36 A,a	56.53 ± 1.83 C,c	45.12 ± 2.41 D,c	
	30–45 cm	66.72 ± 0.72 B,b	56.62 ± 1.49 D,a	76.70 ± 1.74 A,b	63.12 ± 2.72 C,b	58.07 ± 2.36 D,a	
p-value	<0.0001 ***						
MC (%)	0–15 cm	7.06 ± 0.55 B,a	28.83 ± 0.75 A,a	5.78 ± 0.44 B,a	4.97 ± 1.01 B,a	1.64 ± 0.25 C,c	<0.0001 ***
	15–30 cm	5.14 ± 0.62 B,b	19.40 ± 0.79 A,a	3.80 ± 0.81 C,c	4.79 ± 1.08 C,b	2.55 ± 0.52 C,b	
	30–45 cm	3.34 ± 0.40 B,c	23.88 ± 0.96 A,a	3.75 ± 0.80 B,b	4.60 ± 0.88 B,b	3.45 ± 0.52 B,c	
p-value	<0.0001 ***						

Different capital letters showed significant variation among regions at $p < 0.05$ (Duncan's test), with df 4 for the region and 4 for soil layers, respectively. Different small letters revealed significant differences among soil layers (0–15 cm, 15–30 cm, and 30–45 cm). Capital letters indicate the significance of regions and small letters soil layers, EC: electrical conductivity, OM: organic matter, MC: moisture content, *$p < 0.05$, ** $p < 0.01$, *** $p < 0.001$, and "ns" for $p > 0.05$.

Regarding soil texture, the sand content is highest in Salwa for all three layers and the lowest in silt content, while the Qaseem region had the lowest values of sand and the highest of silt for all three soil layers. Clay content was highest in Qaseem for the top layer (16.80%), while it was highest in the Jizan region for the layer of 15–30 cm (13.58%) and the layer of 30–45 cm (12.32%). Moisture content showed a significant difference ($p < 0.0001$) among layers, and it was highest in the Qaseem region for all three layers compared to other regions' respective layers. Overall, the results of soil analyses showed that the highest soil characteristic values trend from top to bottom layer (0–15 > 15–30 > 30–45) for all the regions, with some minor exceptions.

3.2. Morphological Traits Variations among the Studied Regions of *A. lagopoides*

By comparing the five studied regions of *A. lagopoides*, significant differences were determined for all tested parameters of the morphological and reproductive traits as well as for the biomass allocation, except for the leaf thickness, where no significant difference was observed (Table 2).

Table 2. Single-factor analysis of variance (ANOVA) showing the effect of different saline regions on plant functional traits and biomass allocation of *A. lagopoides* having a degree of freedom of the studied regions.

Functional Traits	Unit	SS	MS	*F* Value	*p* Value
Shoot length	cm	1901.38	475.34	40.56	<0.0001 ***
Shoot fresh weight	g	2221.49	555.37	5.04	0.0019 **
Shoot dry weight	g	1182.11	295.52	5.91	0.0006 ***
Number of tillers/plant	No.	2251.37	562.84	42.67	<0.0001 ***
Number of stolons/plant	No.	151.02	37.75	7.10	0.0002 ***
Average stolon length	cm	22701.03	5675.25	25.97	<0.0001 ***
Root length	cm	773.21	193.30	14.38	<0.0001 ***
Root fresh weight	g	92.62	23.15	6.25	0.0004 ***
Root dry weight	g	63.22	15.80	6.89	0.0002 ***
Root area	cm^2	5335.64	1333.91	8.10	0.0001 ***
Leaf fresh weight	g	36182.34	9045.58	13.46	<0.0001 ***
Leaf dry weight	g	24768.00	6192.00	675.49	<0.0001 ***
Number of leaves/plant	No.	342601.02	85650.25	8.28	<0.0001 ***
Specific leaf area	cm^2/g	0.05	0.01	7.53	0.0001 ***
Leaf dry matter content	g	39.38	9.84	6.63	0.0003 ***
Leaf thickness	µm	223.61	55.90	1.28	0.2902 NS
Number of spikes/plant	No.	12999.52	3249.88	6.07	0.0005 ***
Average. spike length	cm	2.10	0.52	3.35	0.0174 *
Spike fresh weight	g	0.03	0.01	14.42	<0.0001 ***
Spike dry weight	g	0.01	0.01	4.05	0.0068 **
Number of seeds/plant	No.	75×10^7	1.87	48.56	<0.0001 ***

SS (Sum of Squares), MS (Mean Square), * $p < 0.05$, ** $p < 0.01$, *** $p < 0.001$, and "NS" for $p > 0.05$.

3.2.1. Shoot Traits

A highly significant difference in the shoot length and biomass was observed among regions ($p < 0.001$). Moreover, the shoot length of *A. lagopoides* growing in the Jizan and Salwa regions was the highest, while it was lowest in Qareenah and Jouf regions (Figure 2). For shoot biomass (fresh and dry weight), Qassem and Salwa regions attained the highest values, while the lowest values of shoot fresh and dry weight were assessed in the Qareenah region (4.68 g and 3.05 g, respectively). The number of tillers per plant, stolon number per plant, and stolon length showed highly significant differences ($p < 0.001$) among regions of *A. lagopoides*. The number of tillers per plant was higher in the Jouf region, while the number and length of stolon were higher in Jizan and Qaseem regions (Figure 2). Jouf region attained the lowest values of stolon measurements.

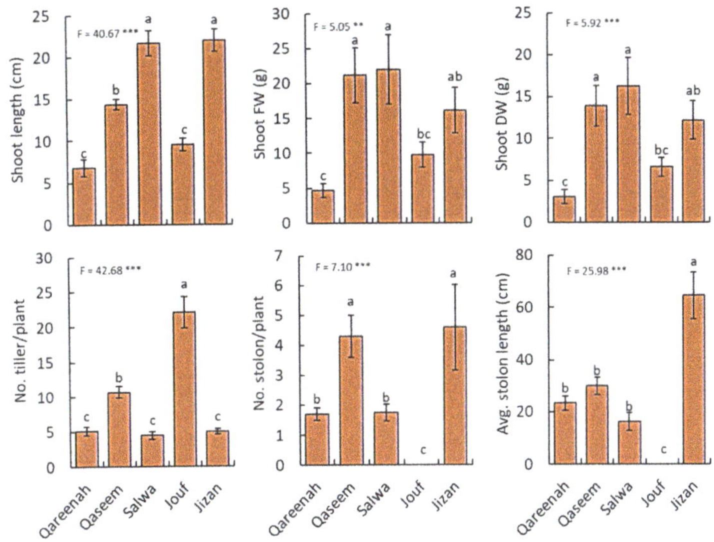

Figure 2. Comparison of shoot traits of *Aeluropus lagopoides* growing in different saline flat regions of Saudi Arabia. Values are average (n = 75), and the bar represents the standard error. Different letters among regions showed significant differences at $p < 0.05$ after Duncan's test. ** $p < 0.01$*** $p < 0.001$.

3.2.2. Root Traits

All studied root traits of the samples showed highly significant differences ($p < 0.001$) among the five studied regions (Figure 3). For root length, Qaseem, Salwa, and Jizan regions attained the highest values, while Qareenah and Jouf had the lowest root length. For biomass, *A. lagopoides* growing in the Qaseem region attained the highest root fresh and dry weight. Moreover, the highest value of the root area was determined for the *A. lagopoides* growing in the Qaseem region (45.27 cm^2), followed by Salwa (35.71 cm^2) and Jizan (32.31 cm^2) regions (Figure 3).

3.2.3. Leaf Traits

All leaf traits of *A. lagopoides* plants showed highly significant variation among regions ($p < 0.001$), except for leaf thickness which did not vary significantly (p = 0.29) from one region to another (Figure 4).

The Jizan region attained the highest values of leaf biomass (fresh and dry weight), while the Qareenah region had the lowest values of both leaf fresh and dry weight. The number of leaves per plant was highest for the population of the Jouf region, while the Qareenah region attained the lowest number of leaves. In contrast, the Qareenah region attained the highest values of specific leaf area (0.16 cm^{-2} g^{-1}), and the Jizan region showed the highest value of leaf dry matter content (90.16%). However, no significant difference in leaf thickness was found in all studied regions.

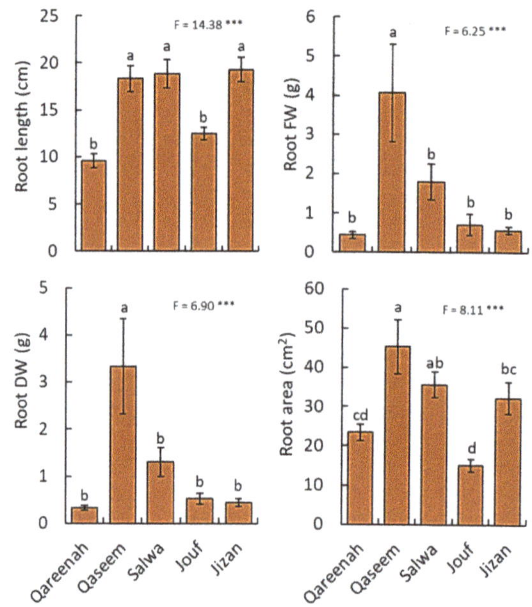

Figure 3. Measured root traits for *Aeluropus lagopoides* growing in different saline flat regions of Saudi Arabia. Values are average ($n = 75$), and the bar represents the standard error. Different letters among regions showed significant differences at $p < 0.05$ after Duncan's test. *** $p < 0.001$. FW: fresh weight, DW: dry weight.

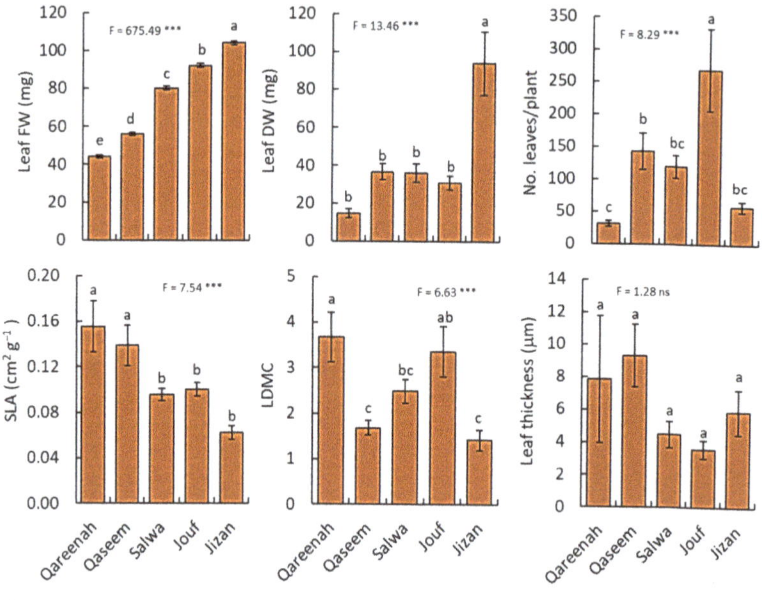

Figure 4. Comparison of leaf traits for *Aeluropus lagopoides* growing in different saline flat regions of Saudi Arabia. Values are average ($n = 10$), and the bar represents standard error. Different letters among regions showed significant differences at $p < 0.05$ after Duncan's test. SLA: specific leaf area, LDMC: leaf dry matter content. FW: fresh weight, DW: dry weight. *** $p < 0.001$, and "ns" for $p > 0.05$.

3.2.4. Reproductive Traits

All the measured reproductive traits of *A. lagopoides* showed a highly significant difference ($p < 0.001$) among the studied regions (Figure 5). Regarding spike numbers, the populations of the Jouf region showed the maximum production of spikes and seeds per plant, while Qareenah and Jizan attained the lowest values (Figure 5). The average spike length of the Qaseem regions was the highest, while Qareenah and Salwa regions attained the lowest values of the spike length. For spike biomass (fresh and dry weight), the Jizan region attained the highest values (0.09 and 0.06 mm, respectively), while Qareenah and Salwa regions showed the lowest biomass among the studied regions.

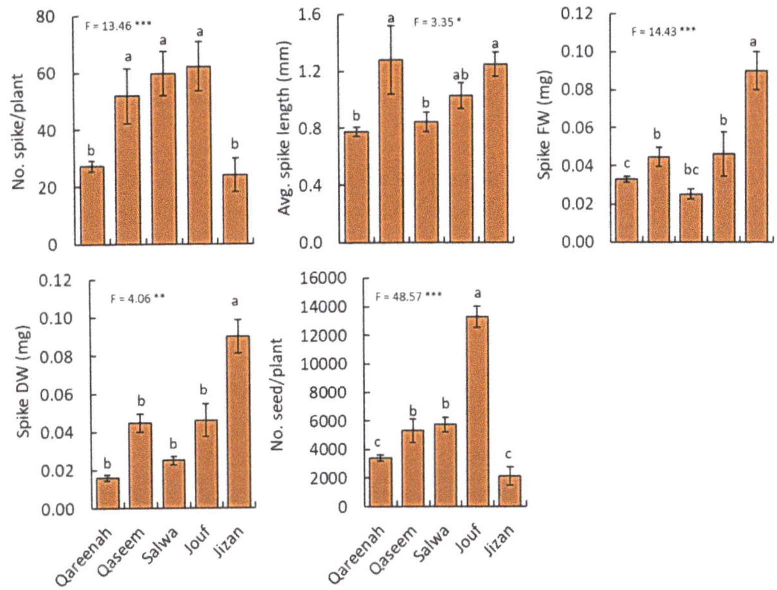

Figure 5. Reproductive traits for *Aeluropus lagopoides* growing in different saline flat regions of Saudi Arabia. Values are average ($n = 75$), and the bar represents the standard error. Different letters among regions showed significant differences at $p < 0.05$ after Duncan's test. * $p < 0.05$, ** $p < 0.01$, *** $p < 0.001$. FW: fresh weight, DW: dry weight.

3.3. *Variation in Ecophysiological Parameters of A. lagopoides among Different Regions*

The photosynthetic pigments (chlorophyll *a*, chlorophyll *b*, total chlorophyll) varied slightly among different regions but had no significance (Figure 6).

Similarly, there is no significant difference in Fv/Fm, and all values were under 0.79. On the other hand, stomatal conductance showed a highly significant difference among the regions ($p < 0.001$), as the Jizan region had the highest stomatal conductance (52.88 mmole m^{-2} s^{-1}), while the Jouf region had the lowest value (20.46 mmole m^{-2} s^{-1}). Moreover, a highly significant difference was observed among the studied regions for the *A. lagopoides* leaf temperature.

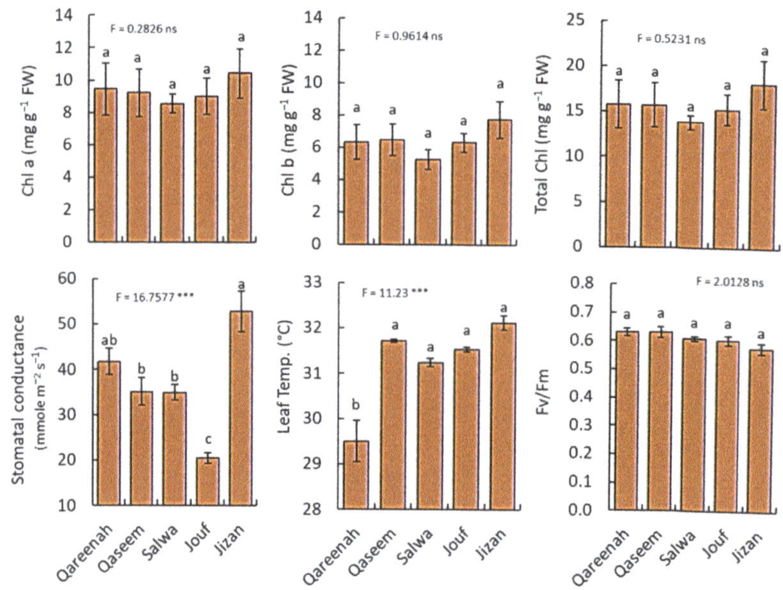

Figure 6. Ecophysiological parameters of *Aeluropus lagopoides* growing in different saline flat regions of Saudi Arabia. Values are average ($n = 75$), and the bar represents the standard error. Different letters among regions showed significant differences at $p < 0.05$ after Duncan's test. *** $p < 0.001$, and "ns" for $p > 0.05$. FW: fresh weight.

3.4. Biomass Allocation of A. lagopoides in Response to Different Habitats

The biomass proportion of plant parts in *A. lagopoides* was significantly different ($p < 0.001$) among the studied regions (Figure 7). Concerning root:shoot ratio, the population of the Qaseem region showed the highest value (0.23), followed by Qareenah (0.16), while the population of the Jizan region attained the lowest values of root/shoot ratio (Figure 7). Similarly, the root mass fraction showed the same pattern, where *A. lagopoides* of the Qaseem region showed the highest root mass fraction (0.18), followed by Qareenah (0.13), while *A. lagopoides* of the Jizan region attained the lowest value (0.04). In contrast, the population of the Jizan region showed the highest shoot mass fraction (0.96), while the population of the Qaseem and Qareenah regions attained the lowest values.

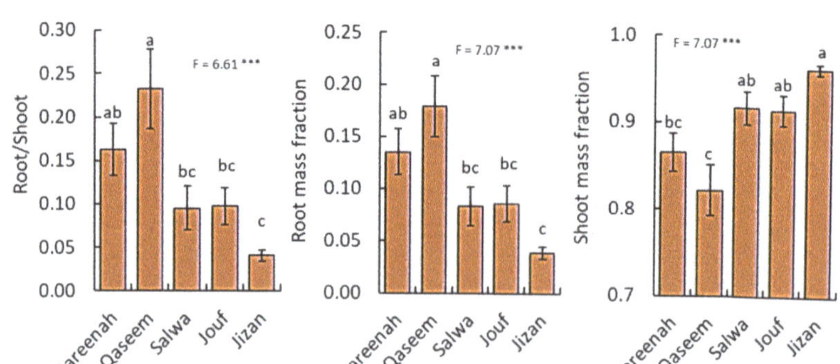

Figure 7. Biomass allocation among *Aeluropus lagopoides* collected from different saline flat regions of Saudi Arabia, based on dry matter. Values are average ($n = 75$), and the bar represents the standard error. Different letters among regions showed significant differences at $p < 0.05$ after Duncan's test. *** $p < 0.05$.

3.5. Correlation Analysis among Functional Plant Traits, Regions, and Soil Variables

3.5.1. Plant Functional Traits-Regions Correlations

The principal component analysis (PCA) revealed the existence of a close correlation between different morphological traits of *A. lagopoides* and the studied regions (Figure 8). The PCA revealed that the number of spikes and number of seeds per plant correlate with the number of tillers and leaves per plant (Figure 8). The *A. lagopoides* growing in Qaseem and Salwa regions are closely correlated and showed a positive correlation with root and shoot biomass. The root traits (root biomass, root area, and root length) are separated on the upper-left side of the PCA biplot, where it showed correlations to each other as well as with shoot biomass. Spike length showed a correlation with the shoot length. The *A. lagopoides* population of the Jizan region showed a substantial correlation with the leaf biomass, spike biomass, and stolon length, where spike biomass revealed a correlation with the leaf biomass as well as the stolon length (Figure 8). The Jouf region showed a significant positive correlation with specific leaf area, number of leaves per plant, number of tillers per plant, and number of seeds per plant (Figure 8). However, leaf dry matter content was the only morphological trait closely correlated to the Qareenah region.

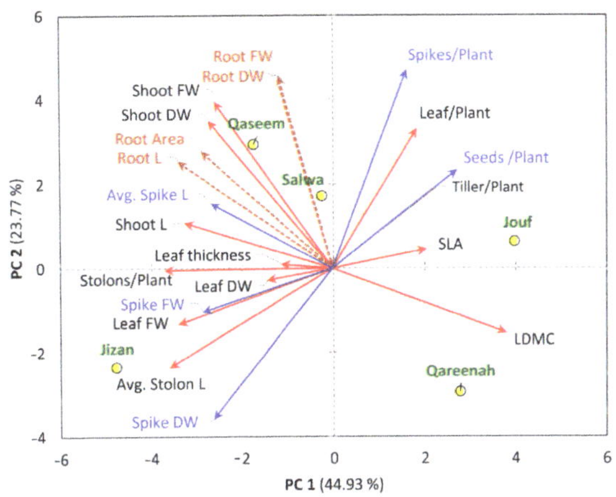

Figure 8. Principal component analysis (PCA) of the measured traits (shoot, represented with red arrows, root represented with brown arrows, reproductive traits represented with blue arrows of *Aeluropus lagopoides* within different saline flat regions (represented with yellow circle) of Saudi Arabia.

3.5.2. Correlations among Soil Variables, Plant Functional Traits, and Regions

The data of soil variables of each layer of each region and the functional traits were correlated using CCA (Figure 9). In general, Qassem and Salwa regions show a close correlation to each other, while Jizan is segregated alone on the lower-right side of the CCA biplot. Jouf region was different from other regions for the soil profile of the three layers (upper, middle, and lower) and separated on the lower-left side of the CCA biplot. Finally, the Qareenah region was separated at the center of the CCA biplot, revealing no specific correlation to any parameters.

Regarding the top layer of the soil (0–15 cm), the Qassem and Salwa regions showed a close correlation to most of the soil parameters that are correlated together, such as moisture content, pH, salinity, organic matter, Na, Cl, Mg, K, and Ca. (Figure 9a). Jizan region showed a close correlation to sulfate content, where it showed a correlation to leaf and spike biomass traits. The soil of the top layer in the Jouf region is different and separated on the lower-left side of the CCA biplot, where it showed a correlation to calcium carbonate and sand contents, and it showed a negative correlation with all morphological traits (Figure 9a).

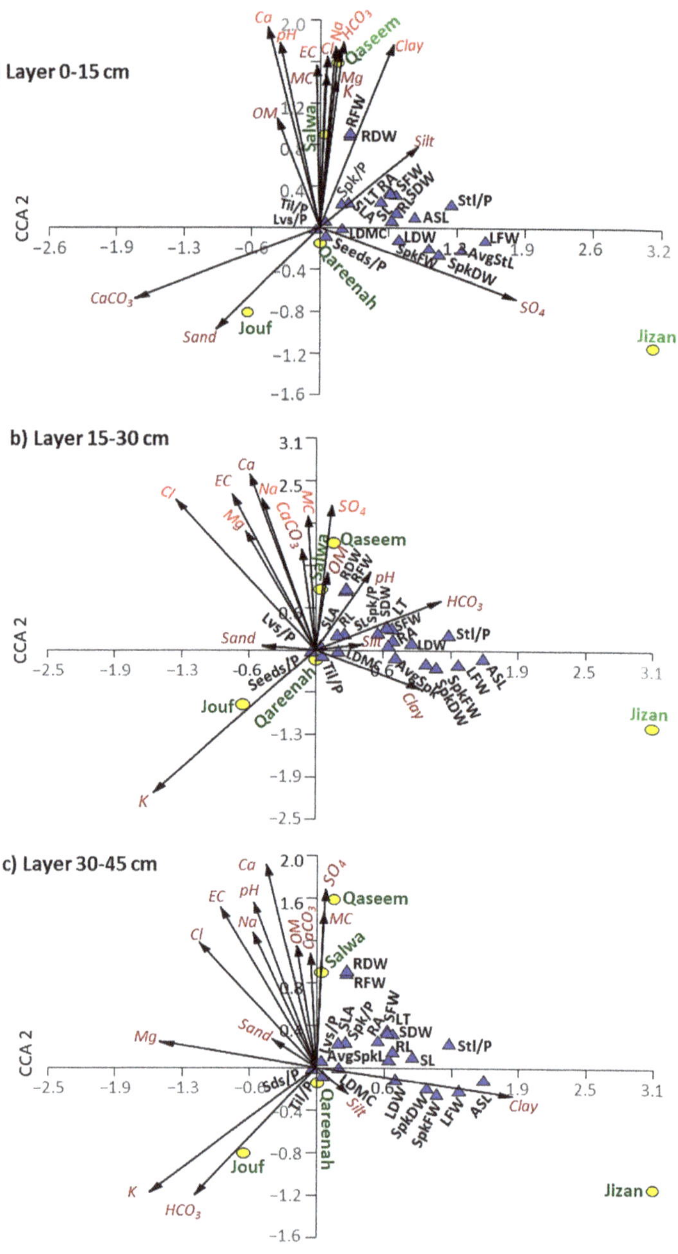

Figure 9. Canonical correspondence analysis (CCA) showing the correlations among the soil variables of different layers separately ((**a**): 0–15 cm, (**b**): 15–30 cm, and (**c**): 30–45 cm layers), regions, and morphological traits of *A. lagopoides*. SFW: shoot fresh weight, SDW: shoot dry weight, RFW: root fresh weight, RDW: root dry weight, lvs/*p*; number of leaves per plant, SLA: specific leaf area, Spk/*p*: number of spikes per plant, AvgSpkL: average spike length, RA: root area, LT: leaf thickness, SL: shoot length, RL: root length, Stl/*p*: number of stolon per plant, LDMC: leaf dry matter content, LFW: leaf fresh weight, LFW: leaf fresh weight, LDW: leaf dry weight, SpkFW: spike fresh weight, SpkDW: spike dry weight, ASL: average stolon length.

The heatmap correlation analysis, based on the soil data of the top layer, revealed that root biomass (root fresh weight and root dry weight) has a significant correlation with Ca^{2+}, Na^{+}, Cl^{-}, Clay, and moisture content, while root area showed a significant correlation to only clay content (Figure S3). On the other hand, specific leaf area showed a significant positive correlation with organic matter, while leaf dry matter content showed a correlation to calcium carbonates.

The leaf thickness revealed a significant positive correlation to Na and K contents. The spike biomass showed a significant correlation with sulfate content. However, leaf dry weight showed a significant negative correlation with K, bicarbonates, and organic matter (Figure S3).

For the middle layer of the soil (15–30 cm), the Qaseem and Salwa regions again showed a close correlation with moisture content, salinity, organic matter, Na, and Ca, while the Jizan region showed a correlation to the clay content. However, the Jouf region showed a close correlation to potassium ions but a negative correlation with all studied plant traits (Figure 9b). Pearson's correlation heatmap of the middle layer showed that root biomass significantly correlates with moisture content and sulfate (Figure S3). Moreover, the specific leaf area showed a significant positive correlation with organic matter like the top layer. The number of tillers and seeds per plant showed a significant correlation with the potassium ion. However, leaf biomass revealed a significant negative correlation with organic matter and calcium carbonates (Figure S3).

The PCA analysis of the lower layer of the soil (30–45 cm) revealed a different pattern compared to the upper and middle layers (Figure 9c). In this soil layer, the plant traits did not show a positive correlation to any soil parameters, except for clay, which correlated to the leaf biomass, spike biomass, and average stolon length of the *A. lagopoides* growing in Jizan (Figure 9c). Pearson's correlation heatmap of the lower layer showed a similar pattern to the middle layer (Figure S3).

3.6. Cluster Analysis of Regions Based on Soil and Plant Functional Traits

The hierarchical clustering for soil variables showed that Qareenah and Salwa regions are quite similar and showed a little pit correlation to the Jouf region (Figure 10A). However, the Jizan region differs in its soil characteristics from other regions. Regarding the plant functional traits, the Qareenah and Jizan regions are closely related, while Qaseem and Salwa regions showed a close correlation to each other (Figure 10C). However, the Jouf region is unique in the functional traits of *A. lagopoides*.

The combination of clustering with heatmap analysis revealed that the *A. lagopoides* populations growing in the Jouf Region showed a negative correlation with all the soil variables except Na^{+1} and pH (Figure 10B), while the Salwa region showed a positive correlation with organic matter and chloride. On the other hand, the heatmap analysis of morphological traits revealed that *A. lagopoides* populations growing in the Qareenah region showed a negative correlation with spike dry weight and fresh weight. In contrast, the Jizan region was positively correlated with root length and fresh weight. The Salwa region revealed a positive correlation with spike length (Figure 10D).

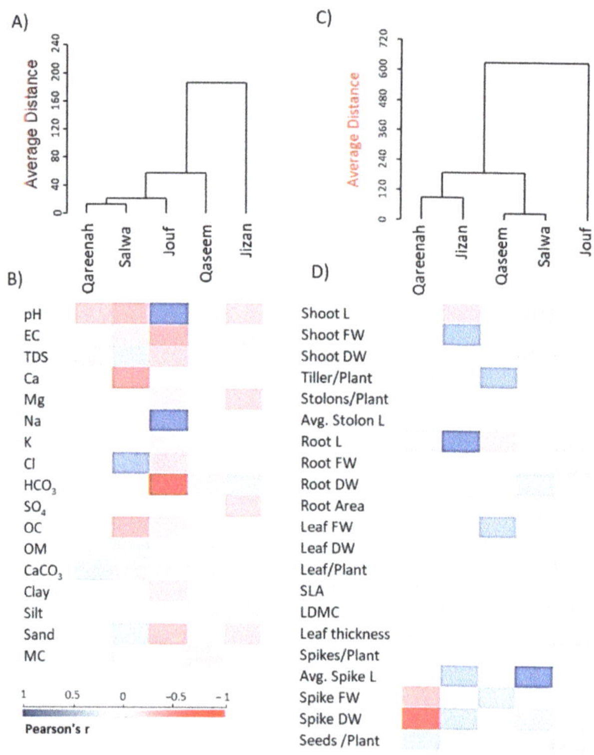

Figure 10. Agglomerative hierarchical clustering (AHC) and heatmaps of the studied parameters within different saline flat regions of *Aeluropus lagopoides*. (**A**) AHC and (**B**) heatmap based on the soil variables, (**C**) AHC and (**D**) heatmap based on the morphological and reproductive traits. EC: electrical conductivity, OM: organic matter, SLA: specific leaf area, LDMC: leaf dry matter content.

4. Discussion

Desert vegetation faces various ecological constraints like high temperature, soil salinity, and low soil moisture due to low precipitation, making the desert region a challenging environment for plants to grow [67]. Under stressful environments, desert grasses show specific structural and functional modifications in morphological and physiological characteristics to thrive well in such harsh environments [68]. The present study revealed that the various saline flat areas inhabited by *A. lagopoides* differed significantly in soil physicochemical characteristics from region to region as well as with soil depth (up to 30 cm), i.e., among layers (Table 1). These edaphic factors shaped these study sites' community structure and species association [55]. Salt stress can cause a reduction in water potential in soil and can induce osmotic stress in plants [69]. The structural and functional mechanisms of differently adapted populations of a desert halophyte (*A. lagopoides*) were studied for its survival and growth in hyper-arid-saline environments. When species undergo specific drought events and variable edaphic factors like soil salinity and high pH, they restrict their growth by utilizing energy for survival rather than further growth and development [70,71]. The continuity of environmental effects and the distance between the studied regions may support the likelihood for specific characteristics to become fixed in this grass over time.

Soil physical and chemical parameters of the habitats of all these five studied regions were significantly different, indicating the adaptive potential of *A. lagopoides* to cope with variable environmental conditions. Thus *A. lagopoides* plant faces the dual environmental stress of salt and water scarcity. Most of the soil physio-chemical characteristics values like salinity, pH, organic matter, cations, and some anions like Cl, and HCO_3, in all three soil layers of the inland saline flat area of the Qaseem region, were high, followed by the inland saline flat area of the Qareenah region (Table 1). The coastal Salwa and Jouf regions were moderately saline, and the least saline was the coastal saline flat area of the Jizan region (Table 1). The soil salinity, pH, cations, and anions were highest in the top soil layer and decreased significantly with depth, where this observation could be attributed to the high evaporation rate [72]. However, the cations and anions increase in value as soil depth increases in the Jouf region which could be ascribed to waterlogging at the site of the Jouf region [73].

Based on these soil characteristic variations, *A. lagopoides* evolved independently to these different salt levels among regions and responded quite differently relating to their morphological and physiological parameters (Figures 2–6). In this study, the highly saline Qaseem region (25.95 dS·m^{-1}) has the stunted growth of aerial parts of *A. lagopoides* (shoot length, stolon length, leaf biomass). This could be due to the high salinity level in the soil of this region. The same less shoot height was reported in *Aeluropuslittoralis* under different salinity [74]. This growth restriction of the aerial part of the plant is an essential morphological adaptation because a short-statured plant may conserve the energy required for vital metabolic processes [75]. In contrast, *A. lagopoides* growing in an inland saline flat area of the Qaseem region had improved root development, such as increased root length, high root fresh weight, root dry weight, and more root area than the low saline Jizan region. The investment in root development is an essential line of defense against salt stress [76] and determines the capacity of the plants to obtain water and nutrients [77,78]. Generally, root parameters increase under salinity in most halophytic species [79], while the opposite is true for glycophytic and less salt-tolerant species [80]. *A. lagopoides*, an indicator species of highly saline soils, grow well and uses Na^+ for many physiological processes [81,82]. The well-developed root system of *A. lagopoides* in highly saline habitats may have provided additional benefits to this plant under physiological drought in extracting moisture from the deeper soil layer, a common phenomenon in plants subjected to limited water availability [83]. This observation is supported by data on biomass allocation, where the *A. lagopoides* growing in the Qaseem region attained the highest root/shoot ratio as well as the root mass fraction (Figure 7). This reflects that when *A. lagopoides* is subjected to more salinity, it invests more energy in root development compared to shoot.

On the other side, most aerial parts like shoot length, shoot fresh weight, and shoot dry weight of the Jouf region were stunted, and the soil was moderately less saline than in the Qaseem region (Figure 2). This may be due to the combined effect of salinity and drought (low soil moisture content). Previous studies also reported reduced stem elongation of *Abies alba* [84] under saline and drought-prone environments. The number of tillers in *A. lagopoides* in the Jouf region is significantly more than in other regions. This, again, may be due to the low soil salinity of the Jouf region. The high osmotic stress of the salt outside the roots reduces the formation rate of new leaves and tiller productions [85] in moderate to high saline regions. *A. lagopoides* tends to produce thick leaves with low fresh and dry weight and low LDMC in the Qaseem region (Figure 4). This may be due to the highest soil salt content and the strongest degree of salinization in this habitat. The leaves of *A. lagopoides* became fleshy and developed a lot of water storage palisade tissues and water transport tissues [86]. Firstly, a fleshy leaf structure can also dilute the cell salt ion concentration to avoid its toxic effect. Secondly, it can also increase vacuole concentration and decrease water potential via ion regionalization, thus alleviating the water stress caused by salt stress [87]. Therefore, forming thick leaves in highly saline habitats may be a survival strategy for inland salt marsh plants to adapt to the harsh environment for a longer time.

Regarding reproductive traits, *A. lagopoides* showed a highly significant difference among the studied regions. Under the more stressful condition, like in Qaseem and Salwa, *A. lagopoides* produce more spikes while showing low spikes in biomass and seeds per plant, compared to the lowest saline habitats, such as Jouf (Figure 5). This could be explained as the plant, under stressful conditions, invests more in seed production and does not make an effort to produce spikes, which is a strategy to maintain more seeds that is more important to the survival of the species in harsh environments [88]. The spike number and inflorescence biomass of *Spartina alterniflora* have been reported to be decreased with increasing salinity [89]. However, the Jouf region showed maximum production of spikes and seeds per plant, while Jizan, in spite of the lowest production, attained the highest value for fresh and dry weight. This could be due to high spike and seed size and mass which the plants tradeoff for spike number and seeds based on the resources available in the habitats they adapt. Plasticity in reproductive investment is also an important trait in varying environments because changes in spike length, seed number, weight, and size directly influence plant fitness [90].

Different levels of salinity adversely influence stomatal conductance (g_s). *A. lagopoides* exhibited declined stomatal conductance (g_s) with increasing salinity (Figure 6). It depicts that stomatal conductance is an effective strategy to prevent water loss for maintaining the normal function of photosynthetic activity under saline conditions. However, stomatal conductance decreased in the low saline and low moisture content Jouf region. The decrease in stomatal conductance could reduce water loss, which is an adaptation mechanism by plants in dry soil conditions [91]. The highest efficiency of the PSII photochemistry (*Fv/Fm*) method has been extensively used to detect plant stress differences in response to environmental challenges and, consequently, to screen tolerance levels to environmental stress [92]. The *Fv/Fm* of *A. lagopoides* did not show a significant difference among the studied regions, while all values were under 0.79 (Figure 6), meaning that plants are under stress conditions.

Under limited resources, *A. lagopoides* improves its fitness by balancing biomass allocation between aboveground and belowground plant parts and synergistic morphological variation between shoot and root systems. In variable environments, plants' developmental traits and biomass allocation strategies are responses toward morphological characteristics of a plant's location adaptation to resource heterogeneity [93]. Vegetative (especially leaves) and roots are essential for plants to acquire resources. Plant morphologically changes with the environmental gradient to obtain most of the resources and strategies ecologically to adapt to environmental changes [94]. Under both water and salt stress in the Qaseem region, the roots of *A. lagopoides* adopted a strategy of root development and expansion to obtain resources to improve their adaptive ability. Thus *A. lagopoides* formed a good root architecture and produced a well-developed network of fibrous roots by increasing the root area and increasing root biomass. Our results are in agreement with the conclusion that an increase in soil salt concentration increased root development [95].

Overall, the present results demonstrated that the morphological architecture and biomass allocation of *A. lagopoides* are significantly affected in different saline flat area regions based on habitat heterogeneity vis a vis moisture content and salinity as a strategy for adaptation to harsh environments.

5. Conclusions

The morphological, reproductive, and physiological traits of *A. lagopoides* in the present study show plasticity with the change in the environmental conditions of saline flat area habitats. The Regions with high salinity, such as Qaseem and Salwa, showed the highest values of most of the shoot and root traits. However, the population of *A. lagopoides* in Qaseem and Salwa showed more spikes and lower spikes in biomass and seeds per plant compared to the lowest saline habitats, such as Jouf. Under stressful conditions, i.e., high salinity, the grass produces more seeds instead of spike biomass or other morphological traits. This plasticity reflects the strategy of *A. lagopoides* to cope with the harsh/saline environment. The ability of *A. lagopoides* to change its morphology with the variations in the environmental conditions enables it to colonize, dominate, and shape the community structure within the salt marsh habitat of different regions. The data on biomass allocation in the present study revealed that *A. lagopoides* invests more energy toward roots than shoots under stressful conditions. Due to the extensive fibrous root network, this plant could be a promising candidate as a soil stabilizer in saline flat areas during the summer season. Based on our data, we can conclude that the population of *A. lagopoides* shows great potential to rehabilitate the saline habitats of inland and coastal saline flat regions, taking saline agriculture, saline soil remediation, and stabilization into consideration, particularly this grass flourished in the dry summer season when these habitats are devoid of forage vegetation. Further study is recommended to evaluate the transplantation of this promising forage grass on a large scale in saline rangeland habitats degraded due to heavy grazing of a few palatable halophytic species.

Supplementary Materials: The following supporting information can be downloaded at: https://www.mdpi.com/article/10.3390/biology12040553/s1, Table S1: Geographical addresses of distinct patches of *A. lagopoides* populations of the studied regions of Saudi Arabia, along with yearly climatic data. Figure S1: Map of Saudi Arabia showing sampled regions of *Aeluropus lagopoides* populations; Figure S2: Pairwise geographical distance between the studied regions; Figure S3. Pearson's Correlation heatmap between the top, middle, and bottom layer soil parameters and the different morphological and reproductive traits of *Aeluropus lagopoides* within different saline flat regions.

Author Contributions: Conceptualization, A.M.A., B.A.D. and A.M.A.-E.; formal analysis, A.M.A., B.A.D. and A.M.A.-E.; investigation, A.M.A., A.M.A.-E., S.L.A.-R., A.A.A.-D., J.A.M. and B.A.D.; writing—original draft preparation, B.A.D. and A.M.A.-E.; writing—review and editing, A.M.A., B.A.D., S.L.A.-R., A.A.A.-D., J.A.M. and A.M.A.-E.; visualization, A.M.A.-E. and B.A.D. All authors have read and agreed to the published version of the manuscript.

Funding: This research was supported by The Researchers Supporting Project number (RSPD2023R676) King Saud University, Riyadh, Saudi Arabia.

Institutional Review Board Statement: Not applicable.

Informed Consent Statement: Not applicable.

Data Availability Statement: Not applicable.

Acknowledgments: The authors extend their appreciation to The Researchers Supporting Project number (RSPD2023R676) King Saud University, Riyadh, Saudi Arabia.

Conflicts of Interest: The authors declare no conflict of interest.

References

1. Nicotra, A.B.; Atkin, O.K.; Bonser, S.P.; Davidson, A.M.; Finnegan, E.J.; Mathesius, U.; Poot, P.; Purugganan, M.D.; Richards, C.L.; Valladares, F. Plant phenotypic plasticity in a changing climate. *Trends Plant Sci.* **2010**, *15*, 684–692. [CrossRef] [PubMed]
2. Hufford, K.M.; Mazer, S.J. Plant ecotypes: Genetic differentiation in the age of ecological restoration. *Trends Ecol. Evol.* **2003**, *18*, 147–155. [CrossRef]
3. Kawecki, T.J.; Ebert, D. Conceptual issues in local adaptation. *Ecol. Lett.* **2004**, *7*, 1225–1241. [CrossRef]
4. Ren, L.; Guo, X.; Liu, S.; Yu, T.; Guo, W.; Wang, R.; Ye, S.; Lambertini, C.; Brix, H.; Eller, F. Intraspecific variation in *Phragmites australis*: Clinal adaption of functional traits and phenotypic plasticity vary with latitude of origin. *J. Ecol.* **2020**, *108*, 2531–2543. [CrossRef]
5. Scheiner, S.M.; Barfield, M.; Holt, R.D. The genetics of phenotypic plasticity XVII. Response to climate change. *Evol. Appl.* **2020**, *13*, 388–399. [CrossRef]
6. Savolainen, O.; Pyhäjärvi, T.; Knürr, T. Gene flow and local adaptation in trees. *Annu. Rev. Ecol. Evol. Syst.* **2007**, *38*, 595–619. [CrossRef]
7. Schmitt, J. Reaction norms of morphological and life-history traits to light availability in *Impatiens capensis*. *Evolution* **1993**, *47*, 1654–1668.
8. Dudley, S.A.; Schmitt, J. Testing the adaptive plasticity hypothesis: Density-dependent selection on manipulated stem length in *Impatiens capensi*. *Am. Nat.* **1996**, *147*, 445–465. [CrossRef]
9. Sultan, S.E. Phenotypic plasticity in plants: A case study in ecological development. *Evol. Dev.* **2003**, *5*, 25–33. [CrossRef]
10. Crick, J.; Grime, J. Morphological plasticity and mineral nutrient capture in two herbaceous species of contrasted ecology. *New Phytol.* **1987**, *107*, 403–414. [CrossRef]
11. Hester, M.W.; Mendelssohn, I.A.; McKee, K.L. Intraspecific variation in salt tolerance and morphology in the coastal grass *Spartina patens* (Poaceae). *Am. J. Bot.* **1996**, *83*, 1521–1527. [CrossRef]

12. Florin, A.-B.; Höglund, J. Absence of population structure of turbot (*Psetta maxima*) in the Baltic Sea. *Mol. Ecol.* **2007**, *16*, 115–126. [CrossRef]
13. Hong-bo, S. Plant gene regulatory network system under abiotic stress. *Acta Biol. Szeged.* **2006**, *50*, 1–9.
14. Gulzar, S.; Khan, M.A.; Ungar, I.A. Effects of salinity on growth, ionic content, and plant–water status of *Aeluropus lagopoides*. *Commun. Soil Sci. Plant Anal.* **2003**, *34*, 1657–1668. [CrossRef]
15. Matesanz, S.; Gianoli, E. Global change and the evolution of phenotypic plasticity in plants. *Ann. N. Y. Acad. Sci.* **2014**, *1206*, 35–55. [CrossRef]
16. Valladares, F.; Gianoli, E.; Gómez, J.M. Ecological limits to plant phenotypic plasticity. *New Phytol.* **2007**, *176*, 749–763. [CrossRef]
17. Chevin, L.-M.; Hoffmann, A.A. Evolution of phenotypic plasticity in extreme environments. *Philos. Trans. R. Soc. B Biol. Sci.* **2017**, *372*, 20160138. [CrossRef]
18. Jacquemyn, H.; De Meester, L.; Jongejans, E.; Honnay, O. Evolutionary changes in plant reproductive traits following habitat fragmentation and their consequences for population fitness. *J. Ecol.* **2012**, *100*, 76–87. [CrossRef]
19. Gurkok, T.; Dolarslan, M.; Gul, E.; Turktas Erken, M. Adaptive Roles of Plant Development and Genome Structure against Soil Composition. *Fresenius Environ. Bull.* **2021**, *30*, 4031–4042.
20. Petrik, P.; Petek, A.; Konopkova, A.; Bosela, M.; Fleischer, P.; Frydl, J.; Kurjak, D. Stomatal and leaf morphology response of European Beech (*Fagus sylvatica* L.) provenances transferred to contrasting climatic conditions. *Forests* **2020**, *11*, 1359. [CrossRef]
21. Leimu, R.; Fischer, M. A meta-analysis of local adaptation in plants. *PLoS ONE* **2008**, *3*, e4010. [CrossRef] [PubMed]
22. Youssef, A.M. Salt tolerance mechanisms in some halophytes from Saudi Arabia and Egypt. *Res. J. Agric. Biol. Sci.* **2009**, *5*, 191–206.
23. Stanton-Geddes, J.; Shaw, R.G.; Tiffin, P. Interactions between soil habitat and geographic range location affect plant fitness. *PLoS ONE* **2012**, *7*, e36015. [CrossRef] [PubMed]
24. Bhatt, A.; Santo, A. Effects of photoperiod, thermoperiod, and salt stress on *Gymnocarpos decandrus* seeds: Potential implications in restoration ecology activities. *Botany* **2017**, *95*, 1093–1098. [CrossRef]
25. Elnaggar, A.; El-Keblawy, A.; Mosa, K.A.; Navarro, T. Adaptive drought tolerance during germination of *Salsola drummondii* seeds from saline and nonsaline habitats of the arid Arabian deserts. *Botany* **2019**, *97*, 123–133. [CrossRef]
26. El-Keblawy, A.; Al-Shamsi, N.; Mosa, K. Effect of maternal habitat, temperature and light on germination and salt tolerance of *Suaeda vermiculata*, a habitat-indifferent halophyte of arid Arabian deserts. *Seed Sci. Res.* **2018**, *28*, 140–147. [CrossRef]
27. Miner, B.G.; Sultan, S.E.; Morgan, S.G.; Padilla, D.K.; Relyea, R.A. Ecological consequences of phenotypic plasticity. *Trends Ecol. Evol.* **2005**, *20*, 685–692. [CrossRef]
28. Rúa, M.A.; Antoninka, A.; Antunes, P.M.; Chaudhary, V.B.; Gehring, C.; Lamit, L.J.; Piculell, B.J.; Bever, J.D.; Zabinski, C.; Meadow, J.F. Home-field advantage? Evidence of local adaptation among plants, soil, and arbuscular mycorrhizal fungi through meta-analysis. *BMC Evol. Biol.* **2016**, *16*, 1–15. [CrossRef]
29. Diekmann, M.; Müller, J.; Heinken, T.; Dupré, C. Wiederansiedlungen von Gefäßpflanzenarten in Deutschland—Eine Übersicht und statistische Auswertung. *Tuexenia* **2015**, *35*, 249–265.
30. Bolnick, D.I.; Amarasekare, P.; Araújo, M.S.; Bürger, R.; Levine, J.M.; Novak, M.; Rudolf, V.H.W.; Schreiber, S.J.; Urban, M.C.; Vasseur, D.A. Why intraspecific trait variation matters in community ecology. *Trends Ecol. Evol.* **2011**, *26*, 183–192. [CrossRef]
31. Chapin, F.S. Integrated responses of plants to stress. *BioScience* **1991**, *41*, 29–36. [CrossRef]
32. Shipley, B.; De Bello, F.; Cornelissen, J.H.C.; Laliberté, E.; Laughlin, D.C.; Reich, P.B. Reinforcing loose foundation stones in trait-based plant ecology. *Oecologia* **2016**, *180*, 923–931. [CrossRef]
33. Poorter, H.; Niklas, K.J.; Reich, P.B.; Oleksyn, J.; Poot, P.; Mommer, L. Biomass allocation to leaves, stems and roots: Meta-analyses of interspecific variation and environmental control. *New Phytol.* **2012**, *193*, 30–50. [CrossRef]
34. Pallas, B.; Da Silva, D.; Valsesia, P.; Yang, W.; Guillaume, O.; Lauri, P.-E.; Vercambre, G.; Génard, M.; Costes, E. Simulation of carbon allocation and organ growth variability in apple tree by connecting architectural and source–sink models. *Ann. Bot.* **2016**, *118*, 317–330. [CrossRef]
35. Mensah, S.; Kakaï, R.G.; Seifert, T. Patterns of biomass allocation between foliage and woody structure: The effects of tree size and specific functional traits. *Ann. For. Res.* **2016**, *59*, 49–60. [CrossRef]
36. Yin, Q.; Tian, T.; Han, X.; Xu, J.; Chai, Y.; Mo, J.; Lei, M.; Wang, L.; Yue, M. The relationships between biomass allocation and plant functional trait. *Ecol. Indic.* **2019**, *102*, 302–308. [CrossRef]
37. Ivanova, K.; Geneva, M.; Anev, S.; Georgieva, T.; Tzvetkova, N.; Stancheva, I.; Markovska, Y. Effect of soil salinity on morphology and gas exchange of two Paulownia hybrids. *Agrofor. Syst.* **2019**, *93*, 929–935. [CrossRef]
38. Hobohm, C.; Schaminée, J.; van Rooijen, N. Coastal Habitats, Shallow Seas and Inland Saline Steppes: Ecology, Distribution, Threats and Challenges. In *Perspectives for Biodiversity and Ecosystems*; Hobohm, C., Ed.; Springer International Publishing: Cham, Switzerland, 2021; pp. 279–310.
39. Goodall, T.M.; Al-Belushi, J.D. A Glossary of Arabic Desert Terminology Used in Southeastern Arabia. In *Quaternary Deserts and Climatic Change*; CRC Press: Boca Raton, FL, USA, 2020; pp. 611–619.
40. Lõhmus, K.; Balke, T.; Kleyer, M. Spatial and temporal patterns of initial plant establishment in salt marsh communities. *J. Veg. Sci.* **2020**, *31*, 1122–1132. [CrossRef]
41. Westerband, A.C.; Funk, J.L.; Barton, K.E. Intraspecific trait variation in plants: A renewed focus on its role in ecological processes. *Ann. Bot.* **2021**, *127*, 397–410. [CrossRef]

42. Weigelt, A.; Mommer, L.; Andraczek, K.; Iversen, C.M.; Bergmann, J.; Bruelheide, H.; Fan, Y.; Freschet, G.T.; Guerrero-Ramírez, N.R.; Kattge, J. An integrated framework of plant form and function: The belowground perspective. *New Phytol.* **2021**, *232*, 42–59. [CrossRef]
43. Pennings, S.C.; Richards, C.L. Effects of wrack burial in salt-stressed habitats: *Batis maritima* in a southwest Atlantic salt marsh. *Ecography* **1998**, *21*, 630–638. [CrossRef]
44. Pennings, S.C.; Bertness, M.D. Salt marsh communities. *Mar. Community Ecol.* **2001**, *11*, 289–316.
45. Richards, C.L.; Pennings, S.C.; Donovan, L.A. Habitat range and phenotypic variation in salt marsh plants. *Plant Ecol.* **2005**, *176*, 263–273. [CrossRef]
46. Flowers, T.J.; Colmer, T.D. Plant salt tolerance: Adaptations in halophytes. *Ann. Bot.* **2015**, *115*, 327–331. [CrossRef]
47. Gul, B.; Ansari, R.; Flowers, T.J.; Khan, M.A. Germination strategies of halophyte seeds under salinity. *Environ. Exp. Bot.* **2013**, *92*, 4–18. [CrossRef]
48. Zahran, M.A.; Murphy, K.J.; Mashaly, I.A.; Khedr, A.A. On the ecology of some halophytes and psammophytes in the Mediterranean coast of Egypt. *Verh.-Ges. Fur Okologie* **1996**, *25*, 133–146.
49. Podda, L.; Santo, A.; Leone, C.; Mayoral, O.; Bacchetta, G. Seed germination, salt stress tolerance and seedling growth of Opuntia *ficus-indica* (Cactaceae), invasive species in the Mediterranean Basin. *Flora* **2017**, *229*, 50–57. [CrossRef]
50. Azad, M.S.; Mollick, A.S.; Ranon, R.J.K.; Khan, M.N.I.; Kamruzzaman, M. Plasticity of leaf morphology of *Bruguiera sexangula* to salinity zones in Bangladesh's Sundarbans. *J. For. Res.* **2022**, *33*, 1857–1866. [CrossRef]
51. Yao, B.; Zhao, C.; Deng, J.; Zhou, H.; Zhao, X.; Liu, J. Phenotypic plasticity of *Thellungiella salsaginea* in response to saline stress. *Evol. Ecol. Res.* **2013**, *15*, 829–846.
52. Cope, T. *Flora of Pakistan, No. 143: Poaceae*; Pakistan Agricultural Research Council Islamabad and University of Karachi: Islamabad, Pakistan, 1982; p. 678.
53. Mandaville, J.P. *Flora of Eastern Saudi Arabia*; Routledge: Abingdon, UK, 2013.
54. Alfarhan, A.H.; Al-Turki, T.A.; Basahy, A.Y. Flora of Jizan region. *Final Rep. Support King Abdulaziz City Sci. Technol.* **2005**, *1*, 545.
55. Dar, B.A.; Assaeed, A.M.; Al-Rowaily, S.L.; Al-Doss, A.A.; Abd-ElGawad, A.M. Vegetation composition of the halophytic grass *Aeluropus lagopoides* communities within coastal and inland sabkhas of Saudi Arabia. *Plants* **2022**, *11*, 666. [CrossRef]
56. Abd-ElGawad, A.M.; Assaeed, A.M.; Al-Rowaily, S.L.; Dar, B.M.; Malik, J.A. Moisture and Salinity Drive the Vegetation Composition of Wadi Hargan, Riyadh, Saudi Arabia. *Diversity* **2021**, *13*, 587. [CrossRef]
57. Phondani, P.C.; Bhatt, A.; Elsarrag, E.; Alhorr, Y.M.; El-Keblawy, A. Criteria and indicator approach of global sustainability assessment system for sustainable landscaping using native plants in Qatar. *Ecol. Indic.* **2016**, *69*, 381–389. [CrossRef]
58. Kumar, A.; Kumar, A.; Lata, C.; Kumar, S. Eco-physiological responses of *Aeluropus lagopoides* (grass halophyte) and *Suaeda nudiflora* (non-grass halophyte) under individual and interactive sodic and salt stress. *South Afr. J. Bot.* **2016**, *105*, 36–44. [CrossRef]
59. Bowles, J.E. *Engineering Properties of Soils and Their Measurement*; McGraw-Hill, Inc.: New York, NY, USA, 1992.
60. Rowell, D. *Soil Science: Methods and Applications*; Longman Group: Harlow, UK, 1994.
61. Rhoades, J. Methods of Soil Analysis. Part 2, Chemical and Microbiological Properties. In *Soluble Salts*; Amer Society of Agronomy: Madison, Wisconsin, 1982; pp. 167–179.
62. Vile, D.; Garnier, E.; Shipley, B.; Laurent, G.; Navas, M.-L.; Roumet, C.; Lavorel, S.; Díaz, S.; Hodgson, J.G.; Lloret, F. Specific leaf area and dry matter content estimate thickness in laminar leaves. *Ann. Bot.* **2005**, *96*, 1129–1136. [CrossRef]
63. Perez-Harguindeguy, N.; Díaz, S.; Garnier Eric, L.S.; Poorter, H.; Jaureguiberry, P.; Bret-Harte, M.; Cornwelf, W.; Craine, J.; Gurvich, D.; Urcelay, C. New handbook for standardized measurement of plant functional traits worldwide. *Aust. J. Bot.* **2013**, *61*, 167–234. [CrossRef]
64. Maxwell, K.; Johnson, G.N. Chlorophyll fluorescence—A practical guide. *J. Exp. Bot.* **2000**, *51*, 659–668. [CrossRef]
65. Lichtenthaler, H.K.; Wellburn, A.R. Determinations of total carotenoids and chlorophylls a and b of leaf extracts in different solvents. *Biochem. Soc. Trans.* **1983**, *11*, 591–592. [CrossRef]
66. Kovach, W. *MVSP-A Multivariate Statistical Package for Windows, ver. 3.1*; Kovach Computing Services: Pentraeth, Anglesey, 1999; p. 137.
67. Weber, D. Adaptive mechanisms of halophytes in desert regions. In *Salinity and Water Stress*; Springer: Dordrecht, Netherlands, 2009; pp. 179–185.
68. Mansoor, U.; Fatima, S.; Hameed, M.; Naseer, M.; Ahmad, M.S.A.; Ashraf, M.; Ahmad, F.; Waseem, M. Structural modifications for drought tolerance in stem and leaves of *Cenchrus ciliaris* L. ecotypes from the Cholistan Desert. *Flora* **2019**, *261*, 151485. [CrossRef]
69. Kang, S.; Zhang, J. Controlled alternate partial root-zone irrigation: Its physiological consequences and impact on water use efficiency. *J. Exp. Bot.* **2004**, *55*, 2437–2446. [CrossRef]
70. Nalina, M.; Saroja, S.; Chakravarthi, M.; Rajkumar, R.; Radhakrishnan, B.; Chandrashekara, K. Water deficit-induced oxidative stress and differential response in antioxidant enzymes of tolerant and susceptible tea cultivars under field condition. *Acta Physiol. Plant.* **2021**, *43*, 1–17. [CrossRef]
71. Ji, Y.; Zhou, G.; New, T. Abiotic factors influencing the distribution of vegetation in coastal estuary of the Liaohe Delta, Northeast China. *Estuaries Coasts* **2009**, *32*, 937–942. [CrossRef]
72. Grunstra, M.; Van Auken, O. Using GIS to display complex soil salinity patterns in an inland salt marsh. *Dev. Environ. Sci.* **2007**, *5*, 407–431.

73. Lu, S.G.; Tang, C.; Rengel, Z. Combined effects of waterlogging and salinity on electrochemistry, water-soluble cations and water dispersible clay in soils with various salinity levels. *Plant Soil* **2004**, *264*, 231–245. [CrossRef]
74. Barzegargolchini, B.; Movafeghi, A.; Dehestani, A.; Mehrabanjoubani, P. Morphological and anatomical changes in stems of Aeluropus littoralis under salt stress. *J. Plant Mol. Breed.* **2017**, *5*, 40–48.
75. Munns, R.; Tester, M. Mechanisms of salinity tolerance. *Annu. Rev. Plant Biol.* **2008**, *59*, 651–681. [CrossRef]
76. Khan, W.-U.-D.; Tanveer, M.; Shaukat, R.; Ali, M.; Pirdad, F. An Overview of Salinity Tolerance Mechanism in Plants. In *Salt and Drought Stress Tolerance in Plants: Signaling Networks and Adaptive Mechanisms*; Hasanuzzaman, M., Tanveer, M., Eds.; Springer International Publishing: Cham, Switzerland, 2020; pp. 1–16.
77. Rubio, G.; Walk, T.; Ge, Z.; Yan, X.; Liao, H.; Lynch, J.P. Root gravitropism and below-ground competition among neighbouring plants: A modelling approach. *Ann. Bot.* **2001**, *88*, 929–940. [CrossRef]
78. Wang, W.; Ding, G.-D.; White, P.J.; Wang, X.-H.; Jin, K.-M.; Xu, F.-S.; Shi, L. Mapping and cloning of quantitative trait loci for phosphorus efficiency in crops: Opportunities and challenges. *Plant Soil* **2019**, *439*, 91–112. [CrossRef]
79. Khan, M.A.; Ungar, I.A.; Showalter, A.M. Effects of salinity on growth, ion content, and osmotic relations in *Halopyrum mucronatum* (L.) Stapf. *J. Plant Nutr.* **1999**, *22*, 191–204. [CrossRef]
80. Horie, T.; Karahara, I.; Katsuhara, M. Salinity tolerance mechanisms in glycophytes: An overview with the central focus on rice plants. *Rice* **2012**, *5*, 1–18. [CrossRef]
81. Pujol, J.A.; Calvo, J.F.; Ramírez-Díaz, L. Seed germination, growth, and osmotic adjustment in response to NaCl in a rare succulent halophyte from southeastern Spain. *Wetlands* **2001**, *21*, 256–264. [CrossRef]
82. Shabala, S.; Mackay, A. Ion transport in halophytes. In *Advances in Botanical Research*; Elsevier: New York, NY, USA, 2011; Volume 57, pp. 151–199.
83. Liu, H.-S.; Li, F.-M.; Xu, H. Deficiency of water can enhance root respiration rate of drought-sensitive but not drought-tolerant spring wheat. *Agric. Water Manag.* **2004**, *64*, 41–48. [CrossRef]
84. Todea, I.M.; Gonzalez-Orenga, S.; Boscaiu, M.; Plazas, M.; Sestras, A.F.; Prohens, J.; Vicente, O.; Sestras, R.E. Responses to water deficit and salt stress in silver fir (*Abies alba* Mill.) seedlings. *Forests* **2020**, *11*, 395. [CrossRef]
85. Munns, R.; James, R.A. Screening methods for salinity tolerance: A case study with tetraploid wheat. *Plant Soil* **2003**, *253*, 201–218. [CrossRef]
86. Naz, N.; Hameed, M.; Nawaz, T.; Batool, R.; Ashraf, M.; Ahmad, F.; Ruby, T. Structural adaptations in the desert halophyte *Aeluropus lagopoides* (Linn.) Trin. ex Thw. under high salinity. *J. Biol. Res.* **2013**, *19*, 150–164.
87. Zhao, K.F.; Li, F.Z.; Zhang, F.S. *Chinese Halophytes*, 2nd ed.; China Science Press: Beijing, China, 2013; pp. 71–74.
88. Hussain, S.; Shaukat, M.; Ashraf, M.; Zhu, C.; Jin, Q.; Zhang, J. Salinity stress in arid and semi-arid climates: Effects and management in field crops. In *Climate Change and Agriculture*; Hussain, S., Ed.; IntechOpen: London, UK, 2019; Volume 13.
89. Xiao, Y.; Tang, J.; Qing, H.; Zhou, C.; An, S. Effects of salinity and clonal integration on growth and sexual reproduction of the invasive grass *Spartina alterniflora*. *Flora* **2011**, *206*, 736–741. [CrossRef]
90. Volis, S. Plasticity, its cost, and phenotypic selection under water and nutrient stress in two annual grasses. *Biol. J. Linn. Soc.* **2009**, *97*, 581–593. [CrossRef]
91. Ilyas, M.; Nisar, M.; Khan, N.; Hazrat, A.; Khan, A.H.; Hayat, K.; Fahad, S.; Khan, A.; Ullah, A. Drought tolerance strategies in plants: A mechanistic approach. *J. Plant Growth Regul.* **2021**, *40*, 926–944. [CrossRef]
92. Baker, N.R.; Rosenqvist, E. Applications of chlorophyll fluorescence can improve crop production strategies: An examination of future possibilities. *J. Exp. Bot.* **2004**, *55*, 1607–1621. [CrossRef]
93. Kleyer, M.; Minden, V. Why functional ecology should consider all plant organs: An allocation-based perspective. *Basic Appl. Ecol.* **2015**, *16*, 1–9. [CrossRef]
94. Zhao, G.; Liu, M.; Shi, P.; Zong, N.; Zhang, X.; Zhang, X. Variation of leaf and root traits and ecological adaptive strategies along a precipitation gradient on Changtang Plateau. *Acta Ecol. Sin.* **2020**, *40*, 295–309.
95. Tan, J.; Ben-Gal, A.; Shtein, I.; Bustan, A.; Dag, A.; Erel, R. Root structural plasticity enhances salt tolerance in mature olives. *Environ. Exp. Bot.* **2020**, *179*, 104224. [CrossRef]

biology

MDPI

Article

Adaptability of Wild-Growing Tulips of Greece: Uncovering Relationships between Soil Properties, Rhizosphere Fungal Morphotypes and Nutrient Content Profiles

Fotis Bilias [1,*], Anastasia-Garyfallia Karagianni [1], Ioannis Ipsilantis [1], Ioulietta Samartza [2], Nikos Krigas [2,*], Georgios Tsoktouridis [2,3] and Theodora Matsi [1]

1 Soil Science Laboratory, School of Agriculture, Aristotle University of Thessaloniki, 54124 Thessaloniki, Greece
2 Institute of Plant Breeding and Genetic Resources, Hellenic Agricultural Organization Demeter, P.O. Box 60458, 57001 Thessaloniki, Greece
3 Theofrastos Fertilizers, Irinis & Filias, Examilia Korithias, 20100 Korinthos, Greece
* Correspondence: fbilias@agro.auth.gr (F.B.); nikoskrigas@gmail.com (N.K.); Tel.: +30-2310-471110 (N.K.)

Simple Summary: Greek wild-growing tulips are protected plants, about which there is scarce knowledge regarding their natural nutrient status and rhizosphere fungal morphotypes. In this study, we collected plant (above-ground and bulb material) and soil samples from 13 tulip species across three phytogeographical units in Greece, and we assessed the tulips' nutrient content and soil properties to determine their interrelationships. We found that soil variables significantly influenced tulip nutrient content, with up to 67% of the detected variability explained by soil properties. Correlation patterns were also found between tulips' essential nutrients. Our study revealed clear distinctions in nutrient content among tulip species from different spatial (phytogeographic) units. The findings shed light on Greek tulips' adaptability and resilience in their natural habitats and may facilitate their domestication in artificial settings.

Citation: Bilias, F.; Karagianni, A.-G.; Ipsilantis, I.; Samartza, I.; Krigas, N.; Tsoktouridis, G.; Matsi, T. Adaptability of Wild-Growing Tulips of Greece: Uncovering Relationships between Soil Properties, Rhizosphere Fungal Morphotypes and Nutrient Content Profiles. *Biology* **2023**, *12*, 605. https://doi.org/10.3390/biology12040605

Academic Editors: Daniel Puppe, Panayiotis Dimitrakopoulos and Baorong Lu

Received: 1 March 2023
Revised: 13 April 2023
Accepted: 14 April 2023
Published: 16 April 2023

Abstract: Wild-growing Greek tulips are protected plants but almost nothing is known about their natural nutrient status and rhizosphere fungal morphotypes in the wild, thus no insight is currently available into their growth and adaptation to their natural environment or artificial settings. To this end, several botanical expeditions were conducted with a special collection permit, and 34 tulip and soil samples were collected, representing 13 species from two phytogeographical regions of Greece (North Aegean Islands, Crete Island) and seven regions of mainland Greece. The tulips' content in essential macro- and micro-nutrients, respective physicochemical soil properties, and rhizosphere fungal morphotypes were assessed across samples, and all parameters were subjected to appropriate statistical analysis to determine their interrelationships. The results showed that soil variables played a significant role in shaping tulips' nutrient content, explaining up to 67% of the detected variability as in the case of phosphorus (P) in the above-ground plant tissue. In addition, significant correlations were observed (with an r value of up to 0.65, $p < 0.001$) between essential nutrients in the tulips, such as calcium (Ca) and boron (B). The principal component analysis (PCA) revealed that between the three spatial units examined, the total variability of tulips' nutrient content produced a clear distinction among sampled species, while the first two PCA axes managed to explain 44.3% of it. This was further confirmed by the analysis of variance (ANOVA) results which showed corresponding significant differences (at $p < 0.05$) in both the tulips' nutrient content and the studied soil properties as well (mean values of N, P, and K in the North Aegean Islands tulips' nutrient content, up to 53%, 119%, and 54% higher compared to those of the Crete Island, respectively). Our study sheds light on Greek tulips' adaptability and resilience in their original habitats, facilitating at the same time the undertaken efforts regarding their conservation and potential domestication in artificial settings.

Keywords: *Tulipa* spp.; elemental variability; plant nutrients; phytogeographical units; edaphic variations; arbuscular mycorrhizal fungi

1. Introduction

Tulips are famous ornamental plants worldwide with high economic importance, therefore, the demand for new botanical tulips with interesting features is constantly high [1]. For example, the high demand for Greek botanical tulips has been satisfied to date by 24 nurseries located in four countries (Greece excepted), and the relatively high prices of the traded materials may showcase an online global market [1]. In addition, there are 41 botanic gardens around the world involved in the ex situ conservation of almost half of the Greek tulip species [1]. In Greece, all wild-growing tulips are protected by the Greek Presidential Decree 67/1981, occurring as wild-growing plants either in the Greek mainland or some of the Greek islands. These phytogenetic resources include 15 different species of the genus *Tulipa*. Among them, six are restricted to Greece (single-country endemics), while five are local sub-endemics in the Balkan or the Aegean regions extending to neighboring countries, and three are considered naturalized alien species originating from Asia [2]. Seven of these species are considered as threatened according to the criteria established by the International Union for Conservation of Nature (IUCN); two of them are assessed as critically endangered (CR), three as endangered (EN), and two as vulnerable (VU), all suffering mainly due to over-collection, habitat degradation and land use changes [1,3–5].

In terms of their specific habitat requirements, the Greek tulips have evolved various adaptations to their local environments in the wider spatial (phytogeographical) units of the country where they thrive. Likewise, distinct morphological and ecological species-specific characteristics have also been developed due to the above-mentioned adaptation process [1,2]. Such adaptation strategies are supported by the combined action of multiple elements (nutrients) which, once they are taken up by the plant root system, are involved in various biochemical reactions dictated by species-specific genotypes with individual variations [6]. Among the latter, macro-nutrients such as nitrogen (N), phosphorus (P), potassium (K), calcium (Ca), and magnesium (Mg) are considered necessary components for performing various metabolic processes, for protecting plants from abiotic and biotic stresses, or for developing their body structure [6–8]. Similarly, micro-nutrients such as iron (Fe), manganese (Mn), copper (Cu), zinc (Zn), and boron (B), or beneficial elements such as sodium (Na) are involved in various plant functions such as osmoregulation and electrochemical reactions [6–8].

As in any plant species, the particularities reflected in the nutrient status of Greek tulip species or the balance of essential macro- and micro-nutrients in their above-ground tissues or bulbs can provide valuable insights into their growth regimes, development patterns, and adaptation strategies to their local environments [9]. For example, it has been well established that different plant species thriving in similar or even identical soil environments may strongly differ in their uptake, transport, or accumulation strategies of nutrients due to the high specificity that each species develop in terms of its ionomic homeostasis [9]. Accordingly, the differences that might arise between several biotic or abiotic factors of different ecological environments may largely influence the ionome or the nutrient status between different species or those of different plant individuals belonging to the same species [7,10].

In this line, among several environmental conditions that exist in the immediate vicinity of any terrestrial growing plant (tulip plants as well), soil properties are considered critical factors since they play a crucial role in determining the availability of plant nutrients, and eventually in regulating their adaptation strategy and reproduction prerequisites dictated by each genotype [11]. Thus, differences that might be observed in properties such as soil texture, $CaCO_3$ content, soil pH, electrical conductivity, or cation exchange capacity may reflect different patterns of nutrients accumulation or balance in different parts of the plants, since they generally play a critical role in shaping their overall ionome [12].

In addition to these abiotic factors, certain biotic factors such as microbes also impact the variations of elemental concentrations in plants thriving in their micro-environments [13–15]. Some of the most typical processes affected or triggered by soil microbial activity include the mineralization process of soil organic matter and the fixation of nitrogen through symbiotic microorganisms as well as the phosphorus or nitrogen acquisition by rhizosphere mycorrhizal fungi [8,16,17].

The domestication of wild plants with attractive characteristics is highly valued in the ornamental horticultural sector, especially when they possess rarity or endemism (uniqueness) [1,18]. Thus, a concerted effort to acquire specialized knowledge about the nutritional requirements of wild-growing Greek tulip species and the mycorrhizal fungi present in their rhizosphere could further improve their cultivation potential.

Since maintaining biodiversity is important for the overall health and sustainability of ecosystems [19], understanding the interaction of different species with biotic and abiotic factors in their environment is of crucial concern. To this end, nutrient cycling plays a major role in ecosystem functioning [20] and can act as an indicator of the health status of different ecosystems [21]. In this regard and considering that both natural and artificial landscapes are threatened by the ongoing and ever-increasing ecological crisis [19], it is essential to identify the properties of the habitats of different species for conservation management. Determining both biotic and abiotic conditions of various habitats can offer knowledge about the species' needs in terms of nutrients including interrelationships thereof with other living factors such as arbuscular mycorrhizal fungi or microbes [19,21]. Thus, establishing species-specific profiles could possibly provide insight into distribution patterns followed regarding species and/or habitats of conservation concern.

This study is focused on the protected wild-growing tulips of Greece (some of which are also threatened with extinction) for which almost nothing is known about their natural nutrient status, thus no insight is currently available into their growth and adaptation to their natural environment or artificial settings. Therefore, it should be considered that constitutes a research gap that needs to be investigated.

Our initial hypothesis was that the status of essential macro- and micro-nutrients in wild-growing tulips of Greece could offer valuable insights into their growth, reproduction, and stress tolerance experienced in their natural habitats. Furthermore, we examined how this overall variance is affected by different functional groups of tulips and analyzed their distribution across different spatial (phytogeographical) units. Therefore, we aimed to identify patterns or groupings that could enhance our understanding of their adaptation strategies to their local environment or specific nutritional requirements for different tulip groups, while we consider that this information brings novelty to conservation management research.

The objectives of our study were (i) to explore the nutrient content profiles of wild-growing Greek tulips analyzing different plant parts (above-ground biomass, and subterranean bulbs); (ii) to investigate their interrelationships with respective soil properties (physicochemical and microbial) as proxies of ecological conditions under which these tulips naturally thrive in the wild; and (iii) to decipher the role that different phytogeographical units may play in shaping differences on the nutritional status of various Greek tulip species. Based on the above-mentioned, we aimed to shed light on Greek tulips' adaptability and resilience in their original habitats, facilitating at the same time the undertaken efforts regarding their conservation and potential domestication in artificial settings.

2. Materials and Methods

2.1. Focal Greek Tulips

In this study, 13 (86.7%) of the Greek *Tulipa* species are included (Figures 1 and 2, Table 1). Plant nomenclature of the Greek tulip species studied herein follow the Flora of Greece web version IV (https://portal.cybertaxonomy.org/flora-greece/intro, accessed on 22 February 2023).

(**A**)—*T. undulatifolia* Boiss.

(**B**)—*T. doerfleri* Gand.

(**C**)—*T. hageri* Heldr.

(**D**)—*T. orhanidea* Boiss. ex Heldr.

(**E**)—*T. clusiana* Redouté

(**F**)—*T. aegenensis* DC.

(**G**)—*T. raddii* Reboul

(**H**)—*T. raddii* overcollected

(**I**)—*T. australis* Link

(**J**)—*T. saxatilis* Spreng.

(**K**)—*T. cretica* Boiss. & Heldr.

(**L**)—*T. bakeri* A.D. Hall

Figure 1. *Cont.*

(**M**)—*T. bithynica* Baker

(**N**)—*T. goulimyi* Sealy & Turrill (Peloponnese)

(**O**)—*T. goulimyi* (Kithyra Island)

Figure 1. Flowering individuals of mainly segetal (**A–H**) and rock-dwelling or non-segetal (**I–O**) wild-growing tulips of Greece employing different biogeographical groups such as Greek endemic ones (*T. bakeri*, *T. cretica*, *T. doerfleri*, *T. goulimyi*, *T. hageri*, *T. orphanidea*), sub-endemic ones extending to Turkey and/or the Balkans (*T. bithynica*, *T. saxatilis*, *T. undulatifolia*), Mediterranean (*T. australis*) and Asiatic ones naturalized in Greece (*T. aegenensis*, *T. clusiana*, *T. raddii*).

(**a**)

(**b**)

(**c**)

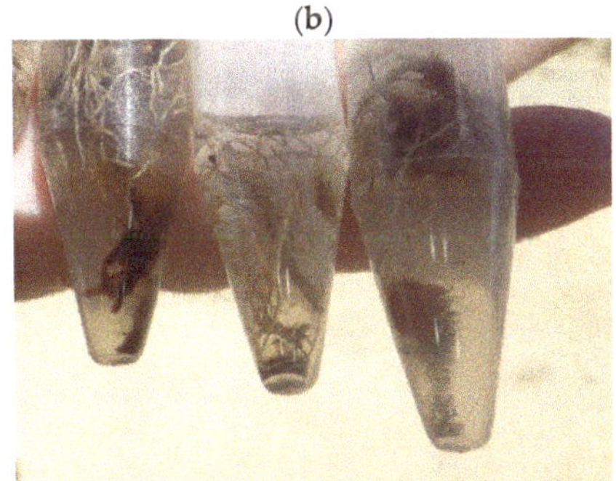

(**d**)

Figure 2. Examples of sampling materials from wild-growing tulip individuals of Greece collected with permission and studied herein: (**a**) propagation materials for ex situ conservation, soil and leaf samples for nutrient diagnostics, and root samples for mycorrhiza analysis collected from *Tulipa orphanidea*; (**b**) leaf samples of *Tulipa orphanidea*; (**c**) bulb samples of *Tulipa goulimyi*; and (**d**) Mycorrhiza samples from different individuals of *Tulipa bakeri*.

Table 1. Collection details and IPEN (International Plant Exchange Network) accession numbers associated with the living material samples regarding the *Tulipa* spp. from Greece studied herein. Different *Tulipa* spp. are arranged alphabetically. For each species, the letter in the parentheses denotes different tulip samples collected (1–6 samples per species).

Tulipa Species	IPEN Accession	Altitude (m)	Phytogeographical Unit of Greece	Latitude (North)	Longitude (East)
T. agenensis (a)	GR-BBGK-1-22,2	101	North Aegean Islands	38.33376	26.081543
T. agenensis (b)	GR-BBGK-1-22,15	99	North Aegean Islands	38.333713	26.081833
T. australis (a)	GR-BBGK-1-21,115	1554	Mainland	40.096552	21.114582
T. australis (b)	GR-BBGK-1-21,116	1856	Mainland	40.096552	21.114582
T. australis (c)	GR-BBGK-1-22,54	477	Mainland	37.809379	23.941191
T. bakeri (a)	GR-BBGK-1-21,8	1057	Crete	35.330758	23.896977
T. bakeri (b)	GR-BBGK-1-22,28	1065	Crete	35.331772	23.907389
T. bithynica (a)	GR-BBGK-1-22,19	870	North Aegean Islands	39.161205	26.066789
T. clusiana (a)	GR-BBGK-1-22,1	422	North Aegean Islands	38.304548	26.056388
T. cretica (a)	GR-BBGK-1-21,79	12	Crete	34.957198	25.099697
T. cretica (b)	GR-BBGK-1-21,83	57	Crete	35.549880	24.150415
T. cretica (c)	GR-BBGK-1- 22,7	131	Crete	35.551054	24.147271
T. cretica (d)	GR-BBGK-1-22,9	158	Crete	35.222128	26.212745
T. cretica (e)	GR-BBGK-1-22,21	974	Crete	34.957373	25.100580
T. doerfleri (a)	GR-BBGK-1-21,106	743	Crete	35.207935	24.560692
T. doerfleri (b)	GR-BBGK-1-22,33	773	Crete	35.213769	24.567137
T. goulimyi (a)	GR-BBGK-1-21,39	415	Mainland *	36.16800	22.96600
T. goulimyi (b)	GR-BBGK-1-21,4	411	Mainland *	36.2531000	22.9677000
T. goulimyi (c)	GR-BBGK-1-21,41	616	Mainland	36.8283000	22.9475000
T. goulimyi (d)	GR-BBGK-1-22,25	177	Crete	35.503031	23.754020
T. hageri (a)	GR-BBGK-1-21,67	516	Mainland	40.799011	23.389939
T. hageri (b)	GR-BBGK-1-21,102	428	Mainland	40.421873	23.694688
T. hageri (c)	GR-BBGK-1-22,55	544	Mainland	38.190514	23.794406
T. orphanidea	GR-BBGK-1-21,44	876	Mainland	37.30900	22.42200
T. raddii (a)	GR-BBGK-1-21,285	183	North Aegean Islands	38.28275	26.07461
T. raddii (b)	GR-BBGK-1-22,3	381	North Aegean Islands	38.319157	26.055302
T. saxatilis (a)	GR-BBGK-1-21,105	721	Crete	35.213210	24.564025
T. saxatilis (b)	GR-BBGK-1-22,8	518	Crete	35.419784	23.740193
T. saxatilis (c)	GR-BBGK-1-22,14	95	Crete	35.074496	25.806127
T, saxatilis (d)	GR-BBGK-1-22,2	599	Crete	35.176982	24.997673
T. saxatilis (e)	GR-BBGK-1-22,27	531	Crete	35.419784	23.740193
T. saxatilis (f)	GR-BBGK-1-22,43	830	Crete	35.1696440	25.4882520
T. undulatifolia (a)	GR-BBGK-1-22,4	25	North Aegean Islands	38.203085	26.030258
T. undulatifolia (b)	GR-BBGK-1-22,11	445	Mainland	38.090000	23.230000

* Kythira island is situated between mainland Greece and Crete but here is classified into mainland Greece due to stronger floristic similarities.

2.2. Collection of Samplings

Samples from about 5–10 wild-growing individuals of each studied species were collected during botanical expedition in 2020 and 2021 in the frame of the research project TULIPS.GR (Figure 2, Table 1). The samples included specimens for ex situ conservation, soil samples, above-ground leaf samples, and below-ground bulbs for nutrient analysis as well as samples for mycorrhiza analysis (Figure 2). All collections were performed using a special permission issued yearly (182336/879 of 16 May 2019, 64886/2959 of 6 July 2020 and 26895/1527 of 21 April 2021) by the national competent authority, namely the Greek Ministry of Environment and Energy.

2.3. Classification of Studied Species into Functional Groups

The sampled wild-growing Greek tulip species were classified into different functional groups (Table 2) based on species-specific biological data [2], such as taxonomic identity and phytogeographical distribution, chorological origin, ecological preferences in terms of habitat types and altitudinal range and categories thereof, and threatened status according to IUCN (International Union for the Conservation of Nature) extinction risk assessments [3,4]. The following functional groups of Greek tulips were outlined:

1. Tulips of mainland Greece and insular tulips of Greece (North Aegean tulips and Cretan tulips);
2. Alien naturalized tulips and wild-growing Greek native tulips;
3. Greek endemic tulips, sub-endemic tulips, and tulips of wider distribution range;
4. Non-threatened and threatened tulips (critically endangered, endangered and vulnerable);
5. Tulips of small altitudinal range (lowland tulips, tulips of intermediate altitudes), tulips of wide altitudinal range (tulips of lowland to intermediate altitudes, tulips of intermediate to high altitudes) and tulips of very wide altitudinal range (tulips occurring from lowlands to high altitudes);
6. Tulips associated with agricultural habitat types, tulips associated with natural habitat types, and tulips occurring in both agricultural and natural habitat types;
7. Rock-dwelling tulips and tulips not occurring in rocky habitats;
8. Segetal tulips and tulips not occurring in agricultural habitats;
9. Early flowering tulips, mid-spring flowering tulips, and late flowering tulips.

2.4. Soil and Plant Analysis

A total of 34 surface soil samples (0–30 cm) were collected in the sampling area in which the respective plant material (tulip above-ground biomass, or bulb) was also collected. Soil samples were air-dried and passed through a 2-mm sieve. Then, they were analyzed in triplicate for the properties described hereafter. Determination of soil texture and respective distribution in sand, silt, and clay particles was conducted by the hydrometer method [22]. Organic carbon (C) was determined by the wet oxidation method [23] and $CaCO_3$ was assessed using a calcimeter. The pH was measured in a 1:2 (*w/v*) water suspension, while the electrical conductivity of the soil solution was determined in the saturation extract (EC_{se}). The sodium absorption ratio (SAR), representing an intensity factor of Na in the soil solution, relative to Ca and Mg, was calculated as follows:

$$SAR = \frac{Na^{+}}{\sqrt{\frac{Ca^{2+}+Mg^{2+}}{2}}} \quad (1)$$

where each chemical element symbol indicates a concentration in millimoles of charge per liter ($mmol_c\ L^{-1}$) [24]. The hexamminecobalt(III) chloride ($[Co(NH_3)_6]Cl_3$) method (ISO 23470) was used to assess the cation exchange capacity (CEC).

Table 2. Overview of the functional groups of the sampled wild-growing Greek tulip species (N = 13) based on biological (taxonomical, phytogeographical, chorological, and ecological) data [2] and threatened status according to IUCN (International Union for the Conservation of Nature) extinction risk assessments [3,4]: CR—critically endangered; EN—endangered; VU—vulnerable.

Tulips (*Tulipa* spp.)	Phytogeographical Status	Chorological Status	Threatened Status	Altitudinal Class (Altitudinal Range in m)	Habitat Types	Rock-Dwelling	Segetal	Flowering
T. agenensis	Naturalized alien	Irano-Turanian (Chios Island, Greece)	No	Lowland (0–300)	Agricultural [1]	No	Yes	Early
T. australis	Wild-growing native	Mediterranean-SW Asiatic	No	Intermediate to high (500–2000)	Natural [2]	Yes	No	Late
T. bakeri	Wild-growing native	Greek endemic (Crete)	Yes (CR)	Intermediate (700–1300)	Agricultural and natural [3]	Yes	No	Mid-spring
T. bithynica	Wild-growing native	Subendemic (Lesvos, Greece-Anatolia)	No	Lowland (200–800)	Agricultural and natural [4]	No	No	Mid-spring
T. clusiana	Naturalized alien	Irano-Turanian (Chios Island, Greece)	No	Lowland (100–600)	Agricultural [5]	No	No	Early
T. cretica	Wild-growing native	Greek endemic (Crete)	Yes (EN)	Lowland to high (0–2100)	Natural [6]	Yes	No	Early
T. doerfleri	Wild-growing native	Greek endemic (Crete)	Yes (CR)	Lowland (400–800)	Agricultural [7]	No	Yes	Mid-spring
T. goulimyi	Wild-growing native	Greek endemic (Peloponnese, nearby islands to Crete)	Yes (VU)	Lowland (0–900)	Natural [8]	No	No	Early
T. hageri	Wild-growing native	Greek endemic (Sterea Hellas, Peloponnese and north Greece)	Yes (EN)	Lowland to intemediate (100–1200)	Agricultural and natural [9]	No	Yes	Mid-spring
T. orphanidea	Wild-growing native	Greek endemic (Sterea Hellas, Peloponnese)	Yes (EN)	Lowland to intemediate (700–1600)	Agricultural and natural [10]	No	Yes	Early
T. raddii	Naturalized alien	East Mediterranean (Chios island)	No	Lowland (0–400)	Agricultural [11]	No	Yes	Early
T. saxatilis	Wild-growing native	Subendemic (South Aegean, Anatolia)	No	Lowland to intemediate (200–1300)	Agricultural and natural [12]	Yes	No	Early
T. undulatifolia	Wild-growing native	Subendemic (Balkan-Anatolia)	Yes (VU)	Lowland (100–800)	Agricultural [13] and natural	No	Yes	Early

[1]: Clay cereal fields and terraced olive groves. [2]: Rocky flats, screes, meadows, open woodland, verges of mountain roads, and rocky slopes. [3]: Mountain plateau field margins, in scrub or by gravelly stream sides. [4]: Terraced olive groves and open *Castanea* woodland. [5]: Seasonally wet sites with cultivated and fallow fields. [6]: Open habitats and phrygana with rich terra rossa or rocky and stony mountain slopes. [7]: Cultivated and fallow fields and olive groves. [8]: Terra rossa with phrygana. [9]: Xeric Mediterranean phrygana and grassland or in agricultural habitats. [10]: Meadows and formerly cultivated land in dolines. [11]: Deep clay terraced olive and mastic groves as well as in cereal fields. [12]: Rock-dweller in rocky limestone hills and flats. [13]: Soil pockets on rocky and stony slopes with phrygana or open woodland or as a weed in cultivated and fallow fields and olive groves.

As far as the available macro- and micro-nutrients are concerned, soil available P was extracted using 0.5 M $NaHCO_3$, pH 8.5, and was measured by the molybdenum blue-ascorbic acid method [25]. Extraction with 1 M KCl was used to determine both NO_3-N and NH_4-N, whereas the measurement was conducted with UV-Vis spectrometry and the sodium salicylate-sodium nitroprusside method, respectively [26]. Exchangeable cations (K, Ca, Mg, Na) were extracted with 1 M ammonium acetate (CH_3COONH_4), pH 7 [27]; K and Na were measured with flame photometry, while Ca and Mg were measured by atomic absorption spectrometry. The DTPA method [28] was used for Cu, Zn, Fe, and Mn extraction, which were measured by atomic absorption spectrometry as well. Boron was extracted with hot water and the determination was carried out with the azomethine-H method by UV-Vis spectrometry [29].

Sub-samples of the above-ground biomass or bulbs of each tulip species collected at the flowering stage were ashed at 500 °C for a four-hour minimum [30]. The ash was then dissolved in 2 M HCl following filtration, while the filtrate was used for the determination of P, K, Ca, Mg, Cu, Zn, Fe, Mn, and B employing the analytical methods described previously for soil analysis. In addition, the above-ground biomass and bulbs were analyzed for total N by the Kjeldahl method [31]. All plant samples were also analyzed in triplicate and their values are presented as their mean.

The wet sieving and decanting method with density gradient centrifugation described by [32] was used for arbuscular mycorrhizal fungi (AMF) spore extraction from 50 g soil samples. The spores were then counted and observed under a dissecting scope at 20–35× magnification and were grouped into morphotypes.

2.5. Statistical Analysis

For each plant or soil parameter determined among different species and sampling sites, descriptive statistics were applied, while the three spatial (phytogeographical) regions of Greece (North Aegean Islands, Crete Island, and mainland Greece) were set as a factor with three levels, and analysis of variance (ANOVA) was conducted. The protected least significant difference (LSD) test was used for mean comparisons at $p \leq 0.05$ to investigate differences between plant or soil parameters whereas correlation analysis and principal component analysis (PCA) was also applied to tulips' nutrients content variables. In addition, linear or linearized single and multiple regression models were fitted to investigate the effects of soil properties on Greek tulips' above-ground macro- and micro-nutrients content variability. All analyses were conducted using the Statgraphics software (STATGRAPHICS, CENTURION 18, version 18.1.12, STATPOINT TECHNOLOGIES, Inc., The Plains, VA, USA), and the PCA results were visualized using R software (V.4.2.2) with "ggplot2" package.

3. Results

3.1. Soil Properties of the Studied Samples

The descriptive statistics of the selected physicochemical properties of the soils are presented in Table 3. The coefficient of variation expressed as a percentage (CV, the ratio of the standard deviation to mean) for the general properties (Table 3) was in the range of 11.2–170.4%, indicating that for variables such as $CaCO_3$, OC, EC_{se}, total N, or clay content and CEC, there was a greater dispersion.

Table 3. Descriptive statistics of the studied physicochemical soil properties and the respective soil available concentrations of macro- and micro-nutrients. SD: standard deviation; CV: coefficient of variation; OC: organic carbon; CEC: cation exchange capacity; EC_{se}: electrical conductivity of the saturation extract; Kex, Naex, Caex, Mgex: exchangeable amounts of K, Na, Ca, Mg, extracted with the ammonium acetate method; and B-HW: Boron extracted with the hot water method.

Variable	Min.	Max.	Median	Mean	SD	CV%
Sand (%)	20.8	78.2	49.6	48.9	13.8	28.2%
Silt (%)	10.4	53.2	26.7	28.1	9.3	33.1%
Clay (%)	6.4	47.0	18.2	23.0	12.8	55.4%
pH	5.20	7.90	7.65	7.21	0.81	11.2%
$CaCO_3$ (%)	0.0	62.4	2.1	9.0	15.4	170.4%
OC (%)	0.64	12.12	1.39	2.39	2.30	96.3%
Total N (%)	0.08	0.80	0.16	0.22	0.16	72.6%
C/N	6.50	15.30	9.60	9.60	2.00	20.8%
CEC ($cmol_c$ kg^{-1})	7.2	56.9	24.4	24.3	12.5	51.7%
EC_{se} (ds m^{-1})	0.14	2.64	0.45	0.57	0.50	84.7%
Kex (mg kg^{-1})	34	900	258	275	214	77.8%
Naex (mg kg^{-1})	12.7	96.3	49.2	47.9	20.5	42.9%
Caex (mg kg^{-1})	616	8472	3856	3663	1981	54.1%
Mgex (mg kg^{-1})	97	884	225	357	256	71.6%
NO_3-N (mg kg^{-1})	1.1	98.7	7.3	13.0	17.5	134.5%
NH_4-N (mg kg^{-1})	2.6	48.3	11.8	12.5	8.2	65.1%
P-Olsen (mg kg^{-1})	0.90	45.20	4.45	6.16	7.95	129.0%
Cu-DTPA (mg kg^{-1})	0.64	2.77	1.26	1.42	0.60	39.4%
Zn-DTPA (mg kg^{-1})	0.31	7.09	0.71	1.33	1.50	114.4%
Fe-DTPA (mg kg^{-1})	3.6	95.0	20.5	27.7	23.7	85.7%
Mn-DTPA (mg kg^{-1})	5.4	115.2	20.0	27.4	21.7	79.0%
B-HW (mg kg^{-1})	0.25	1.81	0.56	0.67	0.40	53.6%

As far as soils' fertility status is concerned, NO_3-N, P, and K were among the soil available macro-nutrients with the largest variability, whereas the same was also the case with the micro-nutrients Zn, Fe, and Mn. On the contrary, concentration values of soil available Ca, NH_4-N, as well as Cu and B turned out to be more stable with lower variations.

3.2. Nutritional Status of the Studied Greek Tulip Species

The nutritional status of the above-ground biomass of the sampled Greek tulips was examined and the results are presented as boxplots in Figure 3, while the respective procedure was also followed for bulb samples (Supplementary Material Figure S1). As regards the above-ground biomass samples, the concentrations of macro- and micro-nutrients generally varied among species, while *T. undulatifolia* recorded the highest values for N, K, and Mg, *T. orphanidea* for P, whereas *T. saxatilis* and *T. cretica* for Ca, and N, accordingly. On the contrary, for the same elements (with the sequence of N, P, K, Ca, Mg, and Na, respectively), the lowest values were recorded for *T. australis*, *T. saxatilis*, *T. bakeri*, *T. hageri*, *T. clusiana*, and *T. orphanidea*.

Regarding the cationic micro-nutrient concentration, *T. doerfleri* was found to contain the highest values in Fe and Mn, while the same species recorded the lowest values in Cu. *T. australis* on the other hand, was found high in Zn above-ground biomass concentration while *T. agenensis* in Cu concentration. The lowest values in Zn, Fe, and Mn were recorded in *T. hageri, T. clusana, and T. raddii*, respectively. However, B concentration of above-ground biomass among species did not follow large variation patterns, and the distribution of its variance proved to be the most stable of all micro-nutrients studied.

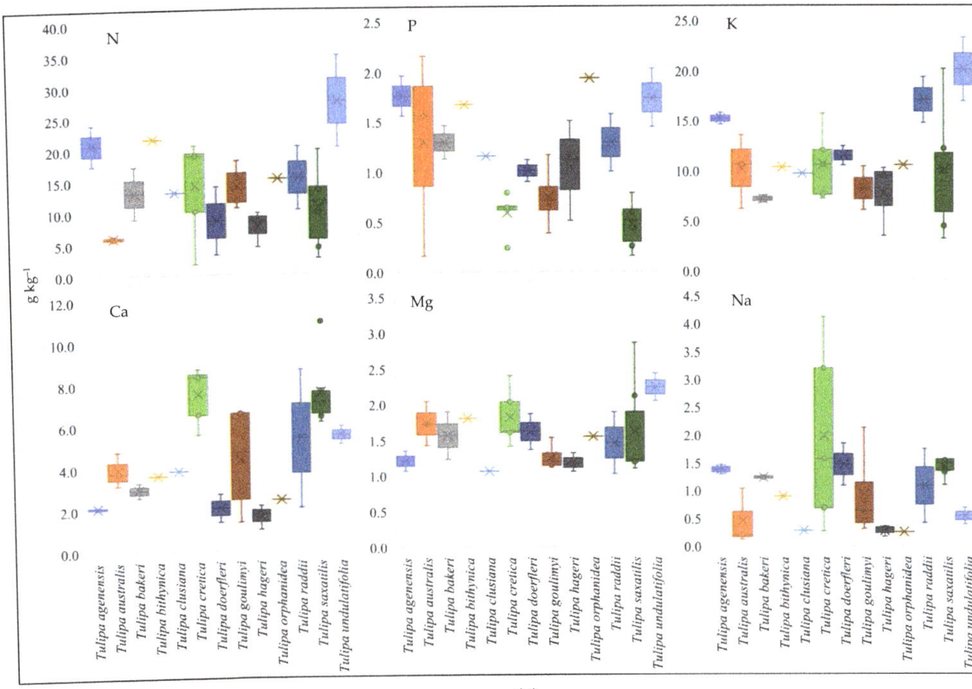

(a)

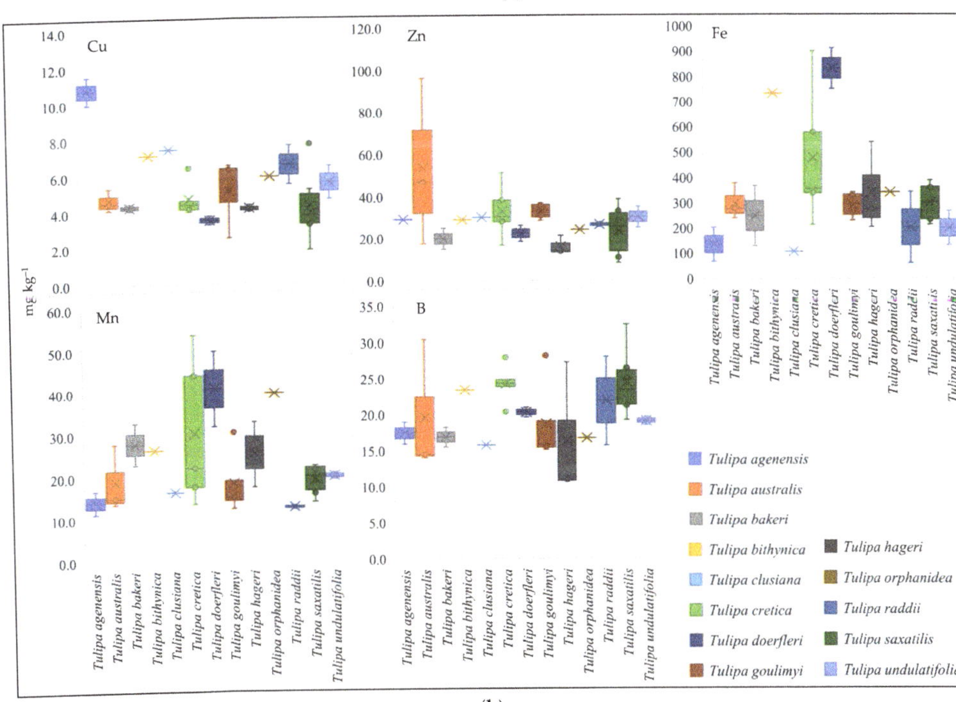

(b)

Figure 3. Boxplots showing concentrations of (**a**) macro-nutrients (N, P, K, Ca, and Mg) and beneficial element Na, and (**b**) micro-nutrients (Cu, Zn, Fe, Mn, and B) in above-ground biomass of wild-growing Greek tulip species. Box colors represent different *Tulipa* species.

3.3. Interrelationships between Tulips' Essential Macro- and Micro-Nutrients Content

Pearson correlations between the set of nutrient content variables examined in the above-ground biomass of tulips revealed significant positive or negative correlations in each case and the results are shown in Table 4. Among them, worth noting are the positive correlations between the three major macro-nutrients (N, P, and K), while the same also applied as expected for Ca and Mg ($r = 0.49$, $p \leq 0.01$). In addition, B was found to be positively correlated with Ca, Mg, and Na (with r value ranging up to 0.65, $p \leq 0.001$), whereas negative correlations on contrary were observed between B and P ($r = -0.65$, $p \leq 0.001$). The latter was also negatively correlated with Ca and Na, while as far as the micro-nutrients are concerned, strong and positive correlations were also found between Fe and Mn ($r = 0.62$, $p \leq 0.001$).

Table 4. Pearson correlation matrix with respective r values between above-ground biomass concentrations of the studied macro- and micro-nutrients of the wild-growing Greek tulip species. Asterisks (*, **, ***) indicate significant r values at $p < 0.05$, 0.01, and 0.001, respectively; $n = 34$.

	N	K	Na	Ca	Mg	Cu	Zn	Fe	Mn	P
K	0.31 [a]									
Na	0.21	−0.04								
Ca	−0.17	0.09	0.38 *							
Mg	−0.22	−0.09	0.15	0.49 **						
Cu	0.35 *	0.11	−0.17	−0.14	0.10					
Zn	0.44 **	0.19	0.07	−0.01	−0.24	0.09				
Fe	−0.28	−0.36 *	0.10	0.14	0.39 *	−0.34	−0.10			
Mn	0.01	−0.31 [a]	0.11	−0.10	0.17	−0.23	0.07	0.62 ***		
P	0.39 *	0.40 *	−0.40 *	−0.57 ***	−0.13	0.46 **	0.22	−0.27	0.02	
B	−0.19	−0.16	0.57 ***	0.65 ***	0.49	−0.20	−0.04	0.37 *	0.15	−0.65 ***

[a] $p = 0.08$.

3.4. Relationships between Rhizosphere's Arbuscular Mycorrhizal Fungi (AMF) Spore Morphotypes, Soil Parameters, and Tulips Nutrients Content

The results from Pearson correlations concerning the arbuscular mycorrhizal fungi (AMF) spore morphotypes showed that among all studied nutrients of tulips' above-ground biomass, they were significantly and positively correlated only with N ($r = 0.4$, $p \leq 0.05$). As regards the soil parameters, on the contrary, a negative correlation was found with soil available NH_4-N ($r = -0.38$, $p \leq 0.05$), while the same was also the case with soil Mn and Fe extracted with DTPA, respectively ($r = -0.36$, $p \leq 0.05$, $r = -0.44$, $p \leq 0.01$). A negative correlation was found with regard to the soil texture parameter between AMF spore morphotypes and the percentage of sand ($r = -0.34$, $p \leq 0.05$).

3.5. Nutrients Variability of Wild-Growing Greek Tulips and Functional Types Distribution

The PCA analysis (Figure 4a) revealed that the major portion of the total variance (almost 80%) of the studied variables (nutrients' content of above-ground biomass, and fungi morphotypes) was grouped between five components, and the two of them explained 44.3% of it (Figure 4b). The variables that most contributed in the first two PCA axes were the elements B (15.7%) and Ca (13.6%), P (12.7%), N (9.8%), and K (9.2%), whereas Mn and Fe contributed mostly on the third component (30% and 26.7%, respectively).

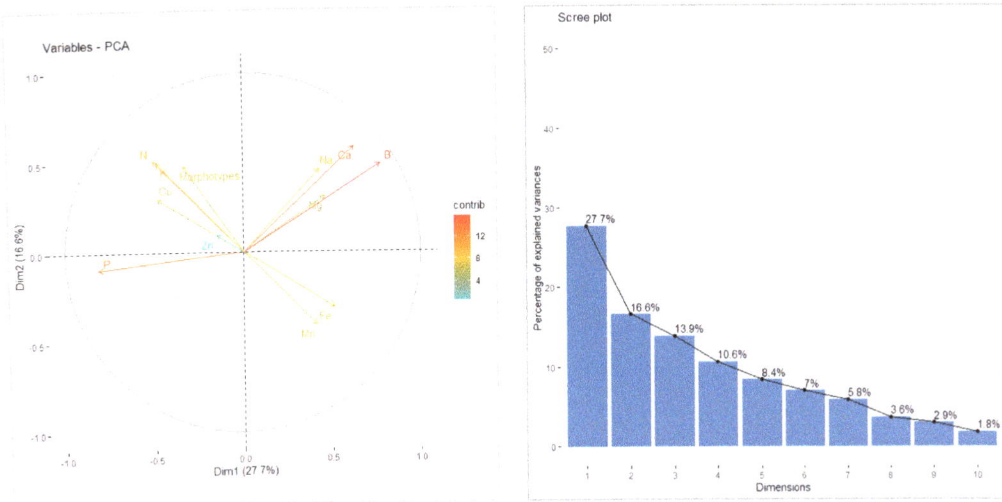

(**a**) (**b**)

Figure 4. Principal component analysis (PCA) of (**a**) the studied macro- and micro-nutrients of tulips' above-ground biomass (vectors), and (**b**) the respective contribution of each component to their total variability.

The distribution of individuals of the PCA analysis (sampled tulips species) as scattered between the first two components was further evaluated using the sampling location as categorial factor as well as the tulips' functional properties to explore potential distribution similarity patterns. The results presented in Figure 5 reveal that between the three spatial (phytogeographical) units examined (North Aegean Islands, Crete Island, mainland Greece), the total nutrients' variability produced a clear distinction among sampled species. More specifically, the samples of the North Aegean Islands were scattered in the upper left quadrant of the PCA plot (Figure 5), indicating a strong association with the respective variables of N, K, and Cu, and with the AMF spores morphotypes (Figure 4a). On the contrary, most of the Cretan samples were scattered in the upper right quadrant (Figure 5), showing respective association with B, Ca, Na, and Mg (Figure 4a). A corresponding association was also observed between the Cretan samples scattered in the lower wright quadrant with Fe and Mn variables, respectively.

Respective though much less distinct patterns were also observed regarding the other functional groups of the studied tulip species outlined in Section 2.3 (Figure 6). For example, the rock-dwelling Greek tulips were mostly scattered in the positive side of the first component axis whereas all other tulip species occupied the negative side. Similarly, the mid-spring flowering Greek tulips were polarized in the most negative side of the second component axis compared with the early or late flowering ones. Concerning the altitudinal class, the lowland Greek tulip species primarily represented by the samples of the North Aegean Islands were scattered in the upper left quadrant of the PCA plot. The latter indicated a strong association of this group of tulips with the respective variables of N, K, and Cu, and with the AMF spores morphotypes as well. The distribution pattern of the chorological status was rather indistinct, denoting that the variability of the tulips' nutrients should not be considered a major factor driving differences between these groups.

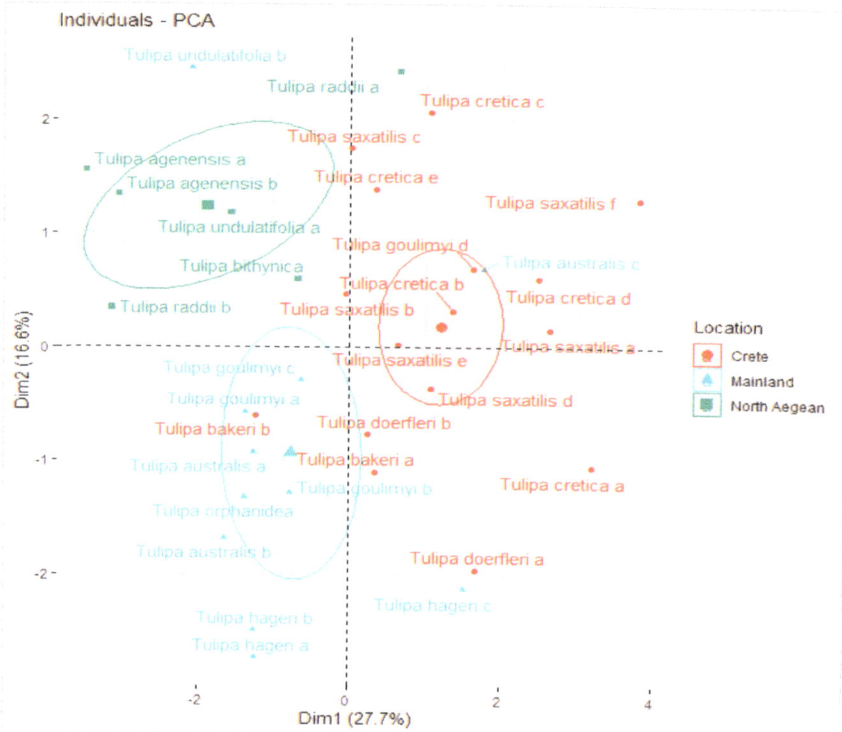

Figure 5. Distribution of each sampled tulip species as scattered among the first two components of the PCA (95% confidence ellipses), and respective grouping in terms of their specific spatial (phytogeographical) units (location) namely North Aegean Islands, Crete Island, and mainland Greece.

3.6. Effects of Soil Properties on Greek Tulips' Above-Ground Macro- and Micro-Nutrients Content Variability

Linear or linearized single and multiple regression models were fitted among the dependent variables of tulips' above-ground biomass elements concentrations, setting as independent variables the soil physicochemical properties. The results presented in Table 5 revealed significant relationships for all studied macro- and micro-nutrients, while parameters such as EC_{se} and CEC in some cases managed to explain almost up to 70% of nutrient's concentration total variance (dependent variable, leaf P).

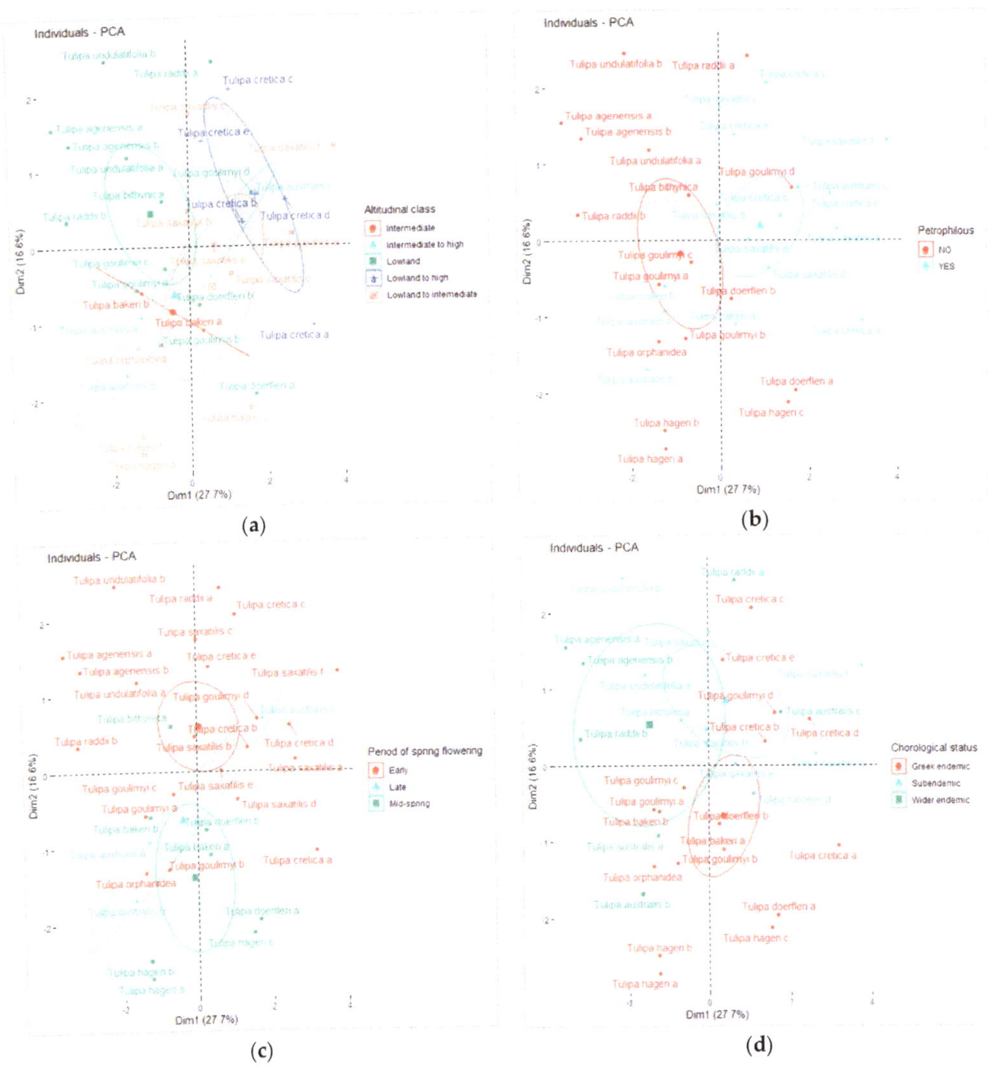

Figure 6. Distribution of each sampled tulip species as scattered betweem the first two components of the PCA (95% confidence ellipses), and respective grouping in terms of (**a**) different altitudinal classes; (**b**) rock-dwelling habit or not (petrphilous); (**c**) period of spring flowering; and (**d**) chorological status.

Except for the CEC and EC_{se} parameters, pH, $CaCO_3$, clay, and OC content were also among the soil properties that contributed the most, while notably, respective soil available concentrations played a significant role in explaining their variance only in the case of Na, Cu, and Zn plants' concentrations.

Table 5. Single and multiple regression models between the tulips' above-ground biomass elements concentrations (Y) and the respective soil physicochemical properties (X). r^2: coefficient of determination; OC: organic carbon; CEC: cation exchange capacity; EC_{se}: electrical conductivity of the saturation extract; and Naex: exchangeable amounts of Na extracted with the ammonium acetate method; n = 34.

Dependent Variable (Y)	Independent Variable (X)	Model	r^2 %	p-Value
N	Clay	Simple linear	32.2	0.0005
P	EC_{se}, CEC	Multiple reciprocal-Y	67.1	0.0000
K	CaCO3, OC%, CEC	Multiple linear	58.2	0.0000
Ca	pH, EC_{se}	Multiple linear	49.6	0.0000
Mg	EC_{se}	Double squared	43.1	0.0000
Na	Naex	Double reciprocal	56.4	0.0000
Cu	Cu-DTPA	Double squared	22.1	0.0050
Zn	Zn-DTPA, pH, EC_{se}	Multiple linear	57.5	0.0000
Fe	pH, OC%	Multiple linear	21.6	0.0258
Mn	pH, OC%	Multiple linear	47.5	0.0001
B	EC_{se}	Squared-Y square root-X	38.3	0.0001

3.7. Evaluating Differences in Nutrients Profile and in Soil Properties between Different Spatial Units

The classification of tulip species in the three spatial (phytogeographical) units of North Aegean Islands, Crete Island and mainland Greece was set as a factor in the analysis of variance (ANOVA) that was conducted. The latter was chosen because of the clear grouping patterns that the PCA analysis provided (Figure 5) in conjunction with the relationships that were shown between tulips' above-ground biomass elements concentrations and the soil physicochemical properties. Thus, potential statistical differences were investigated for three sets of variables: (a) macro- and micro-nutrient content of the tulips' above-ground biomass alongside with the assessed number of fungal morphotypes found in each sample, (b) macro- and micro-nutrient content of bulbs, and (c) soil physicochemical properties.

The results presented in Table 6 showed that regarding the nutrient content of the above-ground biomass of Greek tulip species, statistically significant differences were found for most of the cases (except Mg, and Zn concentration mean values). More specifically, North Aegean tulips were associated with the higher concentrations in the three of the major macro-nutrients (N, P, K) alongside Cu as compared with the respective elements measured in tulips sampled from Crete. The same was also the case with the fungal morphotypes while North Aegean tulips presented the lowest mean values in Ca, Na, Fe, and Mn compared to the Cretan ones. On the contrary, the mean values of the above-mentioned variables recorded in the tulips from mainland Greece were either polarized with the respective values of North Aegean tulips (P, Ca, and Na) or with those in tulips from Crete (N, K, Cu, and fungi morphotypes) or they were clustered in the middle without differing significantly with none of them (Fe, and Mn).

Interestingly, agreement with the above results was only noted for exchangeable K and Cu extracted with DTPA in terms of available macro- and micro-nutrients in soil (Table 7), while even contradictory results were recorded as in the case of soil available Ca. More particularly, even though its concentration in soils sampled from the North Aegean Islands was found high (as exchangeable Ca), the corresponding values of the above-ground biomass were significantly lower compared with those of Crete. In addition, it is noteworthy that unlike the above-ground biomass, nutrient content in tulip bulbs did not show corresponding differences between location groups linked with sample collections (data not shown). On the other hand, among several soil properties studied (Table 7), CEC values were found to be significantly higher in North Aegean samples as compared with those from Crete or mainland Greece, whereas this trend also stood for the clay content parameter, in which, however, significant differences occurred only between North Aegean Islands and Crete. In addition, the mean values of $CaCO_3$ content differed significantly between the North Aegean Islands and the mainland (with higher values in the North Aegean).

Table 6. Mean values of macro- and micro-nutrients' concentrations in above-ground biomass of the studied tulip species or fungi morphotypes in terms of their specific phytogeografical unit (Location) namely North Aegean Islands, Crete Island, and mainland Greece. SE: standard error; NS: non-significant. Within each element or arbuscular mycorrhizal fungi (AMF) spores morphotype, different letters indicate significant differences among means, employing the protected LSD test, at $p \leq 0.05$.

Location	Mean	SE	Mean	SE	Mean	SE	Mean	SE	Mean	SE	Mean	SE
	N		P		K		Ca		Mg		Na	
	(g kg^{-1})											
North Aegean Islands	18.3 a	(1.8)	1.49 a	(0.12)	15.4 a	(1.8)	3.94 b	(0.91)	1.47 a	(0.17)	0.93 b	(0.22)
Crete Island	12.0 b	(1.6)	0.68 b	(0.09)	10.0 b	(1.0)	6.27 a	(0.65)	1.65 a	(0.13)	1.63 a	(0.23)
Mainland Greece	10.3 b	(1.5)	1.24 a	(0.20)	9.4 b	(1.1)	3.27 b	(0.56)	1.52 a	(0.13)	0.43 b	(0.09)
p F-test	0.021		<0.001		0.011		0.007		NS		<0.001	
	Cu		Zn		Fe		Mn		B		AMF spore morphotypes	
	(mg kg^{-1})											
North Aegean Islands	7.74 a	(0.91)	27.9 a	(0.8)	154 b	(44)	16.6 b	(2.0)	19.5 ab	(1. 8)	4.50 a	(0.62)
Crete Island	4.24 b	(0.35)	26.5 a	(2.8)	414 a	(62)	27.1 a	(3.1)	23.1 a	(1.1)	2.75 b	(0.49)
Mainland Greece	5.23 b	(0.31)	33.0 a	(7.1)	316 ab	(28)	19.6 ab	(2.2)	17.3 b	(1.9)	2.18 b	(0.46)
p F-test	<0.001		NS		0.023		0.048		0.022		0.044	

Table 7. Mean values of macro- and micro-nutrients' soil available concentrations and respective soil properties of the sampled soil sites in terms of the specific phytogeographical unit (location) namely North Aegean Islands, Crete Island, and mainland Greece. SE: standard error; NS: non-significant. Within each soil available element, or soil property, different letters indicate significant differences among means, employing the protected LSD test, at $p \leq 0.05$.

Location	Mean	SE	Mean	SE	Mean	SE	Mean	SE	Mean	SE	Mean	SE	Mean	SE
	Clay (%)		pH		$CaCO_3$ (%)		OC (%)		N-Total (%)		CEC (cmol$_c$ kg^{-1})		EC$_{se}$ (ds m^{-1})	
North Aegean Islands	31.9 a	(4.5)	7.51 a	(0.22)	21.6 a	(7.1)	1.86 a	(0.62)	0.18 a	(0.04)	35.2 a	(1.6)	0.61 a	(0.11)
Crete Island	17.1 b	(2.6)	7.29 a	(0.17)	8.78 ab	(4.03)	2.81 a	(0.73)	0.25 a	(0.05)	21.6 b	(3.6)	0.71 a	(0.15)
Mainland Greece	25.9 ab	(3.9)	6.90 a	(0.31)	1.40 b	(0.64)	2.10 a	(0.48)	0.21 a	(0.04)	21.1 b	(3.0)	0.33 a	(0.06)
p F-test	0.020		NS		0.020		NS		NS		0.029		NS	
Location	Mean	SE	Mean	SE	Mean	SE	Mean	SE	Mean	SE	Mean	SE		
	NO_3-N		NH_4-N		P-Olsen		K-ex		Ca-ex		Mg-ex			
	(mg kg^{-1})													
North Aegean Islands	13.6 a	(4.3)	7.81 b	(1.14)	6.10 a	(2.55)	473 a	(114)	5103 a	(222)	627 a	(104)		
Crete Island	15.4 a	(5.9)	10.9 ab	(1.5)	6.81 a	(2.71)	227 b	(42)	3665 ab	(563)	215 c	(25)		
Mainland Greece	9.23 a	(2.62)	14.5 a	(1.3)	5.25 a	(0.58)	218 b	(43)	2744 b	(501)	392 b	(81)		
p F-test	NS		0.020		NS		0.043		<0.001		0.012			
	Na-ex		Cu-DTPA		Zn-DTPA		Fe-DTPA		Mn-DTPA		B-HW			
	(mg kg^{-1})													
North Aegean Islands	54.0 a	(5.1)	1.87 a	(0.24)	1.15 a	(0.30)	17.2 a	(5.3)	14.1 b	(2.9)	0.74 a	(0.18)		
Crete Island	52.9 a	(4.62)	1.26 b	(0.09)	1.46 a	(0.38)	27.2 a	(5.6)	27.8 ab	(3.4)	0.69 a	(0.08)		
Mainland Greece	31.9 b	(5.9)	1.39 ab	(0.20)	1.26 a	(0.59)	35.0 a	(8.7)	35.5 a	(9.7)	0.60 a	(0.11)		
p F-test	<0.001		0.049		NS		NS		NS (0.123)		NS			

4. Discussion

Our study aimed to provide for the first time a comprehensive overview of the wild-growing Greek tulip species focusing on their nutritional profiles and their interrelationships with the specific soil conditions that prevail in the small environmental scale of their original habitats.

As seen from the respective results concerning the soil properties of the studied samples, large variance patterns were observed among several soil parameters such as $CaCO_3$, OC%, EC_{se}, and total N, or clay content and CEC. The latter indicates that the Greek tulip species are wild-growing plants in habitats characterized by diverse soil conditions, while this natural spatial variability of soil properties might adversely affect the soil availability of nutrients required for plant growth. For instance, it is well established that the use of CEC parameter as a covariate significantly improves prediction models of K availability in soils [33,34], whereas EC_{Se} has also been suggested that can reduce the experimental error variance and thus to increase the precision of estimates of available soil P [35,36].

Similar impacts on the availability of soil nutrients derived by respective soil parameters could also be reflected in the tulips' nutrients content profile (Figure 3) to the extent that their variability is explained by such impacts. Previous studies investigating leaf ionomic responses of different plant species grown to diverse edaphic sites [9] have found that the influence of soil parameters to the respective element variations in plants were strong but limited within the phylogenetic preset; however, for some cases such as leaf P concentration, for example, soil condition played a more important role compared to phylogenetic factors. Similar conclusions have also been reached by other researchers suggesting that ionomic variations are mainly driven by environmental conditions [37,38]. Based on our results, we consider that the above assumptions are more evident herein, given that we studied different members (species) of the same genus in which phylogenetic variance is strongly reduced. More specifically, our study showed that significant relationships occurred (from 21.6% up to 67.1%) between the above-ground concentrations of all investigated macro- and micro-nutrients and the respective soil parameters. In addition, corroborating with previous findings [9], tulips P concentration did appear to be strongly and significantly affected by soil parameters, with the EC_{se} playing key role in this effect.

Some of the specific correlation patterns among nutrients content of above-ground biomass (Table 4) have also been reported by other researchers, although such trends could be attributed to several factors. For instance, previous studies have reported that the common source for the element pairs could explain positive correlations among elements in the plants' biomass [39]. The above assumption could stand in the case of the observed Fe–Mn positive correlations detected herein given the fact that these elements were also significantly influenced by specific soil properties such as pH and OC (Table 5).

Hence, similar solubility rates driven from respective soil pH variations, as well as similar decomposition rates of soil organic matter releasing proportionally equal amounts of the respective micro-nutrients probably bound in soils' organic matter could provide a possible explanation mechanism for their positive correlations in tulips' plant tissue content. In addition, the tendency of elements with similar physicochemical properties to share or compete for pathways or transport systems accumulating them in leaves [9,40] might partially explain the observed negative B-P correlations in this study, or the positive Ca–Mg interactions. In this line, the positive correlations between B and Ca could be also attributed to the fact that B tends to keep Ca in a soluble form within the plant [41].

The correlation results regarding the AMF spore morphotypes and their relationships with soil or plant nutrients parameters are also of interest. It is well known that the AMF symbiosis is important for plant P nutrition, and that higher P levels suppress the symbiosis [42]. It is also known that AMF may also contribute to plant N nutrition [43,44], and the positive correlation between AMF spore morphotypes with tulip tissue N points to that direction. On the other hand, previous studies have shown that elevated soil N may reduce AMF spore richness and abundance [45], which is in line with the negative correlation detected in this study between AMF spore morphotypes and the soil available NH_4-N. However, N effects may also be related to N form and plant species [46]. Based on the tissue and soil N correlation, we would expect N fertilization to reduce AMF spore morphotypes and the importance of AMF in tissue N.

The micro-nutrients Fe and Mn are important in many physiological roles in plants, and previous studies have found that N addition may increase foliar Mn suppressing photosynthetic rates [47]. On the other hand, Fe and Mn may function as soil redox indicators, being more mobile at reduced soil conditions, which could also be detrimental for AMF and may offer explanation for the negative correlation of AMF spore morphotypes with these two micro-nutrients in this study. Soil texture is another important soil parameter for AMF spores, with clay rather than sand favoring their presence [48]. Other factors such as climate or elevation that were found to be important for AMF spores [49] exceeded the purposes of the present study and thus they were not examined.

The PCA results in conjunction with the respective ANOVA findings indicated that the distinction observed among sampled tulip species between the three spatial (phytogeographical) units examined (North Aegean Islands, Crete Island, and mainland Greece) might be ascribed to a large extent to differences found in specific soil properties. However, our results indicate that the detected variance in all variables studied regarding mainland Greece was expectedly high due to the comparatively large geographic scale and concomitant spatial heterogeneity of the latter spatial unit including seven phytogeographical regions. Thus, to avoid scale-dependent biases, the discussion made hereafter has been limited to comparisons between Crete and the North Aegean Islands since both spatial units are insular and of comparable size and are referred to as specific phytogeographical regions in the context of the Greek flora (see https://portal.cybertaxonomy.org/flora-greece/annotations, accessed 25 March 2023). In this way, the samples of tulip species from the North Aegean Islands clustered in the upper left side of the PCA plot were related to high levels of nutrients such as N, K, or Cu, a fact which was also documented by the ANOVA results that showed statistically significant higher values of these nutrients in tulip species of this phytogeographical unit. In the same line, the above was also the case for some elements such as Ca, Mg, Na, Fe, and Mn originated from tulips sampled from Crete Island (Table 6).

Considering that specific soil properties significantly influenced the variability of tulips nutrient content (Table 7), the respective ANOVA results concerning differences in soil properties among the studied spatial (phytogeographical) units may provide some interpretation of the above variance. For example, the soil samples of the phytogeographical unit of the North Aegean Islands are characterized by high values in CEC and clay content, respectively, a fact also reflected in the corresponding amounts of exchangeable soil K which were found to be correspondingly high as well.

We consider that the above trend is reflected in the significantly higher amounts of leaf K found in the North Aegean tulip samples whereas the same could also be the case with the correspondence that appeared between the significantly higher levels of foliar Cu and soil available Cu, respectively. The above remarks could fall into a category of cases in which a direct relation between soil available nutrients and their respective concentration in plant tissues occurs, as this has also been reported by several correlation studies and agricultural field experiments [50].

Nevertheless, the above rationale cannot explain the case of the significantly higher concentrations in N or P found in tulips sampled in the North Aegean Islands as compared with Crete, for example, since the physicochemical soil properties did not seem to have any effect in these variations. This leaves open the possibility that tulip species of the North Aegean Islands may be more efficient in P or N uptake, and/or they develop more efficient symbiosis with AMF. The higher number of the AMF spore morphotypes found in the samples of the North Aegean Islands indicated that the latter may be the case [51].

In the same line, the contradictory results that were recorded between the higher tulips' plant Ca content in samples of the North Aegean Islands compared with those from Crete for example (although with non-recorded significant differences in soil available Ca) could not be explained based on the direct soil-plant connection concept, as previously offered for leaf K. Instead, species factors and respective adaptation mechanisms developed to respond to soil environmental specificities might provide some explanation. For instance, it is well established in the literature that when the presence of K is dominant in soil and in plants, antagonistic relationships may occur between other basic cations like Mg or Ca [52,53], a fact which might be true in our case, considering the significantly higher amounts of soil and plant K detected in the North Aegean samples.

Concerning the role that the functional groups of tulips outlined herein might play into potential differences detected on their nutrients content profile, the results presented in Figure 6 did not provide as clear information as the grouping between spatial (phytogeographical) units did. Nevertheless, the combined reading of the overall results of the present work can significantly contribute to a more comprehensive insight into the mechanisms of adaptation of the specific tulip species to their local environments. For example, PCA findings referring to the samples of the North Aegean may be interpreted as follows. These particularly lowland, non-petrophilous (rock-dwelling) and early flowering species (Figure 6), which naturally thrive in clayey soils with high levels of CEC showed higher number of the AMF spore morphotypes, while they were also associated with high levels of nutrients such as N, K, and Cu (Figure 4a). Thus, we consider that the above-mentioned combined information is crucial not only for providing conservation management insights of wild-growing tulips, but also for contributing on future domestication efforts as well.

Overall, the functioning of ecosystems is critical for the maintenance of biodiversity, the provision of ecosystem services, and the sustainability of human societies [19]. The current study has shed light on the nutrient acquisition patterns of 13 wild-growing Greek tulips originating from different phytogeographical units of this small but extremely diverse country in relation to soil properties and AMF morphotypes. Given that natural ecosystems are multifaceted and dynamic [54], these results are important for future conservation efforts and possible sustainable utilization plans concerning the wild-growing Greek tulip species and their original habitats. All the data furnished herein can be further exploited in attempts to cultivate ex situ in a sustainable way the diversity of the wild-growing Greek tulip species (n = 15) and finally domesticate them in artificial settings. Facilitating the latter is a complex and multi-dimensional process given that each of these wild-growing tulips originates from different environmental conditions, has different ecological preferences and has developed different natural adaptations [1,55]. To this end, the data generated herein can be used in the future to formulate species-specific fertilization guidelines to be followed in artificial settings which may benefit to a large extent from the insight into the species-specific nutritional needs, elemental uptake potential, and transport as detected in the original wild habitats of the wild-growing Greek tulips studied herein.

5. Conclusions

Despite the economic interest in tulip hybrids and botanical tulips worldwide [1], no insight is practically available to date into the protected wild-growing tulips of Greece, especially in terms of their growth and adaptation to their natural environment or artificial settings. Based on the latter research gap, our study showed that soil variables played a significant role in shaping wild-growing Greek tulips' nutrient content, explaining up to 67% of the detected variability as in the case of the above-ground plant tissue P. In addition, respective correlation patterns were found between tulips' essential nutrients. Among them, the observed Fe–Mn positive correlations could be attributed to similar solubility rates from respective soil pH variations, while the observed negative B–P correlations or the positive Ca–Mg interactions could be attributed to the tendency of elements with similar physicochemical properties to share or compete for common pathways or transport systems within the plants. The PCA revealed that between the three spatial (phytogeographical) units examined (North Aegean Islands, Crete Island, and mainland Greece), the total variability of tulips' nutrient content produced a clear distinction among sampled species.

In addition, the samples of the North Aegean Island were associated with nutrients such as N, K, and Cu, and with high numbers of AMF morphotypes, whereas the samples of the Crete Island were associated with nutrients such as B, Ca, Na, Mg, Fe, and Mn. This was further confirmed by the ANOVA results which showed corresponding statistically significant differences in both the tulips' nutrient content and the studied soil properties as well. Among them, the significantly higher amounts of leaf K found in the North Aegean tulip samples could be attributed to the high values in CEC and clay content, respectively, a fact also reflected in the corresponding high amounts of exchangeable soil K. In addition, the significantly higher concentrations in N or P found in tulips sampled in the North Aegean Islands might be ascribed to a more efficient AMF symbiosis of the corresponding species, as has been indicated by the higher number of the AMF spore morphotypes found in the North Aegean samples. Therefore, our study sheds light on Greek tulips' adaptability and resilience in their original habitats, facilitating at the same time the undertaken efforts regarding their potential domestication in artificial settings.

Supplementary Materials: The following supporting information can be downloaded at https://www.mdpi.com/article/10.3390/biology12040605/s1: Supplementary Material Figure S1: boxplots showing $\log_{10}$ concentrations of (a) macro-nutrients (N, P, K, Ca, and Mg) and beneficial element Na, and (b) micro-nutrients (Cu, Zn, Fe, Mn, and B) in subterranean bulb biomass of the studied wild-growing Greek tulip species.

Author Contributions: G.T., N.K., T.M. and F.B.: conceived the presented idea. G.T., T.M. and N.K.: coordinated and supervised the field and laboratory work and acquired research funding. F.B., I.I., A.-G.K., I.S. and T.M.: prepared samples and conducted soil, plant analysis, and AMF spore isolation and observation. F.B., I.S., I.I. and T.M.: worked on the data collection and analysis. F.B. and N.K.: visualized all figures and tables. F.B., I.I., I.S., T.M. and N.K.: wrote the first complete draft of the manuscript: F.B., T.M., I.I., I.S., N.K., G.T., A.-G.K. and A.-G.K.: revised this manuscript. All authors have read and agreed to the published version of the manuscript.

Funding: This research was co-financed by the European Union and Greek national funds through the Operational Program Competitiveness, Entrepreneurship and Innovation, under the call RESEARCH-CREATE-INNOVATE (project code: T2EDK- 05115; acronym: TULIPS.GR), entitled "Value chain for Greek native tulips: development of documented propagation material and integrated conservation for sustainable exploitation".

Institutional Review Board Statement: Not applicable.

Informed Consent Statement: Not applicable.

Data Availability Statement: All data are included in this article (and its supplementary information files).

Conflicts of Interest: The authors declare no conflict of interest.

References

1. Krigas, N.; Lykas, C.; Ipsilantis, I.; Matsi, T.; Weststrand, S.; Havström, M.; Tsoktouridis, G. Greek Tulips: Worldwide Electronic Trade over the Internet, Global Ex Situ Conservation and Current Sustainable Exploitation Challenges. *Plants* **2021**, *10*, 580. [CrossRef]
2. Strid, A. *Atlas of the Aegean Flora, Part 1: Text & Plates; Part 2: Maps; Englera, Volume 33*; Botanic Garden and Botanical Museum Berlin, Freie Universität Berlin: Berlin, Germany, 2016; ISBN 978-3-921800-97-3 (Volume 1); 978-3-921800-98-0 (Volume 2).
3. Kougioumoutzis, K.; Kokkoris, I.P.; Panitsa, M.; Strid, A.; Dimopoulos, P. Extinction Risk Assessment of the Greek Endemic Flora. *Biology* **2021**, *10*, 195. [CrossRef]
4. Phitos, D.; Strid, A.; Snogerup, S.; Greuter, W. *The Red Data Book of Rare and Threatened Plants of Greece*; World Wide Fund for Nature: Gland, Switzerland, 1995; ISBN 9789607506047.
5. Phitos, D.; Constantinidis, T.H.; Kamari, G. *The Red Data Book of Rare and Threatened Plants of Greece*; Hellenic Botanical Society: Patra, Greece, 2009; Volume II, ISBN 978-960-9407-09-0.
6. Zhang, J.; Ren, T.; Yang, J.; Xu, L.; Li, M.; Zhang, Y.; Han, X.; He, N. Leaf Multi-Element Network Reveals the Change of Species Dominance Under Nitrogen Deposition. *Front. Plant Sci.* **2021**, *12*, 580340. [CrossRef] [PubMed]
7. Marschner, H. *Marschner's Mineral Nutrition of Higher Plants*, 3rd ed.; Marschner, H., Marschner, P., Eds.; Elsevier: London, UK; Academic Press: Waltham, MA, USA, 2012; ISBN 978-0-12-384905-2.

8. Morgan, J.; Connolly, E. Plant-soil interactions: Nutrient uptake. *Nat. Educ. Knowl.* **2013**, *4*, 2.
9. Zhang, C.; Hiradate, S.; Kusumoto, Y.; Morita, S.; Koyanagi, T.F.; Chu, Q.; Watanabe, T. Ionomic Responses of Local Plant Species to Natural Edaphic Mineral Variations. *Front. Plant Sci.* **2021**, *12*, 614613. [CrossRef] [PubMed]
10. Watanabe, T.; Azuma, T. Ionomic Variation in Leaves of 819 Plant Species Growing in the Botanical Garden of Hokkaido University, Japan. *J. Plant. Res.* **2021**, *134*, 291–304. [CrossRef]
11. Gupta, N.; Yadav, K.K.; Kumar, V.; Kumar, S.; Chadd, R.P.; Kumar, A. Trace Elements in Soil-Vegetables Interface: Translocation, Bioaccumulation, Toxicity and Amelioration—A Review. *Sci. Total Environ.* **2019**, *651*, 2927–2942. [CrossRef] [PubMed]
12. Likar, M.; Vogel-Mikuš, K.; Potisek, M.; Hančević, K.; Radić, T.; Nečemer, M.; Regvar, M. Importance of Soil and Vineyard Management in the Determination of Grapevine Mineral Composition. *Sci. Total Environ.* **2015**, *505*, 724–731. [CrossRef]
13. Zhou, W.; Liu, G.; Xing, W. Variations of Multi-Elements in Wetland Plants on the Tibetan Plateau Are Mainly Determined by Environmental Factors. *Ecol. Indic.* **2023**, *146*, 109807. [CrossRef]
14. Sadiqi, S.; Hamza, M.; Ali, F.; Alam, S.; Shakeela, Q.; Ahmed, S.; Ayaz, A.; Ali, S.; Saqib, S.; Ullah, F.; et al. Molecular Characterization of Bacterial Isolates from Soil Samples and Evaluation of their Antibacterial Potential against MDRS. *Molecules* **2022**, *27*, 6281. [CrossRef] [PubMed]
15. Khan, M.; Ali, S.; Al Azzawi, T.N.I.; Saqib, S.; Ullah, F.; Ayaz, A.; Zaman, W. The Key Roles of ROS and RNS as a Signaling Molecule in Plant–Microbe Interactions. *Antioxidants* **2023**, *12*, 268. [CrossRef]
16. Cole, D.W. Soil Nutrient Supply in Natural and Managed Forests. *Plant Soil* **1995**, *168–169*, 43–53. [CrossRef]
17. Hestrin, R.; Hammer, E.C.; Mueller, C.W.; Lehmann, J. Synergies between Mycorrhizal Fungi and Soil Microbial Communities Increase Plant Nitrogen Acquisition. *Commun. Biol.* **2019**, *2*, 233. [CrossRef] [PubMed]
18. Krigas, N.; Tsoktouridis, G.; Anestis, I.; Khabbach, A.; Libiad, M.; Megdiche-Ksouri, W.; Ghrabi-Gammar, Z.; Lamchouri, F.; Tsiripidis, I.; Tsiafouli, M.A.; et al. Exploring the Potential of Neglected Local Endemic Plants of Three Mediterranean Regions in the Ornamental Sector: Value Chain Feasibility and Readiness Timescale for Their Sustainable Exploitation. *Sustainability* **2021**, *13*, 2539. [CrossRef]
19. Johnson, C.N.; Balmford, A.; Brook, B.W.; Buettel, J.C.; Galetti, M.; Guangchun, L.; Wilmshurst, J.M. Biodiversity Losses and Conservation Responses in the Anthropocene. *Science* **2017**, *356*, 270–275. [CrossRef] [PubMed]
20. Hooper, D.U.; Chapin, F.S.; Ewel, J.J.; Hector, A.; Inchausti, P.; Lavorel, S.; Lawton, J.H.; Lodge, D.M.; Loreau, M.; Naeem, S.; et al. Effects of biodiversity on ecosystem functioning: A consensus of current knowledge. *Ecol. Monogr.* **2005**, *75*, 3–35. [CrossRef]
21. Bhaduri, D.; Sihi, D.; Bhowmik, A.; Verma, B.C.; Munda, S.; Dari, B. A Review on Effective Soil Health Bio-Indicators for Ecosystem Restoration and Sustainability. *Front. Microbiol.* **2022**, *13*, 938481. [CrossRef]
22. Bouyoucos, G.J. Hydrometer method improved for making particle size analysis of soils. *Agron. J.* **1962**, *54*, 464–465. [CrossRef]
23. Walkley, A.; Black, I.A. An examination of the Degtjareff method for determining soil organic matter, and a proposed modification of the chromic acid titration method. *Soil Sci.* **1934**, *37*, 29–38. [CrossRef]
24. Rhoades, J.D. Salinity: Electrical conductivity and total dissolved solids. In *Methods of Soil Analysis, Part 3: Chemical Methods*; SSSA Book Series 5; Sparks, D.L., Ed.; Soil Science Society of America, American Society of Agronomy: Madison, WI, USA, 1996; pp. 417–435.
25. Kuo, S. Phosphorus. In *Methods of Soil Analysis, Part 3: Chemical Methods*; SSSA Book Series 5; Sparks, D.L., Ed.; Soil Science Society of America, American Society of Agronomy: Madison, WI, USA, 1996; pp. 869–919.
26. Mulvaney, R.L. Nitrogen-Inorganic Forms. In *Methods of Soil Analysis, Part 3: Chemical Methods*; SSSA Book Series, 5; Sparks, D.L., Ed.; Soil Science Society of America, American Society of Agronomy: Madison, WI, USA, 1996; pp. 1123–1184.
27. Thomas, G.W. Exchangeable Cations. In *Agronomy Monographs*; Page, A.L., Ed.; American Society of Agronomy, Soil Science Society of America: Madison, WI, USA, 2015; pp. 159–165.
28. Lindsay, W.L.; Norvell, W.A. Development of a DTPA soil test for zinc, iron, manganese, and copper. *Soil Sci. Soc. Am. J.* **1978**, *42*, 421–428. [CrossRef]
29. Keren, R. Boron. In *Methods of Soil Analysis, Part 3: Chemical Methods*; SSSA Book Series, 5; Sparks, D.L., Ed.; Soil Science Society of America, American Society of Agronomy: Madison, WI, USA, 1996; pp. 603–626.
30. Mills, A.H.; Benton, A.J.; Jones, J.B., Jr. *Plant Analysis Handbook II: A Practical Sampling, Preparation, Analysis and Interpretation Guide*; MicroMacro Publishing, Inc.: Athens, GA, USA, 1996.
31. Bremner, J.M. Nitrogen-total. In *Methods of Soil Analysis, Part 3: Chemical Methods*; SSSA Book Series, 5; Sparks, D.L., Ed.; Soil Science Society of America, American Society of Agronomy: Madison, WI, USA, 1996; pp. 1085–1121.
32. Sylvia, D.M. Vesicular-Arbuscular Mycorrhizal Fungi. In *Methods of Soil Analysis: Part 2 Microbiological and Biochemical Properties*; SSSA Book, Series; Weaver, R.W., Angle, S., Bottomley, P., Bezdicek, D., Smith, S., Tabatabai, A., Wollum, A., Eds.; Soil Science Society of America: Madison, WI, USA, 2018; pp. 351–378, ISBN 978-0-89118-865-0.
33. Cox, A.E.; Joern, B.C.; Brouder, S.M.; Gao, D. Plant-Available Potassium Assessment with a Modified Sodium Tetraphenylboron Method. *Soil Sci. Soc. Am. J.* **1999**, *63*, 902–911. [CrossRef]
34. Bilias, F.; Barbayiannis, N. Potassium Availability: An Approach Using Thermodynamic Parameters Derived from Quantity-Intensity Relationships. *Geoderma* **2019**, *338*, 355–364. [CrossRef]
35. Wallor, E.; Kersebaum, K.-C.; Lorenz, K.; Gebbers, R. Soil State Variables in Space and Time: First Steps towards Linking Proximal Soil Sensing and Process Modelling. *Precision Agric.* **2019**, *20*, 313–334. [CrossRef]

36. Kravchenko, A.N.; Harrigan, T.M.; Bailey, B.B. Soil electrical conductivity as a covariate to improve the efficiency of field experiments. *Trans. ASAE* **2005**, *48*, 1353–1357. [CrossRef]
37. Zhao, N.; Liu, H.; Wang, Q.; Wang, R.; Xu, Z.; Jiao, C.; Zhu, J.; Yu, G.; He, N. Root Elemental Composition in Chinese Forests: Implications for Biogeochemical Niche Differentiation. *Funct. Ecol.* **2018**, *32*, 40–49. [CrossRef]
38. Zhang, S.-B.; Zhang, J.-L.; Slik, J.W.F.; Cao, K.-F. Leaf Element Concentrations of Terrestrial Plants across China Are Influenced by Taxonomy and the Environment: Leaf Elements of Chinese Biomes. *Glob. Ecol. Biogeogr.* **2012**, *21*, 809–818. [CrossRef]
39. Ibourki, M.; Ait Bouzid, H.; Bijla, L.; Sakar, E.H.; Asdadi, A.; Laknifli, A.; El Hammadi, A.; Gharby, S. Mineral Profiling of Twenty Wild and Cultivated Aromatic and Medicinal Plants Growing in Morocco. *Biol. Trace Elem. Res.* **2022**, *200*, 4880–4889. [CrossRef] [PubMed]
40. Watanabe, T.; Urayama, M.; Shinano, T.; Okada, R.; Osaki, M. Application of Ionomics to Plant and Soil in Fields under Long-Term Fertilizer Trials. *SpringerPlus* **2015**, *4*, 781. [CrossRef]
41. Tariq, M.; Mott, C.J.B. Effect of Boron on the Behavior of Nutrients in Soil-Plant Systems—A Review. *Asian J. Plant Sci.* **2006**, *6*, 195–202. [CrossRef]
42. Smith, S.E.; Read, D.J. *Mycorrhizal Symbiosis*, 3rd ed.; Academic Press: Amsterdam, The Netherlands; Boston, MA, USA, 2008.
43. Miransari, M. Arbuscular Mycorrhizal Fungi and Nitrogen Uptake. *Arch. Microbiol.* **2011**, *193*, 77–81. [CrossRef]
44. Karagiannidis, N.; Nikolaou, N.; Ipsilantis, I.; Zioziou, E. Effects of Different N Fertilizers on the Activity of *Glomus Mosseae* and on Grapevine Nutrition and Berry Composition. *Mycorrhiza* **2007**, *18*, 43–50. [CrossRef]
45. Egerton-Warburton, L.M.; Allen, E.B. Shifts in arbuscular mycorrhizal communities along an anthropogenic nitrogen deposition gradient. *Ecol. Appl.* **2000**, *10*, 484–496. [CrossRef]
46. Dias, T.; Stürmer, S.L.; Chaves, S.; Fidalgo, C.; Tenreiro, R.; Correia, P.; Carvalho, L.; Martins-Loução, M.-A.; Sheppard, L.J.; Cruz, C. Species of Arbuscular Mycorrhizal Fungal Spores Can Indicate Increased Nitrogen Availability in Mediterranean-Type Ecosystems. In *Nitrogen Deposition, Critical Loads and Biodiversity*; Sutton, M.A., Mason, K.E., Sheppard, L.J., Sverdrup, H., Haeuber, R., Hicks, W.K., Eds.; Springer: Dordrecht, The Netherlands, 2014; pp. 259–266, ISBN 978-94-007-7938-9.
47. Zheng, Z.; Ma, P.; Li, J.; Ren, L.; Bai, W.; Tian, Q.; Sun, W.; Zhang, W. Arbuscular Mycorrhizal Fungal Communities Associated with Two Dominant Species Differ in Their Responses to Long-term Nitrogen Addition in Temperate Grasslands. *Funct. Ecol.* **2018**, *32*, 1575–1588. [CrossRef]
48. Karaarslan, E.; Uyanöz, R.; Doğu, S. Morphological Identification of Vesicular-Arbuscular Mycorrhiza on Bulbous Plants (Taurus Mountain in Turkey). *Arch. Biol. Sci.* **2015**, *67*, 411–426. [CrossRef]
49. Melo, C.D.; Walker, C.; Krüger, C.; Borges, P.A.V.; Luna, S.; Mendonça, D.; Fonseca, H.M.A.C.; Machado, A.C. Environmental Factors Driving Arbuscular Mycorrhizal Fungal Communities Associated with Endemic Woody Plant Picconiaazorica on Native Forest of Azores. *Ann. Microbiol.* **2019**, *69*, 1309–1327. [CrossRef]
50. Black, C.A. *Soil Fertility Evaluation and Control*, 1st ed.; CRC Press: Boca Raton, FL, USA, 1993; ISBN 978-0-429-15979-4.
51. van der Heijden, M.G.A.; Klironomos, J.N.; Ursic, M.; Moutoglis, P.; Streitwolf-Engel, R.; Boller, T.; Wiemken, A.; Sanders, I.R. Mycorrhizal Fungal Diversity Determines Plant Biodiversity, Ecosystem Variability and Productivity. *Nature* **1998**, *396*, 69–72. [CrossRef]
52. Jakobsen, S.T. Interaction between Plant Nutrients: IV. Interaction between Calcium and Phosphate. *Acta Agric. Scand. B Soil Plant Sci.* **1993**, *43*, 6–10. [CrossRef]
53. Xie, K.; Cakmak, I.; Wang, S.; Zhang, F.; Guo, S. Synergistic and Antagonistic Interactions between Potassium and Magnesium in Higher Plants. *Crop J.* **2021**, *9*, 249–256. [CrossRef]
54. Smeti, E.; Tsirtsis, G.; Skoulikidis, N.T. Geology Can Drive the Diversity–Ecosystem Functioning Relationship in River Benthic Diatoms by Selecting for Species Functional Traits. *Biology* **2023**, *12*, 81. [CrossRef]
55. Hatzilazarou, S.; Pipinis, E.; Kostas, S.; Stagiopoulou, R.; Gitsa, K.; Dariotis, E.; Avramakis, M.; Samartza, I.; Plastiras, I.; Kriemadi, E.; et al. Influence of Temperature on Seed Germination of Five Wild-Growing Tulipa Species of Greece Associated with Their Ecological Profiles: Implications for Conservation and Cultivation. *Plants* **2023**, *12*, 1574. [CrossRef]

biology MDPI

icle

rop–Weed Introgression Plays Critical Roles in Genetic ifferentiation and Diversity of Weedy Rice: A Case Study of Iuman-Influenced Weed Evolution

ng-Xing Cai, Zhi Wang, Ye Yuan, Li-Hao Pang, Ying Wang and Bao-Rong Lu *

Ministry of Education Key Laboratory for Biodiversity Science and Ecological Engineering, School of Life Sciences, Fudan University, Shanghai 200438, China; xxcai@fudan.edu.cn (X.-X.C.); 18110700017@fudan.edu.cn (Z.W.); 20110700002@fudan.edu.cn (L.-H.P.); wang_y@fudan.edu.cn (Y.W.)
* Correspondence: brlu@fudan.edu.cn

Simple Summary: To generate knowledge on how human activities influence plant evolution in agroecosystems, we analyzed allelic introgression from *japonica* rice varieties into the *indica* type of weedy rice, and the impact of crop-to-weed introgression on the genetic differentiation and diversity of the weedy populations in Jiangsu Province of China, based on InDel (insertion/deletion) and SSR (simple sequence repeat) molecular fingerprints. Results from these analyses indicated a positive correlation between increased introgression from *japonica* rice varieties and genetic differentiation in weedy rice. In addition, increased crop-to-weed introgression formed a parabola pattern of dynamic genetic diversity in weedy rice. Our case study indicated that human activities such as the frequent change in crop varieties can influence the evolution of their conspecific weeds through crop-to-weed introgression, which promotes their genetic differentiation and dynamics of genetic diversity in agroecosystems.

Abstract: As an important driving force, introgression plays an essential role in shaping the evolution of plant species. However, knowledge concerning how introgression affects plant evolution in agroecosystems with strong human influences is still limited. To generate such knowledge, we used InDel (insertion/deletion) molecular fingerprints to determine the level of introgression from *japonica* rice cultivars into the *indica* type of weedy rice. We also analyzed the impact of crop-to-weed introgression on the genetic differentiation and diversity of weedy rice, using InDel (insertion/deletion) and SSR (simple sequence repeat) molecular fingerprints. Results based on the STRUCTURE analysis indicated an evident admixture of some weedy rice samples with *indica* and *japonica* components, suggesting different levels of introgression from *japonica* rice cultivars to the *indica* type of weedy rice. The principal coordinate analyses indicated *indica–japonica* genetic differentiation among weedy rice samples, which was positively correlated with the introgression of *japonica*-specific alleles from the rice cultivars. In addition, increased crop-to-weed introgression formed a parabola pattern of dynamic genetic diversity in weedy rice. Our findings based on this case study provide evidence that human activities, such as the frequent change in crop varieties, can strongly influence weed evolution by altering genetic differentiation and genetic diversity through crop–weed introgression in agroecosystems.

Keywords: agroecosystem; genetic diversification; *indica–japonica* rice; molecular fingerprint; natural hybridization; *Oryza sativa*

Citation: Cai, X.-X.; Wang, Z.; Yuan, Y.; Pang, L.-H.; Wang, Y.; Lu, B.-R. Crop–Weed Introgression Plays Critical Roles in Genetic Differentiation and Diversity of Weedy Rice: A Case Study of Human-Influenced Weed Evolution. *Biology* **2023**, *12*, 744. https://doi.org/10.3390/biology12050744

Academic Editor: Valeria Terzi

Received: 23 March 2023
Revised: 15 May 2023
Accepted: 16 May 2023
Published: 19 May 2023

1. Introduction

Evolution is one of the most important concepts in biology, in which four essential forces, including mutation or genetic variation, selection, gene flow or introgression, and genetic drift, drive the evolutionary process [1–10]. As an important driving force, introgression plays an essential role in shaping the evolution of plant species [5–13]. Novel alleles can be transferred from one plant species/population into another genetically diverged distinct species/population through introgression [5–8,10]. This process can cause consid able changes in allele frequencies and consequently influence the evolution of the recipie species/populations [5,6,8,13]. When the cross-compatible plant species/populations con into contact, bidirectional or unidirectional introgression is likely to occur naturally throu pollen-mediated gene flow [5,8]. It is estimated that at least 25% of plant species are i volved in introgression from the same or different species [9]. Studies have also indicate that introgression is more effective in bringing genetic variation into a recipient pla population than mutations [5,6,10]. In addition, introgression can alter the genetic diversi of a recipient population in two ways: increases in genetic diversity by the transfer of nov alleles from other species/populations [5–7,10,12], or in contrast, losses of genetic diversi by genetic swamping [6,13].

Domesticated plant species, which are commonly referred to as crops, and the phylogenetically close wild relatives provide excellent cases for studying the influence of introgression on evolution, because of their intimate relationships, allowing natura introgression to occur among crops and their wild relatives [7,8,12,14–17]. Hübner et a sequenced the total genomes of cultivated and wild sunflowers to determine the origi of genetic variation in cultivated sunflowers through wild-to-crop introgression, wher ~1.5% genetic variation in cultivated sunflowers had its origin from introgression [14 In addition, the estimate of introgression and its evolutionary impact involving cro species and their conspecific weed is very practical because of the following advantage firstly, as the same biological species, natural introgression between a crop species and it conspecific weed occurs frequently without reproductive barriers [8,12,14–18]; secondly unidirectional introgression from a crop to its conspecific weed occurring in the same agricultural field can be easily detected through the identification of proper molecula fingerprints [8,15,19,20]; and thirdly, the cultivation histories and farming styles of a crop species are readily available [17,18]. Altogether, these advantages provide an ideal system to trace the evolutionary process of weedy species under the influence of human activities

In fact, a great number of studies have confirmed frequent and extensive introgres sion of crop-specific alleles into weedy populations in agroecosystems, including weedy sorghum (*Sorghum halepense* L.), weedy sunflowers (*Helianthus annus*), and weedy rice (*Oryza sativa* f. *spontanea*) [8,15–17,20]. Such introgression had considerable influences on genetic diversity and adaptation of the recipient weedy populations, which created more weed problems, and also on the extinction of small populations of wild relatives [7,13,16,20] Therefore, it is possible to reveal the consequences and underlying mechanisms of crop-to-weed introgression and its evolutionary impact by studying a conspecific weed and its cooccurring crop species. Crop-specific alleles can accumulate in weedy populations by introgression from different crop varieties through time, which may result in genetic differentiation and increased adaptability of the weedy populations [18–24]. However, the knowledge on how crop-to-weed introgression influences genetic differentiation and diversity of the receipt weedy populations is still limited.

Weedy rice is a typical conspecific weed in the genus *Oryza* (*Poaceae*), infesting world-wide cultivated rice (*Oryza sativa* L.) fields and causing considerable rice yield losses around the world [25–27]. Similar to its cultivated counterpart, weedy rice is also differentiated into *indica* and *japonica* types, depending on its cooccurring rice varieties at the large scales of its geographical distribution [21,23,25,27]. Generally, the *indica* type of weedy rice infests the tropical or subtropical rice cultivation regions, such as in southern China, South, and Southeastern Asian countries [27–30], whereas the *japonica* type of weedy rice occurs in the temperate rice cultivation regions, such as in northern and northeastern China, the United States, and southern Europe [21,27,29–32]. Weedy rice populations with the origin from de-domestication are genetically similar to their cooccurring cultivated rice varieties, showing *indica–japonica* genetic differentiation [22,32]. Usually, long-term cooccurring of weedy rice populations with rice varieties has resulted in their similar morphological and physiological characteristics, most likely through natural gene flow or introgression [27,28].

Therefore, weedy rice provides an ideal system for studying the evolutionary impact of crop-to-weed introgression, because human activities such as the change in rice varieties and cultivation styles happen rapidly in a particular rice planting region [17–22].

Jiangsu Province (JS) is an important rice planting and production region in China, where *indica* rice varieties were prominently cultivated traditionally [33–37]. The gradual replacement of rice seedling transplanting by direct seeding in JS for the past decades largely promoted the emergence and infestation of weedy rice in rice fields, even though farmers have used commercial and certified rice seeds, and applied herbicides to control weeds [29,33,34]. Historically, no natural distribution of wild *Oryza* species has been reported in this province. Studies have indicated that JS weedy rice has its de-domestication origins, meaning that JS weedy rice has evolved from the domesticated rice varieties [38–41], such as the origins of weedy rice from other rice planting regions (e.g., northeastern China) where no wild *Oryza* species are distributed either [21,29,31]. Given that no wild rice species are naturally distributed in these regions, weedy rice there is genetically similar to the cultivated rice, rather than wild *Oryza* species (e.g., *O. rufipogon* and *O. nivara*) [34,38–42]. Similar to the characteristics of their co-occurring rice varieties, weedy rice populations found in JS were essentially the *indica* types [23,34]. Since the end of the 1950s, *japonica* rice varieties were gradually cultivated in this province due to the favorable taste and high yielding of *japonica* rice varieties [35,36]. Now, *japonica* rice varieties are prominently cultivated in JS [33,37]. As a consequence, some weedy rice individuals with *japonica* characteristics are identified in the rice planting regions of this province [23,34,42]. Obviously, the introgression of *japonica*-specific alleles from *japonica* rice varieties into weedy rice with the *indica* genetic background has played an important role in changing *indica–japonica* characteristics of weedy rice in JS where the rapid change from planting *indica* to *japonica* rice varieties has taken place. Apart from observing the *japonica* type of weedy rice in JS, our knowledge concerning *japonica*-crop-to-*indica*-weed introgression in the evolution of weedy rice is still limited. Therefore, investigating the cultivation history of *indica* to *japonica* rice varieties associated with introgression, genetic differentiation, and diversity of weedy rice may provide a deep insight into the role of crop-to-weed introgression in plant evolution.

In this case study, we used the InDel (insertion/deletion) molecular markers [43,44] to determine the genetic differentiation of weedy rice associated with the extent of crop-to-weed introgression in JS. We also applied microsatellite (simple sequence repeats, SSR) molecular markers to examine the genetic diversity pattern of weedy rice associated with the extent of crop-to-weed introgression. The primary objectives of the study were to (1) determine the historical changes in rice cultivation associated with *indica* to *japonica* varieties in JS; (2) examine whether introgression of *japonica* alleles from cultivated rice affects the genetic structure of weedy rice; (3) and assess the influences of crop-to-weed introgression on genetic differentiation of weedy rice and on the patterns of genetic diversity in weedy rice. The generated knowledge will facilitate our understanding of the impact of human-influenced evolution on weedy species in agroecosystems.

2. Materials and Methods

2.1. Historical Changes in Rice Cultivation from Indica to Japonica Varieties in Jiangsu Province

The change in planting areas from *indica* to *japonica* rice varieties in Jiangsu Province (JS) was estimated mainly based on the published statistical data, including Jiangsu Agricultural Statistics (1949–1987), Economic Data of Rural Jiangsu Province (1988–1998), and Statistical Yearbook of Rural Jiangsu Province (2000–2017) [45–47]. In addition, the published books, including The Science of Rice Cultivation in Jiangsu Province, The History of Agricultural Development in Jiangsu Province, and The Documents of Jiangsu Provincial Annals (Agricultural) [48–50], were also used to estimate the historical change in rice varieties cultivated in JS. To demonstrate the pattern of the spatial temporal changes in *japonica* rice cultivation in JS, we calculated the average areas (by 1000 hectares, kha) of *japonica rice* cultivation for every 10 years, based on each county as a unit from the 1950s to 2020s.

The obtained average data of *japonica* rice cultivation areas were visually presented in GIS maps with the different units in JS, using the software ArcGIS ver.10.2 [51].

2.2. Collection of Plant Materials

Mature seed samples were collected from a total of 36 natural weedy rice populations during October 2020~2021 across the rice planting regions in JS (Figure 1 and Table S1). The density of weedy rice occurring in the JS rice fields was about 0.5–1.0 plant per 100 m^2. For sample collection, we included 30~32 randomly selected individuals (samples) from each weedy rice population with the spatial distances >10 m in a sampling rice field (about 6000 m^2). In comparison, mature seed samples of the accompanying rice varieties in the same fields were also collected. The spatial distances between the collecting sites (fields) for each weedy rice population were >10 km. In addition, mature seed samples from three weedy rice populations each in Guangdong Province (GD) and northeast China (NEC) were collected to represent the *indica* and *japonica* types of weedy rice, respectively (Table S1). Furthermore, mature seeds of 13 typical *indica* and 13 typical *japonica* rice varieties from various sources identified by "InDel molecular index" [43] were also included as the references for further analyses (Table S1).

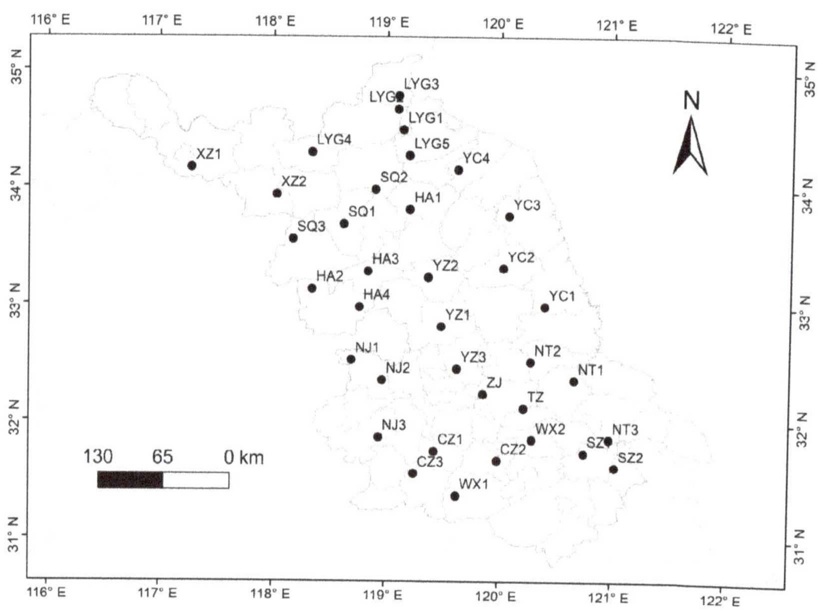

Figure 1. Sampling sites of weedy rice populations (black dots) in Jiangsu Province of China. The identification of the weedy rice populations is provided in Table S1 (Population ID in the 4th column).

2.3. DNA Extraction, Amplification, and Genotyping

Seeds of weedy rice and cultivated rice were germinated in an illuminated incubator (Percival Scientific, Perry, IA, USA) (25 ± 3 °C) with alternating light/dark (16/8 h). The total genomic DNA was extracted from the 14-day-old fresh seedlings following a modified CTAB protocol [52].

Thirty-eight *indica–japonica* specific InDel primer pairs (Table S2) [43,44], distributed on both arms of each of the 12 rice chromosomes, were selected to determine *indica* and *japonica* characteristics of all the weedy and cultivated rice samples. In addition, 47 SSR primer pairs (Table S2) from the rice genome distributed across the 12 chromosomes were selected from the Gramene Markers Database (https://archive.gramene.org/markers/, accessed on 20 April 2023) [53] to analyze genetic diversity of weedy rice populations. All the forward primers of the selected InDel and SSR primer pairs were labeled with one of the following fluorescent dyes: FAM (blue), HEX (green), ROX (red), and TAMRA (black), respectively.

Polymerase chain reactions (PCR) were performed in a total volume of 20 μL containing 1 × PCR buffer (with Mg^{2+}), 200 μM dNTPs, 4 μM of each primer, 0.5 U Taq polymerase, 40 ng template DNA, and ddH_2O to the final volume. Reaction conditions comprised an initial denaturation step for 4 min at 94 °C, followed by 30 cycles of 30 s at 94 °C, 30 s at 52~58 °C, and 30 s at 72 °C, and a final extension step at 72 °C for 10 min. According to the size of the PCR products, labeled PCR products of 3~5 InDel or SSR primer pairs were mixed together in a ratio of FAM:HEX:ROX:TAMRA = 1:3:1:3, then were electrophoresis separated on ABI 3730xl Analyzer (Applied Biosystems, Waltham, MA, USA). For genotyping, the separated InDel or SSR fragments of each sample were scored using the software GeneMapper version 4.1 (Applied Biosystems, Waltham, MA, USA).

2.4. Data Analysis

2.4.1. *Indica–Japonica* Characterization of Weedy Rice and Cultivated Rice

The frequency of *japonica*-specific alleles (F_j) of each weedy and cultivated rice sample was calculated based on the "InDel molecular index" [43]. For each InDel locus, the typical *indica* rice variety (93–11) and the *japonica* rice variety (Nipponbare) were used as the reference to determine the homozygote *indica* (II), *japonica* (JJ), and heterozygote (IJ) genotypes, respectively (for details, see Lu et al. [43]). In this study, the *indica–japonica* characterization of all weedy and cultivated rice samples was determined as the *indica* type ($F_j < 0.25$), intermediate type ($0.25 \leq F_j < 0.75$), and *japonica* type ($F_j \geq 0.75$), respectively, based on the average values of F_j or "InDel molecular index" of the 38 InDel loci [43,44].

2.4.2. Estimate of Crop-to-Weed Introgression Using the Frequency of *Japonica*-Specific Alleles (F_j)

Introgression of the *japonica*-specific alleles was examined based on the InDel data matrix, including JS weedy rice that originally had an *indica* genetic background and the reference rice samples (26 typical *indica* and *japonica* verities), using the software STRUCTURE v.2.3.4 [54]. The STRUCTURE analysis was conducted with the admixture model and the correlated allele frequency model among groups, with an initial burn-in run of 100,000 steps followed by 200,000 MCMC iterations. The group number (K) was set from 2 to 8 with 10 runs for each K value. The appropriate K value was determined by calculating the value of ΔK described in Evanno et al. [55] using the Structure Harvester (https://taylor0.biology.ucla.edu/structureHarvester/ accessed on 20 April 2023).

To estimate the possible application of the frequency of *japonica*-specific alleles (F_j) for measuring the levels of crop-to-weed introgression from *japonica* rice varieties, we analyzed the correlation between the F_j values [43] and the proportion of introgressed *japonica* genetic components (alleles) derived from the "inferred ancestry of individuals" cluster data matrix generated in the STRUCTURE analysis [54]. Results of the correlation between the F_j values and the proportion of introgressed *japonica* genetic components (alleles) were visualized by a linear model of Pearson's correlation method using Prism v.8.0.2 [56].

2.4.3. Correlation between Genetic Differentiation and Crop-to-Weed Introgression in JS Weedy Rice

To determine the *indica–japonica* genetic differentiation pattern of JS weedy rice, we performed a principal coordinate analysis (PCoA) of the genotypic data matrix based on the 38 InDel molecular fingerprints, using the software GenAlEx v.6.5 [57]. The data matrix included weedy rice samples from JS, GD, and NEC, in addition to the typical *indica* and *japonica* rice varieties as references. Results from the first two principal coordinates of all studied samples were graphed in a 2-dimensional scatterplot to illustrate the genetic differentiation of JS weedy rice samples.

To estimate correlations between genetic differentiation and crop-to-weed allelic introgression of weedy rice, we established 30 ideally sampled weedy rice populations (ISWPs), each containing 16 samples randomly drawn from the JS weedy rice pool with a total of 1116 samples. These populations represented the low ($F_j < 0.25$), middle ($0.25 \leq F_j < 0.75$), and high ($F_j \geq 0.75$) levels of introgression, respectively. Consequently, there were 10 low-introgression, 10 middle-introgression, and 10 high-introgression ISWPs established for further correlation analyses. In addition, we randomly selected eight naturally sampled weedy rice populations (NSWPs) from the field sampled JS weedy rice populations based on their F_j values, ranging from 0.05 (lowest) to 0.34 (highest), to analyze the relationships between genetic differentiation of weedy rice and crop-to-weed allelic introgression revealed by ISWPs.

The pairwise Wright's F_{st} values [58] were calculated to represent the genetic differentiation (F_{st}) between ISWPs based on the genotyping data matrixes of both InDel and SSR molecular fingerprints, respectively, using the software GenAlEx v.6.5 [57]. Moreover, the pairwise differences in the F_j values were calculated to represent the differences in introgression ($F_{j\text{-}d}$) between ISWPs based on the same data matrixes. The correlation between genetic differentiation and crop-to-weed allelic introgression was calculated based on the obtained F_{st} and $F_{j\text{-}d}$ values of the 30 ISWPs and eight NSWPs, respectively. The obtained correlation was visualized using Pearson's correlation method (linear model) packaged in the software Prism v.8.0.2 [56].

2.4.4. Correlation between Genetic Diversity and Crop-to-Weed Introgression in JS Weedy Rice

To estimate the correlation of genetic diversity (represented by Nei's expected heterozygosity, H_e and Shannon's information index, I) [59,60] with the level of crop-to-weed introgression (F_j), we calculated the two genetic diversity parameters using the same ISWPs and NSWPs (see the Section 2.4.3), respectively, based on the genotyping data matrices of both InDel and SSR molecular fingerprints. The association of the genetic diversity parameters (H_e and I) with F_j values was obtained using the quadratic equation fitting regression analysis for the ISWPs. The obtained results were visualized using a nonlinear model packaged in the software Prism v.8.0.2 [56]. In addition, we also analyzed the association of the genetic diversity parameters (H_e and I) with the F_j values in the eight NSWPs using Pearson's correlation method (linear model) packaged in the software Prism v.8.0.2 [56] to confirm the correlation pattern obtained based on ISWPs.

3. Results

3.1. *Rapid Alteration of Rice Cultivation from Indica to Japonica Varieties in Jiangsu Province*

The historical archives indicated that the tradition of rice cultivation in Jiangsu Province (JS) was essentially *indica* varieties until the early 1950s, when *japonica* rice varieties were introduced to this province due to their favorable commercial quality (tastes) and high yield. Statistical data between the 1950s and 2010s evidently demonstrated a relatively rapid change in rice cultivation from *indica* to *japonica* rice varieties in this province (Table S3 and Figure 2). Starting from the end of the 1950s, the cultivation areas for *japonica* rice varieties gradually expanded into the areas where the *indica* varieties were originally cultivated in this province within only a few decades. Consequently, *indica* rice varieties were almost completely replaced by *japonica* rice varieties by the end of the 1990s, particularly in some rice cultivation areas, such as the northeast, east, and south part of this province (Figure 2).

The statistical data also indicated that the alteration of cultivation from *indica* to *japonica* rice varieties started in the southern part of JS. Then, the cultivation areas of *japonica* rice varieties were gradually extended northward to cover the rice cultivation areas in the whole province. The cultivation areas for *japonica* rice varieties increased from only 3.4 kha (1000 hectares) in the 1950s to 2015.3 kha in the 2010s. In contrast, the cultivation areas for *indica* rice varieties dramatically decreased from 1988.5 kha in the 1950s to 290.0 kha in the 2010s (Table S3). Therefore, the different durations of rice cultivation for *japonica* varieties in different areas could cause the different levels of *japonica*-specific allelic introgression into the *indica* type of weedy rice cooccurring with the *japonica* rice varieties in these areas.

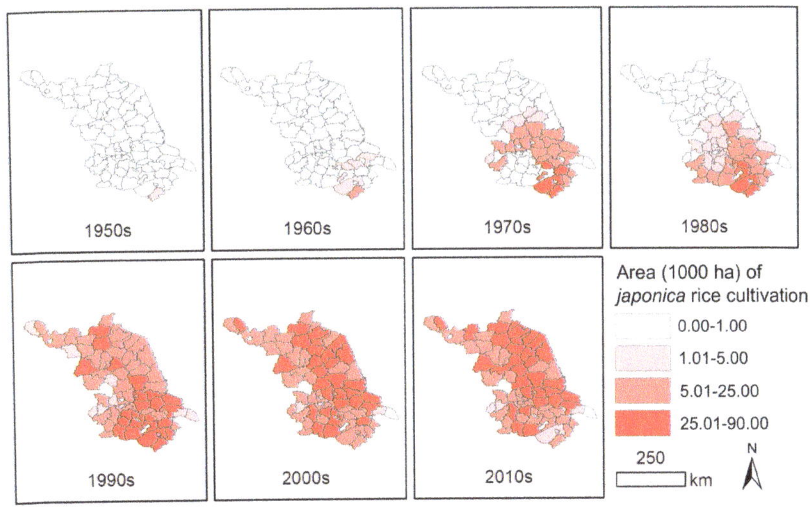

Figure 2. The rapid increases in cultivation areas of *japonica* rice varieties from *indica* rice varieties in Jiangsu Province during the 1950s–2010s. The lines represent the borders of the counties. The color intensity (4 levels) indicates the 10-year average cultivation areas in 1000 hectares (kha) of *japonica* rice varieties in each county: white, 0.00–1.00 kha; light-pink, 1.01–5.00 kha; pink, 5.01–25.00 kha, and red, 25.01–90.00 kha.

3.2. Patterns of Introgression from Japonica Rice Varieties to Weedy Rice

In general, results based on the calculated "InDel molecular index" clearly indicated that each of the examined cultivated rice and weedy rice samples from Guangdong Province (GD), northeast China (NEC), and JS had their distinct *indica*, *japonica*, or *indica–japonica* (determined as intermediate) characteristics (Table S1). This characterization was based on the "InDel molecular index" or examined the frequency of *japonica*-specific alleles (F_j) at the 38 InDel (insertion/deletion) loci across the rice genome. The F_j values of the typical *indica* rice varieties and weedy rice samples from GD were equal to 0 or close to zero, whereas the F_j values of the typical *japonica* rice varieties and weedy rice samples from NEC were equal to one or close to one (Table S1). These results suggest that the reference cultivated rice and weedy rice samples from GD or NEC rice cultivation regions had identical and unique *indica* or *japonica* characteristics, respectively, which set up an ideal standard for a comparison of the *indica–japonica* characteristics of JS weedy rice samples. However, the F_j values of the 1116 weedy rice samples from JS ranged from 0 to 0.97 (Table S1); although, most of the samples still showed their *indica* characteristics (Table S1). In the JS weedy rice pool, ~91% were the *indica* type ($F_j < 0.25$), ~7% were the intermediate type ($0.25 \leq F_j < 0.75$), and ~2% were the *japonica* type ($F_j \geq 0.75$) (Table 1). In addition, most (~81%) cooccurring rice cultivars from JS belonged to the *japonica* type (Table S1). These results clearly indicated the introgression of *japonica*-specific alleles from *japonica* rice varieties into the original *indica* type of weedy rice populations in the JS rice cultivation region.

To determine the level of introgression of *japonica*-specific alleles into weedy rice in JS rice fields, where weedy rice was originally composed of the *indica* genetic background, we conducted the STRUCTURE analysis using the admixture model based on the data matrices of the InDel molecular fingerprints. Results from the STRUCTURE analysis demonstrated the distinct genetic components of the typical *indica* (red and blue) and *japonica* (green) rice varieties (references) at the most optimal K value (K = 3) and their neighboring K values (K = 2, K = 4). The findings suggested the distinct genetic components and substantially diverged genetic relationships of the *indica* and *japonica* cultivated rice samples used as references (Figure 3). The distinguishable *indica* and *japonica* genetic components set up an excellent standard for studying the introgression of *japonica*-specific alleles (green) into the *indica* genetic background (red and blue) of weedy rice samples/populations.

Table 1. Identification of the *indica–japonica* characteristics of weedy rice samples from Guangdong and Jiangsu Provinces, and northeast China, based on the InDel molecular index. F_j indicates the average values and standard deviation (in parenthesis).

Origin	F_j	Type [1]	Percent [2]	Introgression [3]
Guangdong	0.07 (0.021)	*Indica*	-	Low
Jiangsu	0.07 (0.001)	*Indica*	91.85%	Low
Jiangsu	0.38 (0.012)	Intermediate	6.72%	Middle
Jiangsu	0.86 (0.015)	*Japonica*	1.43%	High
Northeast	0.94 (0.058)	*Japonica*	-	Low

[1] Classified based on the InDel index (F_j) following the method of Lu et al. [43]; F_j <0.25, *indica;* $0.25 \leq F_j < 0.75$, intermediate; *japonica*, $F_j \geq 0.75$. [2] Only weedy rice samples from Jiangsu Province were counted. [3] Expected levels of *japonica* allelic introgression.

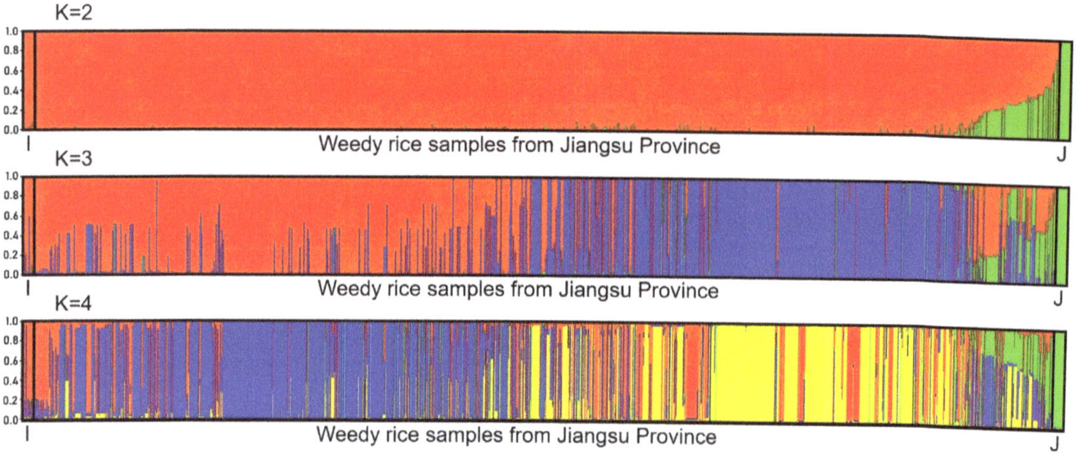

Figure 3. Bar plots indicating genetic components of typical *indica* and *japonica* rice varieties (references) and weedy rice samples from Jiangsu Province based on the STRUCTURE analysis of the InDel molecular fingerprints [43,54] at the most optimal K value (K = 3) and their neighboring K values. Each sample is represented by a single vertical line, proportional to the sample's estimated ancestry of genetic components from *indica* or/and *japonica*. Many weedy rice samples showed the admixture of *indica* and *japonica* genetic components, particularly at K = 3 and K = 4.

As expected, weedy rice samples collected from different locations in JS showed their distinct genetic components (Figure 3). Compared with the genetic components of the references (*indica*: red and blue, *japonica*: green), the weedy rice samples from JS showed both *indica* and *japonica* genetic components (Figure 3). Apparently, most JS weedy rice samples had *indica* genetic components, suggesting their close genetic affinity with their originally cultivated *indica* rice varieties (Figure 3). However, some JS weedy rice samples exhibited an admixture of *indica–japonica* genetic components, indicating the different degree of introgression from *japonica* rice varieties to the *indica* type of weedy rice. In addition, the admixture genetic components of the *japonica* weedy rice also ruled out the possible contamination of *japonica* weedy rice seeds that should not have admixture components. Noticeably, the typical *japonica* rice reference varieties showed a distinctly unique genetic component (green) with nearly no admixture; although, the typical *indica* rice reference varieties showed somehow other genetic components with an extremely low level of admixture at the K = 3 and K = 4 (Figure 3). These results provide opportunities to analyze the level of introgression of the *japonica*-specific alleles in the JS weedy rice samples, most of which had *indica* genetic components.

To determine whether the frequency of the *japonica*-specific alleles (F_j) obtained based on the "InDel molecular index" could be directly used for estimating the level of allelic introgression from *japonica* varieties into the *indica* type of weedy rice, we analyzed the correlation between the obtained F_j values and the ratios of the introgressed *japonica*-component of weedy rice samples extracted from the "ancestry of individuals" cluster data matrix in the STRUCTURE analysis. Results from the correlation analysis indicated a significant positive correlation ($R^2 = 0.94$, $p < 0.001$) between the F_j values and the ratios of introgressed *japonica* components (Figure 4). This finding suggests that the F_j values calculated based on the "InDel molecular index" (Table S1) could be used to estimate the level of *japonica*-specific allelic introgression for further analyses, particularly the relationships of crop-to-weed introgression with genetic differentiation and genetic diversity in JS weedy rice based on the F_j values.

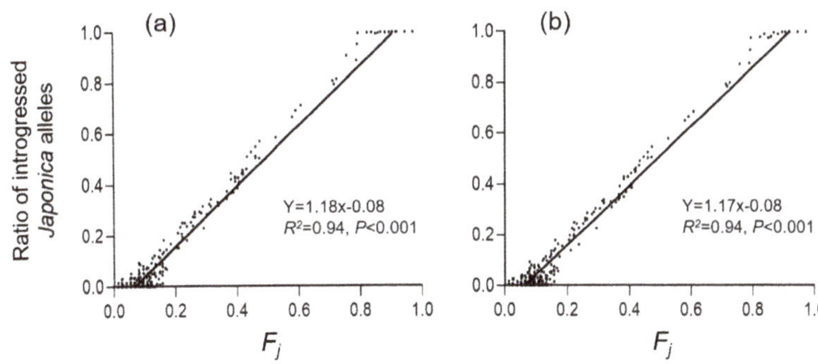

Figure 4. Correlation between the frequency of *japonica*-specific alleles (F_j) based on the InDel molecular index [43] and the ratios of introgressed *japonica*-alleles (components) obtained from the STRUCTURE analysis in weedy rice populations from Jiangsu Province in China, when K = 2 (**a**) and K = 3 (**b**). Black dots represent weedy rice samples, black lines are regression lines.

3.3. Genetic Differentiation in Weedy Rice Associated with Crop-to-Weed Introgression

To investigate the patterns of genetic differentiation in JS weedy rice samples, we conducted the principal coordinate analysis (PCoA) based on the InDel molecular fingerprints, using typical *indica* and *japonica* rice varieties and weedy rice samples from GD and NEC as references. The PCoA results demonstrated evident genetic differentiation of JS weedy rice samples into *indica* and *japonica* types (Figure 5). Obviously, most weedy rice samples from JS were scattered and were closely associated with the typical *indica* rice varieties and the weedy rice samples from GD, at the negative loads of the first principal coordinate (left in Figure 5). Therefore, these weedy rice samples were likely the *indica* type, which was supported by their low F_j values (<0.25, Table 1). A small proportion of weedy rice samples from JS was scattered among the typical *japonica* rice varieties and the weedy rice samples from NEC (references), at the positive loads of the first principal coordinate (right in Figure 5). Therefore, these weedy rice samples were most likely the *japonica* type, which was also supported by their high F_j values (>0.75, Table 1). Noticeably, some of the JS weedy rice samples were scattered between the typical *indica* and *japonica* types along the first principal coordinate (middle in Figure 5), which were determined as the *indica–japonica* intermediate types with the F_j value between 0.25 and 0.75 (Table 1).

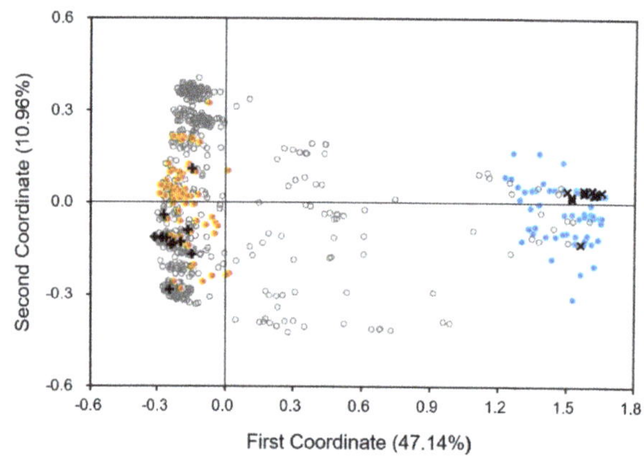

Figure 5. The scatterplot showing genetic relationships of the 1116 weedy rice samples from Jiangsu Province (JS), in addition to the reference samples including typical *indica* and *japonica* rice varieties, and weedy rice from Guangdong Province (GD) and northeast China (NEC) based on the principal coordinate analysis of the InDel dataset. Grey empty dots, JS weedy rice; +, typical *indica* rice varieties; ×, typical *japonica* rice varieties; orange dots, weedy rice from GD; blue dots, weedy rice from NEC.

In general, the PCoA results clearly indicate the genetic differentiation of JS weedy rice samples that are scattered between the typical reference *indica* and *japonica* rice varieties. This finding was supported by the gradually increased *japonica*-specific allelic frequency (F_j) in the JS weedy rice samples that should originally be the *indica* type.

To estimate the correlation between the extent/level of genetic differentiation and crop-to-weed introgression, we calculated the pairwise genetic differentiation (F_{st}) and differences in allelic frequency ($F_{j\text{-}d}$), based on the data matrices of InDel (Figures 6a and 7a) and SSR (Figures 6b and 7b) molecular fingerprints. The correlation analysis was conducted based on the 30 ideally sampled weedy rice populations (ISWPs) with their respective low, middle, and high levels of introgression (Figure 6), and eight natural weedy rice populations (NSWPs) (Figure 7) for both InDel and SSR molecular fingerprints. Results based on the Pearson's correlation analysis showed a significantly positive correlation between genetic differentiation as measured by the pairwise F_{st} values and crop-to-weed introgression as estimated by the pairwise $F_{j\text{-}d}$ values of ISWPs ($R^2 = 0.94$–0.96, $p < 0.001$) and NSWPs ($R^2 = 0.35$–0.73, $p < 0.01$). These results generated from the ISWPs and NSWPs of JS weedy rice suggest that crop allelic introgression would cause considerable genetic differentiation of its conspecific weed.

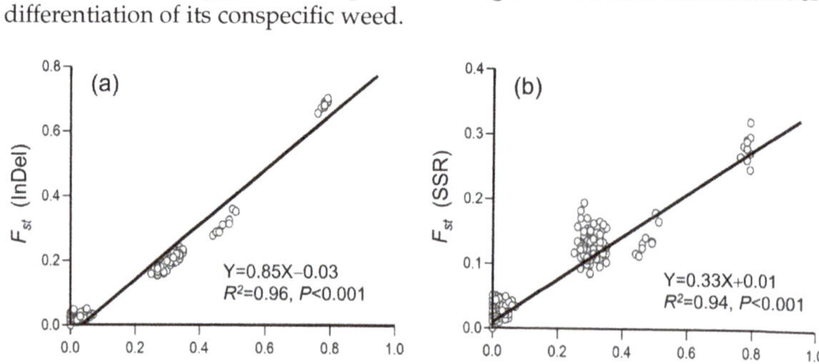

Figure 6. Correlation between the level of pairwise genetic differentiation (F_{st}) and differences in crop-to-weed introgression ($F_{j\text{-}d}$) of the ideally sampled weedy rice populations (empty dots) based on insertion/deletion (InDel, (**a**)) and simple sequence repeat (SSR, (**b**)) molecular fingerprints. Black lines are regression lines.

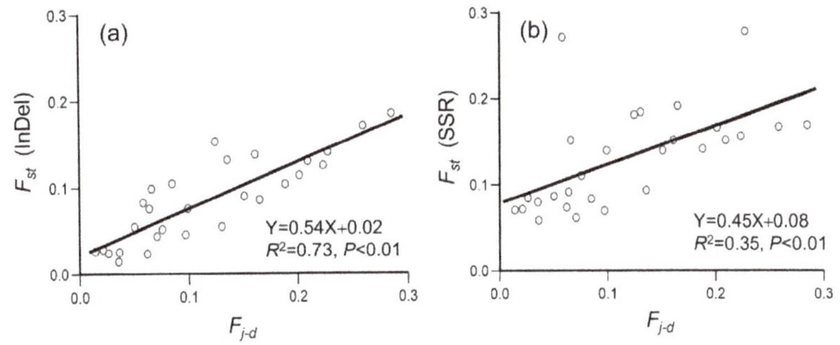

Figure 7. Correlation between the level of pairwise genetic differentiation (F_{st}) and differences in crop-to-weed introgression ($F_{j\text{-}d}$) of the naturally sampled weedy rice populations (empty dots), based on insertion/deletion (InDel, (**a**)) and simple sequence repeat (SSR, (**b**)) molecular fingerprints. Black lines are regression lines.

3.4. Genetic Diversity of Weedy Rice Associated with Crop-to-Weed Introgression

To estimate the correlation/relationship between the level of genetic diversity and crop-to-weed introgression, we calculated Nei's expected heterozygosity (H_e) and Shannon's information index (I) to represent genetic diversity, based on the data matrices of InDel and SSR molecular fingerprints. In addition, we used the frequency of *japonica*-specific alleles (F_j) to represent the level of crop-to-weed introgression. The correlation analysis was conducted based on the 30 ISWPs with their respective low, middle, and high levels of introgression (Figure 8) for both InDel and SSR molecular fingerprints. Results based on the quadratic equation fitting regression analysis indicated a high degree of fitting (R^2 = 0.98 for H_e and 0.99 for I) for the InDel molecular fingerprints (Figure 8a,c). Similarly, the quadratic equation fitting regression analysis indicated a relatively high degree of fitting (R^2 = 0.81 for H_e and 0.78 for I) for the SSR molecular fingerprints (Figure 8b,d).

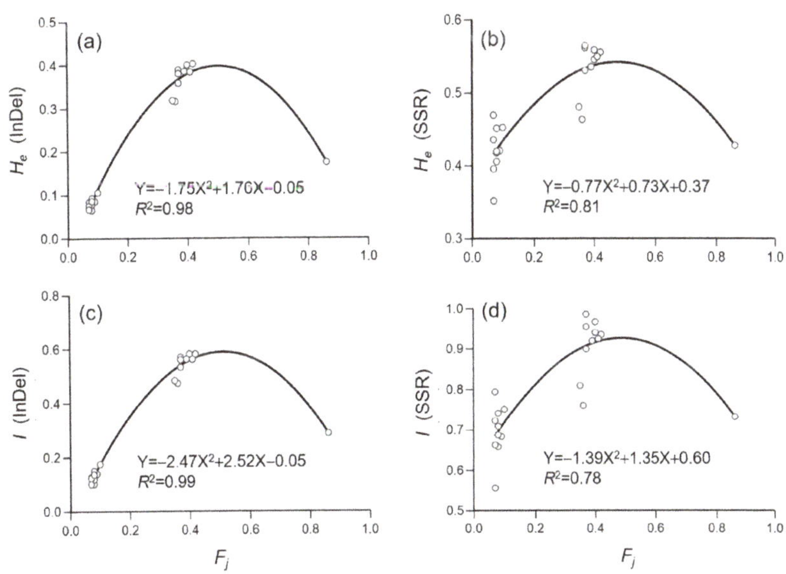

Figure 8. Results of regression between genetic diversity represented by Nei's expected heterozygosity (H_e) and Shannon's information index (I) [59,60] and the frequency of *japonica* specific alleles (F_j) [43], based on the insertion/deletion (InDel, (**a**,**c**)) and simple sequence repeat (SSR, (**b**,**d**)) data matrices of the ideally sampled weedy rice populations (empty dots). Black curves are regression lines.

In addition, the correlation between the level of genetic diversity (H_e and I) and crop-to-weed introgression (F_j) was also analyzed based on the eight NSWPs (Figure 9). Results showed that the level of genetic diversity (H_e and I) significantly increased with the increases in the F_j values for the InDel molecular fingerprints (Figure 9a,c), when F_j varied between 0.05 and 0.34. However, no significant correlations were observed between the level of genetic diversity (H_e and I) and the F_j values for the SSR molecular fingerprints (Figure 9b,d). The relationship between the level of genetic diversity and crop-to-weed introgression (represented by the F_j values) generated based on these results from NSWPs generally agreed with those revealed from ISWPs, particularly for InDel molecular fingerprints.

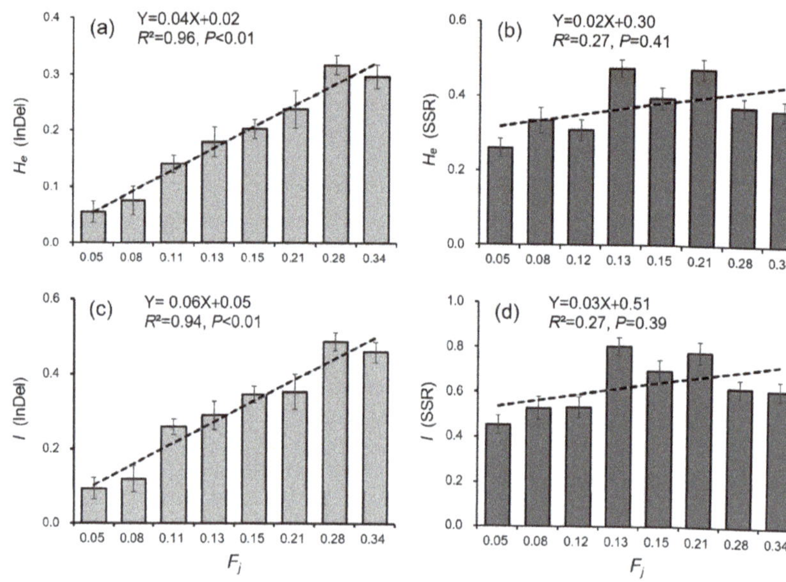

Figure 9. Correlations as represented by dashed lines (regression) between the genetic diversity parameters (H_e, I) [59,60] and the F_j values [43] of the eight naturally sampled weedy rice populations (NSWPs) from Jiangsu Province. The F_j values and the genetic diversity parameters were calculated based on the data matrixes from the InDel (**a**,**c**) and SSR (**b**,**d**) molecular fingerprints, respectively. Each column represents an NSWP (n = 31), and the error bars indicate standard errors.

4. Discussion

4.1. The Change in Rice Varieties Greatly Influences Indica–Japonica Characteristics of Weedy Rice through Crop-to-Weedy Introgression

Our results in this study based on the historical archives clearly demonstrated the pattern of traditional rice cultivation in Jiangsu Province (JS), where *indica* varieties were grown essentially until the early 1950s. Different *japonica* rice varieties were gradually introduced into this province at the beginning of the 1960s because of their favorable commercial quality (taste) and high yield [35–37]. After three decades by the 1990s, the *indica* rice varieties were almost completely replaced by *japonica* rice varieties in different rice cultivation areas in this province. As a consequence, weedy rice with *japonica* characteristics was reported to appear frequently in the rice fields of this province [23,33,42]. Apparently, the changing pattern of rice cultivation from *indica* to *japonica* varieties has considerable influences on the genetic compositions of weedy rice as indicated by its *indica* and *japonica* characteristics. Such influences are hypothetically through gene flow or allelic introgression from *japonica* rice varieties to weedy rice individuals.

In fact, our results showed that most examined JS weedy rice samples (individuals) were the *indica* type (~91%) and the rest of the weedy rice samples were the intermediate and *japonica* types, based on the InDel molecular index [43] or the average frequency of the *japonica*-specific alleles (F_j). These results were obtained based on a relatively large number of weedy rice samples (1116 individuals) collected from 36 populations across the JS rice cultivation areas. This finding indicates the change in characteristics of JS weedy rice gradually from the original *indica* type to the *japonica* and *indica–japonica* intermediate types, associated with the rapid changes in rice cultivation patterns from *indica* to *japonica* varieties in this province. In other words, the rapid alteration of rice cultivation from *indica* to *japonica* varieties promoted the divergence of JS weedy rice characteristics, essentially through crop-to-weed introgression: although, other factors, such as direct-seeding and mechanic harvesting, cannot be completely excluded. The changes in the *indica–japonica* characteristics of JS weedy rice found in this study are similar to those revealed either by phenotypical characterization [42,61] or molecular fingerprinting [23,34], in which weedy rice was composed of the *indica*, *japonica*, and intermediate types, although predominated by the *indica* type. All results clearly indicate a wide range of variation regarding the *indica–japonica* characteristics of JS weedy rice, owing to the rice cultivation change from *indica* to *japonica* varieties.

In this study, we also found that the reference weedy rice populations from NEC were essentially the *japonica* type that was associated strongly with the typical *japonica* rice varieties, whereas those from GD were essentially the *indica* type that was associated intimately with the typical *indica* rice varieties. Our finding is consistent with the previous reports in which weedy rice is genetically closely associated with its cultivated counterparts co-occurring in the same regions [21,23,27,31]. Obviously, the JS weedy rice samples with different genetic backgrounds identified in this study should be the result of successive gene flow or introgression with different types of *indica* and *japonica* rice varieties cultivated during a historical period of time in this province. Successive crop-to-weed gene flow or introgression, coupled with the independent assortment and genetic recombination in the self-pollination process, promoted the admixture of *indica* and *japonica* genotypes in JS weedy rice.

Results from the STRUCTURE analysis based on the InDel molecular fingerprints in this study clearly demonstrated the change in genetic compositions in JS weedy rice from the original *indica* type to the current *japonica* types with evident admixture genetic components, as indicated by the different bar-plots with three K values (Figure 3). The presence of the *indica–japonica* admixture types of weedy rice in the JS rice cultivation areas confirmed crop-to-weed allelic introgression, because of the accumulation of *japonica* alleles in the admixture types of JS weedy rice samples. The presence of *indica–japonica* admixture types in the JS weedy rice samples can also exclude the possible contamination of these samples as *japonica* weedy rice seeds. This is because the mixed *japonica* weedy rice samples, if any, in the certified commercial *japonica* cultivar seeds, should be present as the pure *japonica* genotype in the STRUCTURE analysis. Somehow, there is a possibility of the contamination of a few weedy rice seeds with the newly introduced *japonica* rice varieties in the actual rice production. Given that the opportunities for inter-crosses between the distantly scattered weedy rice plants are very low, the gradual accumulation of *japonica* alleles in weedy rice is more likely through crop-to-weed introgression [62–68] from *japonica* rice varieties, rather than by spontaneous mutations or through seed contamination. In addition, the gradual change in JS weedy rice genetic components from the original *indica* type to the admixture and *japonica* types matches the pattern of JS rice variety change in cultivation for the past seven decades (Figure 2). The relatively low proportion of JS weedy rice samples with *japonica*-specific alleles from their cooccurring cultivars can be explained by the inbreeding feature of weedy rice with a relatively low outcrossing rate (~1%) [62–68]. Thus, we propose that crop-to-weed introgression has played an important role not only in changing the *indica–japonica* characteristics and genetic components of JS weedy rice, but also in shaping the evolution of JS weedy rice as an important driving force. These observations agree with the previous reports concerning weedy rice genetic diversity and evolution [17–22].

4.2. Crop-to-Weed Introgression Impacts Genetic Differentiation and Genetic Diversity in Weedy Rice through Accumilated Crop-Specific Alleles

Our results clearly indicate that the level of gene flow or crop-to-weed introgression can be measured by the frequency of the *japonica*-specific alleles (F_j) of the weedy rice samples with considerably high confidence (Figure 4). Results based on the PCoA analysis of the InDel molecular fingerprints also demonstrated substantial genetic differentiation of JS weedy rice, which was somehow associated with crop-to-weed introgression caused by the change in rice cultivation for *japonica* varieties. Given that the F_j value of each weedy rice sample can be accurately calculated based on the InDel molecular fingerprinting [43], we analyzed the impact of crop-to-weed introgression on genetic differentiation and genetic diversity of weedy rice. The analysis of such impact can be realized through calculating the correlation between the F_j values and the level of genetic differentiation (F_{st}), as well as genetic diversity (H_e and I), using the ideally sampled weedy rice populations (ISWPs) established according to their levels (low, middle, and high) of introgression determined by the F_j values.

Further analyses in this study indicated that the level of *japonica*-specific allelic (crop-to-weed) introgression was highly significantly correlated (R^2 = 0.94, $p < 0.001$) with the level of F_j values of the examined weedy rice samples. Therefore, the F_j values can represent *japonica*-specific allelic introgression and be directly utilized to determine the relationships between the level of crop-to-weed introgression and the genetic differentiation of weedy rice. Results obtained based on ISWPs with low, middle, and high levels of introgression, using both the InDel and SSR molecular fingerprints, clearly demonstrated the correlation patterns, in which crop-to-weed introgression substantially prompted the genetic differentiation of JS weedy rice. Interestingly, similar correlation patterns were obtained based on the randomly selected naturally sampled weedy rice populations (NSWPs) with low to middle levels of introgression in JS, although with a much lower level of correlation for the results obtained using the SSR molecular fingerprints. Therefore, we can conclude based on all results from this study that crop-to-weed introgression can significantly promote the genetic differentiation of weedy rice populations with the original *indica* genotype through accumulating *japonica*-specific alleles from its co-existing *japonica* rice varieties. In other words, the gradual introgression of different crop-specific alleles into weedy rice populations can considerably cause their within- and between-population genetic differentiation.

Our results based on the InDel molecular fingerprints also suggest that the level of F_j values was significantly associated with the two independent genetic diversity parameters (Nei's expected heterozygosity, H_e, and Shannon's information index, I) in both ISWPs and NSWPs (Figures 8 and 9). Noticeably, the polymorphic SSR molecular markers did not show a significant correlation between genetic diversity and introgression in NSWPs, in addition to a comparably lower level of correlation than the dimorphic (*indica–japonica*) InDel molecular markers. The differences in genetic diversity revealed by SSR and InDel molecular markers can easily be explained by the reasons that the formation of genetic diversity in weedy rice is not only determined by crop-to-weed introgression involving *indica–japonica* alleles or characteristics, but also by other types of alleles that are not associated with the *indica* and *japonica* characteristics. Therefore, we propose that, based on this case study, the F_j values that can represent *japonica*-specific allelic introgression can also be used to determine the relationships between the level of crop-to-weed introgression and the dynamics of genetic diversity in weedy rice populations. This conclusion is supported by previous studies, in which an increased level of crop-to-weed introgression can promote rapid changes in genetic diversity patterns in weedy populations [19,20,23,27].

4.3. Human Activities Can Accelerate the Evolution of Conspecific Weeds in Agroecosystems

It is well-known that weedy rice infests worldwide rice fields, causing considerable losses in the grain yield and quality of cultivated rice [25–27]. As a conspecific weed of cultivated rice, weedy rice evolved rapidly in rice ecosystems to adapt to the weed control and the environmental changes associated with human activities around the globe [18,25–27].

Weedy rice has become a great weed problem for rice cultivation around the world, including in the rice-planting regions in China (e.g., Jiangsu Province), which threatens sustainable rice production [33,42,69]. In some regions, a rapid change in the rice ecosystems has taken place, such as the shift from rice transplanting to direct seeding, the application of farming machinery, and the quick change in rice varieties [35–37,69]. These human activities in the rice ecosystems could have imposed a strong impact on the evolutionary processes of weeds, including weedy rice, to adapt to the changing environment. This case study that we completed in Jiangsu Province with documented changes in rice cultivation from *indica* to *japonica* varieties for the past few decades reveals the human-influenced evolution of weedy rice in agroecosystems.

Our results indicated the presence of the *japonica*-specific alleles in JS weedy rice, most likely through crop-to-weed allelic introgression based on the InDel molecular fingerprints. Such introgression is closely associated with the rapid replacement of *indica* rice varieties by *japonica* rice varieties in JS. This finding evidently indicates that human-influenced cultivation changes in rice varieties (from *indica* to *japonica*) alone have already altered the genetic components of the co-occurring weedy rice populations. Our results from the analysis of InDel and SSR molecular fingerprints based on the ISWPs and NSWPs further demonstrated a positive correlation of crop-to-weed introgression, as measured by the F_j values, with the genetic differentiation of weedy rice. In addition, these results also confirmed the close association between the levels of crop-to-weed introgression and genetic diversity both in ISWPs and NSWPs, using the two sets of molecular fingerprints.

Altogether, these findings demonstrated that human activities or disturbances can significantly influence the genetic differentiation and genetic diversity of weedy rice—a conspecific weed—only through consecutive allelic introgression from its cooccurring cultivated counterparts that have constantly been improved by humans at different periods in time. This conclusion is supported by the large number of *japonica*-specific alleles detected in JS weedy rice, which is associated with the rapid change in rice cultivation from *indica* to *japonica* varieties. Genetic differentiation and genetic diversity are two important elements that are closely associated with the evolution of plant species. Therefore, we consider that human activities or disturbances can promote the rapid adaptive evolution of conspecific weedy rice in rice ecosystems through the change in these elements, which makes the control and management of weedy rice very difficult.

Such human-influenced rapid evolution of agricultural weeds as revealed in this case study may also be frequently found in many other weedy plant species in agroecosystems [8,15,16,70,71]. Previous studies reported that crop–weed introgression promoted an increase in genetic diversity in cooccurring weed populations [23,28,71,72], due to the frequent changes in newly developed crop varieties around the world [23,24,73]. Very often, these new varieties contained many new alleles/genes, with even transgenes having great evolutionary potential [17,24,73,74]. The results presented in this study provide a convincing case to explain how a conspecific weed can evolve rapidly by accumulated crop alleles from diverse crop varieties through crop-to-weed introgression to promote its genetic differentiation and diversity. Such an impact on the adaptive evolution of conspecific weeds imposes a great challenge for the control and management of these weeds.

5. Conclusions

In this study, we detected many *japonica*-specific alleles in JS weedy rice that should originally be the *indica* type with the *indica* genetic background, based on the insertion/deletion (InDel) molecular fingerprints. The presence of the *japonica*-specific alleles in the *indica* type of weedy rice is most likely the result of gene flow or introgression of alleles from *japonica* rice varieties; although, in the practical rice production, a very low frequency of *japonica* weedy rice contamination might also happen. Such gradual introgression of *japonica* alleles is closely associated with the change in rice cultivation from *indica* to *japonica* varieties in the past few decades. Our results further indicate that crop-to-weed allelic introgression has considerably changed the genetic components of the cooccurring JS-weedy rice populations. Further analyses based on InDel and SSR molecular fingerprinting indicate a significant positive correlation between the levels of crop-to-weed introgression and genetic differentiation in JS weedy rice. Similarly, increased crop-to-weed introgression promoted a change in genetic diversity in weedy rice with a parabola correlation pattern. Altogether, the above findings demonstrate that human activities, such as the change in cultivated rice varieties, can impose a considerable impact on the evolution of its conspecific weed by promoting genetic differentiation and diversity in rice ecosystems. A similar pattern of crop-to-weed introgression promoting genetic differentiation and genetic diversity is likely found in other crops and their conspecific weeds. Therefore, we conclude that, based on this case study, human activities or disturbances may accelerate the adaptive evolution of conspecific weeds through crop-to-weed introgression in agroecosystems, which may impose a great challenge for the control and management of these weeds.

Supplementary Materials: The following supporting information can be downloaded at: https://www.mdpi.com/article/10.3390/biology12050744/s1, Table S1, The localities, GPS readings, *indica–japonica* characteristics (F_j) of weedy rice and cultivated rice samples used in this study; Table S2, DNA sequences of InDel and SSR primer pairs; Table S3, Average planting areas of *indica* and *japonica* rice varieties for every 10 years (khm^2).

Author Contributions: Conceptualization, B.-R.L.; methodology, X.-X.C. and Z.W.; investigation, X.-X.C. and Y.Y.; data curation, X.-X.C., L.-H.P. and Y.W.; writing—original draft preparation, X.-X.C. and B.-R.L. All authors have read and agreed to the published version of the manuscript.

Funding: This research was funded by the National Natural Science Foundation of China (32070245).

Institutional Review Board Statement: Not applicable.

Informed Consent Statement: Not applicable.

Data Availability Statement: Not applicable.

Conflicts of Interest: The authors declare no conflict of interest.

References

1. Slarkin, M. Gene flow in natural populations. *Annu. Rev. Ecol. Syst.* **1985**, *16*, 393–430. [CrossRef]
2. Hartl, D.L.; Clark, A.G. *Principles of Population Genetics*, 3rd ed.; Sinauer Associates: Sunderland, MA, USA, 1997; pp. 5–307.
3. Templeton, A.R. *Population Genetics and Microevolutionary Theory*, 2nd ed.; John Wiley & Sons, Inc.: Hoboken, NJ, USA, 2021; pp. 77–474.
4. Slatkin, M. Gene flow and the geographic structure of natural populations. *Science* **1987**, *236*, 787–792. [CrossRef] [PubMed]
5. Rieseberg, L.H.; Wendel, J.F. Introgression and its consequences in plants. In *Hybrid Zones and the Evolutionary Process*, 1st ed.; Harrison, R.G., Ed.; Oxford University Press, Inc.: New York, NY, USA, 1993; pp. 70–110.
6. Ellstrand, N.C. Is gene flow the most important evolutionary force in plants? *Am. J. Bot.* **2014**, *101*, 737–753. [CrossRef] [PubMed]
7. Ellstrand, N.C.; Prentice, H.C.; Hancock, J.F. Gene flow and introgression from domesticated plants into their wild relatives. *Annu. Rev. Ecol. Syst.* **1999**, *30*, 539–563. [CrossRef]
8. Morrell, P.L.; Williams Coplin, T.D.; Lattu, A.L.; Bowers, J.E.; Chandler, J.M.; Paterson, A.H. Crop-to-weed introgression has impacted allelic composition of johnsongrass populations with and without recent exposure to cultivated sorghum. *Mol. Ecol.* **2005**, *14*, 2143–2154. [CrossRef]
9. Mallet, J. Hybridization as an invasion of the genome. *TREE* **2005**, *20*, 229–237. [CrossRef] [PubMed]
10. Abbott, R.; Albach, D.; Ansell, S.; Arntzen, J.W.; Baird, S.J.; Bierne, N.; Boughman, J.; Brelsford, A.; Buerkle, C.A.; Buggs, R.; et al. Hybridization and speciation. *J. Evol. Biol.* **2013**, *26*, 229–246. [CrossRef]
11. Nettel, A.; Dodd, R.S.; Afzal-Rafii, Z.; Tovilla-Hernández, C. Genetic diversity enhanced by ancient introgression and secondary contact in East Pacific black mangroves. *Mol. Ecol.* **2008**, *17*, 2680–2690. [CrossRef]
12. Hegstad, J.M.; Nelson, R.L.; Renny-Byfield, S.; Feng, L.; Chaky, J.M. Introgression of novel genetic diversity to improve soybean yield. *Theor. Appl. Genet.* **2019**, *132*, 2541–2552. [CrossRef]
13. Rhymer, J.M.; Simberloff, D. Extinction by hybridization and introgression. *Annu. Rev. Ecol. Syst.* **1996**, *27*, 83–109. [CrossRef]
14. Hübner, S.; Bercovich, N.; Todesco, M.; Mandel, J.R.; Odenheimer, J.; Ziegler, E.; Lee, J.S.; Baute, G.J.; Owens, G.L.; Grassa, C.J. Sunflower pan-genome analysis shows that hybridization altered gene content and disease resistance. *Nat. Plants* **2019**, *5*, 54–62. [CrossRef]

15. Snow, A.A.; Culley, T.M.; Campbell, L.G.; Sweeney, P.M.; Hegde, S.G.; Ellstrand, N.C. Long-term persistence of crop alleles in weedy populations of wild radish (*Raphanus raphanistrum*). *New Phytol.* **2010**, *186*, 537–548. [CrossRef] [PubMed]
16. Ellstrand, N.C.; Meirmans, P.; Rong, J.; Bartsch, D.; Ghosh, A.; De Jong, T.J.; Haccou, P.; Lu, B.-R.; Snow, A.A.; Neal Stewart, C., Jr.; et al. Introgression of crop alleles into wild or weedy populations. *Annu. Rev. Ecol. Evol. Syst.* **2013**, *44*, 325–345. [CrossRef]
17. Wedger, M.J.; Roma-Burgos, N.; Olsen, K.M. Genomic revolution of US weedy rice in response to 21st century agricultural technologies. *Commun. Biol.* **2022**, *5*, 885. [CrossRef] [PubMed]
18. Reagon, M.; Thurber, C.S.; Olsen, K.M.; Jia, Y.; Caicedo, A.L. The long and the short of it: SD1 polymorphism and the evolution of growth trait divergence in U.S. weedy rice. *Mol. Ecol.* **2011**, *20*, 3743–3756. [CrossRef]
19. Xia, H.B.; Wang, W.; Xia, H.; Zhao, W.; Lu, B.-R. Conspecific crop-weed introgression influences evolution of weedy rice (*Oryza sativa* f. *spontanea*) across a geographical range. *PLoS ONE* **2011**, *6*, e16189. [CrossRef]
20. Jiang, Z.; Xia, H.B.; Basso, B.; Lu, B.-R. Introgression from cultivated rice influences genetic differentiation of weedy rice populations at a local spatial scale. *Theor. Appl. Genet.* **2012**, *124*, 309–322. [CrossRef]
21. Sun, J.; Qian, Q.; Ma, D.R.; Xu, Z.J.; Liu, D.; Du, H.B.; Chen, W.F. Introgression and selection shaping the genome and adaptive loci of weedy rice in northern China. *New Phytol.* **2013**, *197*, 290–299. [CrossRef]
22. Song, B.K.; Chuah, T.S.; Tam, S.M.; Olsen, K.M. Malaysian weedy rice shows its true stripes: Wild Oryza and elite rice cultivars shape agricultural weed evolution in Southeast Asia. *Mol. Ecol.* **2014**, *23*, 5003–5017. [CrossRef]
23. Song, Z.J.; Wang, Z.; Feng, Y.; Yao, N.; Yang, J.; Lu, B.-R. Genetic divergence of weedy rice populations associated with their geographic location and coexisting conspecific crop: Implications on adaptive evolution of agricultural weeds. *J. Syst. Evol.* **2015**, *53*, 330–338. [CrossRef]
24. Ellstrand, N.C. Does introgression of crop alleles into wild and weedy living populations create cryptic in situ germplasm banks? *Mol. Ecol.* **2018**, *27*, 38–40. [CrossRef] [PubMed]
25. Delouche, J.C.; Burgos, N.R.; Gealy, D.R.; de San Martín, G.Z.; Labrada, R.; Larinde, M.; Rosell, C. *Weedy Rices-Origin, Biology, Ecology and Control*, 1st ed.; FAO: Rome, Italy, 2007; pp. 1–131.
26. Chauhan, S.B. Strategies to manage weedy rice in Asia. *Crop Prot.* **2013**, *48*, 51–56. [CrossRef]
27. Nadir, S.; Xiong, H.-B.; Zhu, Q.; Zhang, X.-L.; Xu, H.-Y.; Li, J.; Dongchen, W.; Henry, D.; Guo, X.-Q.; Khan, S.; et al. Weedy rice in sustainable rice production. A review. *Agron. Sustain. Dev.* **2017**, *37*, 46. [CrossRef]
28. Ishikawa, R.; Toki, N.; Imai, K.; Sato, Y.I.; Yamagishi, H.; Shimamoto, Y.; Ueno, K.; Morishima, H.; Sato, T. Origin of weedy rice grown in Bhutan and the force of genetic diversity. *Genet. Resour. Crop Evol.* **2005**, *52*, 395–403. [CrossRef]
29. Zhang, L.; Dai, W.; Wu, C.; Song, X.; Qiang, S. Genetic diversity and origin of *japonica*- and *indica*-like rice biotypes of weedy rice in the Guangdong and Liaoning provinces of China. *Genet. Resour. Crop Evol.* **2012**, *59*, 399–410. [CrossRef]
30. Zhang, S.L.; Li, J.; Lee, D.S.; Xu, H.; Zhang, L.D.; Dongchen, W.H.; Xiong, H.B.; Zhu, Q.; Zhang, X.; Lu, B.-R.; et al. Genetic differentiation of Asian weedy rice revealed with InDel markers. *Crop Sci.* **2014**, *54*, 2499–2508. [CrossRef]
31. Cao, Q.J.; Lu, B.-R.; Xia, H.; Rong, J.; Sala, F.; Spada, A.; Grassi, F. Genetic diversity and origin of weedy Rice (*Oryza sativa* f. *spontanea*) populations found in north-eastern China revealed by simple sequence repeat (SSR) markers. *Ann. Bot.* **2006**, *98*, 1241–1252. [CrossRef]
32. Li, L.-F.; Olsen, K.M. Population genomics of weedy crop relatives: Insights from weedy rice. In *Population Genomics: Crop Plants*, 1st ed.; Rajora, O.P., Ed.; Springer Nature: Cham, Switzerland, 2020; pp. 1–25.
33. Zhang, B.H.; Dong, E.J.; Zhang, H.; Shi, Z.H.; Song, X.L.; Qiang, S.; Dai, W.M. Use of hybrid rice and the proliferation of weedy rice: A case in Jiangsu, China. *Crop Sci.* **2016**, *56*, 673–681. [CrossRef]
34. Shao, J.; Dai, W.M.; Zhang, L.J.; Song, X.L.; Qiang, S. Genetic diversity and origin of weedy rice in central region of Jiangsu Province, China. *Acta Agron. Sin.* **2011**, *37*, 1324–1332. (In Chinese) [CrossRef]
35. Gong, J.; Xing, Z.; Hu, Y.; Zhang, H.; Dai, Q.; Huo, Z.; Xu, K.; Wei, H.; Gao, H. Relative advantages of "*indica* to *japonica*" and production development strategies. *China Rice* **2013**, *19*, 1–6. (In Chinese)
36. Xu, D.; Xu, X.; Li, C. Investigation and study on the shift of "*indica* to *japonica*" rice cultivation and the extension of Nongken 58 in the southern rice regions in the middle of the 20th century. *Ancient Mod. Agric.* **2016**, *1*, 1–17. (In Chinese)
37. Deng, J.P.; Du, Y.L. The production situation and development countermeasures of *japonica* in Jiangsu province. *China Rice* **2006**, *4*, 8–11. (In Chinese)
38. Li, L.-F.; Li, Y.-L.; Jia, Y.; Caicedo, A.L.; Olsen, K.M. Signatures of adaptation in the weedy rice genome. *Nat. Genet.* **2017**, *49*, 811–814. [CrossRef] [PubMed]
39. Qiu, J.; Zhou, Y.; Mao, L.; Ye, C.; Wang, W.; Zhang, J.; Yu, Y.; Fu, F.; Wang, Y.; Qian, F.; et al. Genomic variation associated with local adaptation of weedy rice during de-domestication. *Nat. Commun.* **2017**, *8*, 15323. [CrossRef]
40. Qiu, J.; Jia, L.; Wu, D.; Weng, X.; Chen, L.; Sun, J.; Chen, M.; Mao, L.; Jiang, B.; Ye, C.; et al. Diverse genetic mechanisms underlie worldwide convergent rice feralization. *Genome Biol.* **2020**, *21*, 70. [CrossRef]
41. Zhu, Y.-Q.; Fang, J.; Wang, Y.; Pang, L.-H.; Lu, B.-R. Key roles of de-domesticaiton and novel mutation in origin and diversification of global weedy rice. *Biology* **2021**, *10*, 828. [CrossRef]

42. Chen, X.; Qiang, S.; Yang, J.; Zhang, B.H.; Zhang, Z.; Song, X.L.; Dai, W.M. Hierarchical clustering and *indica–japonica* classification: Uncover mutual spread and *indica–japonica* differentiation for weedy rice in Jiangsu Province. *Chin. J. Rice Sci.* **2015**, *29*, 82–90. (In Chinese)
43. Lu, B.-R.; Cai, X.X.; Jin, X. Efficient *indica* and *japonica* rice identification based on the InDel molecular method: Its implication in rice breeding and evolutionary research. *Prog. Nat. Sci.* **2009**, *19*, 1241–1252. [CrossRef]
44. Liu, P.; Cai, X.X.; Lu, B.-R. Single-seeded InDel fingerprints in rice: An effective tool for *indica–japonica* rice classification and evolutionary studies. *J. Syst. Evol.* **2012**, *50*, 1–11. [CrossRef]
45. Agriculture and Forestry Department of Jiangsu Province. *Agricultural Statistics of Jiangsu Province*; Agriculture and Forestry Department of Jiangsu Province: Nanjing, China, 1982; pp. 1–220. (In Chinese)
46. Jiangsu Provincial Bureau of Statistics; Agriculture and Forestry Department of Jiangsu Province. *The Year Book of Jiangsu Rural Economic Statistics*; Jiangsu Provincial Bureau of Statistics: Nanjing, China, 1998. (In Chinese, one book per year from 1988–1998)
47. Liu, G.P.; Kang, C.J. *The Yearbook of Jiangsu Rural Statistical*; Jiangsu Provincial Bureau of Statistics: Nanjing, China, 2018. (In Chinese, one book per year from 2000–2018)
48. Yang, L.J.; Cui, J.L.; Tang, Y.G. *The Science of Rice Cultivation in Jiangsu Province*; Jiangsu Science & Technology Press: Nanjing, China, 1990; pp. 5–188. (In Chinese)
49. Wang, Q.Q.; Zhang, Y.B.; Shi, Y.; Xia, Y.L.; Chen, Z.W.; Chen, D.; Huang, S.S.; Lu, J.D.; Zhang, Y.Y.; Zhao, H.Y.; et al. *The History of Agricultural Development in Jiangsu Province*; Jiangsu Science & Technology Press: Nanjing, China, 1992; pp. 8–132. (In Chinese)
50. Jiangsu Provincial Local Records Compilation Committee. *Jiangsu Provincial Annals (Agricultural Annals)*; Jiangsu Classics Publishing House: Nanjing, China, 1997; pp. 10–120.
51. Esri. Available online: https://www.esri.com/en-us/home (accessed on 20 April 2023).
52. Doyle, J.J.; Doyle, J.L. Isolation of plant DNA from fresh tissue. *Focus* **1990**, *12*, 13–15.
53. Gramene Markers Database. Available online: https://archive.gramene.org/markers/ (accessed on 20 April 2023).
54. Pritchard, J.K.; Stephens, M.; Donnelly, P. Inference of population structure using multilocus genotype data. *Genetics* **2000**, *155*, 945–959. [CrossRef] [PubMed]
55. Evanno, G.; Regnaut, S.; Goudet, J. Detecting the number of clusters of individuals using the software STRUCTURE: A simulation study. *Mol. Ecol.* **2005**, *14*, 2611–2620. [CrossRef] [PubMed]
56. GraphPad. Available online: https://www.graphpad.com/ (accessed on 20 April 2023).
57. Peakall, R.; Smouse, P.E. GenAlEx 6.5: Genetic analysis in Excel. Population genetic software for teaching and research: An update. *Bioinformatics* **2012**, *28*, 2537–2539. [CrossRef] [PubMed]
58. Wright, S. The genetical structure of populations. *Ann. Eugen.* **1951**, *15*, 323–354. [CrossRef]
59. Nei, M. Estimation of average heterozygosity and genetic distance from a small number of individuals. *Genetics* **1978**, *89*, 583–590. [CrossRef]
60. Shannon, C.E. A mathematical theory of communication. *Bell Syst. Tech. J.* **1948**, *27*, 379–423. [CrossRef]
61. Yang, J.; Wang, J.; Cao, Q.; Chen, Z.D.; Tang, L.H.; Wang, Y.P.; Fang, X.W.; Wang, C.L.; Zhong, W.G. *Indica–japonica* differentiation of chloroplast DNA of weedy rice in the Changjiang and Huaihe River valley of China. *Chin. J. Rice Sci.* **2009**, *4*, 391–397. (In Chinese)
62. Chen, L.J.; Lee, D.S.; Song, Z.P.; Suh, H.S.; Lu, B.-R. Gene flow from cultivated rice (*Oryza sativa*) to its weedy and wild relatives. *Ann. Bot.* **2004**, *93*, 67–73. [CrossRef]
63. Zuo, J.; Zhang, L.J.; Song, X.L.; Dai, W.M.; Qiang, S. Innate factors causing differences in gene flow frequency from transgenic rice to different weedy rice biotypes. *Pest Manag. Sci.* **2011**, *67*, 677–690. [CrossRef]
64. Gealy, D.R.; Burgos, N.R.; Yeater, K.M.; Jackson, A.K. Outcrossing potential between US blackhull red rice and indica rice cultivars. *Weed Sci.* **2015**, *63*, 647–657. [CrossRef]
65. Shivrain, V.K.; Burgos, N.R.; Anders, M.M.; Rajguru, S.N.; Moore, J.; Sales, M.A. Gene flow between Clearfield™ rice and red rice. *Crop Prot.* **2007**, *26*, 349–356. [CrossRef]
66. Shivrain, V.K.; Burgos, N.R.; Gealy, D.R.; Moldenhauer, K.A.K.; Baquireza, C.J. Maximum outcrossing rate and genetic compatibility between red rice (*Oryza sativa*) biotypes and Clearfield™ rice. *Weed Sci.* **2008**, *56*, 807–813. [CrossRef]
67. Goulart, I.C.G.R.; Menezes, V.G.; Bortoly, E.D.; Kupas, V.; Merotto, A. Detecting gene flow from ALS-resistant hybrid and inbred rice to weedy rice using single plant pollen donors. *Exp. Agric.* **2015**, *52*, 237–250. [CrossRef]
68. Nam, K.-H.; Kim, D.Y.; Moon, Y.S.; Pack, I.S.; Jeong, S.-C.; Park, K.W.; Kim, C.-G. Gene flow from transgenic PPO-inhibiting herbicide-resistant rice to weedy rice, and agronomic performance by their hybrids. *J. Plant Biol.* **2019**, *62*, 286–296. [CrossRef]
69. Dai, W.M.; Song, X.L.; Wu, C.; Zhang, L.J.; Zuo, R.L.; Zhang, Z.; Li, S.S.; Cao, D.; Zuo, J.; Yang, L.; et al. Investigation of weedy rice (*Oryza sativa* f. *spontantea*) occurrence in Jiangsu Province. *Jiangsu J. Agric. Sci.* **2009**, *25*, 712–714. (In Chinese)
70. Barnaud, A.; Deu, M.; Garine, E.; Chantereau, J.; Bolteu, J.; Koïda, E.O.; Mckey, D.; Joly, H.I. A weed-crop complex in sorghum: The dynamics of genetic diversity in a traditional farming system. *Am. J. Bot.* **2009**, *96*, 1869–1879. [CrossRef] [PubMed]
71. Campbell, L.G.; Lee, D.; Shukla, K.; Waite, T.A.; Bartsch, D. An ecological approach to measuring the evolutionary consequences of gene flow from crops to wild or weedy relatives. *Appl. Plant Sci.* **2016**, *4*, 1500114. [CrossRef]
72. Muller, M.H.; Latreille, M.; Tollon, C. The origin and evolution of a recent agricultural weed: Population genetic diversity of weedy populations of sunflower (*Helianthus annuus* L.) in Spain and France. *Evol. Appl.* **2011**, *4*, 499–514. [CrossRef]

73. Valérie, L.C.; Siol, M.; Vigouroux, Y.; Tenaillon, M.I.; Délye, C. Adaptive introgression from maize has facilitated the establishment of teosinte as a noxious weed in Europe. *Proc. Natl. Acad. Sci. USA* **2020**, *117*, 25618–25627.
74. Pandolfo, E.C.; Presotto, A.; Carbonell, F.T.; Ureta, S.; Poverene, M.; Cantamutto, M. Transgene escape and persistence in an agroecosystem: The case of glyphosate-resistant *Brassica rapa* L. in central Argentina. *Environ. Sci. Pollut. Res.* **2018**, *25*, 6251–6264. [CrossRef]

biology

MDPI

Article

Genome Study of α-, β-, and γ-Carbonic Anhydrases from the Thermophilic Microbiome of Marine Hydrothermal Vent Ecosystems

Mohammad Sadegh Gheibzadeh [1], Colleen Varaidzo Manyumwa [2], Özlem Tastan Bishop [2], Hossein Shahbani Zahiri [1], Seppo Parkkila [3,4] and Reza Zolfaghari Emameh [1,*]

1 Department of Energy and Environmental Biotechnology, National Institute of Genetic Engineering and Biotechnology (NIGEB), Tehran 14965/161, Iran; m_gheibzadeh@nigeb.ac.ir (M.S.G.); shahbani@nigeb.ac.ir (H.S.Z.)
2 Research Unit in Bioinformatics (Rubi), Department of Biochemistry and Microbiology, Rhodes University, Grahamstown 6140, South Africa; colleen.manyumwa06@gmail.com (C.V.M.); o.tastanbishop@ru.ac.za (Ö.T.B.)
3 Faculty of Medicine and Health Technology, Tampere University, 33520 Tampere, Finland; seppo.parkkila@tuni.fi
4 Fimlab Ltd., Tampere University Hospital, 33520 Tampere, Finland
* Correspondence: zolfaghari@nigeb.ac.ir; Tel.: +98-21-44787476

Simple Summary: Hydrothermal vents are regions such as hot springs found on the seafloor in the mid-ocean and near tectonic plates. They contain fluids with highly enriched carbon dioxide, which is the central element of life on Earth. Many organisms live in this environment and can survive in extreme conditions (extremophiles), such as up to 400 °C or higher, low pH, and high pressure. All organisms need the carbonic anhydrase (CA) enzyme to handle the acid-base imbalance through the hydration of carbon dioxide and the production of bicarbonate necessary for pH homeostasis and many cellular functions. The CAs have been categorized into eight families. In this study, we focused on α-, β-, and γ-CAs from the thermophilic microbiome of marine hydrothermal vents. Microorganisms in this environment need CA to capture CO_2, which is an important contribution to marine hydrothermal vent ecosystem functioning. Previously, we showed the transfer of β-CA gene sequences from prokaryotes to protozoans, insects, and nematodes via horizontal gene transfer (HGT). HGT is not only the transfer and movement of genetic information between organisms but is also a powerful tool in natural biodiversity. If the CA coding gene is transferred horizontally between microorganisms in hydrothermal vents, it is hypothesized that CA is essential for survival in these environments and one of the key players in the carbon cycle in the ocean.

Citation: Gheibzadeh, M.S.; Manyumwa, C.V.; Tastan Bishop, Ö.; Shahbani Zahiri, H.; Parkkila, S.; Zolfaghari Emameh, R. Genome Study of α-, β-, and γ-Carbonic Anhydrases from the Thermophilic Microbiome of Marine Hydrothermal Vent Ecosystems. *Biology* **2023**, *12*, 770. https://doi.org/10.3390/biology12060770

Academic Editors: Daniel Puppe, Panayiotis Dimitrakopoulos and Baorong Lu

Received: 23 March 2023
Revised: 17 May 2023
Accepted: 17 May 2023
Published: 25 May 2023

Abstract: Carbonic anhydrases (CAs) are metalloenzymes that can help organisms survive in hydrothermal vents by hydrating carbon dioxide (CO_2). In this study, we focus on alpha (α), beta (β), and gamma (γ) CAs, which are present in the thermophilic microbiome of marine hydrothermal vents. The coding genes of these enzymes can be transferred between hydrothermal-vent organisms via horizontal gene transfer (HGT), which is an important tool in natural biodiversity. We performed big data mining and bioinformatics studies on *α-*, *β-*, and *γ-CA* coding genes from the thermophilic microbiome of marine hydrothermal vents. The results showed a reasonable association between thermostable α-, β-, and γ-CAs in the microbial population of the hydrothermal vents. This relationship could be due to HGT. We found evidence of HGT of α- and β-CAs between *Cycloclasticus* sp., a symbiont of *Bathymodiolus heckerae*, and an endosymbiont of *Riftia pachyptila* via Integrons. Conversely, HGT of β-CA genes from the endosymbiont Tevnia jerichonana to the endosymbiont Riftia pachyptila was detected. In addition, *Hydrogenovibrio crunogenus* SP-41 contains a *β-CA* gene on genomic islands (GIs). This gene can be transferred by HGT to *Hydrogenovibrio* sp. MA2-6, a methanotrophic endosymbiont of *Bathymodiolus azoricus*, and a methanotrophic endosymbiont of *Bathymodiolus puteoserpentis*. The endosymbiont of *R. pachyptila* has a *γ-CA* gene in the genome. If *α-* and *β-CA* coding genes have been derived from other microorganisms, such as endosymbionts of

T. jerichonana and *Cycloclasticus* sp. as the endosymbiont of *B. heckerae*, through HGT, the theory of the necessity of thermostable CA enzymes for survival in the extreme ecosystem of hydrothermal vents is suggested and helps the conservation of microbiome natural diversity in hydrothermal vents. These harsh ecosystems, with their integral players, such as HGT and endosymbionts, significantly impact the enrichment of life on Earth and the carbon cycle in the ocean.

Keywords: big data mining; carbonic anhydrase; extreme ecosystems; horizontal gene transfer; hydrothermal vents; mobile genetic elements; thermophilic microbiome

1. Introduction

Deep-sea hydrothermal vents are one of the best environments for evolutionary studies. Hydrothermal vents are regions such as hot springs found on the seafloor. These are located in the mid-ocean and near tectonic plates initially discovered in 1977 at a depth of 2.5 km around a hot spring on the Galápagos volcanic rift (spreading ridge) off the coast of Ecuador [1,2]. Based on their characteristics, deep-sea hydrothermal vents are called either black smokers or white smokers [3]. Black smokers' fluid temperature goes up to 400 °C or above and has a low pH, but white smokers have an alkaline pH, and their temperature is approximately 40–75 °C [3]. Hydrothermal vents contain fluids with highly enriched carbon dioxide (CO_2), which are discharged into the deep sea by these vents [4]. CO_2 is a very stable form of carbon, the central element of life on Earth, and consists of a carbon atom covalently double-bonded to two oxygen atoms. Carbonic acid (H_2CO_3) is derived from the reaction of CO_2 and water molecules, so the product is an unstable compound that spontaneously splits into bicarbonate (HCO_3^-) and protons (H^+).

Many organisms live in this environment, especially bacterial and archaeal species that can survive in extreme conditions such as high temperatures and pressure. The organisms adapted to this habit are called extremophiles. All organisms need carbonic anhydrases (CAs) to handle the large amount of CO_2 and, consequently, the related acid-base imbalance [5–7]. CA is the metalloenzyme that catalyzes the reversible hydration of CO_2 to HCO_3 and H^+ as follows:

$$CO_2 + H_2O \leftrightarrow HCO_3^- + H^+$$

CAs are encoded by eight evolutionarily divergent gene families, including alpha (α), beta (β), gamma (γ), delta (δ), zeta (ζ), eta (η), theta (θ), and iota (ι) CA. α-CA has been reported in vertebrates, prokaryotes, fungi, algae, protozoa, and plants [7]. β-CA is expressed in prokaryotes, plants, fungi, protozoa, arthropods, and nematodes [8–13]. γ-CA is present in many plants, fungi, and prokaryotes. δ-CA and ζ-CA are present in marine diatoms [7,12]. η-CA was identified in the causative agent of malaria, *Plasmodium* spp., and θ-CA was identified in marine diatoms [7,14,15]. Iota(ι)-CA was recently reported to be expressed in diatoms and bacteria [16]. In this study, we focused on α-, β-, and γ-CAs from the thermophilic microbiome of marine hydrothermal vents. These metalloenzymes have an active site containing a Zn(II) metal ion cofactor [17], while Co(II) and Fe(II) can be included in α- and γ-CA, respectively [7]. The structures of α-CAs are frequently monomers and rarely dimers [18]; β-CAs are dimers, tetramers, or octamers [19]; and γ-CAs are trimers [20].

A previous study showed that β-CA gene sequences could be transferred from prokaryotes to protozoans, insects, and nematodes via HGT [21]. Additionally, the involvement of bacterial β-CA gene sequences in the gastrointestinal tract and their horizontal transfer to their host during evolution has been demonstrated [22]. HGT, also called lateral gene transfer (LGT), is the transfer and movement of genetic information between organisms and thus is differentiated from the vertical transmission of genes from parent to the next generations [23]. HGT plays a crucial role in natural biodiversity as a general mechanism [24,25], and it often causes dramatic changes in the ecological and pathogenic properties of bacterial species, thereby promoting microbial diversification and speciation [25]. HGT may occur via mobile genetic elements (MGEs) such as integrons, genomic islands (GIs), integrative conjugative elements (ICEs), transposable elements (TEs), plasmids, and phages [26–30]. MGEs are parts of DNA that encode enzymes and other proteins that interpose the transfer of DNA in HGT within genomes (intracellular mobility) or between bacterial cells (intercellular mobility) [26]. Intercellular transfer of DNA takes three forms in prokaryotes: transformation, conjugation, and transduction [31].

Integrons are MGEs that allow the capture and expression of exogenous genes. Integrons have three essential core features: intI, attI, and Pc [32,33]. IntI is the gene encoding an enzyme for catalyzing recombination between incoming gene cassettes called integron integrase (IntI) [32]. AttI is an integron-associated recombination site [33], and Pc is an integron-associated promoter that is expressed once a gene cassette is recombined [34]. The length of GIs is more than 10 kb, a part of a chromosome, recognized as discrete DNA segments, and can be different from closely related strains, and transposase is a primary tool for HGT through GIs [32–34]. Another family of MGEs is the integrative conjugative element (ICE), called the conjugative transposon. ICEs have two features: first, they are integrated into a host genome, and second, they encode a type IV secretion system (functional conjugation system) [28,35]. TEs are DNA sequences that can move from one location to another in the genome [30]. TEs fall into two classes: retrotransposons (Class I) or RNA transposons [36] and true transposons (Class II) or DNA transposons that consist of a transposase gene with two terminal inverted repeats (TIRs) on either side [30]. Additionally, insertion sequences (IS) are small MGEs that carry more than one or two transposase genes [37]. The CA genes may be transferred between organisms living in hydrothermal vents and their endosymbionts via HGT. Endosymbiotic bacteria are located in the trophosome of the host, which contains animal cells, so-called bacteriocytes [38]. For instance, one of the important living organisms living in deep-sea hydrothermal vents is the giant tubeworm *Riftia pachyptila*, which lives with its symbiont bacteria. Nitrate, oxygen, hydrogen sulfide, and inorganic carbon are taken up from the environment and it feeds its symbiotic bacteria with these substances in an organ known as the trophosome [39].

In addition, about one-third of the added carbon from atmospheric CO_2 uptake into the ocean increases dissolved CO_2 in seawater [40]. The accompanying acidification may reduce the seawater saturation of calcite, thus affecting marine calcifications. CA helps the concentration of inorganic carbon in the fluid from which calcium carbonate is sedimented and directly affects the calcification in some calcifiers, such as gastropods, oysters, and giant clams as well as coral calcification. The calcification can be reduced by 40%, which has been affected by high atmospheric CO_2 levels. Even a modest impact on producing carbonate shells and skeletons may have important consequences on the global carbon cycle [41].

Microorganisms in this environment need CA to capture CO_2, which is an important contribution to marine hydrothermal vent ecosystem functioning [42]. It has been suggested that α-CA evolution may contribute to the vulnerability to environmental changes of bivalves and their diversity [43] since HGT would create a large variability acted on by natural selection [39]. If the coding gene of this enzyme is transferred horizontally between hydrothermal vent microorganisms, it is hypothesized that CA is essential for survival and for preserving natural biodiversity in this extreme ecosystem. For this purpose, we investigated the evolutionary relationship and the possibility of HGT in the hydrothermal vent ecosystem. We conducted a large data mining and bioinformatics study focusing on the HGT of *α-*, *β-* and *γ-CA* genes in the microbial population of deep-sea hydrothermal vents.

2. Materials and Methods

2.1. Identification of α-, β-, and γ-CA Sequences

We collected the names of all microbial populations from hydrothermal vents based on the literature. The protein and DNA sequences of CA candidates were retrieved from databases that were previously annotated in these databases after performing genomics and proteomics studies (Tables S1–S3). We retrieved their α-, β-, and γ-CA protein sequences from UniProt (http://www.uniprot.org/, 1 March 2023). In addition, we utilized a Position-Specific Iterated BLAST (PSI-BLAST) in the National Center for Biotechnology Information (NCBI) for two iterations to identify sequences that were homologous to the query sequences from organisms originating from hydrothermal vents. Each CA family has a defined conserved amino acid sequence to retrieve other CAs from the relevant CA family. α-CAs have three conserved histidine residues [21] (Figure 1A) that can be used as a pattern for identifying bacterial α-CAs. β-CAs have two highly conserved motifs; the first motif includes three residues of cysteine, aspartic acid, and arginine (CxDxR); the second highly conserved motif includes histidine and cysteine residues (HxxC) [21] (Figure 1B). γ-CAs have three histidine residues as well as asparagine and glutamine residues (NxQxxxxxH) and (HxxxxH) [44,45] (Figure 1C).

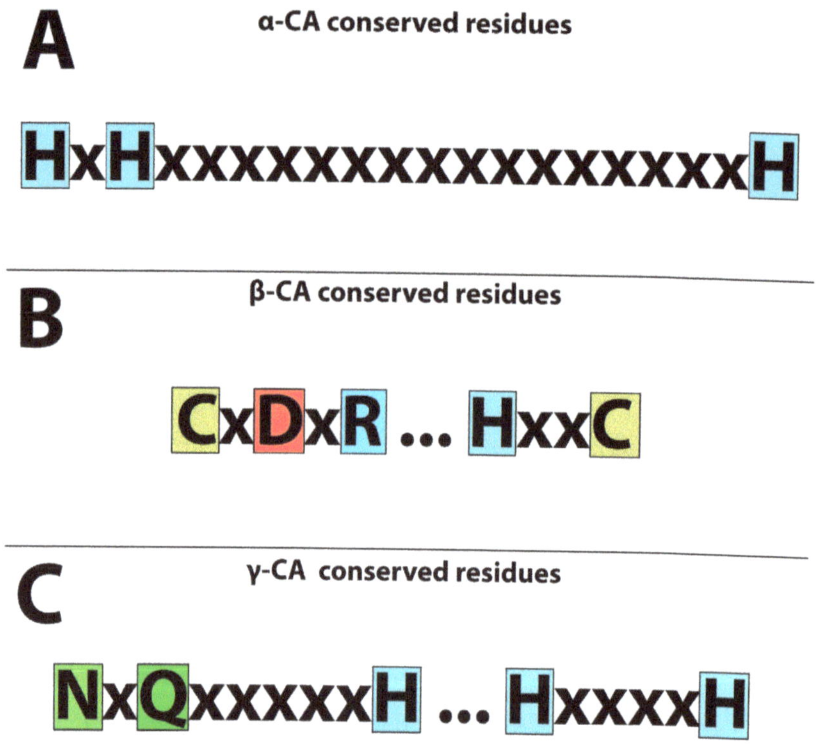

Figure 1. Conserved residues of CAs in the catalytic active sites. (**A**) Three "H" histidines are highly conserved in α-CAs. (**B**) Two cysteines "C", one histidine "H", one aspartic acid "D", and one arginine "R" are highly conserved amino acids in β-CAs. (**C**) Three histidine residues "H", one asparagine "N", and one glutamine "Q" are highly conserved amino acids in γ-CAs.

In addition, α-, β-, and γ-CA proteins from the microbiome of marine hydrothermal vent ecosystems with taxonomic classifications have been listed in Table S1 [46–62], Table S2 [63–95], and Table S3 [96–118], respectively.

Multiple sequence alignment (MSA) was performed using the Tree-based Consistency Objective Function for Alignment Evaluation (T-Coffee) [119] for the identification of conserved residues in α-, β-, and γ-CA protein sequences. Additionally, we analyzed these MSA results in Jalview2 software [120]. Then, we made a dataset for each organism (the whole genome, if available) from the NCBI database (https://www.ncbi.nlm.nih.gov/nuccore) (Access date: 1 March 2023) and apperceived the α-, β-, and γ-CA gene positions on our bacterial genomes from the Ensembl Bacteria (https://bacteria.ensembl.org) (Access date: 1 March 2023) and KEGG (https://www.genome.jp/kegg/) (Access date: 1 March 2023) databases. We annotated our integrons via Geneious prime version: 2021.0.3 software with default parameters.

2.2. Phylogenetic Analysis

We retrieved the Tax ID of all microbiomes from marine hydrothermal vents containing α-, β-, and γ-CA from the UniProt database (https://www.uniprot.org/taxonomy/) (Access date: 1 March 2023) and NCBI database (https://www.ncbi.nlm.nih.gov/taxonomy/) (Access date: 1 March 2023) for more accuracy. Phylogenetic trees were constructed for evolutionary study using maximum likelihood, and models with the lowest Bayesian Information Criterion (BIC) scores were considered to best describe the substitution pattern [121] via MEGA X software [122] and annotated in FigTree V1.4.4 software for all protein sequences. Then, we generated a heatmap based on the pairwise sequence identity between them using GraphPad Prism version 8.00 software for Windows (www.graphpad.com, 1 March 2023).

2.3. Identification of *α*-, *β*-, and *γ*-CA Genes on the MGEs

2.3.1. Integrons

Integrons have three essential core features: intI, attI, and *Pc* [32–34], so we tried to find these features in our dataset. Integrons gain new genes as part of gene cassettes [123]. In addition to these features, we needed to find cassettes as simple structures consisting of a single open reading frame (ORF) bounded by a cassette-associated recombination site called a 59-base element or *attC* [124]. Gene cassettes exist in a circular free state and are integrated into *attI* [125,126]. Integron integrase mediates the integration of circular gene cassettes by site-specific recombination between *attI* and *attC* reversibly and excises [126–128]. For the identification of the mentioned features, we used the Integron finder. Integron Finder has two forms: a standalone program (https://github.com/gem-Pasteur/Integron_Finder) (Access date: 1 March 2023) and a web application (https://galaxy.pasteur.fr/#forms::integronfinder) (Access date: 1 March 2023). Hidden Markov model (HMM) profiles were used for the search of integron-integrase and covariance models for *attC* sites. Pattern matching was also used for other features (such as promoters and *attI* sites) [129]. In this study, we applied the web application of integron finder.

2.3.2. Genomic Islands (GIs)

Prediction of GIs was studied using tools such as SIGI-HMM, IslandPath-DIMOB [130], PAI-IDA [131], and Centroid [132], based on the evaluation of sequence compositions as well as BLAST homology searches and whole-genome sequence alignment for comparative genomics methods [48]. For this purpose, we applied the IslandViewer 4 (http://www.pathogenomics.sfu.ca/islandviewer/) (Access date: 1 March 2023) database using a web server to predict and visualize genomic islands in bacterial and archaeal genomes [133]. After searching all microorganisms in this database, we retrieved their annotations and searched for *α*-, *β*-, and *γ*-*CA* genes on their GIs.

2.3.3. Integrative Conjugative Elements (ICEs)

ICEs comprise the ICE integration and excision module, ICE conjugation module, and ICE regulation module, which are the main genetic modules [134]. ICEs contain integrase- and relaxase-coding genes and/or type IV secretion systems. For the identification of ICEs, we used ICEberg 2.0 (https://db-mml.sjtu.edu.cn/ICEberg/) (Access date: 1 March 2023) [135,136].

2.3.4. Transposable Elements (TEs), Phages, and Plasmids

Insertion sequences (IS) and true transposons (Tn) consist of a transposase gene with two terminal inverted repeats (TIRs) on either side [30]. IS are small mobile elements that carry little more than one or two transposase genes [37]. For the identification of these elements, we used the MobileElementFinder web server (https://cge.cbs.dtu.dk/services/MobileElementFinder/) (Access date: 1 March 2023) [137]. To study phages in our datasets, we needed to find evidence of prophages. Evidence of insertion sites includes alteration of GC content and the presence of tRNA flanking the region [138]. PhageWeb (http://computationalbiology.ufpa.br/phageweb/) (Access date: 1 March 2023) was used to search for this evidence. Utilizing information from a 2018 study by Sousa, A.L.d., et al., we set options to default (BLAST options to identify 80% and six minimum of CDS) in prophage identification [139]. After that, we checked the location of our genes for the position on the chromosome or plasmid.

3. Results

3.1. Identification of α-, β-, and γ-CA and Protein Sequences

This study evaluated 83 previously isolated microorganisms in or around hydrothermal vents (Tables S1–S3). They consisted of bacteria and archaea and were classified into ten groups of bacterial species, including Alphaproteobacteria, Deltaproteobacteria, Epsilonproteobacteria, Gammaproteobacteria, Zetaproteobacteria, Aquificae, Bacilli, Deferribacteres, Deinococci, and Fusobacteria, as well as four groups for archaea, including Archaeoglobi, Methanopyri, Methanococci, and Thermococci. We retrieved 25 α-CA, 55 β-CA, and 47 γ-CA protein sequences from the UniProt database [140]. We must note that we have abbreviated microorganism names for convenience, and they are stated in the Supplementary Materials (Tables S1–S3). It is worth noting that many of these isolated species from hydrothermal vents are endosymbiotic microorganisms.

The results of the MSA for verification of α-, β-, and γ-CA protein sequences are shown in the Supplementary Materials (Figures S1–S3). Many α-CAs from the thermophilic microbiome of marine hydrothermal vents have been studied previously [42]. At first, the MSA of α-CA showed conserved residues (Figure S1) in which three conserved histidine residues (His107, His109, and His126) [21] were visible and coordinated with the Zn^{2+} metal ion cofactor in the enzyme catalytic active site [141]. Next, the MSA of β-CAs showed three conserved residues in the first highly conserved motif (CxDxR), including cysteine, aspartic acid, and arginine, with variation in the residues between them [21]. The second highly conserved motif (HxxC), which contained histidine and cysteine residues with two other residues between them, was also observed [21] (Figure S2). Finally, in the MSA of γ-CAs, we identified three histidine residues, asparagine and glutamine residues, that were highly conserved [44,45] (Figure S3).

3.2. Phylogenetic Analysis

The results of phylogenetic analysis and heatmaps of α-, β-, and γ-CAs from the thermophilic microbiome of hydrothermal vents are shown in Figure 2, Figure 3, and Figure 4, respectively. We highlighted the bacterial CAs with blue and archaea with orange. The evolutionary history was inferred using the maximum-likelihood method and the result of calculating the best model using the Le Gascuel model with discrete gamma distribution and invariable sites (LGGI) [142]. Since the phylogenetic analysis of α-CAs from the thermophilic microbiome of marine hydrothermal vents has been studied previously [42], we analyzed additional species, including *Hydrogenovibrio crunogenus* (XCL-2), *Hydrogenovibrio crunogenus* (SP-41), *Bacillus oceanisediminis, Sulfurivirga caldicuralii, Caldithrix abyssi*, an endosymbiont of *Riftia pachyptila* (vent Ph05), *Bathymodiolus platifrons* as a methanotrophic gill symbiont, and *Cycloclasticus* sp. as symbionts of *Bathymodiolus heckerae, Nitrosophilus alvini, Nitrosophilus labii, Sulfurimonas paralvinellae, Hydrogenimonas urashimensis, Sulfurovum indicum* and *Persephonella atlantica* (Figure 2).

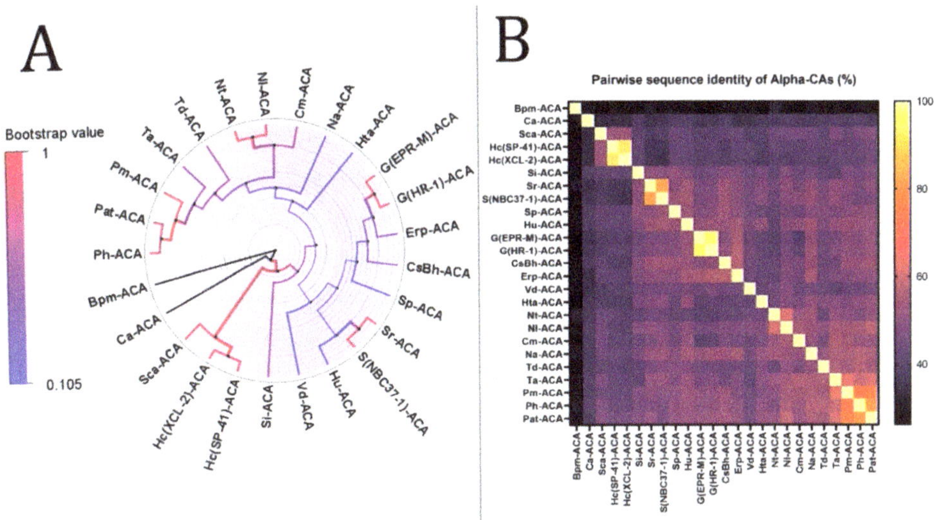

Figure 2. Phylogenetic analysis of α-CAs from the thermophilic microbiome of hydrothermal vents. (**A**) The tree's branches and nodes were colored based on bootstrap values (0–1), and the bacterial CAs and archaea were highlighted with blue and orange, respectively, via FigTree V1.4.4 software. (**B**) α-CA pairwise sequence identity heatmap. The heatmap for the all-versus-all pairwise sequence identity of α-CA calculations was generated using T-Coffee MSA. Pairwise sequence identity values are colored from yellow (highest) to black (lowest).

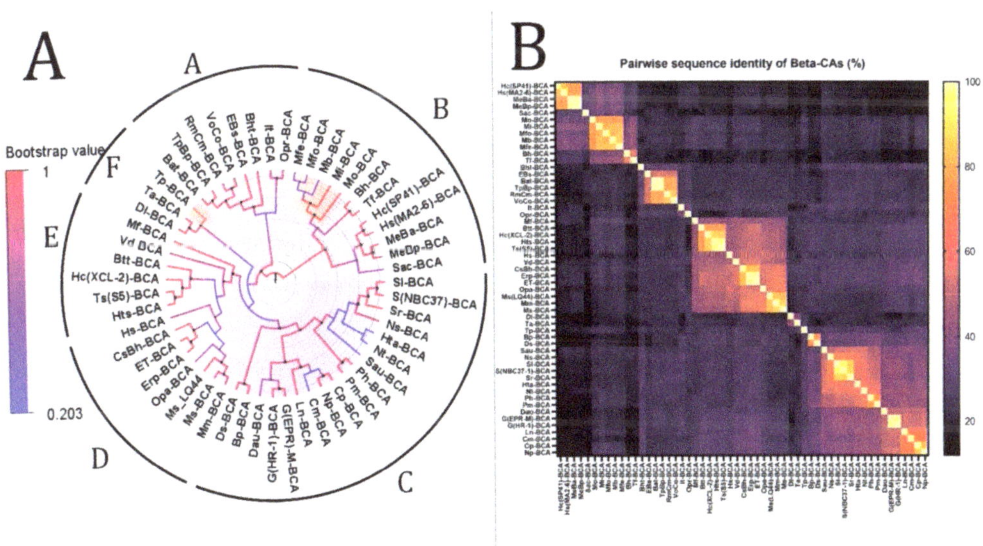

Figure 3. Phylogenetic analysis of β-CAs from the thermophilic microbiome of hydrothermal vents. (**A**) The tree's branches and nodes were colored based on bootstrap values (0–1), and the bacterial CAs and archaea were highlighted with blue and orange via FigTree V1.4.4 software. (**B**) β-CA pairwise sequence identity heatmap. The heatmap for the all-versus-all pairwise sequence identity of β-CA calculations was generated using T-Coffee MSA. Pairwise sequence identity values are colored from yellow (highest) to black (lowest).

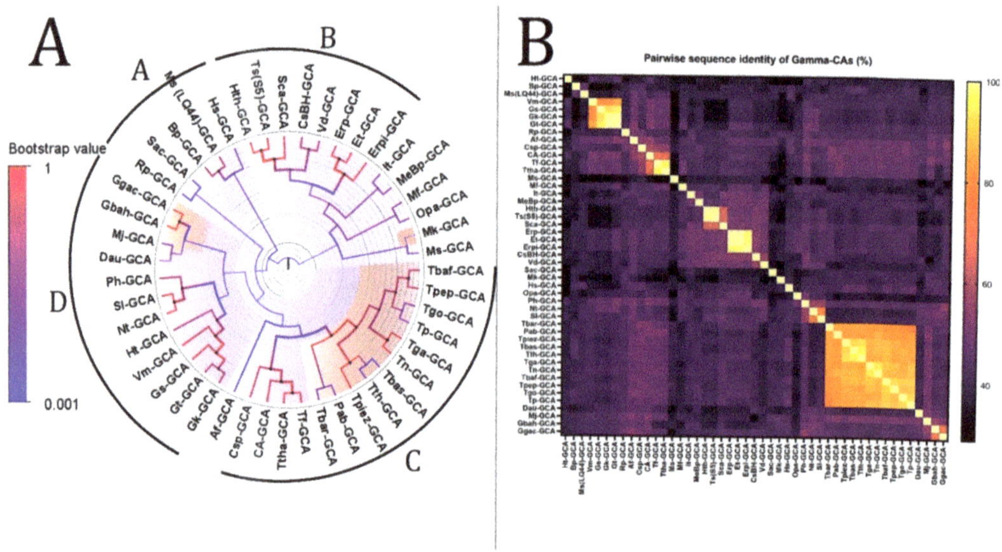

Figure 4. Phylogenetic analysis of γ-CAs from the thermophilic microbiome of hydrothermal vents. (**A**) The tree's branches and nodes were colored based on bootstrap values (0–1), and the bacterial CAs and archaea were highlighted with blue and orange, respectively, via FigTree V1.4.4 software. We did not find any specific items between clades A and B or between clades C and D that can be categorized as separate clades. (**B**) γ-CA pairwise sequence identity heatmap. The heatmap for the all-versus-all pairwise sequence identity of γ-CA calculations was generated using T-Coffee MSA. Pairwise sequence identity values are colored from yellow (highest) to black (lowest).

The phylogenetic tree of α-CAs was performed with the highest log-likelihood (−9636.17). Initial trees for the heuristic search were obtained automatically by applying neighbor-joining and BioNJ algorithms to a matrix of pairwise distances estimated using the Jones-Taylor-Thornton (JTT) model and then selecting the topology with a superior log-likelihood value. A discrete gamma distribution was used to model evolutionary rate differences among sites (two categories (+G, parameter = 1.1935)). The rate variation model allowed some sites to evolve invariably ([+I], 3.92% sites). The tree was drawn to scale, with branch lengths measured in the number of substitutions per site. Twenty-five α-CA amino acid sequences were involved in this analysis. There were a total of 357 positions in the final dataset. The analysis revealed that there is a common ancestor between Hc(XCL-2)-ACA and Hc(SP-41)-ACA; Sr-ACA and S(NBC37-1)-ACA; G(EPR-M)-ACA and G(HR-1)-ACA; Nt-ACA and Nl-ACA; and Pat-ACA and Ph-ACA.

The phylogenetic tree of β-CAs with the highest log-likelihood (−15,146.85) is shown in Figure 3. Initial trees for the heuristic search were obtained automatically by applying neighbor-joining and BioNJ algorithms to a matrix of pairwise distances estimated using the JTT model and then selecting the topology with a superior log-likelihood value. A discrete gamma distribution was used to model evolutionary rate differences among sites (two categories (+*G*, parameter = 2.7462)). The rate variation model allowed some sites to evolve invariably ([+*I*], 0.97% sites). This analysis involved 55 β-CA amino acid sequences. There were a total of 308 positions in the final dataset. Based on bootstrap values and identity, we divided this tree (Figure 3) into six clades from A to F. The analysis revealed that there is a common ancestor between the β-CAs in each clade.

The phylogenetic tree of γ-CAs with the highest log-likelihood (−9786.77) is shown in Figure 4. The initial phylogenetic trees for the heuristic search were automatically obtained by applying Neighbor-Join and BioNJ algorithms to a matrix of pairwise distances estimated using the JTT model. The topology with a superior log-likelihood value was then selected. A discrete gamma distribution was used to model evolutionary rate differences among sites (two categories (+*G*, parameter = 1.6179)). The variation model rate allowed some sites to evolve invariably ([+*I*], 3.65% sites). Forty-seven amino acid sequences were involved in this analysis. There were a total of 219 positions in the final dataset. We divided this tree into four clades from A to D based on bootstrap values and identity, similar to the β-CA phylogenetic analysis. The analysis revealed that there is a common ancestor between the γ-CAs in each clade.

3.3. *Identification of α-, β-, and γ-CA Genes on MGEs*

3.3.1. Integrons

Integrons are divided into complete integrons, In0 elements, and CALINs elements. Complete integrons have an integrase and one *attC* site or more. The In0 elements consist of an integron integrase without *attC* sites, and CALINs have two *attC* sites or more without integron integrases. After searching integron features on our dataset, we found integrons in many microorganisms, which have been mentioned in Table 1. We performed a BLAST analysis on all protein CDS (protein-coding sequences) on integrons. The results showed that only the endosymbiont of *R. pachyptila* and the endosymbiont of *Tevnia jerichonana* have CA-coding genes in their integron area.

Table 1. Integrons in the thermophilic microbiome from hydrothermal vents.

Microorganisms	Integron Type	Integrase	CA Gene
Cycloclasticus sp. symbiont of *Bathymodiolus heckerae*	In0	Intersection tyr intI	-
Endosymbiont of *Riftia pachyptila* (vent Ph05)	Integron 1: CALIN Integron 2: CALIN	-	*α-CA* *β-CA*
Sulfurovum sp. NBC37-1	Integron 1: CALIN Integron 2: CALIN	-	-
Caldithrix abyssi	CALIN	-	-
Hydrogenovibrio crunogenus SP-41	Integron 1: CALIN Integron 2: CALIN Integron 3: Complete Integron 4: CALIN	Intersection tyr intI	-
Thiomicrospira crunogena XCL-2	Integron 1: CALIN Integron 2: CALIN	-	-
Bathymodiolus thermophilus thioautotrophic gill symbiont	Integron 1: CALIN Integron 2: CALIN Integron 3: CALIN	-	-
Endosymbiont of *Tevnia jerichonana*	CALIN	-	*β-CA*
Halomonas sulfidaeris strain SST4	CALIN	-	-
Marinobacter sp. LQ44	Integron 1: CALIN Integron 2: In0	Intersection tyr intI	-
Sulfurimonas autotrophica	Integron 1: In0 Integron 2: CALIN	Intersection tyr intI	-
Endosymbiont of *Bathymodiolus septemdierum*	Integron 1: CALIN Integron 2: CALIN Integron 3: In0 Integron 4: CALIN Integron 5: CALIN	Intersection tyr intI	-
Hydrogenovibrio thermophilus	Complete	Intersection tyr intI	-
Thermococcus barophilus strain CH5	CALIN	-	-
Cycloclasticus sp. symbiont of *Bathymodiolus heckerae*	In0	Intersection tyr intI	-
Endosymbiont of *Riftia pachyptila* (vent Ph05)	Integron 1: CALIN Integron 2: CALIN	-	*α-CA* *β-CA*

Table 1. *Cont.*

Microorganisms	Integron Type	Integrase	CA Gene
Sulfurovum sp. NBC37-1	Integron 1: CALIN Integron 2: CALIN	-	-
Caldithrix abyssi	CALIN	-	-
Hydrogenovibrio crunogenus SP-41	Integron 1: CALIN Integron 2: CALIN Integron 3: Complete Integron 4: CALIN	Intersection tyr intI	-
Thiomicrospira crunogena XCL-2	Integron 1: CALIN Integron 2: CALIN	-	-
Bathymodiolus thermophilus thioautotrophic gill symbiont	Integron 1: CALIN Integron 2: CALIN Integron 3: CALIN	-	-
Endosymbiont of *Tevnia jerichonana*	CALIN	-	*β-CA*
Halomonas sulfidaeris strain SST4	CALIN	-	-
Marinobacter sp. LQ44	Integron 1: CALIN Integron 2: In0	Intersection tyr intI	-
Sulfurimonas autotrophica	Integron 1: In0 Integron 2: CALIN	Intersection tyr intI	-
Endosymbiont of *Bathymodiolus septemdierum*	Integron 1: CALIN Integron 2: CALIN Integron 3: In0 Integron 4: CALIN Integron 5: CALIN	Intersection tyr intI	-
Hydrogenovibrio thermophilus	Complete	Intersection tyr intI	-
Thermococcus barophilus strain CH5	CALIN	-	-

According to data analysis by Integron Finder, the endosymbiont of *Riftia pachyptila* contains two integrons. The first integron has one CDS, and the second has two *attC* sites; the CALIN type has six CDSs, an α-CA gene on the fourth CDS, and a β-CA gene on the sixth CDS (Figure 5A). The endosymbiont of *Tevnia jerichonana* has one integron with two *attC* sites, six CDS is CALIN type, and a β-CA gene on the fourth CDS (Figure 5B).

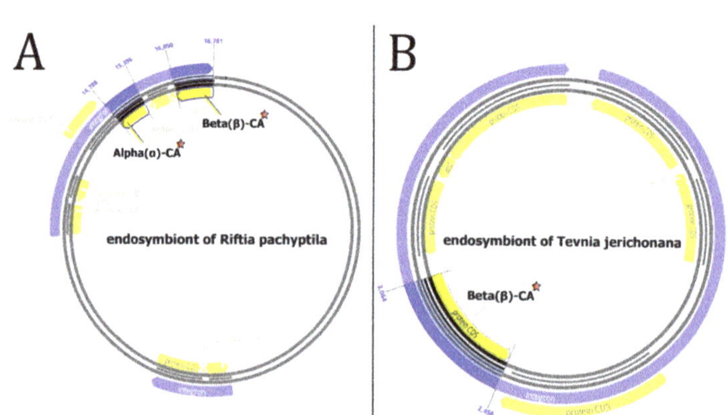

Figure 5. Integrons of endosymbionts of (**A**) *Riftia pachyptila* and (**B**) *Tevnia jerichonana*. The α- and *β-CA* genes are bolded and marked with red stars.

3.3.2. Genomic Islands (GIs)

According to the IslandViewer 4 (http://www.pathogenomics.sfu.ca/islandviewer/) (Access date: 1 March 2023) database, 25 out of 83 of our microorganisms have GIs, and only one of the *Hydrogenovibrio crunogenus* SP-41 GIs carries a *β-CA* gene (*Hc(SP41)-BCA*) (UniProt ID: Q31FD6) and three transposase genes that are primary tools for HGT [143] (Figure 6). This GI is predicted by SIGI-HMM [144] and IslandPath-DIMOB methods [130]. However, the HGT of β-CA genes with GIs between prokaryotes and protists was previously studied [22].

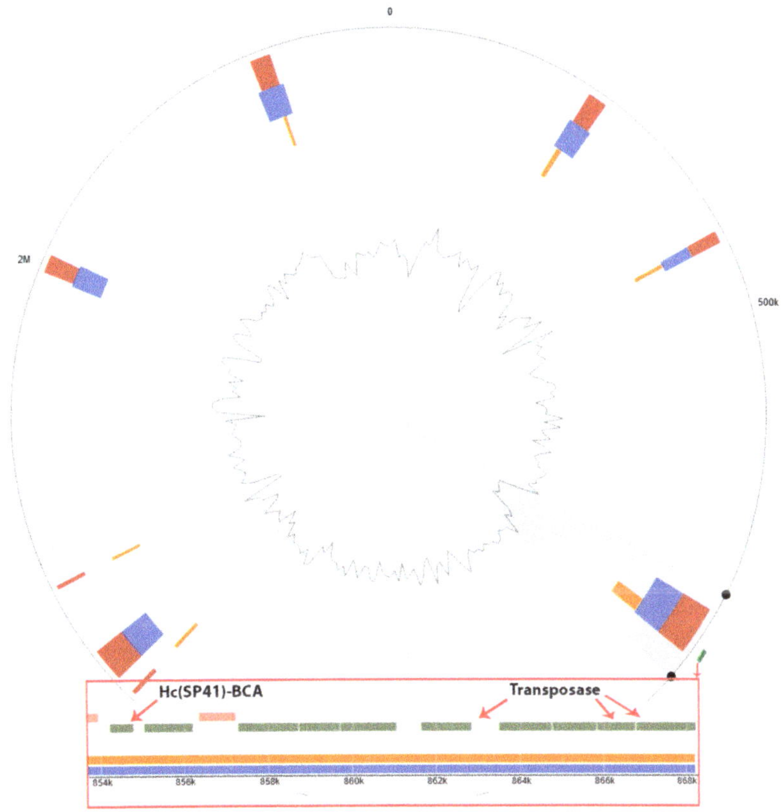

Figure 6. GIs of *Hydrogenovibrio crunogenus* SP-41. *Hc(SP41)-BCA* and transposase genes are marked green and shown in the red box. The GC content is visible at the center of the figure.

3.3.3. Integrative Conjugative Elements (ICEs), Transposable Elements (TEs), Phages, and Plasmids

According to ICEberg 2.0 (https://db-mml.sjtu.edu.cn/ICEberg/) (Access date: 1 March 2023) and MobileElementFinder web server results, we did not find any α-, β-, and γ-CA genes on the ICEs and TEs. Additionally, using PhageWeb (https://github.com/phagewebufpa/API (Access date: 25 April 2019), we did not find any evidence supporting the transfer of α-, β-, and γ-CA genes via phages. Based on the details of our dataset, CA genes were not located on the plasmids from the thermophilic microbiome of hydrothermal vents, and all genes were found on the chromosomes.

4. Discussion

The evolutionary process in hydrothermal vent ecosystems and the role of viruses in the biodiversity in this harsh environment have been studied previously. A study performed by Cheng et al. [145] revealed that bacteriophages are the most predominant viruses across the global hydrothermal vents, while single-stranded DNA viruses, including Microviridae and small eukaryotic viruses, have been located in the next steps. The metagenomics analysis showed that this virome plays a crucial role in the evolution and biodiversity of the microbiome of hydrothermal vents, especially Gammaproteobacteria and Campylobacterota [145]. Although the bacteriophages have no role in the HGT of CA genes in the hydrothermal ecosystems, our previous studies showed the HGT of *β-CA* genes from prokaryotic endosymbionts to their protozoan, insects, and nematodes hosts. In addition, the genomic islands have been shown to have a potential role in the HGT of *β-CA* genes from ancestral prokaryotes to protists. Since then, no further study has been performed on the HGT of CA genes. Since hydrothermal vent ecosystems have been reported as potent environments for HGT and biodiversity, these harsh deep-sea fissures were studied.

According to the heatmap and phylogenetic analysis (Figure 2) of α-CAs, Bpm-ACA and Ca-ACA showed no significant relationship with the other α-CAs. Hc(XLC-2)ACA, Hc(SP41)-ACA, and Sca-ACA clustered together with branch bootstrap values of 1.00, showing significant relationships. Additionally, G(HR-1)-ACA and G(EPR-M)-ACA had a branch bootstrap value of 1.00, indicating a robust evolutionary relationship similar to that between Sr-ACA and S(NBC37-1)-ACA, whose bootstrap value was also 1.00. Similar to a previous study, the branch for Pm-ACA and Ph-ACA was observed to have a high bootstrap value of 0.99. A high branch bootstrap value of 0.86 was observed for CsBh-ACA and Erp-ACA. It is necessary to mention that all the α-CAs above belong to the Proteobacteria phylum except for Pm-ACA and Ph-ACA, which belong to the Aquificae phylum. According to the heatmap and phylogenetic analysis (Figure 3) of β-CAs, in clade A, Opr-BCA and It-BCA have poor relationships with other clade members, showing a branch bootstrap value of 0.37. All members of clade B have the same root, but Mfe-BCA, Mfo-BCA, and Mb-BCA have poor relationships with other clade members. In clade C, a significant relationship between Sr-BCA, S(NBC37)-BCA, and Si-BCA showed a branch bootstrap value of 1.00, in which pairwise sequence identities of more than 88.6% were revealed. Although these three cases with a branch bootstrap value of 0.92 have a significant relationship with Ns-BCA, they have a poor relationship with other members of clade C. In clade D, a relationship between Erp-BCA, Et-BCA, and CsBh-BCA was observed with a 1.00 branch bootstrap value, in which a pairwise sequence identity of more than 83.5% was observed for all three. In clade E, the cluster containing Hts-BCA, Ts-BCA, Hc(XCL-2) –BCA, and Btt-BCA was observed with a branch bootstrap value of 1.00. Clade F with a 0.3 branch bootstrap value did not show a good relationship with other clades, while Di-BCA and Ta-BCA have the same root as Tp-BCA, a member of archaea. According to the heatmap and phylogenetic analysis of γ-CAs (Figure 4), clades A and B, with branch bootstrap values of 0.02 and 0.001, respectively, have a very poor relationship with other clades, including Erpi-GCA, Erp-GCA, and ET-GCA with different branch bootstrap values of more than 0.98 and a pairwise sequence identity value of more than 97.78, which have a significant relationship together. In addition, a meaningful relationship was observed for Hts-GCA, Ts(S5)-GCA, and Sca-GCA with a branch bootstrap value of 0.99. In clade C, archaea and bacteria have the same root, and according to the heatmap, all archaea have high pairwise sequence identity values. In clade D, Gbah-GCA and Ggac-GCA have a good relationship with a branch bootstrap value of 0.99 and a pairwise sequence identity value of 69.18. According to the heatmap of γ-CA (Figure 4B) in clade D, Gs-GCA with a branch bootstrap value of 0.99 and a pairwise sequence identity value of 94.29 had a significant relationship with Gk-GCA.

CALIN elements (Table 1) might have arisen from a missing integrase in a previously complete integron. The *α*- and *β*-CA genes from CALIN may be cut by the integron-integrase and reinserted in the integron at an *attI* site. Since the stable circular form of CALINs can survive in the environment, these genetic elements can be taken up by transformable bacteria through a transformation mechanism [146]. On the other hand, integrons often capture cassettes from CALIN elements [129], so the α- and β-CA genes can be derived from different microorganisms or transferred to other hosts. According to the phylogenetic trees of α- and β-CA (Figures 2A and 3A), Erp-ACA has the highest relationship with CsBh-ACA, with a bootstrap value of 0.40 and a pairwise sequence identity value of 57.61, which is a weak relationship. In addition, it has a relatively weak relationship with G(EPR-M)-ACA and G(HR-1)-ACA twins with a bootstrap value of 0.44 and pairwise sequence identity values of 55.19 and 57.39, respectively. The β-CAs in clades A and B, with 0.02 and 0.001 branch bootstrap values, respectively, did not have a good relationship with other clades. At the same time, Erp-BCA is the highest related compound to Et-BCA and CsBh-BCA, with a 1.00 branch bootstrap value and pairwise sequence identity of 99 and 100, respectively. Moreover, the β-CA gene (*Et-BCA*) from the endosymbiont of *T. jerichonana* is related the highest to Erp-BCA and CsBh-BCA, with a branch bootstrap value of 1.00 and pairwise sequence identity of 99 and 84, respectively, which indicates the possibility of horizontal gene transfer of β-CA coding genes in these microorganisms.

It should be noted that inorganic carbon from CO_2 is first obtained from the environment via diffusion through the plume, a branchial organ [147]. Next, CO_2 is transformed to HCO_3^- and transported to trophosome cells, particularly bicarbonate, at the surrounding branchial plume interface. Then, HCO_3^- is transformed to CO_2 on the body fluids and bacterial cells [148] and adhered via the bacterial symbiont enzyme RuBisCO form II. In the arginine biosynthesis and pyrimidine pathways, carbamylphosphate synthetase uses inorganic HCO_3^- to start the biosynthesis process. Since the metabolic relationship between *R. pachyptila* and its endosymbiont is vital for the survival of each organism, this issue can explain the cause and importance of HGT of CA in these organisms. Furthermore, *R. pachyptila* contains an α-CA gene [149] with UniProt ID: Q8MPH8, which is not similar to Erp-ACA.

Additionally, *T. jerichonana* has no reported CA family. Identification of the β-CA gene beside three transposase genes on one of the GIs of *H. crunogenus* SP-41 could lead to the theory that this gene may be transferred with plasmids and phages or occur through transposon accumulation in recombination sites. Experimental studies have suggested the release of about 1.5 billion symbionts from dead tubeworm clumps into the environment [47], which provides the opportunity for the spread and HGT of CA genes in the environment and preparing the biodiversity condition.

In addition to the β-CA phylogenetic tree, the heatmap showed that the Hc(SP41)-BCA in clade B is closely related to Hs(MA2-6)–BCA with a branch bootstrap value of 0.99 and a pairwise sequence identity value of 82.9. In addition, MeBa-BCA and MeBp-BCA showed a close relationship with Hc(SP41)-BCA with branch bootstrap values of 1.0 and pairwise sequence identity values of 76.56 for both cases. The HGT of hydrogenase-coding genes between *H. crunogenus* SP-41 and *H. crunogenus* XCL-2 was studied previously [150]; however, in this study, *H. crunogenus* SP-41 (*Hc(SP41)-BCA*) had no HGT relationship with *H. crunogenus* XCL-2. *R. pachyptila* has cytosolic α-CA in the trophosome. Although these organisms need secretory CA for their physiological needs and use Erp-ACA, this theory must be experimentally studied.

The significance of this study revealed that there is an evolutionary relationship between Hc(XLC-2)ACA, Hc(SP41)-ACA, and Sca-ACA; G(HR-1)-ACA and G(EPR-M)-ACA; Sr-ACA and S(NBC37-1)-ACA; Pm-ACA and Ph-ACA; and CsBh-ACA and Erp-ACA in α-CAs. In addition, there is an evolutionary relationship between Sr-BCA, S(NBC37)-BCA, and Si-BCA; Erp-BCA, Et-BCA, and CsBh-BCA; and Hts-BCA, Ts-BCA, Hc(XCL-2)–BCA, and Btt-BCA in β-CAs. Additionally, there is an evolutionary relationship between Erpi-GCA, Erp-GCA, and ET-GCA; Hts-GCA, Ts(S5)-GCA, and Sca-GCA; Gbah-GCA and Ggac-GCA; and Gs-GCA and Gk-GCA in γ-CAs.

Elevated CO_2 pressure in seawater can affect marine organisms by disrupting acid-base physiology and decreasing mineralization rates (affecting calcium carbonate saturation and calcification). Ocean uptake of anthropogenic CO_2 and associated changes in seawater chemistry adversely affect biodiversity, other ecosystem processes, and the global carbon cycle [151]. The HGT and distribution of CA genes in the hydrothermal vent area may also help the survival and diversity of the organisms in this environment.

5. Conclusions

According to the results of this big data mining and bioinformatics study, α-, β-, and γ-CAs from the thermophilic microbiome of marine hydrothermal vents have a reasonable evolutionary relationship. The α-, β-, and γ-*CA* genes can be transferred to other microorganism habitats in hydrothermal vents via HGT and cause natural biodiversity in this extreme ecosystem. Given the presence of an integron with an integrase coding gene in the *Cycloclasticus* sp. symbiont of *Bathymodiolus heckerae,* it is highly possible that the α-CA coding gene is transferred between *Cycloclasticus* sp. as the symbiont of *B. heckerae* and endosymbiont of *Riftia pachyptila*. This evolutionary phenomenon can also be applied to β-*CA*-coding genes.

According to the β-CA gene on the endosymbiont of *T. jerichonana* and the endosymbiont of *R. pachyptila* and the evolutionary relationship between them, the HGT of the β-CA gene from the endosymbiont of *T. jerichonana* to the endosymbiont of *R. pachyptila* and conversely is highly possible. In addition, the endosymbiont of *R. pachyptila* has a γ-CA gene on the chromosome; if α- and β-CA coding genes are derived from other microorganisms, such as the endosymbiont of *T. jerichonana* and *Cycloclasticus* sp. as the symbiont of *B. heckerae,* the theory of the necessity of the CA enzyme for survival in this extreme ecosystem and its effect on preserved natural biodiversity is proposed. Despite the presence of the α-CA gene in *R. pachyptila* and the α-, β-, and γ-CA genes in its endosymbiont, this theory is suggested for this giant marine worm. Therefore, the prokaryotic endosymbionts of mussels and giant marine worms have evolutionary relationships through HGT. With more focus on the HGT phenomenon, endosymbionts are integral parts of natural biodiversity and ecosystem functioning of marine hydrothermal vents.

Supplementary Materials: The following supporting information can be downloaded at: https://www.mdpi.com/article/10.3390/biology12060770/s1. Figure S1, Multiple Sequence Alignment of α-CA from the thermophilic microbiome of hydrothermal vents; Figure S2, Multiple Sequence Alignment of β-CAs from the thermophilic microbiome of hydrothermal vents; Figure S3, Multiple Sequence Alignment of γ-CAs from the thermophilic microbiome of hydrothermal vents; Table S1, α-CA proteins from the microbiome of marine hydrothermal vent ecosystems with taxonomic classifications; Table S2, β-CA proteins from the microbiome of marine hydrothermal vent ecosystems with taxonomic classifications; Table S3, γ-CA proteins from the microbiome of marine hydrothermal vent ecosystems with taxonomic classifications.

Author Contributions: M.S.G. performed the study and drafted the first version of this manuscript. R.Z.E., as the supervisor of this study, designed the study and edited the manuscript, figures, and tables. M.S.G., C.V.M., Ö.T.B., H.S.Z., S.P. and R.Z.E. performed further revisions on the manuscript. All authors have read and agreed to the published version of the manuscript.

Funding: The MSc thesis grant from NIGEB supported MSG. Also, RZE has been supported by grant No. 737 from NIGEB. In addition, to perform this study, RZE was supported by grant No. M/75137 from the Ministry of Science, Research, and Technology (MSRT) of the Islamic Republic of Iran based on the collaboration agreement between Iran and South Africa. In addition, OTB was supported by grant No. 111212 from the National Research Foundation (NRF) of South Africa. SP supported this study through his grants from the Jane and Aatos Erkko Foundation and the Academy of Finland, Finland.

Institutional Review Board Statement: Not applicable.

Informed Consent Statement: Not applicable.

Data Availability Statement: Not applicable.

Acknowledgments: We thank the National Institute of Genetic Engineering and Biotechnology (NIGEB) of the Islamic Republic of Iran and the National Research Foundation (NRF) of South Africa for preparing the conditions to perform this study. No funding organizations had any role in the design of the study; in the collection, analysis, or interpretation of data; in the writing of the manuscript; or in the decision to publish the results.

Conflicts of Interest: The authors declare that they have no known competing financial interests or personal relationships that could have appeared to influence the work reported in this paper.

References

1. Lonsdale, P. Clustering of suspension-feeding macrobenthos near abyssal hydrothermal vents at oceanic spreading centers. *Deep Sea Res.* **1977**, *24*, 857–863. [CrossRef]
2. Corliss, J.B.; Dymond, J.; Gordon, L.I.; Edmond, J.M.; von Herzen, R.P.; Ballard, R.D.; Green, K.; Williams, D.; Bainbridge, A.; Crane, K. Submarine thermal springs on the Galapagos Rift. *Science* **1979**, *203*, 1073–1083. [CrossRef] [PubMed]
3. Colín-García, M.; Heredia, A.; Cordero, G.; Camprubí, A.; Negrón-Mendoza, A.; Ortega-Gutiérrez, F.; Beraldi, H.; Ramos-Bernal, S. Hydrothermal vents and prebiotic chemistry: A review. *Bol. Soc. Geol. Mex.* **2016**, *68*, 599–620. [CrossRef]
4. Shitashima, K. CO2 supply from deep-sea hydrothermal systems. *Waste Manag.* **1998**, *17*, 385–390. [CrossRef]
5. Maren, T.H. Carbonic anhydrase: Chemistry, physiology, and inhibition. *Physiol. Rev.* **1967**, *47*, 595–781. [CrossRef]
6. Smith, K.S.; Jakubzick, C.; Whittam, T.S.; Ferry, J.G. Carbonic anhydrase is an ancient enzyme widespread in prokaryotes. *Proc. Natl. Acad. Sci. USA* **1999**, *96*, 15184–15189. [CrossRef]
7. Capasso, C.; Supuran, C.T. An overview of the alpha-, beta-and gamma-carbonic anhydrases from Bacteria: Can bacterial carbonic anhydrases shed new light on evolution of bacteria? *J. Enzym. Inhib. Med. Chem.* **2015**, *30*, 325–332. [CrossRef]
8. Zolfaghari Emameh, R.; Kuuslahti, M.; Näreaho, A.; Sukura, A.; Parkkila, S. Innovative molecular diagnosis of T richinella species based on β-carbonic anhydrase genomic sequence. *Microb. Biotechnol.* **2016**, *9*, 172–179. [CrossRef]
9. Zolfaghari Emameh, R.; Barker, H.R.; Syrjänen, L.; Urbański, L.; Supuran, C.T.; Parkkila, S. Identification and inhibition of carbonic anhydrases from nematodes. *J. Enzym. Inhib. Med. Chem.* **2016**, *31*, 176–184. [CrossRef]
10. Emameh, R.Z.; Kuuslahti, M.; Vullo, D.; Barker, H.R.; Supuran, C.T.; Parkkila, S. Ascaris lumbricoides β carbonic anhydrase: A potential target enzyme for treatment of ascariasis. *Parasites Vectors* **2015**, *8*, 479. [CrossRef]
11. Emameh, R.Z.; Barker, H.; Hytönen, V.P.; Tolvanen, M.E.; Parkkila, S. Beta carbonic anhydrases: Novel targets for pesticides and anti-parasitic agents in agriculture and livestock husbandry. *Parasites Vectors* **2014**, *7*, 403. [CrossRef] [PubMed]
12. Zolfaghari Emameh, R.; Kuuslahti, M.; Nosrati, H.; Lohi, H.; Parkkila, S. Assessment of databases to determine the validity of beta- and gamma-carbonic anhydrase sequences from vertebrates. *BMC Genom.* **2020**, *21*, 352. [PubMed]
13. Zolfaghari Emameh, R.; Hosseini, S.N.; Parkkila, S. Application of beta and gamma carbonic anhydrase sequences as tools for identification of bacterial contamination in the whole genome sequence of inbred Wuzhishan minipig (Sus scrofa) annotated in databases. *Database* **2021**, *2021*, baab029. [PubMed]
14. Lindskog, S. Structure and mechanism of carbonic anhydrase. *Pharmacol. Ther.* **1997**, *74*, 1–20. [CrossRef] [PubMed]
15. Akocak, S.; Supuran, C.T. Activation of α-, β-, γ-δ-, ζ-and η-class of carbonic anhydrases with amines and amino acids: A review. *J. Enzym. Inhib. Med. Chem.* **2019**, *34*, 1652–1659. [CrossRef]
16. Del Prete, S.; Nocentini, A.; Supuran, C.T.; Capasso, C. Bacterial ι-carbonic anhydrase: A new active class of carbonic anhydrase identified in the genome of the Gram-negative bacterium Burkholderia territorii. *J. Enzym. Inhib. Med. Chem.* **2020**, *35*, 1060–1068. [CrossRef]
17. Supuran, C.T. Structure and function of carbonic anhydrases. *Biochem. J.* **2016**, *473*, 2023–2032. [CrossRef]
18. Supuran, C.T.; Capasso, C. An overview of the bacterial carbonic anhydrases. *Metabolites* **2017**, *7*, 56. [CrossRef]
19. Ferraroni, M.; Del Prete, S.; Vullo, D.; Capasso, C.; Supuran, C.T. Crystal structure and kinetic studies of a tetrameric type II β-carbonic anhydrase from the pathogenic bacterium Vibrio cholerae. *Acta Crystallogr. Sect. D Biol. Crystallogr.* **2015**, *71*, 2449–2456. [CrossRef]
20. Iverson, T.M.; Alber, B.E.; Kisker, C.; Ferry, J.G.; Rees, D.C. A closer look at the active site of γ-class carbonic anhydrases: High-resolution crystallographic studies of the carbonic anhydrase from Methanosarcina thermophila. *Biochemistry* **2000**, *39*, 9222–9231. [CrossRef]
21. Zolfaghari Emameh, R.; Barker, H.R.; Tolvanen, M.E.; Parkkila, S.; Hytönen, V.P. Horizontal transfer of β-carbonic anhydrase genes from prokaryotes to protozoans, insects, and nematodes. *Parasites Vectors* **2016**, *9*, 152. [CrossRef] [PubMed]
22. Emameh, R.Z.; Barker, H.R.; Hytönen, V.P.; Parkkila, S. Involvement of β-carbonic anhydrase genes in bacterial genomic islands and their horizontal transfer to protists. *Appl. Environ. Microbiol.* **2018**, *84*, e00771-18.
23. Andersson, J.O. Lateral gene transfer in eukaryotes. *Cell. Mol. Life Sci.* **2005**, *62*, 1182–1197. [CrossRef] [PubMed]
24. Emamalipour, M.; Seidi, K.; Zununi Vahed, S.; Jahanban-Esfahlan, A.; Jaymand, M.; Majdi, H.; Amoozgar, Z.; Chitkushev, L.; Javaheri, T.; Jahanban-Esfahlan, R. Horizontal gene transfer: From evolutionary flexibility to disease progression. *Front. Cell Dev. Biol.* **2020**, *8*, 229. [CrossRef] [PubMed]
25. Dutta, C.; Pan, A. Horizontal gene transfer and bacterial diversity. *J. Biosci.* **2002**, *27*, 27. [CrossRef] [PubMed]

26. Frost, L.S.; Leplae, R.; Summers, A.O.; Toussaint, A. Mobile genetic elements: The agents of open source evolution. *Nat. Rev. Microbiol.* **2005**, *3*, 722–732. [CrossRef]
27. Bellanger, X.; Payot, S.; Leblond-Bourget, N.; Guédon, G. Conjugative and mobilizable genomic islands in bacteria: Evolution and diversity. *FEMS Microbiol. Rev.* **2014**, *38*, 720–760. [CrossRef]
28. Johnson, C.M.; Grossman, A.D. Integrative and conjugative elements (ICEs): What they do and how they work. *Annu. Rev. Genet.* **2015**, *49*, 577–601. [CrossRef]
29. Watson, B.N.; Staals, R.H.; Fineran, P.C. CRISPR-Cas-mediated phage resistance enhances horizontal gene transfer by transduction. *mBio* **2018**, *9*, e02406-17. [CrossRef]
30. Muñoz-López, M.; García-Pérez, J.L. DNA transposons: Nature and applications in genomics. *Curr. Genom.* **2010**, *11*, 115–128. [CrossRef]
31. Thomas, C.M.; Nielsen, K.M. Mechanisms of, and barriers to, horizontal gene transfer between bacteria. *Nat. Rev. Microbiol.* **2005**, *3*, 711–721. [CrossRef] [PubMed]
32. Messier, N.; Roy, P.H. Integron integrases possess a unique additional domain necessary for activity. *J. Bacteriol.* **2001**, *183*, 6699–6706. [CrossRef] [PubMed]
33. Partridge, S.R.; Recchia, G.D.; Scaramuzzi, C.; Collis, C.M.; Stokes, H.; Hall, R.M. Definition of the attI1 site of class 1 integrons. *Microbiology* **2000**, *146*, 2855–2864. [CrossRef] [PubMed]
34. Collis, C.M.; Hall, R.M. Expression of antibiotic resistance genes in the integrated cassettes of integrons. *Antimicrob. Agents Chemother.* **1995**, *39*, 155–162. [CrossRef]
35. Alvarez-Martinez, C.E.; Christie, P.J. Biological diversity of prokaryotic type IV secretion systems. *Microbiol. Mol. Biol. Rev.* **2009**, *73*, 775–808. [CrossRef]
36. Domsic, J.F.; McKenna, R. Sequestration of carbon dioxide by the hydrophobic pocket of the carbonic anhydrases. *Biochim. Biophys. Acta-Proteins Proteom.* **2010**, *1804*, 326–331. [CrossRef]
37. Mahillon, J.; Chandler, M. Insertion sequences. *Microbiol. Mol. Biol. Rev.* **1998**, *62*, 725–774. [CrossRef]
38. Fisher, R.M.; Henry, L.M.; Cornwallis, C.K.; Kiers, E.T.; West, S.A. The evolution of host-symbiont dependence. *Nat. Commun.* **2017**, *8*, 15973. [CrossRef]
39. Hall, R.J.; Whelan, F.J.; McInerney, J.O.; Ou, Y.; Domingo-Sananes, M.R. Horizontal Gene Transfer as a Source of Conflict and Cooperation in Prokaryotes. *Front Microbiol* **2020**, *11*, 1569.
40. Sabine, C.L.; Tanhua, T. Estimation of anthropogenic CO2 inventories in the ocean. *Ann. Rev. Mar. Sci.* **2010**, *2*, 175–198. [CrossRef]
41. De Goeyse, S.; Webb, A.E.; Reichart, G.-J.; De Nooijer, L.J. Carbonic anhydrase is involved in calcification by the benthic foraminifer Amphistegina lessonii. *Biogeosciences* **2021**, *18*, 393–401. [CrossRef]
42. Manyumwa, C.V.; Emameh, R.Z.; Tastan Bishop, Ö. Alpha-carbonic anhydrases from hydrothermal vent sources as potential carbon dioxide sequestration agents: In silico sequence, structure and dynamics analyses. *Int. J. Mol. Sci.* **2020**, *21*, 8066. [CrossRef] [PubMed]
43. Cardoso, J.C.; Ferreira, V.; Zhang, X.; Anjos, L.; Félix, R.C.; Batista, F.M.; Power, D.M. Evolution and diversity of alpha-carbonic anhydrases in the mantle of the Mediterranean mussel (Mytilus galloprovincialis). *Sci. Rep.* **2019**, *9*, 10400. [CrossRef] [PubMed]
44. Parisi, G.; Perales, M.; Fornasari, M.; Colaneri, A.; Schain, N.; Casati, D.; Zimmermann, S.; Brennicke, A.; Araya, A.; Ferry, J. Gamma carbonic anhydrases in plant mitochondria. *Plant Mol. Biol.* **2004**, *55*, 193–207. [CrossRef]
45. Smith, K.S.; Ferry, J.G. Prokaryotic carbonic anhydrases. *FEMS Microbiol. Rev.* **2000**, *24*, 335–366. [CrossRef]
46. Dobrinski, K.P.; Longo, D.L.; Scott, K.M. The carbon-concentrating mechanism of the hydrothermal vent chemolithoautotroph Thiomicrospira crunogena. *J. Bacteriol.* **2005**, *187*, 5761–5766. [CrossRef]
47. Miroshnichenko, M.L.; Kostrikina, N.A.; Chernyh, N.A.; Pimenov, N.V.; Tourova, T.P.; Antipov, A.N.; Spring, S.; Stackebrandt, E.; Bonch-Osmolovskaya, E.A. Caldithrix abyssi gen. nov., sp. nov., a nitrate-reducing, thermophilic, anaerobic bacterium isolated from a Mid-Atlantic Ridge hydrothermal vent, represents a novel bacterial lineage. *Int. J. Syst. Evol. Microbiol.* **2003**, *53*, 323–329. [CrossRef]
48. Voordeckers, J.W.; Starovoytov, V.; Vetriani, C. Caminibacter mediatlanticus sp. nov., a thermophilic, chemolithoautotrophic, nitrate-ammonifying bacterium isolated from a deep-sea hydrothermal vent on the Mid-Atlantic Ridge. *Int. J. Syst. Evol. Microbiol.* **2005**, *55*, 773–779. [CrossRef]
49. Ward, M.E.; Shields, J.D.; Van Dover, C.L. Parasitism in species of Bathymodiolus (Bivalvia: Mytilidae) mussels from deep-sea seep and hydrothermal vents. *Dis. Aquat.* **2004**, *62*, 1–16. [CrossRef]
50. Klose, J.; Polz, M.F.; Wagner, M.; Schimak, M.P.; Gollner, S.; Bright, M. Endosymbionts escape dead hydrothermal vent tubeworms to enrich the free-living population. *Proc. Natl. Acad. Sci. USA* **2015**, *112*, 11300–11305. [CrossRef]
51. L'Haridon, S.; Reysenbach, A.-L.; Tindall, B.; Schönheit, P.; Banta, A.; Johnsen, U.; Schumann, P.; Gambacorta, A.; Stackebrandt, E.; Jeanthon, C. Desulfurobacterium atlanticum sp. nov., Desulfurobacterium pacificum sp. nov. and Thermovibrio guaymasensis sp. nov., three thermophilic members of the Desulfurobacteriaceae fam. nov., a deep branching lineage within the Bacteria. *Int. J. Syst. Evol. Microbiol.* **2006**, *56*, 2843–2852. [CrossRef] [PubMed]
52. Hansen, M.; Perner, M. Reasons for Thiomicrospira crunogena's recalcitrance towards previous attempts to detect its hydrogen consumption ability. *Environ. Microbiol. Rep.* **2016**, *8*, 53–57. [CrossRef]

53. Takai, K.; Nealson, K.H.; Horikoshi, K. Hydrogenimonas thermophila gen. nov., sp. nov., a novel thermophilic, hydrogen-oxidizing chemolithoautotroph within the ε-Proteobacteria, isolated from a black smoker in a Central Indian Ridge hydrothermal field. *Int. J. Syst. Evol. Microbiol.* **2004**, *54*, 25–32. [CrossRef] [PubMed]
54. Mino, S.; Shiotani, T.; Nakagawa, S.; Takai, K.; Sawabe, T. Hydrogenimonas urashimensis sp. nov., a hydrogen-oxidizing chemolithoautotroph isolated from a deep-sea hydrothermal vent in the Southern Mariana Trough. *Syst. Appl. Microbiol.* **2021**, *44*, 126170. [CrossRef] [PubMed]
55. Shiotani, T.; Mino, S.; Sato, W.; Nishikawa, S.; Yonezawa, M.; Sievert, S.M.; Sawabe, T. Nitrosophilus alvini gen. nov., sp. nov., a hydrogen-oxidizing chemolithoautotroph isolated from a deep-sea hydrothermal vent in the East Pacific Rise, inferred by a genome-based taxonomy of the phylum "Campylobacterota". *PLoS ONE* **2020**, *15*, e0241366. [CrossRef]
56. Nakagawa, S.; Takai, K.; Inagaki, F.; Horikoshi, K.; Sako, Y. Nitratiruptor tergarcus gen. nov., sp. nov. and Nitratifractor salsuginis gen. nov., sp. nov., nitrate-reducing chemolithoautotrophs of the ε-Proteobacteria isolated from a deep-sea hydrothermal system in the Mid-Okinawa Trough. *Int. J. Syst. Evol. Microbiol.* **2005**, *55*, 925–933. [CrossRef]
57. François, D.X.; Godfroy, A.; Mathien, C.; Aubé, J.; Cathalot, C.; Lesongeur, F.; L'Haridon, S.; Philippon, X.; Roussel, E.G. Persephonella atlantica sp. nov.: How to adapt to physico-chemical gradients in high temperature hydrothermal habitats. *Syst. Appl. Microbiol.* **2021**, *44*, 126176. [CrossRef]
58. Nakagawa, S.; Takai, K.; Horikoshi, K.; Sako, Y. Persephonella hydrogeniphila sp. nov., a novel thermophilic, hydrogen-oxidizing bacterium from a deep-sea hydrothermal vent chimney. *Int. J. Syst. Evol. Microbiol.* **2003**, *53*, 863–869. [CrossRef]
59. Götz, D.; Banta, A.; Beveridge, T.; Rushdi, A.; Simoneit, B.; Reysenbach, A. Persephonella marina gen. nov., sp. nov. and Persephonella guaymasensis sp. nov., two novel, thermophilic, hydrogen-oxidizing microaerophiles from deep-sea hydrothermal vents. *Int. J. Syst. Evol. Microbiol.* **2002**, *52*, 1349–1359.
60. Waite, D.W.; Vanwonterghem, I.; Rinke, C.; Parks, D.H.; Zhang, Y.; Takai, K.; Sievert, S.M.; Simon, J.; Campbell, B.J.; Hanson, T.E. Comparative genomic analysis of the class Epsilonproteobacteria and proposed reclassification to Epsilonbacteraeota (phyl. nov.). *Front. Microbiol.* **2017**, *8*, 682. [CrossRef]
61. Xie, S.; Wang, S.; Li, D.; Shao, Z.; Lai, Q.; Wang, Y.; Wei, M.; Han, X.; Jiang, L. Sulfurovum indicum sp. nov., a novel hydrogen-and sulfur-oxidizing chemolithoautotroph isolated from a deep-sea hydrothermal plume in the Northwestern Indian Ocean. *Int. J. Syst. Evol. Microbiol.* **2021**, *71*, 004748. [CrossRef] [PubMed]
62. Takai, K.; Suzuki, M.; Nakagawa, S.; Miyazaki, M.; Suzuki, Y.; Inagaki, F.; Horikoshi, K. Sulfurimonas paralvinellae sp. nov., a novel mesophilic, hydrogen-and sulfur-oxidizing chemolithoautotroph within the Epsilonproteobacteria isolated from a deep-sea hydrothermal vent polychaete nest, reclassification of Thiomicrospira denitrificans as Sulfurimonas denitrificans comb. nov. and emended description of the genus Sulfurimonas. *Int. J. Syst. Evol. Microbiol.* **2006**, *56*, 1725–1733. [PubMed]
63. Boutet, I.; Ripp, R.; Lecompte, O.; Dossat, C.; Corre, E.; Tanguy, A.; Lallier, F.H. Conjugating effects of symbionts and environmental factors on gene expression in deep-sea hydrothermal vent mussels. *BMC Genom.* **2011**, *12*, 530. [CrossRef] [PubMed]
64. Poli, A.; Gugliandolo, C.; Spanò, A.; Taurisano, V.; Di Donato, P.; Maugeri, T.L.; Nicolaus, B.; Arena, A. Poly-[Gamma]-Glutamic Acid from Bacillus Horneckiae Strain APA of Shallow Marine Vent Origin with Antiviral and Immunomodulatory Effects against Herpes Simplex Virus Type-2. *J. Mar. Sci. Res. Dev.* **2015**, *5*, 173. [CrossRef]
65. Wu, Y.-H.; Xu, L.; Zhou, P.; Wang, C.-S.; Oren, A.; Xu, X.-W. Brevirhabdus pacifica gen. nov., sp. nov., isolated from deep-sea sediment in a hydrothermal vent field. *Int. J. Syst. Evol. Microbiol.* **2015**, *65*, 3645–3651. [CrossRef]
66. Ponnudurai, R.; Sayavedra, L.; Kleiner, M.; Heiden, S.E.; Thürmer, A.; Felbeck, H.; Schlüter, R.; Sievert, S.M.; Daniel, R.; Schweder, T. Genome sequence of the sulfur-oxidizing Bathymodiolus thermophilus gill endosymbiont. *Stand. Genom. Sci.* **2017**, *12*, 50. [CrossRef]
67. Grosche, A.; Sekaran, H.; Pérez-Rodríguez, I.; Starovoytov, V.; Vetriani, C. Cetia pacifica gen. nov., sp. nov., a chemolithoautotrophic, thermophilic, nitrate-ammonifying bacterium from a deep-sea hydrothermal vent. *Int. J. Syst. Evol. Microbiol.* **2015**, *65*, 1144–1150. [CrossRef]
68. Slobodkina, G.; Kolganova, T.; Chernyh, N.; Querellou, J.; Bonch-Osmolovskaya, E.; Slobodkin, A. Deferribacter autotrophicus sp. nov., an iron (III)-reducing bacterium from a deep-sea hydrothermal vent. *Int. J. Syst. Evol. Microbiol.* **2009**, *59*, 1508–1512. [CrossRef]
69. Cao, J.; Birien, T.; Gayet, N.; Huang, Z.; Shao, Z.; Jebbar, M.; Alain, K. Desulfurobacterium indicum sp. nov., a thermophilic sulfur-reducing bacterium from the Indian Ocean. *Int. J. Syst. Evol. Microbiol.* **2017**, *67*, 1665–1668. [CrossRef]
70. Jiang, L.; Xu, H.; Shao, Z.; Long, M. Defluviimonas indica sp. nov., a marine bacterium isolated from a deep-sea hydrothermal vent environment. *Int. J. Syst. Evol. Microbiol.* **2014**, *64*, 2084–2088. [CrossRef]
71. Hansen, M.; Perner, M. Hydrogenase gene distribution and H2 consumption ability within the Thiomicrospira lineage. *Front. Microbiol.* **2016**, *7*, 99. [CrossRef] [PubMed]
72. Kaye, J.Z.; Marquez, M.C.; Ventosa, A.; Baross, J.A. Halomonas neptunia sp. nov., Halomonas sulfidaeris sp. nov., Halomonas axialensis sp. nov. and Halomonas hydrothermalis sp. nov.: Halophilic bacteria isolated from deep-sea hydrothermal-vent environments. *Int. J. Syst. Evol. Microbiol.* **2004**, *54*, 499–511. [CrossRef] [PubMed]
73. Jiang, L.; Lyu, J.; Shao, Z. Sulfur metabolism of Hydrogenovibrio thermophilus strain S5 and its adaptations to deep-sea hydrothermal vent environment. *Front. Microbiol.* **2017**, *8*, 2513. [CrossRef] [PubMed]

74. Slobodkina, G.B.; Baslerov, R.V.; Novikov, A.A.; Viryasov, M.B.; Bonch-Osmolovskaya, E.A.; Slobodkin, A.I. Inmirania thermothiophila gen. nov., sp. nov., a thermophilic, facultatively autotrophic, sulfur-oxidizing gammaproteobacterium isolated from a shallow-sea hydrothermal vent. *Int. J. Syst. Evol. Microbiol.* **2016**, *66*, 701–706. [CrossRef]
75. Nagata, R.; Takaki, Y.; Tame, A.; Nunoura, T.; Muto, H.; Mino, S.; Sawayama, S.; Takai, K.; Nakagawa, S. Lebetimonas natsushimae sp. nov., a novel strictly anaerobic, moderately thermophilic chemoautotroph isolated from a deep-sea hydrothermal vent polychaete nest in the Mid-Okinawa Trough. *Syst. Appl. Microbiol.* **2017**, *40*, 352–356. [CrossRef]
76. Stewart, L.C.; Jung, J.-H.; Kim, Y.-T.; Kwon, S.-W.; Park, C.-S.; Holden, J.F. Methanocaldococcus bathoardescens sp. nov., a hyperthermophilic methanogen isolated from a volcanically active deep-sea hydrothermal vent. *Int. J. Syst. Evol. Microbiol.* **2015**, *65*, 1280–1283. [CrossRef]
77. McAllister, S.M.; Davis, R.E.; McBeth, J.M.; Tebo, B.M.; Emerson, D.; Moyer, C.L. Biodiversity and emerging biogeography of the neutrophilic iron-oxidizing Zetaproteobacteria. *Appl. Environ. Microbiol.* **2011**, *77*, 5445–5457. [CrossRef]
78. Takai, K.; Nealson, K.H.; Horikoshi, K. Methanotorris formicicus sp. nov., a novel extremely thermophilic, methane-producing archaeon isolated from a black smoker chimney in the Central Indian Ridge. *Int. J. Syst. Evol. Microbiol.* **2004**, *54*, 1095–1100. [CrossRef]
79. Takeuchi, M.; Katayama, T.; Yamagishi, T.; Hanada, S.; Tamaki, H.; Kamagata, Y.; Oshima, K.; Hattori, M.; Marumo, K.; Nedachi, M. Methyloceanibacter caenitepidi gen. nov., sp. nov., a facultatively methylotrophic bacterium isolated from marine sediments near a hydrothermal vent. *Int. J. Syst. Evol. Microbiol.* **2014**, *64*, 462–468. [CrossRef]
80. Wang, H.; Li, H.; Shao, Z.; Liao, S.; Johnstone, L.; Rensing, C.; Wang, G. Genome sequence of deep-sea manganese-oxidizing bacterium Marinobacter manganoxydans MnI7-9. *J. Bacteriol.* **2012**, *194*, 899–900. [CrossRef]
81. Takai, K.; Inoue, A.; Horikoshi, K. Methanothermococcus okinawensis sp. nov., a thermophilic, methane-producing archaeon isolated from a Western Pacific deep-sea hydrothermal vent system. *Int. J. Syst. Evol. Microbiol.* **2002**, *52*, 1089–1095. [PubMed]
82. Zhou, M.; Dong, B.; Shao, Z. Complete genome sequence of Marinobacter sp. LQ44, a haloalkaliphilic phenol-degrading bacterium isolated from a deep-sea hydrothermal vent. *Mar. Genom.* **2020**, *50*, 100697. [CrossRef]
83. Handley, K.M.; Hery, M.; Lloyd, J.R. Marinobacter santoriniensis sp. nov., an arsenate-respiring and arsenite-oxidizing bacterium isolated from hydrothermal sediment. *Int. J. Syst. Evol. Microbiol.* **2009**, *59*, 886–892. [CrossRef] [PubMed]
84. Smith, J.L.; Campbell, B.J.; Hanson, T.E.; Zhang, C.L.; Cary, S.C. Nautilia profundicola sp. nov., a thermophilic, sulfur-reducing epsilonproteobacterium from deep-sea hydrothermal vents. *Int. J. Syst. Evol. Microbiol.* **2008**, *58*, 1598–1602. [CrossRef]
85. Dong, C.; Lai, Q.; Chen, L.; Sun, F.; Shao, Z.; Yu, Z. Oceanibaculum pacificum sp. nov., isolated from hydrothermal field sediment of the south-west Pacific Ocean. *Int. J. Syst. Evol. Microbiol.* **2010**, *60*, 219–222. [CrossRef]
86. Miroshnichenko, M.; L'haridon, S.; Jeanthon, C.; Antipov, A.; Kostrikina, N.; Tindall, B.; Schumann, P.; Spring, S.; Stackebrandt, E.; Bonch-Osmolovskaya, E. Oceanithermus profundus gen. nov., sp. nov., a thermophilic, microaerophilic, facultatively chemolithoheterotrophic bacterium from a deep-sea hydrothermal vent. *Int. J. Syst. Evol. Microbiol.* **2003**, *53*, 747–752. [CrossRef]
87. Newton, I.; Woyke, T.; Auchtung, T.; Dilly, G.; Dutton, R.; Fisher, M.; Fontanez, K.; Lau, E.; Stewart, F.; Richardson, P. The Calyptogena magnifica chemoautotrophic symbiont genome. *Science* **2007**, *315*, 998–1000. [CrossRef]
88. Guo, W.; Zhang, H.; Zhou, W.; Wang, Y.; Zhou, H.; Chen, X. Sulfur metabolism pathways in Sulfobacillus acidophilus TPY, a gram-positive moderate thermoacidophile from a hydrothermal vent. *Front. Microbiol.* **2016**, *7*, 1861. [CrossRef]
89. Inagaki, F.; Takai, K.; Nealson, K.H.; Horikoshi, K. Sulfurovum lithotrophicum gen. nov., sp. nov., a novel sulfur-oxidizing chemolithoautotroph within the ε-Proteobacteria isolated from Okinawa Trough hydrothermal sediments. *Int. J. Syst. Evol. Microbiol.* **2004**, *54*, 1477–1482. [CrossRef]
90. Giovannelli, D.; Chung, M.; Staley, J.; Starovoytov, V.; Le Bris, N.; Vetriani, C. Sulfurovum riftiae sp. nov., a mesophilic, thiosulfate-oxidizing, nitrate-reducing chemolithoautotrophic epsilonproteobacterium isolated from the tube of the deep-sea hydrothermal vent polychaete Riftia pachyptila. *Int. J. Syst. Evol. Microbiol.* **2016**, *66*, 2697–2701. [CrossRef]
91. Vetriani, C.; Speck, M.D.; Ellor, S.V.; Lutz, R.A.; Starovoytov, V. Thermovibrio ammonificans sp. nov., a thermophilic, chemolithotrophic, nitrate-ammonifying bacterium from deep-sea hydrothermal vents. *Int. J. Syst. Evol. Microbiol.* **2004**, *54*, 175–181. [CrossRef] [PubMed]
92. Urios, L.; Cueff, V.; Pignet, P.; Barbier, G. Tepidibacter formicigenes sp. nov., a novel spore-forming bacterium isolated from a Mid-Atlantic Ridge hydrothermal vent. *Int. J. Syst. Evol. Microbiol.* **2004**, *54*, 439–443. [CrossRef] [PubMed]
93. Hensley, S.A.; Jung, J.-H.; Park, C.-S.; Holden, J.F. Thermococcus paralvinellae sp. nov. and Thermococcus cleftensis sp. nov. of hyperthermophilic heterotrophs from deep-sea hydrothermal vents. *Int. J. Syst. Evol. Microbiol.* **2014**, *64*, 3655–3659. [CrossRef]
94. Martins, E.; Santos, R.S.; Bettencourt, R. Vibrio diabolicus challenge in Bathymodiolus azoricus populations from Menez Gwen and Lucky Strike hydrothermal vent sites. *Fish. Shellfish Immunol.* **2015**, *47*, 962–977. [CrossRef] [PubMed]
95. Kuwahara, H.; Yoshida, T.; Takaki, Y.; Shimamura, S.; Nishi, S.; Harada, M.; Matsuyama, K.; Takishita, K.; Kawato, M.; Uematsu, K. Reduced genome of the thioautotrophic intracellular symbiont in a deep-sea clam, Calyptogena okutanii. *Curr. Biol.* **2007**, *17*, 881–886. [CrossRef] [PubMed]
96. Zeng, X.; Zhang, Z.; Li, X.; Zhang, X.; Cao, J.; Jebbar, M.; Alain, K.; Shao, Z. Anoxybacter fermentans gen. nov., sp. nov., a piezophilic, thermophilic, anaerobic, fermentative bacterium isolated from a deep-sea hydrothermal vent. *Int. J. Syst. Evol. Microbiol.* **2015**, *65*, 710–715. [CrossRef] [PubMed]

97. Zammuto, V.; Fuchs, F.M.; Fiebrandt, M.; Stapelmann, K.; Ulrich, N.J.; Maugeri, T.L.; Pukall, R.; Gugliandolo, C.; Moeller, R. Comparing spore resistance of Bacillus strains isolated from hydrothermal vents and spacecraft assembly facilities to environmental stressors and decontamination treatments. *Astrobiology* **2018**, *18*, 1425–1434. [CrossRef]
98. Wery, N.; Moricet, J.-M.; Cueff, V.; Jean, J.; Pignet, P.; Lesongeur, F.; Cambon-Bonavita, M.-A.; Barbier, G. Caloranaerobacter azorensis gen. nov., sp nov., an anaerobic thermophilic bacterium isolated from a deep-sea hydrothermal vent. *Int. J. Syst. Evol. Microbiol.* **2001**, *51*, 1789–1796. [CrossRef]
99. Alain, K.; Pignet, P.; Zbinden, M.; Quillevere, M.; Duchiron, F.; Donval, J.-P.; Lesongeur, F.; Raguenes, G.; Crassous, P.; Querellou, J. Caminicella sporogenes gen. nov., sp. nov., a novel thermophilic spore-forming bacterium isolated from an East-Pacific Rise hydrothermal vent. *Int. J. Syst. Evol. Microbiol.* **2002**, *52*, 1621–1628.
100. Takai, K.; Kobayashi, H.; Nealson, K.H.; Horikoshi, K. Deferribacter desulfuricans sp. nov., a novel sulfur-, nitrate-and arsenate-reducing thermophile isolated from a deep-sea hydrothermal vent. *Int. J. Syst. Evol. Microbiol.* **2003**, *53*, 839–846. [CrossRef]
101. Chao, L.S.L.; Davis, R.E.; Moyer, C.L. Characterization of bacterial community structure in vestimentiferan tubeworm Ridgeia piscesae trophosomes. *Mar. Ecol.* **2007**, *28*, 72–85. [CrossRef]
102. Won, Y.-J.; Hallam, S.J.; O'Mullan, G.D.; Pan, I.L.; Buck, K.R.; Vrijenhoek, R.C. Environmental acquisition of thiotrophic endosymbionts by deep-sea mussels of the genus Bathymodiolus. *Appl. Environ. Microbiol.* **2003**, *69*, 6785–6792. [CrossRef] [PubMed]
103. Slobodkina, G.; Kolganova, T.; Querellou, J.; Bonch-Osmolovskaya, E.; Slobodkin, A. Geoglobus acetivorans sp. nov., an iron (III)-reducing archaeon from a deep-sea hydrothermal vent. *Int. J. Syst. Evol. Microbiol.* **2009**, *59*, 2880–2883. [CrossRef]
104. Maugeri, T.L.; Gugliandolo, C.; Caccamo, D.; Stackebrandt, E. Three novel halotolerant and thermophilic Geobacillus strains from shallow marine vents. *Syst. Appl. Microbiol.* **2002**, *25*, 450–455. [CrossRef] [PubMed]
105. Roalkvam, I.; Bredy, F.; Baumberger, T.; Pedersen, R.-B.; Steen, I.H. Hypnocyclicus thermotrophus gen. nov., sp. nov. isolated from a microbial mat in a hydrothermal vent field. *Int. J. Syst. Evol. Microbiol.* **2015**, *65*, 4521–4525. [CrossRef]
106. Van Dover, C. *The Ecology of Deep-Sea Hydrothermal Vents*; Princeton University Press: Princeton, NJ, USA, 2000.
107. Erauso, G.; Reysenbach, A.-L.; Godfroy, A.; Meunier, J.-R.; Crump, B.; Partensky, F.; Baross, J.A.; Marteinsson, V.; Barbier, G.; Pace, N.R. Pyrococcus abyssi sp. nov., a new hyperthermophilic archaeon isolated from a deep-sea hydrothermal vent. *Arch. Microbiol.* **1993**, *160*, 338–349. [CrossRef]
108. Marteinsson, V.T.; Bjornsdottir, S.H.; Bienvenu, N.; Kristjansson, J.K.; Birrien, J.-L. Rhodothermus profundi sp. nov., a thermophilic bacterium isolated from a deep-sea hydrothermal vent in the Pacific Ocean. *Int. J. Syst. Evol. Microbiol.* **2010**, *60*, 2729–2734. [CrossRef]
109. Miroshnichenko, M.; Gongadze, G.; Rainey, F.; Kostyukova, A.; Lysenko, A.; Chernyh, N.; Bonch-Osmolovskaya, E. Thermococcus gorgonarius sp. nov. and Thermococcus pacificus sp. nov.: Heterotrophic extremely thermophilic archaea from New Zealand submarine hot vents. *Int. J. Syst. Evol. Microbiol.* **1998**, *48*, 23–29. [CrossRef]
110. Marteinsson, V.T.; Birrien, J.-L.; Reysenbach, A.-L.; Vernet, M.; Marie, D.; Gambacorta, A.; Messner, P.; Sleytr, U.B.; Prieur, D. Thermococcus barophilus sp. nov., a new barophilic and hyperthermophilic archaeon isolated under high hydrostatic pressure from a deep-sea hydrothermal vent. *Int. J. Syst. Evol. Microbiol.* **1999**, *49*, 351–359. [CrossRef]
111. Duffaud, G.D.; d'Hennezel, O.B.; Peek, A.S.; Reysenbach, A.-L.; Kelly, R.M. Isolation and characterization of Thermococcus barossii, sp. nov., a hyperthermophilic archaeon isolated from a hydrothermal vent flange formation. *Syst. Appl. Microbiol.* **1998**, *21*, 40–49. [CrossRef]
112. Jolivet, E.; l'Haridon, S.; Corre, E.; Forterre, P.; Prieur, D. Thermococcus gammatolerans sp. nov., a hyperthermophilic archaeon from a deep-sea hydrothermal vent that resists ionizing radiation. *Int. J. Syst. Evol. Microbiol.* **2003**, *53*, 847–851. [CrossRef]
113. Gorlas, A.; Croce, O.; Oberto, J.; Gauliard, E.; Forterre, P.; Marguet, E. Thermococcusnautili sp. nov., a hyperthermophilic archaeon isolated from a hydrothermal deep-sea vent. *Int. J. Syst. Evol. Microbiol.* **2014**, *64*, 1802–1810. [CrossRef] [PubMed]
114. González, J.M.; Kato, C.; Horikoshi, K. Thermococcus peptonophilus sp. nov., a fast-growing, extremely thermophilic archaebacterium isolated from deep-sea hydrothermal vents. *Arch. Microbiol.* **1995**, *164*, 159–164. [CrossRef] [PubMed]
115. Kobayashi, T.; Kwak, Y.S.; Akiba, T.; Kudo, T.; Horikoshi, K. Thermococcus profundus sp. nov., a new hyperthermophilic archaeon isolated from a deep-sea hydrothermal vent. *Syst. Appl. Microbiol.* **1994**, *17*, 232–236. [CrossRef]
116. Dalmasso, C.; Oger, P.; Selva, G.; Courtine, D.; L'haridon, S.; Garlaschelli, A.; Roussel, E.; Miyazaki, J.; Reveillaud, J.; Jebbar, M. Thermococcus piezophilus sp. nov., a novel hyperthermophilic and piezophilic archaeon with a broad pressure range for growth, isolated from a deepest hydrothermal vent at the Mid-Cayman Rise. *Syst. Appl. Microbiol.* **2016**, *39*, 440–444. [CrossRef] [PubMed]
117. Slobodkin, A.; Tourova, T.; Kostrikina, N.; Chernyh, N.; Bonch-Osmolovskaya, E.; Jeanthon, C.; Jones, B. Tepidibacterthalassicus gen. nov., sp. nov., a novel moderately thermophilic, anaerobic, fermentative bacterium from a deep-sea hydrothermal vent. *J. Med. Microbiol.* **2003**, *53*, 1131–1134.
118. L'Haridon, S.; Miroshnichenko, M.; Kostrikina, N.; Tindall, B.; Spring, S.; Schumann, P.; Stackebrandt, E.; Bonch-Osmolovskaya, E.; Jeanthon, C. Vulcanibacillus modesticaldus gen. nov., sp. nov., a strictly anaerobic, nitrate-reducing bacterium from deep-sea hydrothermal vents. *Int. J. Syst. Evol. Microbiol.* **2006**, *56*, 1047–1053. [CrossRef]
119. Notredame, C.; Higgins, D.G.; Heringa, J. T-Coffee: A novel method for fast and accurate multiple sequence alignment. *J. Mol. Biol.* **2000**, *302*, 205–217. [CrossRef] [PubMed]
120. Waterhouse, A.M.; Procter, J.B.; Martin, D.M.; Clamp, M.; Barton, G.J. Jalview Version 2—A multiple sequence alignment editor and analysis workbench. *Bioinformatics* **2009**, *25*, 1189–1191. [CrossRef]

121. Nei, M.; Kumar, S. *Molecular Evolution and Phylogenetics*; Oxford University Press: New York, NY, USA, 2000.
122. Kumar, S.; Stecher, G.; Li, M.; Knyaz, C.; Tamura, K. MEGA X: Molecular evolutionary genetics analysis across computing platforms. *Mol. Biol. Evol.* **2018**, *35*, 1547–1549. [CrossRef]
123. Hall, R.M.; Brookes, D.E.; Stokes, H. Site-specific insertion of genes into integrons: Role of the 59-base element and determination of the recombination cross-over point. *Mol. Microbiol.* **1991**, *5*, 1941–1959. [CrossRef] [PubMed]
124. Rowe-Magnus, D.A.; Guérout, A.-M.; Mazel, D. Super-integrons. *Res. Microbiol.* **1999**, *150*, 641–651. [CrossRef] [PubMed]
125. Rowe-Magnus, D.A.; Guerout, A.-M.; Ploncard, P.; Dychinco, B.; Davies, J.; Mazel, D. The evolutionary history of chromosomal super-integrons provides an ancestry for multiresistant integrons. *Proc. Natl. Acad. Sci. USA* **2001**, *98*, 652–657. [CrossRef] [PubMed]
126. Stokes, H.; O'Gorman, D.; Recchia, G.D.; Parsekhian, M.; Hall, R.M. Structure and function of 59-base element recombination sites associated with mobile gene cassettes. *Mol. Microbiol.* **1997**, *26*, 731–745. [CrossRef]
127. Collis, C.M.; Hall, R.M. Gene cassettes from the insert region of integrons are excised as covalently closed circles. *Mol. Microbiol.* **1992**, *6*, 2875–2885. [CrossRef]
128. Hocquet, D.; Llanes, C.; Thouverez, M.; Kulasekara, H.D.; Bertrand, X.; Plésiat, P.; Mazel, D.; Miller, S.I. Evidence for induction of integron-based antibiotic resistance by the SOS response in a clinical setting. *PLoS Pathog.* **2012**, *8*, e1002778. [CrossRef]
129. Cury, J.; Jové, T.; Touchon, M.; Néron, B.; Rocha, E.P. Identification and analysis of integrons and cassette arrays in bacterial genomes. *Nucleic Acids Res.* **2016**, *44*, 4539–4550. [CrossRef]
130. Hsiao, W.; Wan, I.; Jones, S.J.; Brinkman, F.S. IslandPath: Aiding detection of genomic islands in prokaryotes. *Bioinformatics* **2003**, *19*, 418–420. [CrossRef]
131. Tu, Q.; Ding, D. Detecting pathogenicity islands and anomalous gene clusters by iterative discriminant analysis. *FEMS Microbiol. Lett.* **2003**, *221*, 269–275. [CrossRef]
132. Rajan, I.; Aravamuthan, S.; Mande, S.S. Identification of compositionally distinct regions in genomes using the centroid method. *Bioinformatics* **2007**, *23*, 2672–2677. [CrossRef]
133. Bertelli, C.; Laird, M.R.; Williams, K.P.; Group, S.F.U.R.C.; Lau, B.Y.; Hoad, G.; Winsor, G.L.; Brinkman, F.S. IslandViewer 4: Expanded prediction of genomic islands for larger-scale datasets. *Nucleic Acids Res.* **2017**, *45*, W30–W35. [CrossRef] [PubMed]
134. Boyd, E.F.; Almagro-Moreno, S.; Parent, M.A. Genomic islands are dynamic, ancient integrative elements in bacterial evolution. *Trends Microbiol.* **2009**, *17*, 47–53. [CrossRef] [PubMed]
135. Liu, M.; Li, X.; Xie, Y.; Bi, D.; Sun, J.; Li, J.; Tai, C.; Deng, Z.; Ou, H.-Y. ICEberg 2.0: An updated database of bacterial integrative and conjugative elements. *Nucleic Acids Res* **2019**, *47*, D660–D665. [CrossRef] [PubMed]
136. Bi, D.; Xu, Z.; Harrison, E.M.; Tai, C.; Wei, Y.; He, X.; Jia, S.; Deng, Z.; Rajakumar, K.; Ou, H.-Y. ICEberg: A web-based resource for integrative and conjugative elements found in Bacteria. *Nucleic Acids Res.* **2012**, *40*, D621–D626. [CrossRef]
137. Johansson, M.H.; Bortolaia, V.; Tansirichaiya, S.; Aarestrup, F.M.; Roberts, A.P.; Petersen, T.N. Detection of mobile genetic elements associated with antibiotic resistance in Salmonella enterica using a newly developed web tool: MobileElementFinder. *J. Antimicrob. Chemother.* **2021**, *76*, 101–109. [CrossRef]
138. Campbell, A. Prophage insertion sites. *Res. Microbiol.* **2003**, *154*, 277–282. [CrossRef]
139. Sousa, A.L.D.; Maués, D.; Lobato, A.; Franco, E.F.; Pinheiro, K.; Araújo, F.; Pantoja, Y.; Costa da Silva, A.L.D.; Morais, J.; Ramos, R.T. PhageWeb–Web interface for rapid identification and characterization of prophages in bacterial genomes. *Front. Genet.* **2018**, *9*, 644. [CrossRef]
140. Consortium, T.U. UniProt: The universal protein knowledgebase in 2021. *Nucleic Acids Res.* **2020**, *49*, D480–D489.
141. Domsic, J.F.; Avvaru, B.S.; Kim, C.U.; Gruner, S.M.; Agbandje-McKenna, M.; Silverman, D.N.; McKenna, R. Entrapment of carbon dioxide in the active site of carbonic anhydrase II. *J. Biol. Chem.* **2008**, *283*, 30766–30771. [CrossRef]
142. Le, S.Q.; Gascuel, O. An improved general amino acid replacement matrix. *Mol. Biol. Evol.* **2008**, *25*, 1307–1320. [CrossRef]
143. Peters, J.E.; Fricker, A.D.; Kapili, B.J.; Petassi, M.T. Heteromeric transposase elements: Generators of genomic islands across diverse bacteria. *Mol. Microbiol.* **2014**, *93*, 1084–1092. [CrossRef] [PubMed]
144. Waack, S.; Keller, O.; Asper, R.; Brodag, T.; Damm, C.; Fricke, W.F.; Surovcik, K.; Meinicke, P.; Merkl, R. Score-based prediction of genomic islands in prokaryotic genomes using hidden Markov models. *BMC Bioinform.* **2006**, *7*, 142. [CrossRef] [PubMed]
145. Cheng, R.; Li, X.; Jiang, L.; Gong, L.; Geslin, C.; Shao, Z. Virus diversity and interactions with hosts in deep-sea hydrothermal vents. *Microbiome* **2022**, *10*, 235. [PubMed]
146. Gestal, A.M.; Liew, E.F.; Coleman, N.V. Natural transformation with synthetic gene cassettes: New tools for integron research and biotechnology. *Microbiology* **2011**, *157*, 3349–3360. [PubMed]
147. Goffredi, S.; Childress, J.; Desaulniers, N.; Lee, R.; Lallier, F.; Hammond, D. Inorganic carbon acquisition by the hydrothermal vent tubeworm Riftia pachyptila depends upon high external PCO2 and upon proton-equivalent ion transport by the worm. *J. Exp. Biol.* **1997**, *200*, 883–896. [CrossRef] [PubMed]
148. De Cian, M.C.; Bailly, X.; Morales, J.; Strub, J.M.; Van Dorsselaer, A.; Lallier, F.H. Characterization of carbonic anhydrases from Riftia pachyptila, a symbiotic invertebrate from deep-sea hydrothermal vents. *Proteins Struct. Funct. Bioinf.* **2003**, *51*, 327–339. [CrossRef] [PubMed]
149. De Cian, M.-C.; Andersen, A.C.; Bailly, X.; Lallier, F.H. Expression and localization of carbonic anhydrase and ATPases in the symbiotic tubeworm Riftia pachyptila. *J. Exp. Biol.* **2003**, *206*, 399–409. [CrossRef]

150. Gonnella, G.; Adam, N.; Perner, M. Horizontal acquisition of hydrogen conversion ability and other habitat adaptations in the Hydrogenovibrio strains SP-41 and XCL-2. *BMC Genom.* **2019**, *20*, 339. [CrossRef]
151. Fabry, V.J.; Seibel, B.A.; Feely, R.A.; Orr, J.C. Impacts of ocean acidification on marine fauna and ecosystem processes. *ICES J. Mar. Sci.* **2008**, *65*, 414–432. [CrossRef]

MDPI
St. Alban-Anlage 66
4052 Basel
Switzerland
Tel. +41 61 683 77 34
Fax +41 61 302 89 18
www.mdpi.com

Biology Editorial Office
E-mail: biology@mdpi.com
www.mdpi.com/journal/biology

www.ingramcontent.com/pod-product-compliance
Lightning Source LLC
Chambersburg PA
CBHW061805210926
51850CB00006B/623